Eosinophils in Allergy and Inflammation

CLINICAL ALLERGY AND IMMUNOLOGY

Series Editor

MICHAEL A. KALINER, M.D.

Head, Allergic Disease Section
National Institute of Allergy and Infectious Diseases
National Institutes of Health
Bethesda, Maryland

1. Sinusitis: Pathophysiology and Treatment, *edited by Howard M. Druce*
2. Eosinophils in Allergy and Inflammation, *edited by Gerald J. Gleich and A. Barry Kay*

ADDITIONAL VOLUMES IN PREPARATION

Molecular Biology of the Allergic Immune Response, *edited by Arnold I. Levinson and Yvonne Paterson*

Eosinophils in Allergy and Inflammation

edited by
Gerald J. Gleich
Mayo Clinic and Mayo Foundation
Rochester, Minnesota

A. Barry Kay
National Heart and Lung Institute
London, England

CRC Press
Taylor & Francis Group
Boca Raton London New York

CRC Press is an imprint of the
Taylor & Francis Group, an **informa** business

First published 1994 by Marcel Dekker, Inc.

Published 2019 by CRC Press
Taylor & Francis Group
6000 Broken Sound Parkway NW, Suite 300
Boca Raton, FL 33487-2742

First issued in paperback 2019

ISBN 13: 978-0-367-45599-6 (pbk)
ISBN 13: 978-0-8247-9121-6 (hbk)

Visit the Taylor & Francis Web site at
http://www.taylorandfrancis.com

and the CRC Press Web site at
http://www.crcpress.com

Library of Congress Cataloging-in-Publication Data

Eosinophils in allergy and inflammation / edited by Gerald J. Gleich, A. Barry Kay.
p. cm. -- (Clinical allergy and immunology; 2)
Includes bibliographical references and index.
ISBN 0-8247-9121-5 (alk. paper)
1. Allergy--Pathophysiology. 2. Inflammation--Pathophysiology. 3. Eosinophils. I. Gleich, G. J. (Gerald J.). II. Kay, A. B. III. Series.
[DNLM: 1. Eosinophils--physiology--congresses. 2. Hypersensitivity--etiology--congresses. 3. Inflammation--etiology--congresses. WH 200 E6152 1994
QR188.E57 1994
616.97'071--dc20
DNLM/DLC

93-32584
CIP

Series Introduction

The decision to initiate a series of books on clinical allergy and immunology was based upon the need to create a library of texts useful for both clinicians and scientists in these rapidly enlarging fields. There already are excellent textbooks providing overviews of the fields of allergy and immunology, and the scientific journals attempt to provide concise reviews of selected topics of interest. However, there is no library of books that take relevant topics and expand them into texts, with the express intent of making them of interest to both clinicians and scientists. Thus, this new series.

Clinical Allergy and Immunology will develop into the premier series of texts for our field. The initial book is directed at sinusitis and represents the type of amalgamation of pathophysiology with treatment that will be reflected in most of the books yet to be published. The series will include books focused on areas of allergy, immunology, specific diseases, and important clinical entities, developing areas of research relevant to clinicians, major changes in therapeutic approaches, and research areas that warrant a text. The benefactor of this series should be the patient because physicians will now have a series to seek when searching for concise but authoritative summaries of a field. The other benefactors will be clinicians. Despite the fact that allergy is the single most prevalent chronic disease, only a small number of medical schools incorporate allergy in their curricula, and medical students are notably deficient in the knowledge of these diseases. As allergy is largely an outpatient specialty, few house officers see allergic patients other than asthmatics and patients experiencing anaphylaxis. To clinicians, this series offers the chance to develop an extensive knowledge about selected topics, each chosen for its clinical relevance.

Immunology is the field of the future. Advances in our understanding of immune processes and their relevance to health and disease are increasing at a breakneck pace. As therapeutic approaches become defined, the new relevant immunological knowledge will become incorporated into texts in this series.

The series begins with the publication of the first book, *Sinusitis: Pathogenesis and Treatment*, edited by Howard Druce, MD. The first text represents an extraordinary effort by Dr. Druce, who organized and completed the book in record time. This initial contribution is well written, contains authoritative chapters on all the relevant aspects of sinusitis, and should become the reference for this topic. It is a pleasure to have such a stalwart book initiate the series.

At the moment, some 16 additional books are in the planning and development stages, with about 4 to appear in the next year or so. This commitment involves the combined talents of some 300 scientists and clinicians, and practically guarantees the success of the series.

I accepted the challenge of producing this series with the hope and expectation that clinicians and scientists will find information that will lead to improved understanding of allergic and immunological diseases and their treatment. I hope the readers of these texts agree that we have succeeded.

Eosinophilia is the hallmark of the allergic process and the association between eosinophilia and allergic diseases is so constant that one heralds the other. Moreover, recent work has not only amplified this association, but also has shown that the eosinophil likely plays a critical role in mediating allergic inflammation. Interest in the eosinophil has increased exponentially in the past few years and, with this interest, the number of important observations has exceeded the capacity of the ordinary reader to keep abreast of advances in this area. Thus, it was an easy decision to invite Drs. Gleich and Kay to publish their international symposium in this series.

In this work, the newest understandings of the biology of the eosinophil have been summarized and discussed in a manner that makes the information concisely available to the reader, whether clinician or scientist. As an appropriate extension of the scientific chapters, there are excellent summaries of the roles of the eosinophil in allergic diseases as well. Therefore, this excellent book personifies the type of focus that will be presented in this series of books: current, scientifically sound, clinically relevant, and of use to clinicians and scientists alike.

Michael A. Kaliner

Preface

In 1879 Paul Ehrlich developed a reliable staining method for peripheral blood leukocytes and coined the term eosinophil for a particular type of cell having a strong affinity for acidic dyes. By the end of the last century, the association of peripheral blood eosinophilia with helminth infection and allergic disease was firmly established and, by the early 1900s, the association of eosinophilia (in the blood, sputum, and lung) with bronchial asthma had been made. However, the role of the eosinophil in these diseases remained elusive and it was not until the 1970s that insight into this question was obtained. We now recognize that the eosinophil is armed with a set of powerful toxins contained in its specific granules and is able to generate inflammatory mediators including lipids and partial reductive oxygen molecules. Thus, the eosinophil has the capacity to inflict damage on a variety of targets, and it seems likely (although still not proven) that the eosinophil causes tissue damage in a variety of diseases.

Because the mechanisms by which this presumed damage occurs are at the frontier of our understanding, it seemed reasonable to convene a group of investigators concerned with eosinophils in order to share and evaluate our current knowledge and identify critical questions relating to gaps in our information. Another stimulus for the meeting was (and is) the remarkable increase in the number of publications on eosinophils. The figure plots the annual number of publications on eosinophils from 1965 through 1991 and shows a dramatic increase from the decades of the 1960s and 1970s to the 1980s and early 1990s. Whereas

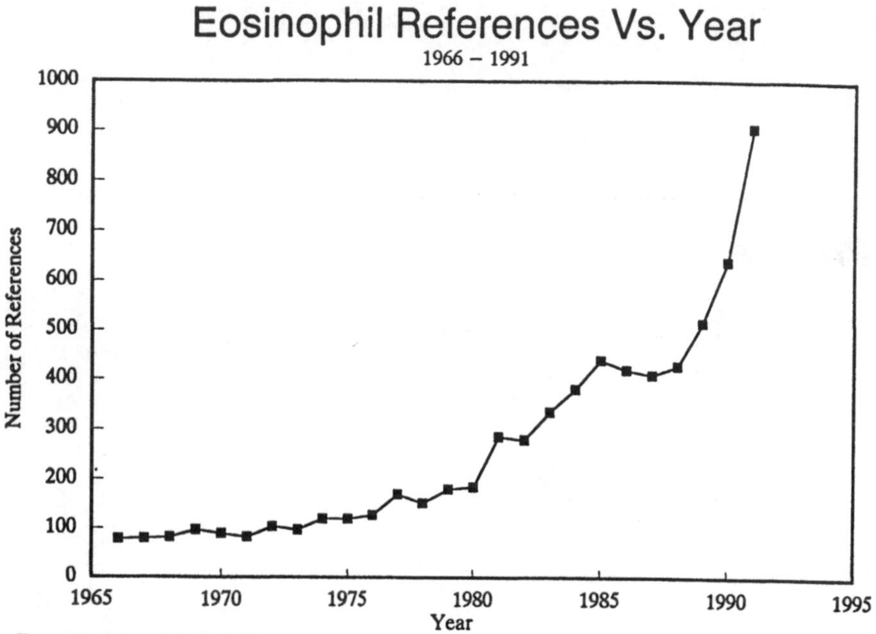

From the Mayo Medical Library

in the 1960s and 1970s only about 100 articles per year citing eosinophils were published, in 1990 almost 1000 such articles were published. This testifies to the surge of interest in the eosinophil and the mechanisms by which it participates in immunity and disease. Finally, because a major international meeting devoted to the eosinophil has not been held since 1979, the time was ripe for a reevaluation.

Fortunately, it was possible to assemble a group of 50 participants from 11 countries at the delightful Adare Manor in County Limerick, Ireland, where for a period of 3 days eosinophils and their activities were deliberated. The result is a marvelous blend of insights ranging from studies of the cell itself to diseases associated with eosinophilia. Included among the topics are studies of eosinophil ultrastructure, eosinophil-associated granule and cytoplasmic proteins, and the ability of the eosinophil to migrate from the bloodstream into the tissues and to become activated, synthesizing mediators, and degranulating. Recent information about the adhesion molecules involved in the interaction of eosinophils with endothelial cells and with extracellular matrix proteins is discussed. Of particular interest are findings that the eosinophil expresses surface receptors that enable it to interact with other cells of the immune system and that it has the ability to synthesize, store, and secrete a number of cytokines, possibly including interleukin (IL)-5. New information about the cytokines regulating eosinophil function is presented, and here IL-5 emerges as a dominant player in the control of eosinophil production by the bone marrow and a major participant in control of eosinophil function as shown by studies of IL-5 transgenic mice, animal experiments using anti-IL-5, and studies in human disease.

A delightful summing-up was done by Dr. Christopher Spry, who served as the conscience of the convention, asking penetrating questions throughout the discussion periods and then cogently and wittily summing up the entire meeting.

We are grateful to UCB Pharma and Pfizer for their generous support of this meeting and, in particular, to Mr. Johan de Clercq and Mr. Willy Dellisse for their personal attention to the planning and conduct of the meeting. A debt of gratitude is also due to the manager and staff of Adare Manor; Ms. Adele Hartnell, Ms. Roma Sehmi, and Dr. Garry Walsh for summarizing so expertly the discussion sections; and Ms. Jennifer Mitchell for her editorial assistance in preparation of this volume.

Gerald J. Gleich
A. Barry Kay

Conference Participants

John S. Abrams DNAX Research Institute of Molecular and Cellular Biology, Inc., Palo Alto, CA

Steven J. Ackerman Infectious Diseases Division, Department of Medicine, Beth Israel Hospital and Harvard Medical School, Boston, MA

A. R. E. Anwar Department of Allergy and Clinical Immunology, National Heart & Lung Institute, London, U.K. (present address: Department of Immunology, Western Infirmary, Glasgow, Scotland, U.K.)

K. Frank Austen Department of Medicine, Harvard Medical School (Department of Rheumatology and Immunology, Brigham and Women's Hospital), Boston, MA

Peter J. Barnes Department of Thoracic Medicine, National Heart & Lung Institute, London, U.K.

Kurt Blaser Swiss Institute for Allergy and Asthma Research, CH-7270 Davos, Switzerland

Bruce S. Bochner Department of Medicine/Clinical Immunology, Johns Hopkins Asthma and Allergy Center, Baltimore, MD

William W. Busse Section of Allergy and Clinical Immunology, Department of Medicine, University of Wisconsin Medical School, Madison, WI

Anthony E. Butterworth Microbiology and Parasitology Division, Department of Pathology, Cambridge University, Cambridge, U.K.

Andre Capron Centre d'Immunologie et de Biologie Parasitaire, Unité Mixte INSERM U 167-CNRS 624, Institut Pasteur, Lille, France

Monique Capron Centre d'Immunologie et de Biologie Parasitaire, Unité Mixte INSERM U 167-CNRS 624, Institut Pasteur, Lille, France

Martin K. Church Clinical Pharmacology Group, Southampton General Hospital, Southampton, Hampshire, U.K.

Robert L. Coffman DNAX Research Institute, Department of Immunology, Palo Alto, CA

Oliver Cromwell Department of Allergy and Clinical Immunology, National Heart & Lung Institute, London, U.K. (present address: Allergopharma, Joachim Ganzer KG, Hamburg, Germany)

Judah Denburg Departments of Medicine and Pathology, McMaster University, Health Sciences Centre, Hamilton, Ontario, Canada

Rene Devos Roche Research Gent, Gent, Belgium

Stephen R. Durham Department of Allergy and Clinical Immunology, National Heart & Lung Institute, London, U.K.

Ann M. Dvorak Electron Microscopy Unit, Department of Pathology, Beth Israel Hospital, Boston, MA

Robert W. Egan Allergy and Inflammation Department, Schering Corporation, Bloomsfield, NJ

Stephen J. Galli Division of Experimental Pathology, Department of Pathology, Boston, MA

Gerald J. Gleich Departments of Immunology and Medicine, Mayo Clinic and Mayo Foundation, Rochester, MN

Robert H. Gundel Department of Pharmacology, Boehringer Ingelheim Pharmaceuticals Inc., Ridgefield, CT

Trevor T. Hansel Swiss Institute for Allergy and Asthma Research, CH-7270 Davos, Switzerland (present address: Human Pharmacology, Sandoz AG, Basel, Switzerland)

Caroline Hebert Cell Biology Department, Genentech, Inc., South San Fransisco, CA

Stephen T. Holgate Immunopharmacology Group, Medicine I and Clinical Pharmacology, Southampton General Hospital, Southampton, Hampshire, U.K.

A. Barry Kay Department of Allergy and Clinical Immunology, National Heart & Lung Institute, London, U.K.

Hirohito Kita Allergic Diseases Laboratory, Mayo Clinic, Rochester, MN

Wolfgang Konig Ruhr-Universitat Bochum, Medizinische Mikrobiologie und Immunologie, Bochum, Germany

Tak H. Lee Department of Allergy and Allied Respiratory Disorders, Guy's Hospital Medical School, London, U.K.

Kristin M. Leiferman Dermatology Research Unit, Mayo Clinic and Mayo Foundation, Rochester, MN

Angel Lopez Division of Human Immunology, Hanson Centre for Cancer Research, Institute of Medical and Veterinary Science, Adelaide, South Australia

Sohei Makino Department of Medicine & Clinical Immunology, Dokkyo University School of Medicine, MIBU, Tochigi-Ken, Japan

Redwan Moqbel Department of Allergy and Clinical Immunology, National Heart & Lung Institute, London, U.K.

John Morley Preclinical Research, Pharma Division, Sandoz Ltd., Basel, Switzerland

William F. Owen, Jr. Department of Medicine, Harvard Medical School (Department of Rheumatology and Immunology, Brigham and Women's Hospital), Boston, MA

Roland Repp Medizinische Universitätsklinik III, Universität Erlangen, D-8520 Erlangen, Germany

Jean-Pierre Rihoux UCB–Pharmaceutical Sector, Braine-l'Alleud, Belgium

Dirk Roos Central Laboratory of the Netherlands Red Cross Blood Transfusion Service and Laboratory of Experimental and Clinical Immunology, University of Amsterdam, Amsterdam, The Netherlands

Colin J. Sanderson Searle, High Wycombe, Bucks, U.K. (present address: Western Australian Research Institute for Child Health, Princess Margaret Hospital for Children, Perth, Australia)

Robert P. Schleimer Johns Hopkins Asthma & Allergy Center, Baltimore, MD

David S. Silberstein Department of Rheumatology and Immunology, Harvard Medical School, Boston, MA

Christopher J. F. Spry Departments of Cellular & Molecular Sciences & Cardiological Sciences, St. George's Hospital Medical School, London, U.K.

Kiyoshi Takatsu Department of Biology, Institute for Medical Immunology, Kumamoto University Medical and Department of Immunology, Institute of Medical Science, University of Tokyo, Tokyo, Japan

Larry L. Thomas Department of Immunology/Microbiology, Rush-Presbyterian-St. Luke's Medical Center, Chicago, IL

Robert G. Townley Pulmonary-Allergy Division, Creighton University School of Medicine, Omaha, NE

Mathew Vadas Hanson Centre for Cancer Research, Division of Human Immunology, Institute of Medical and Veterinary Sciences, Adelaide, South Australia

Per Venge Department of Clinical Chemistry, University Hospital, Uppsala, Sweden

Christine De Vos UCB–Pharmaceutical Sector, Braine-l'Alleud, Belgium

Andrew J. Wardlaw Department of Medicine, University of Leicester School of Medicine, Glenfield General Hospital, Leicester, U.K.

Peter F. Weller Harvard Thorndike Laboratory, Charles A. Dana Research Institute, Harvard Medical School, Department of Medicine, Beth Israel Hospital, Boston, MA

Contents

Contributors

John S. Abrams DNAX Research Institute of Molecular and Cellular Biology, Palo Alto, California

Randa I. Abu-Ghazaleh Department of Biochemistry and Molecular Biology, Mayo Clinic and Mayo Foundation, Rochester, Minnesota

Steven J. Ackerman Infectious Diseases Division, Department of Medicine, Beth Israel Hospital and Harvard Medical School, Boston, Massachusetts

A. R. E. Anwar Department of Allergy and Clinical Immunology, National Heart and Lung Institute, London, England

K. Frank Austen Department of Medicine, Harvard Medical School and Department of Rheumatology and Immunology, Brigham and Women's Hospital, Boston, Massachusetts

Peter J. Barnes Department of Thoracic Medicine, National Heart and Lung Institute, London, England

S. Barry Division of Human Immunology, Institute of Medical and Veterinary Sciences, Hanson Centre for Cancer Research, Adelaide, South Australia

Andrew M. Bentley Department of Allergy and Clinical Immunology, National Heart and Lung Institute, London, England

M. C. Berndt Vascular Biology Laboratory, Baker Medical Research Institute, Prahran, Victoria, Australia

Michela Blom Central Laboratory of the Netherlands Red Cross Blood Transfusion Service and Laboratory of Experimental and Clinical Immunology, University of Amsterdam, Amsterdam, The Netherlands

Bruce S. Bochner Department of Medicine/Clinical Immunology, Johns Hopkins Asthma and Allergy Center, Baltimore, Maryland

William W. Busse Section of Allergy and Clinical Immunology, Department of Medicine, University of Wisconsin Medical School, Madison, Wisconsin

William F. Calhoun Section of Allergy and Clinical Immunology, Department of Medicine, University of Wisconsin Medical School, Madison, Wisconsin

M. G. Campos Immunopharmacology Group, Southampton General Hospital, Southampton, England

André Capron Centre d'Immunologie et de Biologie Parasitaire, Institut Pasteur, Lille, France

Monique Capron Centre d'Immunologie et de Biologie Parasitaire, Institut Pasteur, Lille, France

M. K. Church Immunopharmacology Group, Southampton General Hospital, Southampton, England

Mike A. Clark Allergy Division, Schering-Plough Research, Kennelworth, New Jersey

Robert L. Coffman Department of Immunology, DNAX Research Institute, Palo Alto, California

Christopher J. Corrigan Department of Allergy and Clinical Immunology, National Heart and Lung Institute, London, England

John J. Costa Division of Experimental Pathology, Department of Pathology, Beth Israel Hospital and Harvard Medical School, Boston, Massachusetts

Oliver Cromwell Department of Allergy and Clinical Immunology, National Heart and Lung Institute, London, England

Judah A. Denburg Departments of Medicine and Pathology, McMaster University, Hamilton, Ontario, Canada

Lindsay A. Dent Department of Microbiology, University of South Australia, Adelaide, Australia

Pierre Desreumaux Centre d'Immunologie et de Biologie Parasitaire, Institut Pasteur, Lille, France

Jerry Dolovich Departments of Medicine and Pathology, McMaster University, Hamilton, Ontario, Canada

Stephen R. Durham Department of Allergy and Clinical Immunology, National Heart and Lung Institute, London, England

Ann M. Dvorak Harvard Medical School and Electron Microscopy Unit, Department of Pathology, the Charles A. Dana Research Institute, Beth Israel Hospital, Boston, Massachusetts

P. Dyson Division of Haematology, Institute of Medical and Veterinary Sciences, Hanson Centre for Cancer Research, Adelaide, South Australia

Motohiro Ebisawa Johns Hopkins Asthma and Allergy Center, Baltimore, Maryland

Aram Elovic Department of Oral Medicine and Oral Pathology, Harvard School of Dental Medicine, Boston, Massachusetts

Susetta Finotto Departments of Medicine and Pathology, McMaster University, Hamilton, Ontario, Canada

Stephen J. Galli Division of Experimental Pathology, Department of Pathology, Beth Israel Hospital and Harvard Medical School, Boston, Massachusetts

J. R. Gamble Division of Human Immunology, Institute of Medical and Veterinary Sciences, Hanson Centre for Cancer Research, Adelaide, South Australia

Steve N. Georas Johns Hopkins Asthma and Allergy Center, Baltimore, Maryland

Peter Gibson Departments of Medicine and Pathology, McMaster University, Hamilton, Ontario, Canada

Mark A. Giembycz Department of Thoracic Medicine, National Heart and Lung Institute, London, England

Gerald J. Gleich Departments of Immunology and Medicine, Mayo Clinic and Mayo Foundation, Rochester, Minnesota

Dohn G. Glitz Department of Biological Chemistry, UCLA School of Medicine, University of California, Los Angeles, California

Bastien D. Gomperts Department of Physiology, University College, London, England

John R. Gordon Division of Experimental Pathology, Department of Pathology, Beth Israel Hospital and Harvard Medical School, Boston, Massachusetts

Robert H. Gundel Department of Pharmacology, Boehringer Ingelheim Pharmaceuticals, Inc., Ridgefield, Connecticut

Qutayba Hamid Department of Allergy and Clinical Immunology, National Heart and Lung Institute, London, England

Fred Hargreave Departments of Medicine and Pathology, McMaster University, Hamilton, Ontario, Canada

Adele Hartnell Department of Allergy and Clinical Immunology, National Heart and Lung Institute, London, England

T. Hercus Division of Human Immunology, Institute of Medical and Veterinary Sciences, Hanson Centre for Cancer Research, Adelaide, South Australia

S. T. Holgate Immunopharmacology Group, Medicine I and Clinical Pharmacology, Southampton General Hospital, Southampton, England

T. C. Hunt Immunopharmacology Group, Southampton General Hospital, Southampton, England

Charles G. Irvin Pulmonary Physiology Division, Department of Medicine, National Jewish Center for Immunology and Respiratory Medicine, Denver, Colorado

Manel Jordana Departments of Medicine and Pathology, McMaster University, Hamilton, Ontario, Canada

A. Barry Kay Department of Allergy and Clinical Immunology, National Heart and Lung Institute, London, England

Leo Koenderman Department of Pulmonary Diseases, Academic Hospital, University of Utrecht, Utrecht, The Netherlands

Bouchaïb Lamkhioued Centre d'Immunologie et de Biologie Parasitaire, Institut Pasteur, Lille, France

Kristin M. Leiferman Department of Dermatology, Mayo Clinic and Mayo Foundation, Rochester, Minnesota

L. Gordon Letts Department of Pharmacology, Boehringer Ingelheim Pharmaceuticals, Inc., Ridgefield, Connecticut

Ming-Shi Li Departments of Cellular and Molecular Sciences and Cardiological Sciences, St. George's Hospital Medical School, London, England

Manfred Lindau Max Planck Institute for Medical Research, Heidelberg, Germany

Angel F. Lopez Division of Human Immunology, Institute of Medical and Veterinary Sciences, Hanson Centre for Cancer Research, Adelaide, South Australia

C. M. Lucas Division of Human Immunology, Institute of Medical and Veterinary Sciences, Hanson Centre for Cancer Research, Adelaide, South Australia

Jean Marshall Departments of Medicine and Pathology, McMaster University, Hamilton, Ontario, Canada

Redwan Moqbel Department of Allergy and Clinical Immunology, National Heart and Lung Institute, London, England

Ian Mudway National Institute for Medical Research, London, England

Yoshiyuki Murata Department of Biology, Institute for Medical Immunology, Kumamoto University Medical School, Kumamoto, and University of Tokyo, Tokyo, Japan

Isao Ohno Departments of Medicine and Pathology, McMaster University, Hamilton, Ontario, Canada

William F. Owen, Jr. Department of Medicine, Harvard Medical School and Department of Rheumatology and Immunology, Brigham and Women's Hospital, Boston, Massachusetts

Isabelle Pin Departments of Medicine and Pathology, McMaster University, Hamilton, Ontario, Canada

Douglas S. Robinson Department of Allergy and Clinical Immunology, National Heart and Lung Institute, London, England

Dirk Roos Central Laboratory of the Netherlands Red Cross Blood Transfusion Service and Laboratory of Experimental and Clinical Immunology, University of Amsterdam, Amsterdam, The Netherlands

Colin J. Sanderson Western Australian Research Institute for Child Health, Perth, Australia

Takahiro Satoh Departments of Cellular and Molecular Sciences and Cardiological Sciences, St. George's Hospital Medical School, London, England

Robert P. Schleimer Johns Hopkins Asthma and Allergy Center, Baltimore, Maryland

Julie B. Sedgwick Section of Allergy and Clinical Immunology, Department of Medicine, University of Wisconsin Medical School, Madison, Wisconsin

M-C. Seminario Immunopharmacology Group, Southampton General Hospital, Southampton, England

M. F. Shannon Division of Human Immunology, Institute of Medical and Veterinary Sciences, Hanson Centre for Cancer Research, Adelaide, South Australia

J. K. Shute Immunopharmacology Group, Southampton General Hospital, Southampton, England

Jon E. Silver DNAX Research Institute of Molecular and Cellular Biology, Palo Alto, California

P. Simmons Division of Haematology, Institute of Medical and Veterinary Sciences, Hanson Centre for Cancer Research, Adelaide, South Austalia

M. P. Skinner Vascular Biology Laboratory, Baker Medical Research Institute, Prahran, Victoria, Australia

Christopher J. F. Spry Departments of Cellular and Molecular Sciences and Cardiological Sciences, St. George's Hospital Medical School, London, England

Malcolm Strath National Institute for Medical Research, London, England

Li Sun Departments of Cellular and Molecular Sciences and Cardiological Sciences, St. George's Hospital Medical School, London, England

Satoshi Takaki Department of Biology, Institute for Medical Immunology, Kumamoto University Medical School, Kumamoto, and Department of Immunology, Institute of Medical Science, University of Tokyo, Tokyo, Japan

Kiyoshi Takatsu Department of Biology, Institute for Medical Immunology, Kumamoto University Medical School, Kumamoto, and Department of Immunology, Institute of Medical Science, University of Tokyo, Tokyo, Japan

Daniel G. Tenen Hematology/Oncology Division, Department of Medicine, Beth Israel Hospital and Harvard Medical School, Boston, Massachusetts

Margherita Tomassini Centre d'Immunologie et de Biologie Parasitaire, Institut Pasteur, Lille, France

Akira Tominaga Department of Biology, Institute for Medical Immunology, Kumamoto University Medical School, Kumamoto, and University of Tokyo, Tokyo, Japan

Anton T. J. Tool Central Laboratory of the Netherlands Red Cross Blood Transfusion Service and Laboratory of Experimental and Clinical Immunology, University of Amsterdam, Amsterdam, The Netherlands

Marie-José Truong Centre d'Immunologie et de Biologie Parasitaire, Institut Pasteur, Lille, France

Anne Tsicopoulos Department of Allergy and Clinical Immunology, National Heart and Lung Institute, London, England

Yuan-Po Tu National Jewish Center for Immunology and Respiratory Medicine, Denver, Colorado

Mathew A. Vadas Division of Human Immunology, Institute of Medical and Veterinary Sciences, Hanson Centre for Cancer Research, Adelaide, South Australia

Robert E. Van Dyke Department of Molecular Pharmacology, Arris Pharmaceutical Corporation, South San Francisco, California

Arthur J. Verhoeven Central Laboratory of the Netherlands Red Cross Blood Transfusion Service and Laboratory of Experimental and Clinical Immunology, University of Amsterdam, Amsterdam, The Netherlands

Christine De Vos UCB SA, Pharma Sector, Braine-l'Alleud, Belgium

Garry M. Walsh Department of Allergy and Clinical Immunology, National Health and Lung Institute, London, England

Andrew J. Wardlaw Department of Allergy and Clinical Immunology, National Heart and Lung Institute, London, England

Craig D. Wegner Department of Pharmacology, Boehringer Ingelheim Pharmaceuticals, Inc., Ridgefield, Connecticut

Peter F. Weller Department of Medicine, Beth Israel Hospital and Harvard Medical School, Boston, Massachusetts

Barry K. Wershil Division of Experimental Pathology, Department of Pathology, Beth Israel Hospital and Harvard Medical School, Boston, Massachusetts

David T. W. Wong Department of Oral Medicine and Oral Pathology, Harvard School of Dental Medicine, Boston, Massachusetts

J. Woodcock Division of Human Immunology, Institute of Medical and Veterinary Sciences, Hanson Centre for Cancer Research, Adelaide, South Australia

Sun Ying Department of Allergy and Clinical Immunology, National Heart and Lung Institute, London, England

Zeqi Zhou Infectious Diseases Division, Department of Medicine, Beth Israel Hospital and Harvard Medical School, Boston, Massachusetts

Eosinophils in Allergy and Inflammation

1

Eosinophil Granule Proteins: Structure and Function

Gerald J. Gleich and Randa I. Abu-Ghazaleh
Mayo Clinic and Mayo Foundation, Rochester, Minnesota

Dohn G. Glitz
UCLA School of Medicine, University of California, Los Angeles, California

I. INTRODUCTION

The pioneering studies of white blood cell morphology by Paul Ehrlich established that one granulated leukocyte stained with essentially all acid dyes (and Ehrlich tried over 30 acid dyes), but this same leukocyte was not stained by the basic aniline dyes (1). Because Ehrlich discovered that the acid dye eosin was especially useful for staining this cell, the peripheral blood leukocyte containing the acidophilic granules was named the eosinophil. Ehrlich's observations presumed the existence of basic charge in the eosinophil, and subsequent observations have abundantly substantiated this presumption. Here, information on the principal cationic eosinophil granule proteins will be reviewed with emphasis on the toxicity of the major basic protein (MBP) and the ability of the eosinophil-derived neurotoxin (EDN) and the eosinophil cationic protein (ECP) to elicit the neurotoxic reaction in rabbits referred to as the Gordon phenomenon.

II. EOSINOPHIL GRANULE PROTEINS

Analyses of eosinophils from guinea pigs and humans have established the existence of four predominant cationic proteins, referred to as MBP, the eosinophil peroxidase (EPO), EDN, and ECP.

A. Major Basic Protein

MBP is so named because in the guinea pig it accounts for approximately 55% of granule protein, its isoelectric point is >10, and it is proteinaceous in nature (2–4). Analysis of guinea pig MBP by sodium dodecyl sulfate–polyacrylamide gel electrophoresis revealed a molecular weight of approximately 11,000 and showed that the molecule is rich in arginine and has a marked propensity to polymerize on the basis of the formation of disulfide bonds. Analyses of human MBP showed that it has similar properties (5). Detailed studies of human MBP indicated that it consists of a single polypeptide chain of 117 amino acids, has a molecular weight of about 14,000, and is rich in arginine with a calculated isoelectric point of 10.9 (6,7). The MBP cDNA specifies the existence of a prepromolecule with a 15-amino-acid leader sequence and a 90-amino-acid prosequence followed by the 117-amino-acid sequence for MBP itself (7,8). The 90-amino-acid pro-portion is markedly enriched in acidic amino acids, especially glutamic acid, and has an isoelectric point of 3.9. When the pro-portion of MBP and the mature protein are combined, one obtains a molecule of 270 amino acids with roughly equal numbers of strongly basic and strongly acidic amino acids and an isoelectric point of 6.2. The balance of charge between the pro-portion and MBP itself suggests that the pro-portion serves to neutralize the toxic properties of MBP. In this view, the pro-portion would protect the cell from the toxic effects of MBP during the transport of pro-MBP from the Golgi apparatus to the eosinophil granule. Of interest, the cDNA of barley toxin-α-hordothionin, another toxic cationic protein, codes for a nearly neutral precursor (9). Here, the isoelectric point of the barley toxin-α-hordothionin is 9.6 and the isoelectric point of the pro-portion is 3.6 so that the isoelectric point of the entire molecule is 7.6.

The occurrence of a pro-portion in MBP has also been observed in the guinea pig (10,11). Two types of guinea pig MBP, MBP-1 and MBP-2, have been isolated and partially sequenced. Analyses of cDNA indicated that both types show a prepro-MBP with three domains consisting of a signal peptide, an acidic pro-portion, and mature MBP. Whereas guinea pig MBP-1 and MBP-2 are quite similar and resemble human MBP, the pro-portions of the molecule are not very homologous to the human propiece despite similar pI values. These results indicate that the synthesis of MBP proceeds through a promolecule with a markedly acidic pro-portion in both humans and guinea pigs. Furthermore, while the MBP portions are quite homologous, the pro-portions are less so, suggesting that anionic charge and not sequence homology is critical.

Information regarding MBP localization and MBP function is summarized in Table 1 and reviewed in detail in Ref. 12. Briefly, MBP is localized to the eosinophil granule; it is also present in basophils and placental X cells (13). MBP is a potent toxin and kills parasites, both helminths and protozoa, bacteria, and mammalian cells. It causes histamine release from basophils and rat mast cells, activates neutrophils and platelets, and causes bronchoconstriction and bronchial hyperreactivity when instilled into the lungs of monkeys.

Table 1 Some Properties of Cationic Human Eosinophil Granule Proteins and Their Encoding cDNA and Genes

Protein	Site	M_r (× 10^{-3})	Isoelectric point[a]	Cell content (μg/10^6 eos)	Activities	Molecular biology	
						cDNA	Gene
MBP	Core	14	10.9	9	(1) Potent helminthotoxin and cytotoxin, (2) causes histamine release from basophils and rat mast cells, (3) neutralizes heparin, (4) bactericidal, (5) increases bronchial reactivity to methacholine in primates, (6) unique, strong platelet agonist, (7) provokes bronchospasm in primates, (8) activates neutrophils	~900 nt[b] (prepro-MBP)	3.3 kb 5 introns 6 exons
ECP	Matrix	18–21	10.8	5	(1) Potent helminthotoxin, (2) potent neurotoxin, (3) inhibits cultures of peripheral blood lymphocytes, (4) causes histamine release from rat mast cells, (5) weak RNase activity, (6) bactericidal, (7) neutralizes heparin and alters fibrinolysis	~725 nt (pre-ECP)	~ 1.2 kb 1 intron in UTR
EDN	Matrix	18–19	8.9	3	(1) Potent neurotoxin, (2) inhibits cultures of peripheral blood lymphocytes, (3) potent RNase activity, (4) weak helminthotoxin, (5) identical to eosinophil protein X	~725 nt (pre-EDN)	~ 1.2 kb 1 intron in UTR
EPO	Matrix	66	10.8	12	In the presence of H_2O_2+halide: (1) kills microorganisms and tumor cells, (2) causes histamine release and degranulation from rat mast cells, (3) inactivates leukotrienes, (4) in the absence of H_2O_2+halide kills *Brugia* microfilariae, (5) damages respiratory epithelium, (6) provokes bronchospasm in primates	~2500 nt (210b nt ORF)	12 kb 11 introns 12 exons

[a]Calculated from amino acid sequences deduced from the cDNAs.
[b]nt = nucleotides; kb = kilobases; ORF = open reading frame; UTR = untranslated region; see Ref. 12 for references concerning the molecular biology of these proteins and their genes.
Source: Used with permission from Ref. 12.

B. Eosinophil Peroxidase

EPO differs from neutrophil myeloperoxidase (MPO) in its physicochemical properties and in its deduced amino acid sequence (12). Human EPO consists of two subunits, a heavy chain with M_r 54–58 × 10^3 and a light chain with M_r 10.5–15.5 × 10^3 (Table 1). EPO cDNA has been cloned and consists of an open reading frame of 2106 nucleotides coding for a 381-base-pair prosequence, a 333-base-pair sequence coding for the EPO light chain, a 1392-base-pair sequence coding for the EPO heavy chain, and a 452-base-pair untranslated 3′ region. EPO belongs to the peroxidase multigene family, including EPO, MPO, thyroid peroxidase, and lactoperoxidase.

When armed with H_2O_2 and a halide, such as Br^-, EPO generates HOBr, a potent oxidant. Similar reactions occur with I^- and Cl^-, although EPO prefers Br^-. The resultant hypohalous acids (most experiments used I^-) kill most biological targets and are able to stimulate histamine release from rat mast cells (12). However, the ability of EPO+H_2O_2+halide to generate potent oxidizing hypohalous acids must be reexamined in light of recent information that EPO prefers SCN^- over Br^- by a 100-fold margin (14) and that hypothiocyanous acid, HOSCN, is a weak oxidant. Finally, even in the presence of catalase, which consumes H_2O_2, the cationic EPO is a potent toxin comparable to MBP (12).

C. Eosinophil-Derived Neurotoxin

EDN, initially identified by its ability to cause a neurotoxic reaction in rabbits (and guinea pigs) termed the Gordon phenomenon, is a member of the RNase gene superfamily composed of EDN, ECP, RNase, and angiogenin (Table 1). EDN is almost as potent an RNase as pancreatic RNase and approximately 100-fold more potent as an RNase than ECP, but it is considerably less toxic to parasites and cells than ECP (12).

D. Eosinophil Cationic Protein

Although ECP is also a member of the RNase gene superfamily and a potent neurotoxin, it differs from EDN by its marked basicity, pI~11 (Table 1). ECP is a more potent toxin than EDN and kills bacteria, parasites, and mammalian cells by forming voltage-sensitive ionic pores, as has been shown in planar lipid bilayers (12).

III. THE TOXICITY OF EOSINOPHIL GRANULE MBP

A. Effects of Acidic Polyamino Acids

To test the hypothesis that the acidic pro-portion of pro-MBP inhibits the toxicity of mature MBP, acidic polyamino acids were used as substitutes for the pro-portion of pro-MBP (15). Table 2 shows the properties of the amino acids, polyamino acids, and proteins tested. A variety of polyamino acids were employed, with different degrees of polymerization and different net charge, and the concentrations needed to balance the cationic charge of MBP at 5×10^{-6} M were calculated. In the inhibition experiments, homopolymers of glutamic acid and aspartic acid of differing lengths were added to K562 cells followed by addition of MBP, and cell viabilities were determined after 4 h. Table 3 shows the relative abilities of the polyamino acids to inhibit MBP toxicity. At equimolar concentrations, all of the acidic polyamino acids inhibited MBP toxicity. This inhibition was related to the anionic nature of the polymers because poly-L-asparagine at 5×10^{-6} M did not inhibit MBP toxicity. Further, polymerization was necessary for the inhibitory effect because the acidic amino acid monomers L-aspartic acid and L-glutamic acid did not inhibit MBP toxicity. The acidic polyamino acids themselves were not toxic to the K562 cells at the concentrations tested (Table 3).

As the molar concentrations of the acidic polyamino acids were reduced below that of MBP, inhibition of MBP toxicity to the K562 cells was reduced and then lost. The reduced inhibition was dependent on the molecular weight and, thus, the degree of polymerization of the acidic polyamino acid. Inhibition was not influenced by the type of acidic polyamino acid in that both D and L isomers of glutamic and aspartic acids were active. Furthermore, concentrations of the acidic polyamino acids less than those necessary to balance the cationic charge of MBP showed a drastic reduction in inhibition of MBP (at 5×10^{-6} M) toxicity. For example, as shown in Table 3, poly-L-glutamic acid with a molecular weight of 77,800 was not toxic for K562 cells and completely neutralized the toxicity of MBP to K562 cells at concentrations as low as 5×10^{-7} M; at this concentration, poly-L-glutamic acid provides negative charge sufficient to balance the positive electrical charge of MBP. However, a further 50-fold reduction in the poly-L-glutamic acid concentration produced a total loss in inhibition.

To more closely define the point at which acidic polyamino acids lost their ability to inhibit MBP toxicity, poly-L-glutamic acid (6800 daltons), poly-L-glutamic acid (13,600 daltons), poly-L-aspartic acid (42,500 daltons), and poly-L-glutamic acid (77,800 daltons) were tested at 2.0, 1.0, and 0.5 times the concentration necessary to balance the charge of MBP at 5×10^{-6} M. At a concentration of two times that needed to balance the charge of MBP, all four polymers inhibited MBP toxicity to K562 cells and they were not significantly different from the medium control. At concentrations yielding charges equal to that of MBP (5×10^{-6} M), the polyglutamic acids appeared to protect somewhat better than the polyaspartic acids. Finally, at a concentration of 0.5 times the cationic charge of MBP, the acidic polymers afforded little protection against MBP toxicity.

In the experiments described above, the acidic polyamino acids were added to the K562 cells before MBP. When the acidic polyamino acids were added simultaneously with MBP to the K562 cells, they were also quite effective in

Table 2 Selective Properties of Amino Acids, Polyamino Acids, and Proteins Studied

	Mol wt	Degree of polymerization[a]	Charge (e) at pH 7.0[b]	Concentration for balanced charge[c]
L-Aspartic acid (A-0651)[d]	155	1	−1.1	7.5×10^{-5} M
L-Glutamic acid salt (G-1626)	169	1	−1.1	7.5×10^{-5} M
Poly-(α,β)-DL-aspartic acid (P-3418)	6,800	50	−50.1	1.6×10^{-6} M
Poly (aspartic acid, glutamic acid) 1:1 (P-1408)	9,000	60	−60.0	1.4×10^{-6} M
Poly-L-aspartic acid (P-5387)	11,500	84	−84.0	9.7×10^{-7} M
Poly-L-glutamic acid (P-4636)	13,600	90	−89.9	9.1×10^{-7} M
Polyglutamic acid (P-4761)	36,240	240	−239.0	3.4×10^{-7} M
Poly-D-glutamic acid (P-4033)	41,000	272	−271.0	3.0×10^{-7} M
Poly-L-aspartic acid (P-6762)	42,500	310	−310.0	2.6×10^{-7} M
Poly-D-glutamic acid (P-4637)	66,000	437	−436.0	1.9×10^{-7} M
Poly-L-glutamic acid (P-4886)	77,800	515	−514.0	1.6×10^{-7} M
Poly-L-asparagine (P-8137)	10,400	91	−0.1	
Native MBP	13,801	117	16.3	

[a]Degree of polymerization indicates number of polyamino acid residues per molecule.
[b]Charge (e) calculated by Titrate program (DNAstar, Madison, WI).
[c]Concentraion of acidic polyamino acid necessary to balance charge of cationic MBP at 5×10^{-6} M.
[d]All acids used were purchased as sodium salts; Sigma product code appears in parentheses after the product name.
Source: Revised and used with permission from Ref. 15.

Table 3 Inhibition of MBP Toxicity by Acidic Amino and Polyamino Acids

			Percent viability of K562 cells[a]					
			Native MBP (5×10^{-6} M) added to aa concentration					
	Mol wt	Controls[b]	1×10^{-3} M	5×10^{-6} M	1×10^{-6} M	5×10^{-7} M	1×10^{-7} M
L-Aspartic acid	155	98.5 ± 0.2	3.4 ± 4.7[c]	ND[d]	ND	ND	ND
L-Glutamic acid	169	98.8 ± 0.2	0.0 ± 0.0[c]	ND	ND	ND	ND
Poly-(α,β)-DL-aspartic acid	6,800	97.0 ± 2.7	ND	*90.1 ± 6.7*[e]	1.6 ± 1.5	6.0 ± 3.9[c]	2.9 ± 0.4[c]
Poly (aspartic acid, glutamic acid) 1:1	9,000	98.4 ± 1.4	ND	*95.5 ± 0.9*	22.0 ± 19.7[c]	4.9 ± 2.3[c]	0.0 ± 0.0[c]
Poly-L-aspartic acid	11,500	98.0 ± 2.4	ND	87.5 ± 7.9	*84.3 ± 1.6*[c]	22.7 ± 7.2[c]	2.7 ± 3.8[c]
Poly-L-glutamic acid	13,600	98.9 ± 1.0	ND	89.5 ± 2.0	*72.9 ± 0.7*[c]	4.9 ± 5.9[c]	0.0 ± 0.0[c]
Poly-L-glutamic acid	36,240	94.7 ± 2.6	ND	96.2 ± 6.5	96.9 ± 1.8	*67.5 ± 20.4*	1.3 ± 1.8[c]
Poly-D-glutamic acid	41,000	100.0 ± 0.0	ND	98.0 ± 1.8	99.0 ± 1.7	*96.2 ± 1.4*	11.2 ± 0.9[c]
Poly-L-aspartic acid	42,500	97.4 ± 2.3	ND	100.0 ± 0.0	99.4 ± 0.9	*100.0 ± 0.0*	8.7 ± 1.4[c]
Poly-D-glutamic acid	66,000	93.4 ± 6.1	ND	96.4 ± 1.7	97.3 ± 1.0	*90.6 ± 3.4*	0.0 ± 0.0[c]
Poly-L-glutamic acid	77,800	95.1 ± 3.8	ND	93.9 ± 8.0	95.5 ± 6.4	*95.4 ± 3.2*	2.6 ± 3.5[c]
Poly-L-asparagine	10,400	85.3 ± 5.1	ND	5.9 ± 9.5[c]	1.2 ± 2.1[c]	0.0 ± 0.0[c]	0.0 ± 0.0[c]
Native MBP	13,801	1.0 ± 3.2					
Column buffer		96.7 ± 1.6					
Medium		96.5 ± 3.7					

[a]Values are mean percent viability ± 1 SD of one to three 4-h experiments, each consisting of duplicate wells.
[b]Controls consist of amino acid (aa), acidic poly aa, poly-L-asparagine, native MBP, column buffer, or medium tested alone. All controls at 5×10^{-6} M except L-aspartic and L-glutamic acid monomers, which were tested at 1×10^{-3} M.
[c]Indicates $p < 0.001$ for values tested against appropriate nontoxic aa control using Student's *t*-test; MBP added within 5 min after addition of acidic aa.
[d]ND, not determined.
[e]Values in italics are from acidic poly aa test concentrations closest to but not < concentration for balanced charge (Table 1) for a particular acidic poly aa.
Source: Used with permission from Ref. 15.

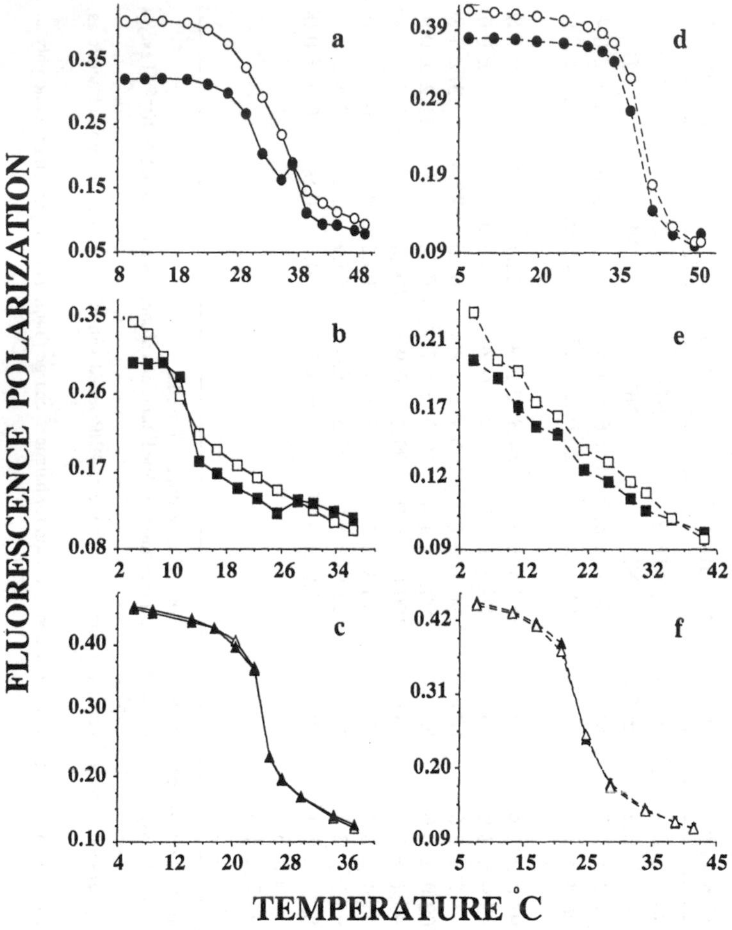

Figure 1 Temperature transition profiles of liposomes as measured by the fluorescence polarization of DPH in liposomes made from different lipids (2 mM) and exposed to MBP (closed symbols) or an equivalent volume of buffer (open symbols). (a, b, and c) The effect of raMBP compared to PBS controls. (d, e, and f) The effect of nMBP compared to acetate buffer controls. Standard deviations are within the symbols. (a) 1,2-Dimyristoyl-*sn*-glycero-3-phosphocholine:1,2-dimyristoyl-*sn*-glycero-3-phosphatidic acid liposomes (circles) with raMBP (10 μM) or PBS. The difference on addition of raMBP is significant ($p \leqslant 0.0001$) below 35°C. (b) 1-Palmitoyl-2-oleoyl-*sn*-glycero-3-phosphatidyl serine liposomes (squares) with raMBP (10 μM) or PBS. The difference on addition of raMBP is significant ($p \leqslant 0.001$) below 10°C. (c) 1,2-Dimyristoyl-*sn*-glycero-3-phosphocholine liposomes (triangles, 2 mM) with raMBP (5 μM) or PBS. (d) 1,2-Dimyristoyl-*sn*-glycero-3-phosphocholine:1,2-dimyristoyl-*sn*-glycero-3-phosphatidic acid liposomes (circles, dashed line) with nMBP (5 μM) or acetate buffer. The difference on addition of nMBP is significant

neutralizing the toxicity of MBP. Further, some protection was evident when the acidic polyamino acids were added within 15 min after exposure of K562 cells to MBP. However, when the acidic polymers were added one or more hours after MBP, little protection was seen (because approximately 90% of cells were killed within 30 min following exposure to MBP).

Thus, the acidic polyamino acids are potent inhibitors of MBP when permitted to interact with target cells before or at the time of exposure to MBP. Some residual inhibitory effect is seen as long as 15 min after exposure of K562 cells to MBP, but no inhibitory effects of the acidic polyamino acids were observed at 1 h and 2 h even at concentrations of polyamino acids up to 5×10^{-5} M. Thus, these results encourage belief that acidic polyamino acids are effective inhibitors and support the hypothesis that the pro-portion of MBP functions to inhibit the toxicity of MBP during the processing of pro-MBP to mature MBP.

B. Interactions of MBP with Synthetic Lipid Bilayers

As noted above, MBP is a potent toxin to a variety of targets including mammalian cells, parasites, and bacteria (16–19), is able to cause histamine release from basophils and mast cells (20), activates neutrophils (21), and induces bronchoconstriction and bronchial hyperreactivity in primates (22). The broad spectrum of MBP's activities suggests that the plasma membrane may be its target. To test this hypothesis, the effect of MBP on liposomes was analyzed (23). First, the temperature transition profiles of liposomes (which indicate their ordered state) made from various lipids were analyzed in the presence and absence of MBP. Liposomes were prepared utilizing a variety of neutral and acidic lipids in the presence of 1,6-diphenyl-1,3,5-hexatriene (DPH), a fluorescent indicator probe. Figure 1 shows the temperature transition profiles of lipids and reveals that both mildly reduced and alkylated MBP (raMBP) and native MBP (nMBP) change the ordered state of only acidic lipids (Fig. 1a, 1b, 1d, 1e). No change in the transition temperature profiles was observed when raMBP or nMBP was added to zwitterionic liposomes made of 1,2-dimyristoyl-*sn*-glycero-3-phosphocholine (Fig. 1c, 1f). Furthermore, MBP did not interact with other zwitterionic liposomes made from

---

($p \leq 0.0001$) below 48°C. (e) 1-Palmitoyl-2-oleoyl-*sn*-glycero-3-phosphatidyl serine liposomes (squares, dashed line) with nMBP (10 μM) or acetate buffer. The difference on addition of nMBP is significant ($p \leq 0.0002$) below 31°C. (f) 1,2-Dimyristoyl-*sn*-glycero-3-phosphocholine liposomes (triangle, dashed line) with nMBP (5 μM) or acetate buffer. Excitation wavelength, 360 nm; emission cutoff filter, KV 450. Liposomes were prepared by standard methods of tip sonication as described (23). DPH was added to the lipids before drying. (Used with permission from Ref. 23.)

1-palmitoyl-2-oleoyl-*sn*-glycero-3-phosphocholine or 1-oleoyl-2-hydroxy-*sn*-glycero-3-phosphocholine (results not shown). Thus, MBP interacts with synthetic bilayers made from acidic, but not neutral, lipids.

Second, the ability of MBP to induce lysis and fusion of liposomes was tested. Preliminary experiments showed that both nMBP and raMBP cause liposome lysis. To analyze liposome fusion and lysis, a mixture of two liposome preparations, one containing the fluorescent molecule calcein and its quencher $CoCl_2$ and the other containing EDTA, was employed. Both raMBP and nMBP induced

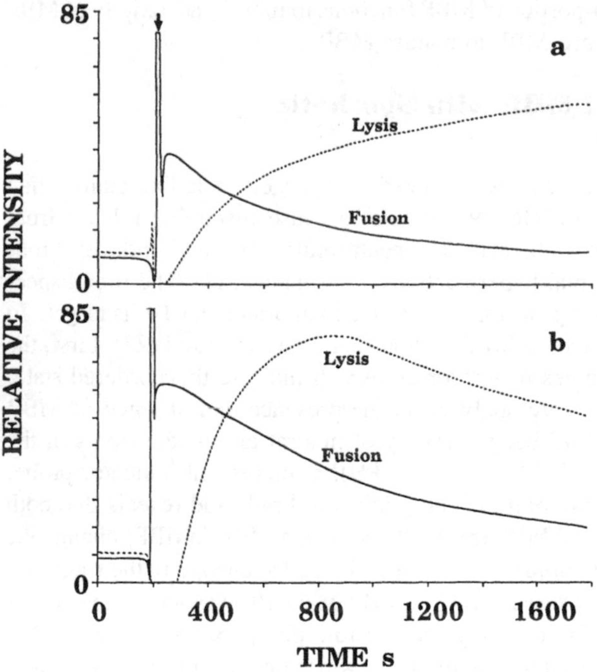

Figure 2 Ability of MBP to induce lysis and fusion of calcein-containing liposomes. Deoxycholate (0.1%) was used as a positive control for lysis and fusion. The data shown are for 5 μM MBP added at the arrow to liposomes [(30 μM) (1,2-dimyristoyl-*sn*-glycero-3-phosphocholine:1,2-dimyristoyl-*sn*-glycero-3-phosphatidic acid)]. The fusion curves (solid lines) were calculated by dividing the corrected time course of the increase in calcein fluorescence (in the presence of $CoCl_2$ and citrate in the buffer) by the time course of the total lysis. The lysis curves (dotted lines) were calculated by subtracting the relative protein-induced fusion from the relative protein-induced lysis and fusion. (a) raMBP, (b) nMBP, excitation wavelength, 490 nm; emission wavelength, 520 nm. (Used with permission from Ref. 23.)

fusion and lysis of the liposomes as shown by the increase in the fluorescence of calcein (Fig. 2). To determine specific fusion of liposomes, $CoCl_2$ was included in the solution bathing the liposomes so that calcein released from lysed liposomes was quenched by the $CoCl_2$. The relative fusion and lysis were calculated from the correlated spectra (Fig. 2). The results indicated that the initial fusion is followed by lysis with leakage of calcein outside of the liposomes.

Figure 3 shows a model for the interaction of MBP and synthetic lipid bilayers. In this model, MBP reacts with the liposome surface by electrostatic and hydrophobic interactions and causes aggregation of liposomes, and subsequently fusion and lysis. Although the results of these experiments implicate surface lipids as the target for MBP activity, the possibility also exists that the cellular glycocalyx and/or acidic proteins are also the targets of MBP action.

IV. THE NEUROTOXICITY OF THE EOSINOPHIL-DERIVED NEUROTOXIN, THE HUMAN LIVER RNASE, AND THE EOSINOPHIL CATIONIC PROTEIN: RELATIONSHIP TO RIBONUCLEASE ACTIVITY

While attempting to identify an etiological agent for Hodgkin's disease 60 years ago, M. H. Gordon discovered that extracts of spleen from patients with Hodgkin's disease caused a neurotoxic reaction when injected intrathecally or intracerebrally into rabbits (24). Subsequently, this response was shown not to be specific for Hodgkin's disease, but rather for eosinophils within the spleen, and subsequently the active factors causing the neurotoxic reaction, now referred to as the Gordon phenomenon, have been identified as EDN and ECP (25,26). Interest in the Gordon phenomenon has increased recently because of the recognition that neurological complications are especially common in patients with the eosinophilia-myalgia syndrome associated with L-tryptophan ingestion (27,28).

As noted above, EDN and ECP are homologous to ribonuclease, possess RNase activity (29), and are members of the ribonuclease superfamily (12). Two major types of human ribonucleases have been defined by structural and functional studies. Human liver ribonuclease is classified as a nonsecretory RNase (30,31) and it is structurally related to, but differs from, human pancreatic RNase, a secretory enzyme (32). EDN appears very similar to human liver RNase in that the amino-terminal portions of liver RNase and EDN are identical to each other and to the terminal portion of the nonsecretory RNase present in human urine (24,30,31). Further, the nucleotide sequence of EDN cDNA (33,34) defines a protein that corresponds to the complete amino acid sequence determined for urinary RNase (31). Therefore, human liver ribonuclease and EDN were compared to determine whether their amino terminal amino acid sequence similarity reflects

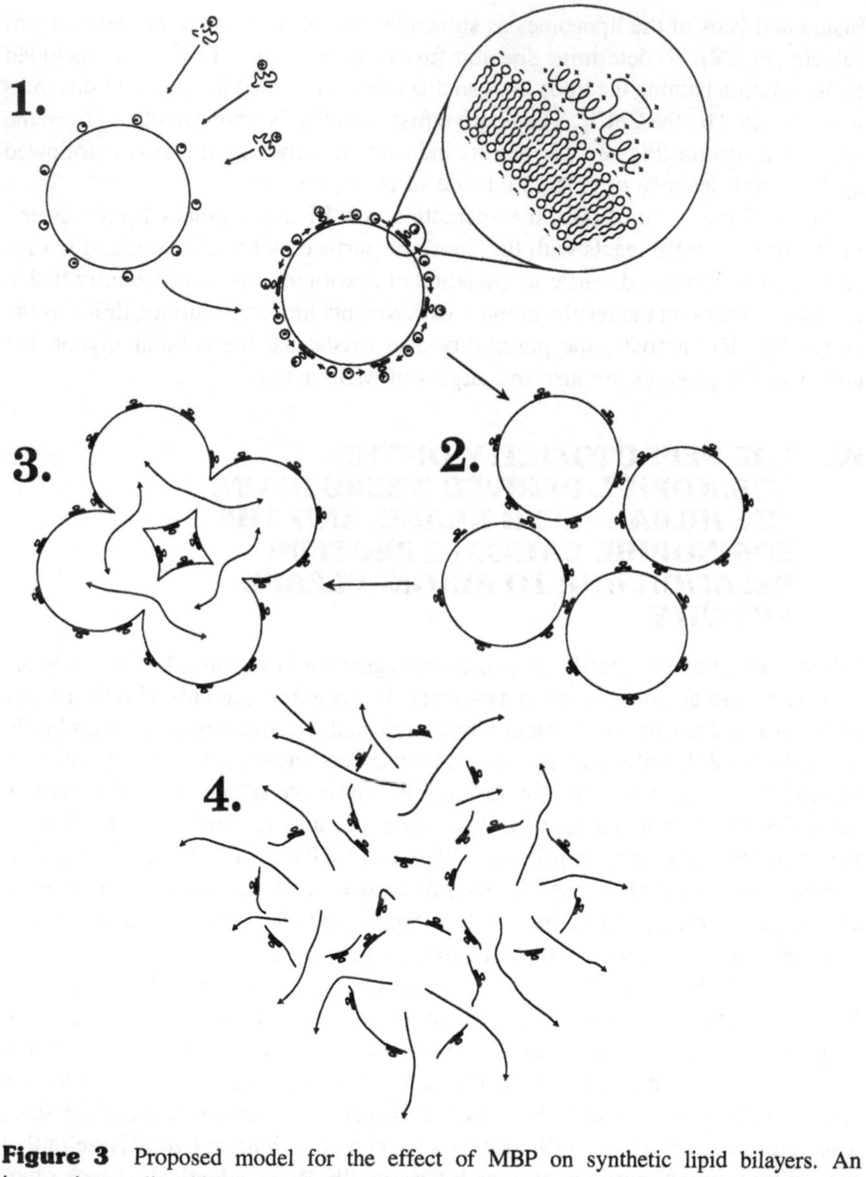

Figure 3 Proposed model for the effect of MBP on synthetic lipid bilayers. An interaction mediated by electrostatic and hydrophobic interactions will cause (1) the binding of MBP to the surface of the bilayer and clustering of the surface charged molecules, (2) aggregation of liposomes and destabilization of the bilayer, inducing (3) fusion of the bilayer, and subsequently (4) lysis. Because no evidence exists supporting the actual penetration of MBP into the bilayer or formation of a channel, the protein is shown on the surface of the membrane. (Used with permission from Ref. 23.)

overall similarity of the functional properties of the molecule, and also whether human liver RNase possesses neurotoxic activity (35).

First, comparison of EDN and liver RNase to bovine pancreatic RNase A showed that EDN and liver RNase were remarkably similar to each other in their substrate preference and they differed from pancreatic RNase (35). EDN and liver RNase both showed a preference for polynucleotides that contain uridine rather than cytidine, whereas the opposite was true for pancreatic RNase. Both liver RNase and EDN showed an identical and generally low activity for simple polymeric RNA substrates and for small defined substrates; in contrast, RNase A was able to hydrolyze substrates, such as cytidine cyclic 2,3-phosphate, uridine cyclic-2,3-phosphate, and corresponding dinucleoside phosphates. The similarity between EDN and liver RNase was further shown by the reactivity of these proteins with antibodies. For example, antibodies to either EDN or liver RNase inhibited the RNase activity of both proteins to an equivalent extent. In contrast, antibodies to EDN or liver RNase did not inhibit human pancreatic RNase. Furthermore, radioiodinated liver RNase bound to anti-EDN and to antiliver RNase and was inhibited to an equivalent extent by these molecules, whereas human pancreatic RNase showed less than 1% cross-reactivity in a similar assay (30). Finally, both EDN and liver RNase lost nuclease activity at an identical rate when incubated at pH 5.5 with iodoacetic acid. Interestingly, the conformations of antigenic determinants of EDN and liver RNase were only moderately affected by carboxymethylation, as judged by the reactivity of the iodoacetate-modified protein with polyclonal antibodies to EDN and liver RNase, whereas fully reduced and carboxymethylated EDN possessed less than 0.01% of its initial immunological activity. For example, carboxymethyl-EDN retained more than 50% of its reactivity.

Analyses of the neurotoxic activity of RNases are shown in Table 4 (35) and indicate that both EDN and liver RNase possess neurotoxic activity (experiment 1). Experiment 2 shows that the neurotoxic activity of both EDN and liver RNase, as well as ECP, is abolished by treatment with iodoacetic acid at pH 5.5. In contrast, EDN that has been exposed to the buffer and incubation conditions of the experiment, but not to iodoacetic acid itself (buffer control), EDN treated with iodoacetic acid in the presence of excess 2-mercaptoethanol (CM-EDN), and EDN that was simply shipped between the laboratories (in Los Angeles and Rochester) with the other samples retained neurotoxic activity. Thus, these results suggest that RNase activity is needed for expression of neurotoxicity associated with EDN, ECP, and human liver RNase. However, experiment 3 of Table 4 shows that RNase activity alone is not sufficient to confer neurotoxicity. For example, neither bovine RNase A nor its glycosylated equivalent RNase B induced the Gordon phenomenon even at doses 10- to 20-fold higher than those that were effective with EDN or liver RNase. Interestingly, human pancreatic RNase was neurotoxic, but only at a dose significantly higher than those needed for EDN or liver RNase. An inconclusive attempt was also made to inhibit EDN neurotoxicity with RNasin, a protein inhibitor of the RNase activity of members of the RNase superfamily. These

Table 4 Neurotoxic Activity of RNases: Induction of the Gordon Phenomenon

Sample assayed	Dose (μg)	Days of observation Total	To symptoms	Positive responses/ total rabbits
Experiment 1				
EDN	1	8	—	0/3
	5	8	3–5	3/3
	10	8	3	3/3
Liver RNase	1	8	—	0/3
	5	8	5–6	2/3
	10	8	4–6	2/3
Phosphate/saline	—	8	—	0/3
Experiment 2				
EDN	10	11	4–6	3/3
	20	9	2–3	3/3
CM-EDN	10	11	—	0/3
	20	9	—	0/3
Sham CM-EDN[a]	20	9	2–3	3/3
Shipping Control[b]	20	9	2–3	3/3
EDN-Buffer Control[c]	20	9	2–3	3/3
Liver RNase	10	11	4–6	3/3
CM-Liver RNase	10	11	—	0/3
ECP	10	11	3–5	3/3
CM-ECP	10	11	—	0/3
Phosphate/saline	—	9–11	—	0/6
Experiment 3				
EDN	10	7–10	2–4	8/8
	20	10	2–3	2/2
ECP	10	10	4–5	2/2
	20	10	4	1/2
Bovine RNase A	10	10	—	0/3
	33	10	—	0/3
	100	10	—	0/3
	200	7	—	0/3
Bovine RNase B	10	10	—	0/3
	100	10	—	0/3
Human pancreatic RNase				
Unfractionated	10	7	—	0/3
	35	7	3–4	2/3
Fraction A (unglycosylated)	30	8	3–5	3/4
Fraction B (glycosylated)	30	8	—	0/2
Phosphate/saline	—	8–10	—	0/9

[a]Glutathione present during iodoacetate treatment.
[b]Protein shipped between laboratories but not included in experimental manipulations.
[c]Iodoacetate omitted from reaction mixture but sample carried through the experimental procedure.
Source: Revised and used with permission from Ref. 35.

experiments were not successful. However, measurement of the stability of EDN-RNasin complexes showed a half-life of only about 10 h, indicating that almost all of the RNase was released early in the 3-day period before expression of the Gordon phenomenon. Prior experiments showed that RNasin inhibited EDN toxicity toward *Trypanosoma cruzi* trypomastigotes (36); however, this experiment required only a 2- to 6-h exposure of the EDN-RNasin complex to the trypomastigotes. Therefore, the inability of RNasin to inhibit the neurotoxic activity of EDN may be due to the short half-life of the RNasin-EDN complex, as well as the necessity of more concentrated preparations of active RNasin than we were able to obtain.

Overall, these results show that EDN and human liver RNase are indistinguishable from each other, but distinct from pancreatic ribonucleases on the basis of their substrate preference and immunological activity. Further, both EDN and liver RNase are neurotoxic, as judged by their ability to cause the Gordon phenomenon, and interestingly human pancreatic RNase was neurotoxic, albeit less so than EDN or human liver RNase. Bovine pancreatic RNase was not neurotoxic even though doses up to 200 μg were tested. The neurotoxicity of EDN, liver RNasc, and ECP was destroyed by incubation with iodoacetic acetic acid pH 5.5, which led to inactivation of their RNase activity. In the case of EDN, the conformation of the moleculc, as judged by reactivity with antibody, was not greatly affected by iodoacetate treatment, and, thus, we conclude that RNase activity is necessary, but not sufficient, to induce neurotoxic activity. The clinical importance of this neurotoxic activity of EDN and liver RNase is not clear; however, EDN and/or ECP could mediate certain of the neurological complications of the eosinophilia-myalgia syndrome, and liver RNase, almost certainly an identical protein, might mediate certain of the symptoms of hepatic encephalopathy.

V. SUMMARY AND CONCLUSIONS

The eosinophil granule contains four cationic proteins that account for the bulk of the granule protein. Presumably, their basicity is the reason for the eosinophil's intense avidity for the acidic dye eosin. These molecules are potent toxins for targets, such as parasites and mammalian cells, and possess numerous biological activities. The toxicity of MBP is neutralized by polymers of glutamic and aspartic acids, and these anionic polymers may be useful as therapeutic agents for eosinophil-associated diseases. Experiments with liposomes suggest that MBP interacts with acidic lipids, causing fusion and lysis of the liposomes.

Liver RNase has the same amino-terminal amino acid sequence as EDN, and both proteins have similar patterns of RNase activity. In addition, both EDN and liver RNase are neurotoxic, and controlled alkylation (which destroys RNase activity and modestly affects RNase confirmation) also abolishes neurotoxicity. Furthermore, human pancreatic RNase (but not bovine pancreatic RNase) possesses neurotoxic activity. These findings suggest that possession of RNase activity is necessary but not sufficient for neurotoxic activity.

ACKNOWLEDGMENTS

This work was supported in part by grants from the National Institutes of Health, AI 09728, AI 15231, AI 31155, and HD 22924; the National Science Foundation, Grant 89-20753; the American Heart Association, Minnesota Affiliate (Dr. Abu-Ghazaleh is a fellow of the American Heart Association); and by the Mayo Foundation.

REFERENCES

1. Ehrlich, P. Methodologische Beiträge zur Physiologie und Pathologie der verschiedenen Formen der Leukocyten. Z Klin Med 1880; 1:553.
2. Gleich GJ, Loegering DA, Maldonado JE. Identification of a major basic protein in guinea pig eosinophil granules. J Exp Med 1973; 137:1459–1471.
3. Gleich GJ, Loegering DA, Kueppers F, Bajaj SP, Mann KG. Physiochemical and biological properties of the major basic protein from guinea pig eosinophil granules. J Exp Med 1974; 140:313–332.
4. Ackerman SJ, Loegering DA, Venge P, Olsson I, Harley JB, Fauci AS, Gleich GJ. Distinctive cationic proteins of the human eosinophil granule: major basic protein, eosinophil cationic protein, and eosinophil-derived neurotoxin. J Immunol 1983; 131: 2977–2982.
5. Gleich GJ, Loegering DA, Mann KG, Maldonado JE. Comparative properties of the Charcot-Leyden crystal protein and the major basic protein from human eosinophils. J Clin Invest 1976; 57:633–640.
6. Wasmoen TL, Bell MP, Loegering DA, Gleich GJ, Prendergast FG, McKean DJ. Biochemical and amino acid sequence analysis of human eosinophil granule major basic protein. J Biol Chem 1988; 263:12559–12563.
7. Barker RL, Gleich GJ, Pease LR. Acidic precursor revealed in human eosinophil granule major basic protein cDNA. J Exp Med 1988; 168:1493–1498.
8. McGrogan M, Simonsen C, Scott R, Griffith J, Ellis N, Kennedy J, Campanelli D, Nathan C, Gabay J. Isolation of a complementary DNA clone encoding a precursor to human eosinophil major basic protein. J Exp Med 1988; 168:2295–2308.
9. Ponz F, Paz-Ares J, Hernandez-Lucas C, Garcia-Olmedo F, Carbonero P. Cloning and nucleotide sequence of a cDNA encoding the precursor of the barley toxin α-hordothionin. Eur J Biochem 1986; 156:131–135.
10. Aoki I, Shindoh Y, Nishida T, Nakai S, Hong Y-M, Mio M, Saito T, Tasaka K. Sequencing and cloning of the complementary DNA of guinea pig eosinophil major basic protein. FEBS Lett 1991; 279:330–334.
11. Aoki I, Shindoh Y, Nishida T, Nakai S, Hong Y-M, Mio M, Saito T, Tasaka K. Comparison of the amino acid and nucleotide sequences between human and two guinea pig major basic proteins. FEBS Lett 1991; 282:56–60.
12. Gleich GJ, Adolphson CR, Leiferman KM. Eosinophils. In: Gallin JI, Goldstein IM, Synderman R, eds. Inflammation: Basic Principles and Clinical Correlates. 2d ed. New York: Raven Press, 1992; 663–700.
13. Wasmoen TL, McKean DJ, Benirschke K, Coulam CB, Gleich GJ. Evidence for

eosinophil granule major basic protein in human placenta. J Exp Med 1989; 170:2051–2063.

14. Slungaard A, Mahoney JR, Jr. Thiocyanate is the major substrate for eosinophil peroxidase in physiologic fluids: implications for cytotoxicity. J Biol Chem 1991; 266: 4903–4910.
15. Barker RL, Gleich GJ, Checkel JL, Loegering DA, Pease LR, Hamann KJ. Acidic polyamino acids inhibit human eosinophil granule major basic protein toxicity: evidence of a functional role for proMBP. J Clin Invest 1991; 88:798–805.
16. Butterworth AE, Wassom DL, Gleich GJ, Loegering DA, David JR. Damage to schistosomula of *Schistosoma mansoni* induced directly by eosinophil major basic protein. J Immunol 1979; 122:221–229.
17. Wassom DL, Gleich GJ. Damage to *Trichinella spiralis* newborn larvae by eosinophil major basic protein. Am J Trop Med Hyg 1979; 28:860–863.
18. Hamann KJ, Gleich GJ, Checkel JL, Loegering DA, McCall JW, Barker RL. In vitro killing of microfilariae of *Brugia pahangi* and *Brugia malayi* by eosinophil granule proteins. J Immunol 1990; 144:3166–3173.
19. Lehrer RI, Szklarez D, Barton A, Ganz T, Hamann KJ, Gleich GJ. Anti-bacterial properties of eosinophil major basic protein (MBP) and eosinophil cationic protein (ECP). J Immunol 1989; 142:4428–4434.
20. O'Donnell MC, Ackerman SJ, Gleich GJ, Thomas LL. Activation of basophil and mast cell histamine release by eosinophil granule major basic protein. J Exp Med 1983; 157:1981–1991.
21. Moy JN, Gleich GJ, Thomas LL. Noncytotoxic activation of neutrophils by eosinophil granule major basic protein: effect on superoxide anion generation and lysosomal enzyme release. J Immunol 1990; 145:2626–2632.
22. Gundel RH, Letts LG, Gleich GJ. Human eosinophil major basic protein induces airway constriction and airway hyperresponsiveness in primates. J Clin Invest 1991; 87:1470–1473.
23. Abu-Ghazaleh RI, Gleich GJ, Prendergast FG. Interaction of eosinophil granule major basic protein with synthetic lipid bilayers: a mechanism for toxicity. J Membrane Biol 1992; 128:153–164.
24. Gordon MH. Remarks on Hodgkin's disease. A pathogenic agent in the glands and its application in diagnosis. Br Med J 1933; 1:641–644.
25. Durack DT, Ackerman SJ, Loegering DA, Gleich GJ. Purification of human eosinophil-derived neurotoxin. Proc Natl Acad Sci USA 1981; 78:5165–5169.
26. Gleich GJ, Loegering DA, Bell MP, Checkel JL, Ackerman SJ, McKean DJ. Biochemical and functional similarities between human eosinophil-derived neurotoxin and eosinophil cationic protein: homology with ribonuclease. Proc Natl Acad Sci USA 1986; 83:3146–3150.
27. Hertzman PA, Blevins WL, Mayer J, Greenfield B, Ting M, Gleich GJ. Association of the eosinophilia-myalgia syndrome with the ingestion of tryptophan. N Engl J Med 1990; 322:869–873.
28. Belongia EA, Hedberg CW, Gleich GJ, White KE, Mayeno AN, Loegering DA, Dunnette SL, Pirie PL, MacDonald KL, Osterholm MT. An investigation of the cause of the eosinophilia-myalgia syndrome associated with tryptophan use. N Engl J Med 1990; 323:357–365.

29. Slifman NR, Loegering DA, McKean DJ, Gleich GJ. Ribonuclease activity associated with human eosinophil-derived neurotoxin and eosinophil cationic protein. J Immunol 1986; 137:2913–2917.
30. Sorrentino S, Tucker GK, Glitz DG. Purification and characterization of a ribonuclease from human liver. J Biol Chem 1988; 263:16125–16131.
31. Beintema JJ, Hofsteenge J, Iwama M, Morita T, Ohgi K, Irie M, Sugiyama RH, Schieven GL, Dekker CA, Glitz DG. Amino acid sequence of the nonsecretory ribonuclease of human urine. Biochemistry 1988; 27:4530–4538.
32. Weickmann JL, Elson M, Glitz DG. Purification and characterization of human pancreatic ribonuclease. Biochemistry 1981; 20:1272–1278.
33. Hamann KJ, Barker RL, Loegering DA, Pease LR, Gleich GJ. Sequence of human eosinophil-derived neurotoxin cDNA: identity of deduced amino acid sequence with human nonsecretory ribonucleases. Gene 1989; 83:161–167.
34. Rosenberg HF, Tenen DG, Ackerman SJ. Molecular cloning of the human eosinophil-derived neurotoxin: a member of the RNase gene family. Proc Natl Acad Sci USA 1989; 86:4460–4464.
35. Sorrentino S, Glitz DG, Hamman KJ, Loegering DA, Checkel JL, Gleich GJ. Eosinophil-derived neurotoxin and human liver ribonuclease: identity of structure and linkage of neurotoxicity to nuclease activity. J Biol Chem 1992; 267:14859–14865.
36. Molina HA, Kierszenbaum F, Hamann KJ, Gleich GJ. Toxic effects produced or mediated by human eosinophil granule components on *Trypanosoma cruzi*. Am J Trop Med Hyg 1988; 38:327–334.

DISCUSSION (Speaker: G. J. Gleich)

Ackerman: What is the relationship between the RNAse activity of EDN and its ability to induce the Gordon phenomenon?

Gleich: The RNAse activity is necessary but not sufficient to induce the Gordon phenomenon.

Sanderson: Are eosinophils susceptible to the toxic effects of eosinophil granule proteins? Angel Lopez showed that eosinophils could not kill antibody-coated eosinophils. They were, however, killed by K cells. This suggests eosinophils may be resistant to their own granule products.

Gleich: We have not tested whether eosinophils can resist attack by their own granule proteins. We have shown that MBP can stimulate eosinophil degranulation (1), and on that basis it seems likely that the more toxic granule proteins (MBP, EPO, ECP) would attack the eosinophil.

Egan: With regard to the effect of MBP on liposome disordering, do polycationic peptides such as polyarginine cause the same phenomenon? That is the case with several other MBP-related phenomena.

Gleich: Dr. Abu-Ghazelah has not tested these molecules on liposomes. Nonetheless one would predict that polyarginine would be as active as the eosinophil cationic toxins on lysosomes, based on earlier comparisons of the two.

Spry: Have you some new thoughts on the relationship between the eosinophil ribonuclease acid syndromes in humans where eosinophils may be involved in central and/or peripheral nervous system disease?

Gleich: The eosinophil RNAses may be active on the human central nervous system, but as yet this has not been shown.

Wardlaw: One of the pillars of the "eosinophils cause asthma" hypothesis is that the basic granule proteins are highly toxic for human bronchial epithelium. However, the literature appears somewhat confusing in that in some studies the effects of granule proteins on epithelium occur only at very high concentrations and the damage is not marked.

Gleich: Our studies of this question were conducted in vitro and showed that MBP concentrations as low as 10 μg/ml caused respiratory epithelial damage, albeit over 48–72 h. Further, sputa from asthma patients requiring hospitalization contained a geometric mean of 8 μg/ml, and it seems likely that MBP concentrations at the tissue were considerably higher. Finally, MBP has been localized to lesions of asthma, and the concentrations required to yield that intensity of fluorescence are $1–5 \times 10^{-5}$ M or 140–700 μg/ml; these concentrations are frankly cytotoxic to respiratory epithelium.

Durham: What is the influence of mucus glycoproteins (highly negatively charged) on eosinophil protein toxicity? May this relate to a protective role in vivo?

Gleich: Dr. Shinji Morojima in our laboratory investigated the ability of mucus glycoproteins to neutralize MBP and found that the mucus was quite active as an inhibitor.

Morley: What, if anything, is the effect of administration of antibodies to MBP to animals prior to induction of allergic bronchospasm and allergic hyperreactivity?

Gleich: We have not performed this experiment. We have shown (2) that polyglutamic acid potently neutralizes the effect of MBP both in vitro and in vivo.

Kay: Are the granule proteins causing damage to the bronchial epithelium in asthma by lysis? What is the evidence?

Gleich: The information supporting the role of the eosinophil in asthma has been well summarized in a recent review (3).

Hansel: MBP is known to activate neutrophils and platelets, but does MBP activate T cells? In particular, does MBP influence the pattern of cytokine production by T cells?

Gleich: We have not tested whether MBP can activate T cells.

Vadas: Is the release of active granule components the inevitable consequence of eosinophil death?

Gleich: We have studied the release of eosinophil granule EDN following culture of eosinophils in vitro in both the presence and absence of IL-5 and have found that dying eosinophils release virtually all of their content of EDN into the medium (4). Furthermore, analyses of the other eosinophil granule proteins in eosinophils cultured with IL-5 suggest that up to 50% of granule proteins are released into the medium by 7 days even though the eosinophils remain viable.

Abrams: In view of literature showing that lysolecithin and phospholipase A_2 can induce liposome fusion, is MBP a phospholipase?

Gleich: No, not that we could detect.

Moqbel: The presence of ubiquitous RNase activity in eosinophils presents investigators with a major problem when investigating mRNA expression for various proteins. From your work can you suggest which RNase inhibitor(s) may be most efficient in blocking the RNase activity in these cells?

Gleich: Both alkylation of EDN/human liver RNase (with iodoacetate) and the placental RNasin potently inhibit the RNase activity of these proteins.

Denburg: Does MBP, or other eosinophil granule proteins, cause eosinophil cytokine release?

Gleich: We have no results on this question.

DISCUSSION REFERENCES

1. Abu Ghazaleh R, Gleich GJ. The effect of major basic protein on eosinophils. Fed Proc 1987; 46:988.
2. Barker RL, Gundel RH, Gleich GJ, Checkel JL, Loegering DA, Pease LR, Hamann KJ. Acidic polyamino acids inhibit human eosinophil granule major basic protein toxicity: evidence of a functional role for proMBP. J Clin Invest 1991; 88:798–805.
3. Hamann KJ, Gleich GJ, Gundel RH, White SR. Interactions between respiratory epithelium and eosinophil granule proteins in asthma: the eosinophil hypothesis. In: Farmer SG, Hay DWP, eds. The Airway Epithelium: Physiology, Pathophysiology, and Pharmacology. New York: Marcel Dekker, 1991; 255–300.
4. Kita H, Weiler DA, Abu-Ghazaleh R, Sanderson CJ, Gleich GJ. Release of granule proteins from eosinophils cultured with IL-5. J Immunol 1992; 149:629–635.

2

Human Eosinophil Lysophospholipase (Charcot-Leyden Crystal Protein): Molecular Cloning, Expression, and Potential Functions in Asthma

Steven J. Ackerman, Zeqi Zhou, and Daniel G. Tenen
Beth Israel Hospital and Harvard Medical School, Boston, Massachusetts

Mike A. Clark
Schering-Plough Research, Kennelworth, New Jersey

Yuan-Po Tu and Charles G. Irvin
National Jewish Center for Immunology and Respiratory Medicine, Denver, Colorado

I. INTRODUCTION

Charcot-Leyden crystals (CLC) were first described in 1853 by Charcot and Robin in the postmortem spleen and blood of a leukemia patient (1); a later report by Leyden described them in the sputum of asthmatics (2). These distinctive hexagonal, bipyramidal crystals have classically been observed in tissues and secretions from sites of eosinophil-associated inflammatory reactions in asthma, myeloid leukemias, allergic, parasitic, and other diseases (3,4). Although CLC are considered a hallmark of the eosinophil, basophils have also been shown to form CLC in vitro (5,6), and CLC have been identified by ultrastructural studies as intragranular inclusions in the large, particle-filled granule of human basophils participating in a variety of diseases (7–9). Eosinophil CLC are formed by a small, markedly hydrophobic polypeptide (CLC protein) of 17,400 daltons (10–12) that exhibits lysophospholipase activity (lysolecithin acylhydrolase EC 3.1.1.5) (12). Chromatographically purified human eosinophil lysophospholipase is immunochemically, physicochemically, and enzymatically indistinguishable from eosinophil CLC protein (13) and crystallizes to form identical hexagonal, bipyramidal crystals (12).

CLC protein comprises the sole protein constituent of both native CLC formed in vivo and CLC prepared from disrupted eosinophils (14) or basophils (Ackerman, SJ, unpublished observations) in vitro. At approximately 8.5 pg/cell, CLC protein is one of the most prominent constituents of the eosinophil (5,11). While purified basophil-derived CLC protein has not been analyzed for lysophospholipase activity, basophils contain comparable amounts (4–6 pg/cell) of CLC protein to the eosinophil, and basophil cell lysates express significant lysophospholipase activity (Ackerman, SJ, unpublished observations).

The tendency of CLC protein to form the distinctive bipyramidal crystals at sites of eosinophil infiltration has generated significant interest in the biochemical properties of this protein. However, although CLC protein comprises an estimated 7–10% of total eosinophil protein (13) and possesses lysophospholipase activity, its role in eosinophil and basophil function remains obscure.

A. Physicochemical Properties

The molecular mass of CLC protein was initially estimated at 13 kDa on the basis of its mobility on SDS-PAGE (10,11) and by molecular sieve chromatography in denaturing 6 M guanidine HCl (10). The multiple bands seen on nondenaturing PAGE gels were assumed to be due to carbohydrate heterogeneity and/or aggregation, as the individual bands were shown to be immunochemically identical (11). The estimate of molecular mass was later revised to 17.4 kDa (13); the erroneously high electrophoretic mobility reported previously was attributed to excessive binding of SDS to the markedly hydrophobic CLC protein (13). A determination of the amino acid content of eosinophil-derived CLC protein (eosinophil lysophospholipase) identified a polypeptide of 117–119 amino acids (10) with a single reactive sulfhydryl group (11). The amino-terminal residue of eosinophil-derived CLC is blocked (12; Ackerman, SJ, unpublished observations); the identity of the blocking substituent and the amino terminal residue are not known. However, preliminary analyses suggest that mature CLC protein may possess an *n*-acetylated serine residue (Zhou, Z, Chin, DT, Ackerman, SJ, unpublished observations) (see below).

B. Biosynthesis

Little has been reported on the biosynthesis of CLC protein. There is one report of an increase in the lysophospholipase content of murine eosinophils cocultured with T cells, macrophages, and *Trichinella spiralis* antigen, suggesting a role for cytokines in lysophospholipase (the putative murine CLC protein analog) biosynthesis (15). Biosynthetic labeling studies of human CLC protein have suggested that myristic acid may be added as a posttranslational modification to the newly synthesized polypeptide (16); neither the number of myristic acid residues nor the points of attachment along the polypeptide chain have been determined.

Pulse-chase experiments of CLC protein biosynthesis in HL-60 promyelocytic leukemia cells and human bone marrow mononuclear cells demonstrated the synthesis of a 15.5- to 16-kDa polypeptide without any evidence for significant posttranslational modification (16,17; Ackerman, SJ, unpublished observations).

C. Crystal Formation

Perhaps the most interesting of the known properties of CLC protein is its ability to form distinctive hexagonal bipyramidal crystals. As noted above, CLC protein is the sole component of these crystals (14), which are seen most commonly at sites of eosinophil infiltration (18) and degranulation (19). Crystals can also form intracellularly in vitro in both eosinophils and basophils when swelled in hypotonic solutions (6) and have been seen in bone marrow eosinophilic promyelocytes (20) and, most recently, in eosinophils in vivo (19). Crystals tend to be resistant to proteolytic digestion (21), but are soluble in hot water, acetic acid, sodium hydroxide, and alcohol (22). Purified CLC protein has been recrystallized by hanging-drop techniques, and an analysis of its crystallographic, three-dimensional structure is in progress (23).

D. Subcellular Localization

Early cell fractionation studies (which screened for lysophospholipase activity) suggested that the eosinophil enzyme was associated with the cell membrane (24); no lysophospholipase activity was found in the granule fraction. In contrast, ultrastructural immunogold localization studies by Dvorak, Ackerman, and colleagues have localized CLC protein to a number of eosinophil subcellular compartments, including a small (<5%) population of large, crystalloid-free, primary granules (25) and the cytosol, vesicles, and nucleus of activated tissue eosinophils (19) and eosinophils derived from interleukin-5-induced umbilical cord blood cultures (26). Localization by indirect immunofluorescence (5,27,28) has also demonstrated additional subcellular sites in nucleus and cytoplasm. In the basophil, CLC protein is localized to the large, particle-filled, histamine- and chondroitin sulfate–containing granule population and to intragranular crystals or crystalloids in activated cells (29). For a more detailed review of CLC subcellular localization, see Chapter 9.

E. Enzymatic Activity

Weller and colleagues (12) first reported the lysophospholipase (lysolecithin acylhydrolase) activity of CLC protein. Several lines of evidence suggested that the CLC (crystal forming) protein was identical to eosinophil lysophospholipase. Purified CLC had lysophospholipase activity with the same K_m as the chromatographically purified enzyme (13), and purified lysophospholipase formed hexagonal bipyramidal crystals (12). The proteins were immunochemically indistinguishable (14), had identical electrophoretic mobilities on SDS-PAGE (12), had identical amino acid compositions, and both had blocked amino termini (12). Recently, we have shown that the gene for eosinophil CLC protein directs the expression of considerable lysophospholipase activity in transiently transfected COS cells and stably transfected CHO cell lines (28; see below), confirming the expression of lysophospholipase activity by this protein.

F. Hypothetical Functions in Allergic and Parasitic Diseases

The biological significance of lysophospholipase activity to eosinophil function remains unclear. Several as yet unsubstantiated hypotheses have been presented. For example, the released fatty acid may have membrane-perturbing effects on the eosinophil's target cell (30). Others have suggested that the lysophospholipase protects eosinophils from lysophospholipids and other toxic parasitic products (21,24,31); another possibility is a direct antiparasitic effect by hydrolyzing lysophospholipids in the cuticle of parasite targets (32). The fact that eosinophils (and basophils) contain such large quantities of this enzyme suggests a role for these cells and CLC protein in the metabolism of lysophosphoglycerides that may be generated in inflammatory foci by neutrophils (33,34) or other agents (35), in anaphylactic reactions by mast cells (36), directly by pathogenic microorganisms (37,38) or by helminth parasites such as *Schistosoma mansoni* (39). Schistosomula of *S. mansoni* secrete considerable quantities of a potent, biologically active lysophosphatide, monopalmitoyllysophosphatidylcholine, that apparently mediates membrane fusion and lysis of cytotoxic effector cells (40); since eosinophils, but not neutrophils, can effectively kill schistosomula in vitro, they may be resistant to the lytic and membrane-perturbing effects of parasite lysophosphatides by virtue of their considerable content of a lysophospholipase. Enzymatically active CLC protein in or on the plasma membrane could inactivate these lysophosphatides, potentially protecting the eosinophil or, as demonstrated experimentally with a bacterial lysophospholipase, inhibiting lysophosphatide-mediated inflammation in vivo (41).

Although it has been suggested that lysophospholipases may catabolize potentially toxic and membrane-perturbatory lysophosphatides, studies of erythrocyte membrane lysophospholipase have shown that it promotes complement-mediated erythrocyte lysis, potentially by altering the stoichiometric balance of phospholipid and free fatty acid (30). These observations suggest a role for lysophospholipases such as CLC protein in altering membrane fluidity and/or integrity and a potential need for intracellular regulation of the enzyme activity of this protein in eosinophils (and basophils). By altering the balance between phospholipids and free fatty acids in membranes, CLC protein might act synergistically to enhance the cytotoxicity or helminthotoxicity of eosinophil granule cationic proteins, such as MBP, ECP, or EPO, which have been shown to alter membrane structure (42). At noncytolytic concentrations, lysophospholipids also possess multiple pro- and anti-inflammatory properties, including potentiation of mast cell histamine release, stimulation or inhibition of membrane-bound enzymes such as adenylate and guanylate cyclases, inhibition of prostaglandin synthesis and platelet aggregation, modification of membrane fluidity, and potentiation of humoral and cell-mediated immune responses (reviewed in 43, 44) and thus could act as modulators of inflammation. Catabolism by CLC protein of lysophosphatides generated at sites of eosinophil-, basophil-, or mast-cell-mediated inflammation could be a physiologically significant activity of this lysophospholipase.

II. MOLECULAR CLONING AND CHARACTERIZATION

We have obtained, sequenced, and characterized a full-length cDNA clone for eosinophil CLC protein, compared the deduced amino acid sequence to other published protein sequences, and analyzed the expression of CLC mRNA in various hematopoietic cells and cell lines (45). A 598-base-pair, full-length cDNA clone for CLC protein was obtained by expression screening of a BCGF-II/IL-5-induced HL-60 eosinophilic subline cDNA library with antidenatured CLC antiserum (45,46). The clone was confirmed as encoding the CLC protein since the predicted amino acid composition matched that determined for the native protein (10,13); more important, the deduced amino acid sequence also matched that of amino-terminal sequence obtained by unblocking the amino terminus with trifluoroacetic acid (45; Zhou, Z, Chin, DT, Ackerman, SJ, unpublished observations). Further, expression of the cDNA in COS and CHO cells yielded immunoreactive CLC protein with both lysophospholipase and crystal-forming activities comparable to the native protein (23; see below). The open reading frame encodes a protein of 16.4 kDa containing 142 amino acids, 36% of which are either hydrophobic or aromatic. The translated amino acid sequence of CLC protein showed similarities to the S-type beta-galactoside-binding animal lectin superfamily (47–49), but not to any known sequences of lysophospholipases (50,51). The cDNA sequence encodes not one, but two cysteines; a hydropathy plot suggests that both the cysteines are engulfed in hydrophobic pockets, and as such could be associated with the active site of the enzyme. In this regard, CLC protein is a sulfhydryl-dependent lysophospholipase (13) and possesses only a single reactive (free) sulfhydryl group (11); of note, the peptide pattern produced by 2-nitro-5-thiocyanobenzoic acid cleavage at cysteine residues in the reduced and denatured protein suggests that the second cysteine contributes the reactive sulfhydryl (Zhou, Z, Ackerman, SJ, unpublished observations) and is thus more likely to be associated with the active site of the enzyme. The mRNA for CLC protein is approximately 900 base pairs and was expressed in peripheral blood eosinophils, in HL-60 cells induced toward eosinophilic differentiation with IL-5, and, surprisingly, in HL-60 cells induced toward neutrophilic differentiation with DMSO; however, no mRNA for CLC protein was expressed in peripheral blood neutrophils. Low levels of mRNA were also expressed in basophils obtained from a patient with CML.

A. Isolation and Sequencing of cDNA Clones

Primary expression screening of an induced HL-60 cDNA lambda gt11 expression library (HL-60 3c5I) with monospecific polyclonal rabbit antisera to denatured CLC protein yielded an initial 318-bp clone (nucleotides 280–598) containing an open reading frame encoding 60 amino acids followed by a stop codon (TAA) and a 3′ untranslated region containing a polyadenylation signal (ATTAAA) and poly A tail. Further screening of the HL-60 3c5I library with this 318-bp cDNA clone yielded a longer clone (nucleotides 14–598), which also included a putative eukaryotic translation initiation sequence (53). Determination of the remainder of the 5′ untranslated sequence of the full-length CLC cDNA (nucleotides 1–13) was facilitated by using the PCR to directly amplify CLC cDNAs from the HL-60 3c5I library using a specific 3′ oligonucleotide and lambda primers as previously described (46); a 529-bp cDNA was obtained by PCR (nucleotides 1–529) that encompassed the 5′ untranslated and coding regions and 64 bp of the 3′ untranslated region of the cDNA. Both the coding and noncoding strands of these clones were sequenced to completion (Fig. 1). The 13 bp of 5′ untranslated sequence obtained from the PCR-amplified cDNA clones has been confirmed by direct sequencing of genomic clones (54). The cDNA contains a Kozak-like translation initiation sequence (nucleotides 31–37) (53), an open reading frame encoding a total of 142 amino acids (nucleotides 34–459), a termination codon (TAA, nucleotides 460–462), and a polyadenylation signal (nucleotides 566–571) with a 15-base spacer preceding the poly A tail. Because the initial 318 bp cDNA clone was obtained by expression screening using polyclonal antiserum to denatured CLC protein, the authenticity of the full-length cDNA and translation was confirmed by a number of criteria; these included (1) an amino acid composition essentially identical to that published for CLC protein (10) and eosinophil lysophospholipase (13) (Table 1), (2) 16 residues of confirmatory amino-terminal protein sequence from the native eosinophil-derived protein (beginning with the serine at +2) obtained by Edman degradation after first unblocking the amino terminus with trifluoroacetic acid (TFA) (55) (Fig. 1), and (3) identity with two additional overlapping peptide amino acid sequences near the carboxy-terminal end of the protein (beginning with serines at residues 92 and 94) that were obtained by limited TFA cleavage at serine residues (Fig. 1). Furthermore, we have

```
                                        *
CAATTCAGAAGAGCCACCCAGAAGGAGACAACAATGTCCCTGCTACCCGTGCCATACACA   60
                                   MetSerLeuLeuProValProTyrThr    9

GAGGCTGCCTCTTTGTCTACTGGTTCTACTGTGACAATCAAAGGGCGACCACTTGTCTGT  120
GluAlaAlaSerLeuSerThrGlySerThrValThrIleLysGlyArgProLeuValCys   29

TTCTTGAATGAACCATATCTGCAGGTGGATTTCCACACTGAGATGAAGGAGGAATCAGAC  180
PheLeuAsnGluProTyrLeuGlnValAspPheHisThrGluMetLysGluGluSerAsp   49

ATTGTCTTCCATTTCCAAGTGTGCTTTGGTCGTCGTGTGGTCATGAACAGCCGTGAGTAT  240
IleValPheHisPheGlnValCysPheGlyArgArgValValMetAsnSerArgGluTyr   69

GGGGCCTGGAAGCAGCAGGTGGAATCCAAGAACATGCCCTTTCAGGATGGCCAAGAATTT  300
GlyAlaTrpLysGlnGlnValGluSerLysAsnMetProPheGlnAspGlyGlnGluPhe   89

GAACTGAGCATCTCAGTGCTGCCAGATAAGTACCAGGTAATGGTCAATGGCCAATCCTCT  360
GluLeuSerIleSerValLeuProAspLysTyrGlnValMetValAsnGlyGlnSerSer  109

TACACCTTTGACCATAGAATCAAGCCTGAGGCTGTGAAGATGGTGCAAGTGTGGAGAGAT  420
TyrThrPheAspHisArgIleLysProGluAlaValLysMetValGlnValTrpArgAsp  129

ATCTCCCTGACCAAATTTAATGTCAGCTATTTAAAGAGATAACCAGACTTCATGTTGCCA  480
IleSerLeuThrLysPheAsnValSerTyrLeuLysArg                       142

AGGAATCCCTGTCTCTACGTGAACTTGGGATTCCAAAGCCAGCTAACAGCATGATCTTTT  540

CTCACTTCAATCCTTACTCCTGCTCATTAAAACTTAATCAAACTTCAAAAAAAAAAAA...  598
```

Figure 1 Complete CLC cDNA sequence and translation. The nucleotide sequence of the 598-bp coding strand of the CLC cDNA (clone 32/1) is shown with translation of the open reading frame (142 amino acids). The Kozak-like initiation sequence (ACAATGT) containing the presumptive start codon* (ATG) is boxed and shaded and the stop codon (TAA) at nucleotides 460–462 is boxed. A 3′ polyadenylation signal (ATTAAA) at nucleotides 566–571 is boldly underlined. A single potential N-linked glycosylation site (Asp-X-Thr/Ser) at amino acids 136–138 is double-underlined. Confirmatory amino acid sequences of three peptides obtained by TFA cleavage at serine residues of the purified crystal-derived eosinophil CLC protein are underlined. These sequence data have been submitted to the EMBL/GenBank Data Libraries under Accession #L01664. [From Ackerman et al. (45).]

expressed the full-length CLC cDNA transiently in transfected COS cells and in stably transfected CHO cell lines; the recombinant CLC protein being shown to possess lysophospholipase activity and to spontaneously crystallize into typical CLC (28; see below).

B. Characterization of the Translated Amino Acid Sequence

The 426-bp open reading frame encoded a 142-amino-acid polypeptide with a predicted mass of 16.5 kDa, isoelectric point of 7.28, and a single potential

Table 1 Predicted Versus Chemically Determined Amino Acid Compositions of CLC Protein

Amino acid	CLC protein[a]	Eosinophil LPLase[b]	Predicted CLC composition[c]	Percent of residues[d]
Ala	4	6	4	3
Arg	6	8	7	5
Asx	10[e]	12[e]	11	
Asp	—	—	6	4
Asn	—		5	4
Cys	2[f]	1[g]	2	1
Glx	19	22	19	
Glu	—	—	10	7
Gln	—	—	9	6
Gly	5	7	6	4
His	2	3	3	2
Ile	4	5	5	4
Leu	9	12	10	7
Lys	7	10	9	6
Met	4	5	6	4
Phe	7	9	9	6
Pro	5	7	7	5
Ser	9	13	13	9
Thr	5	7	7	5
Trp	4	—	2	1
Tyr	5	5	6	4
Val	13	14	16	11
Total	120	146	142	100

[a]Based on mass of 13 kDa for CLC protein (10).
[b]Based on mass of 17.4 kDa for eosinophil lysophospholipase (13).
[c]Calculated mass of 16,481 daltons (from the translated amino acid sequence, Fig. 1).
[d]In the translated amino acid sequence, Fig. 1.
[e]Amino acid analyses did not distinguish between Asn/Asp and Gln/Glu.
[f]As cysteic acid.
[g]As carboxymethylcysteine.
(—) = not determined.
Source: From Ackerman et al. (45.)

N-linked glycosylation site (Asn-Val-Ser) (56) at amino acids 136–138. Comparisons of the predicted amino acid composition from the CLC cDNA clone with that of authentic CLC protein (10) and eosinophil lysophospholipase (13) showed them to be essentially identical (Table 1). Hydrophobic (A, I, L, W, V) and aromatic (F, Y) residues comprise 36.6% of the amino acids, and there are equivalent numbers of strongly basic (K, R) and strongly acidic (D, E) residues. In contrast to the marked number of hydrophobic and aromatic residues, a Kyte and Doolittle (57) hydrophobicity profile showed relatively few markedly hydrophobic domains in the sequence aside from a hydrophobic amino terminus of seven residues that does not appear to resemble a typical cleaved (58) or uncleaved signal sequence (59,60) and hydrophobic regions containing the two cysteine residues in the sequence one of which contributes the reactive thiol group thought to comprise part of the active site of this sulfhydryl-dependent lysophospholipase (13). Secondary structure as evaluated by the method of Chou and Fassman (61) suggests 37% helical, 80% extended, and 21% beta-turn conformations.

C. Expression of CLC mRNA in Hematopoietic Cells and Leukemic Cell Lines

Northern analyses have demonstrated an ~900-bp mRNA encoding CLC protein in an eosinophil-committed subline of HL-60 (3+C-5) (62) that was markedly up-regulated by induction toward the eosinophil lineage with BCGF-II as a source of IL-5. CLC mRNA was present in the eosinophilic subline of HL-60 (3+C-5) (62) induced toward eosinophilic differentiation with BCGF-II, the parental line of HL-60 induced with DMSO, peripheral blood eosinophils from patients with eosinophilia, normal bone marrow aspirates, and basophils isolated from patients with CML; no CLC mRNA was detected in neutrophils, monocytes, T-cell or B-cell lines, or HL-60 cells induced toward monocytic differentiation with vitamin D_3. Separation of peripheral blood leukocytes from a patient with atopic dermatitis and eosinophilia by centrifugation over a cushion of Ficoll-Hypaque showed that CLC mRNA was associated exclusively with the eosinophils that sedimented in the cell pellet and not with the lighter-density mononuclear cell fraction containing primarily lymphocytes and monocytes. While uninduced cells of the eosinophil-committed subline of HL-60 (3+C-5) constitutively expressed small amounts of CLC mRNA, induction of these cells for 72 h with BCGF-II strongly up-regulated the expression of CLC mRNA. As noted above, both eosinophils and basophils from the peripheral blood of patients with HES or CML, respectively, contained mRNA for CLC protein. The presence of mRNA for CLC protein in basophils would suggest that these granulocytes synthesize CLC protein rather than acquire it from extracellular (presumably eosinophil) sources during differentiation in the bone marrow. In this regard, we have also demonstrated CLC protein biosynthesis in the CML basophil preparations that were used in these studies for RNA isolation (Golightly, LM, and Ackerman, SJ, unpublished observations).

D. Sequence Comparisons

The deduced amino acid sequence of CLC protein did not show any significant similarities to any of the published sequences of mammalian lysophospholipases or other lipolytic enzymes, including lysophospholipases of rat (50) and bovine

(51) origin as well as a series of 37-, 38-, and 63-kDa sulfhydryl-dependent lysophospholipases of mouse macrophage origin that have been partially sequenced (Clark, MA, unpublished observations). In contrast, the CLC protein sequence showed 25–30% identity over regions of 100–120 amino acids with the carboxyl-terminal domains of four highly conserved IgE-binding proteins (Fig. 2), thus far isolated from mouse, rat, and human including the 31-kDa rat (63,64) and human (65,66) IgE-binding proteins, the 35-kDa mouse carbohydrate-binding protein (CBP35) (67,68), and Mac-2, the murine macrophage cell surface protein that is identical to CBP-35 (69). These proteins are members of a growing superfamily of β-galactoside (lactose) binding S-type animal lectins (Fig. 3) (47–49,70–72), which includes a large group of smaller-molecular-weight, highly conserved 14- to 16-kDa carbohydrate-binding proteins thus far isolated from diverse species, organs, and tissues including human and rat lung (71–74), bovine heart (75), human placenta (48,67,76), chicken skin (77), mouse fibroblasts (78),

```
CLC PROTEIN ... A A S L S T G S T V T I K G R P L V C F L N E P Y - L Q V D F H T E M    44
                                                                                  : :
hMac-2      ... I V P Y N L P L P G G V V P R M L I T I L G T V K P N A N R I A L D F   149
RIgEBP      ... T V P Y D M T L T G G V M P R M L I T I I G T V K P N A N S I T L N F   161
CBP35/Mac-2 ... T V P Y D L P L P G G V M P R M L I T I M G T V K P N A N R I V L D F   162

CLC PROTEIN     - K E E S D I V F H F Q V C F - - - - G R R V V M - N S R E Y G A W K    73
                    : : :     :       :               :       : :     : : :   : :
hMac-2          - Q R G N D V A F H F N P R F N E N - N R R V I V C N T K L D N N W G   182
RIgEBP          - K K G N D I A F H F N P R F N E N - N R R V I V C N T K Q D N N W G   194
CBP35/Mac-2     - R R G N D V A F H F N P R F N E N - N R R V I V C N T K Q D N N W G   195

CLC PROTEIN     Q Q V E S K N M P F Q D G Q E F E L S I S V L P D K Y Q V M V N G Q S   108
                : :       : : :     : :   :     : :   :           : : :         :
hMac-2          R E E R Q S V F P F E S G K P F K I Q V L V E P D H F K V A V N D A H   217
RIgEBP          R E E R Q S A F P F E S G K P F K I Q V L V E A D H F K V A V N D V H   229
CBP35/Mac-2     K E E R Q S A F P F E S G K P F K I Q V L V E A D H F K V A V N D A H   230

CLC PROTEIN     S Y T F D H R I K P - E A V K M V Q V W R D I S L T K F N V S Y L K R   142
                      : :     :         : : :   :   :         :
hMac-2          L L Q Y N H R V K K L N E I S K L G I S G D I D L T S A S Y T M I       250
RIgEBP          L L Q Y N H R M K N L R E I S Q L G I I G D I T L T S A S H A M I       262
CBP35/Mac-2     L L Q Y N H R M K N L R E I S Q L G I S G D I T L T S A N H A M I       263
```

Figure 2 Comparison of the translated CLC protein amino acid sequence to human, rat, and mouse IgE/carbohydrate-binding proteins. Shown are alignments of the cDNA-derived amino acid sequence (residues 11–142) of CLC protein with the carboxyl-terminal domains of four IgE/carbohydrate (beta-galactoside) binding proteins, including human Mac-2 (hMac-2) (65,66), rat IgE-binding protein (RIgEBP) (63,64), and mouse carbohydrate-binding protein (CBP35) (67,68), which is identical in sequence to mouse Mac-2 (Mac-2) (69). Regions of conserved sequence in comparison to CLC protein are boxed and shaded; colons indicate potentially conservative amino acid changes. Numbering of amino acids is as per the published sequences. Dashes in the CLC sequence indicate gaps inserted for purposes of optimal alignment. CLC protein also contains the boxed three-amino-acid sequence (VPY) at residues 6–8 (Fig. 1). The percent similarity of CLC protein to that of hMac-2, RIgEBP, and CBP35/Mac-2 was 29%, 29%, and 28%, respectively, for the regions of sequence shown. [From Ackerman et al. (45).]

```
CLC    ..A A S L S T G S T V T I K G R P L V C F L N E P Y - L Q V D F H T E M    44
CBP35  ...T V P Y D L P L P G G V M P R M L I T I M G T V K P N A N R I V L D F  161
RIgEBP...T V P Y D M T L T G G V M P R M L I T I I G T V K P N A N S I T L N F
C14    Ac S C Q G P V C T N L G L K P G Q R L T V K G I I A P N A K S F V M N L    35
R14    Ac A C - G L V A S N L N L K P G E C L K V R G E L A P D A K S F V L N L
H14    Ac A C - G L V A S N L N L K P G E C L R V R G E V A P D A K S F V L N L
B14    (M) S C - G L V A S N L N L K P G E C L R V R G E V A A D A K S F L L N L
E14    (S M)N G V V D E R M S F K A G Q N L T V K G V P S I D S T N F A I N V
M14    (M) A C - G L V D Q Q A E S Q T G A M S Q S S G QRW T S D D K S F V L N L

                          53        58               65            72
CLC     - K E E S D I V F H F Q V C F - - - G - R R V V M - N S R E Y G A W K   73
CBP35   - R R G N D V A F H F N P R F N E N - N R R V I V C N T K Q D N N W G  194
RIgEBP  - K K G N D I A F H F N P R F N E N - N R R V I V C N T K Q D N N W G
C14     G K D S T H L G L H F N P R F D A H G D V N L I V C N S K K M E E W G   70
R14     G K D S N N L C L H F N P R F N A H G D A N T I V C N S K D D G T W G
H14     G K D S N N L C L H F N P R F N A H G D A N T I V C N S K D G G A W G
B14     G K D D N N L C L H F N P R F N A H G D V N T I V C N S K D A G A W G
E14     G N S A E D L A L H I N P R F D A H G D Q Q A V V V N S F Q G G N W G
M14     G K D S N N L C L H F N P R F N A H G D A N T I V C N T K E D G T W G

                       82 83    86
CLC     Q Q V E S K N M P F Q D G Q E F E L S I S V L P D K Y Q V M V N G Q S  108
CBP35   K E E R Q S A F P F E S G K P F K I Q V L V E A D H F K V A V N D A V  229
RIgEBP  R E E R Q S A F P F E S G K P F K I Q V L V E A D H F K V A V N D V H
C14     T E Q R E T V F P F Q K G A P I E I T F S I N P S D L T V H L P - G H  104
R14     T E Q R E T A F P F Q P G S I T E V C I T F D Q A D L T I K L P D G H
H14     T E Q R E A V F P F Q P G S V A E V C I T F D Q A N L T V K L P D G Y
B14     A E Q R E S A F P F Q P G S V V E V C I S F N Q T D L T I K L P D G Y
E14     T E Q R E G G F P F K Q G E D F K I Q I T F N S E E F R I I L P D G S
M14     T E H R E P A F P F Q P G S I T E V C I T F D Q A D L T I K L P D G H

CLC     S Y T F D H R I K P - E A V K M V Q V W R D I S L T K F N V S Y L K R  142
CBP35   L L Q Y N H R M K N L R E I S Q L G I S G D I T L T S A N H A M I      262
RIgEBP  L L Q Y N H R M K N L R E I S Q L G I I G D I T L T S A S H A M I
C14     Q F S F P N R L - G L S V F D Y F D T H G D F T L R S V S W E          134
R14     E F K F P N R L - N M E A I N Y M A A D G D F K I K C V A F E
H14     E F K F P N R L - N L E A I N Y M A A D G D F K I K C V A F D
B14     E F K F P N R L - N L E A I N Y L S A G G D F K I K C V A F E
E14     E I H F P N N R - Y M H F E G - E A R I Y S I E I K...
M14     E F...
```

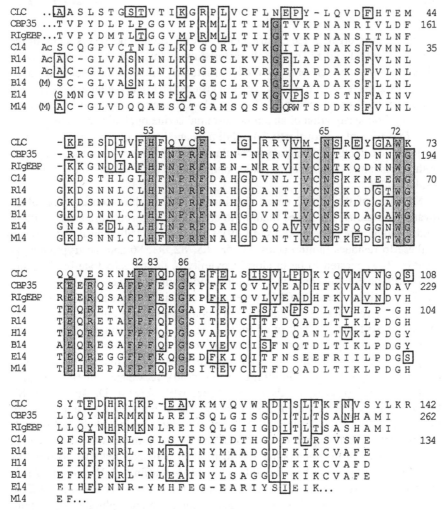

Figure 3 Alignments of the translated CLC protein amino acid sequence with sequences of members of the S-type animal lectin superfamily. Comparison of the translated CLC protein amino acid sequence to the carboxyl-terminal halves of the 31- to 35-kDa IgE-binding proteins from mouse (CBP35) (67,68) and rat (RIgEBP) (63,64), and to the superfamily of β-galactoside binding S-type animal lectins, a group of highly conserved 14-kDa proteins isolated from chicken (C14) (77), rat and human lung (R14, H14) (71–74), bovine heart (B14) (75), mouse fibroblasts (M14) (78), and electric eel (E14) (70) (boxed residues). Members of this superfamily show conservation of 16 invariant residues (boxed and shaded) presumed to comprise the carbohydrate β-galactoside-binding domain of these lectins (49). CLC protein sequence is conserved at seven of these 16 residues (shown by numbers above sequence for amino acids 53, 58, 65, 72, 82, 83, 86). Gaps have been inserted in the sequences for optimal alignment. Numbering at right is from the published sequences. The percent identity of CLC protein to that of the human lung (H14) lectin is 20%. [From Ackerman et al. (45).]

and the electric organ of the electric eel (70). CLC protein possesses many of the physicochemical and biological characteristics of this family of proteins (47). These include the lack of conserved, structurally important cysteine residues (Fig. 3), the presence of free/reactive thiol groups (11,13), apparent lack of cleaved amino-terminal signal sequences (Fig. 3), blocked amino terminus (13) generally resulting from acetylation of an amino-terminal serine or alanine residue (Fig. 3) (48), formation of noncovalently bound dimers and higher oligomers, (11,13), and difficulty in establishing unique subcellular localizations, which routinely include both cytoplasmic (cytosolic), granule, nuclear, and extracellular sites (5,19,26–28,47,67,79).

Members of the S-type animal lectin superfamily show conservation of 16 invariant amino acid residues thought to comprise the carbohydrate recognition/binding domain of these proteins (48,49); CLC protein is also conserved at seven of these 16 residues [amino acids 53, 58, 65, 82, 83, 86 and the tryptophan at position 72 suggested to be in close proximity to the carbohydrate-binding site (70)], with potentially conservative amino acid changes at others (amino acids 55, 64, 75, 81) (Figs. 2 and 3). CLC protein also contains a sequence of three amino acids (VPY, residues 6–8) that is highly conserved at the amino-terminal end of the carboxyl-terminal half of all four of the IgE-binding proteins (Fig. 2) (71). These findings suggest that CLC protein could possess similar carbohydrate (β-galactoside) or IgE-binding activities. In this regard, hypodense activated eosinophils express a low-affinity IgE receptor (80), yet we and others have failed to detect CD23 ($F_c\epsilon RII$) on these cells (Lee, BW, Vercelli, D, Ackerman, SJ, Geha, R, unpublished observations). Intact viable eosinophils have been shown to express CLC protein lysophospholipase activity that is inhibitable by sulfhydryl-group-reactive, cell-impermeant reagents (81). CLC protein, which can have a plasma and cytosolic membrane localization in activated reactive eosinophils (19,26), could therefore be a candidate for this low-affinity IgE receptor if it is shown to bind IgE. The potential IgE-binding or other lectin-like activities of CLC protein and its possible role in aspects of eosinophil and basophil biology require further study.

A number of studies have suggested that the S-type lectins may be involved in cell-cell and cell-matrix interactions and that their appearance is developmentally regulated and temporally associated with the appearance of specific carbohydrate-containing structures in tissues (reviewed in 47). For example, rat lung has been shown to contain a 67-kDa galactoside-binding S-type lectin that is a component of the elastin receptor complex (82), and the 14-kDa rat lung β-galactoside lectin is secreted and localizes extracellularly to elastic tissue fibers of pulmonary blood vessels (47). The similarities of CLC protein to the 31- to 35-kDa IgE-binding proteins, especially mouse Mac-2 (69), the major nonintegrin laminin-binding protein of macrophages (83), suggests that CLC protein could be involved in interactions with extracellular matrix constituents such as laminin or elastin.

Likewise, CLC protein is similar to the human macrophage equivalent of Mac-2, which has recently been cloned and characterized by a number of groups (65,66) and shown to bind to laminin as well (66).

Using ultrastructural immunogold techniques, we have localized CLC protein to a primary granule population in mature eosinophils obtained from peripheral blood (25) and in eosinophils that differentiate in vitro in IL-5-stimulated cultures of human umbilical cord blood (26). Additionally, eosinophilic promyelocytes and myelocytes have been shown to secrete CLC protein in vitro during eosinophil differentiation and colony formation in soft agar, possibly by exocytosis of primary granules (84). There is ample ultrastructural evidence for the degranulation/exocytosis of eosinophil primary granules in the bone marrow during eosinophil development (reviewed in 84); these findings are further supported by the observation that CLC protein levels are significantly elevated in the marrow sinusoidal cavity of normal individuals (84). By analogy, the 15-kDa chicken lectin is concentrated intracellularly in developing embryonic muscle, but becomes mostly extracellular with maturation (85). The observations that the expression and localization of S-type animal lectins are developmentally regulated (reviewed in 47) suggest that CLC protein might play a role in the regulation of eosinophil or basophil development or participate extracellularly in the organization of the hematopoietic microenvironment in the bone marrow. In this regard, a recently characterized murine β-galactoside-binding S-type animal lectin (mGBP) from mouse embryo fibroblasts (86) was recently shown to function constitutively as a cell-growth-regulatory molecule during cell division and as a cytostatic factor, activities that were not attributable to its lectin-like properties and likely involved specific high-affinity receptor-ligand interactions (86).

The possibility that CLC protein is bifunctional with both lysophospholipase- and carbohydrate-binding activities is currently under investigation. The cDNA clones and deduced amino acid sequence of CLC protein will facilitate further studies of this protein's rather unusual hydrophobic properties, unique propensity for crystallization, lipolytic enzyme activity, potential lectin-like and associated activities, and ultimately its intracellular or extracellular role(s) in eosinophil and basophil development or functions in inflammatory reactions.

E. Characterization of the CLC Gene

The structure of the CLC gene has been examined using Southern blot analysis of restriction digests of total genomic DNA isolated from unseparated peripheral blood leukocytes of normal donors (45). Digestions with PstI, EcoRI, and BamHI gave single positive restriction fragments, the smallest of which was ~5–6 kb in length. Although not conclusive, these results are consistent with CLC protein being encoded by a single copy gene in this size range. We have now mapped the gene for human CLC protein to chromosome 19 using Southern analyses of mouse-human somatic cell hybrids (52) and to 19q13.1 by in situ hybridization (Trask, B, Mohrenweiser, H, unpublished observations). Further, we have cloned the CLC gene from a chromosome 19–specific library (54), and a fragment overlapping the transcriptional start site (identified by RNase protection to be 43 bp upstream of the 5′ end of the known cDNA sequence) has been isolated and sequenced. Constructs containing 411 bp of genomic sequence upstream of the transcriptional start site directed reported gene expression in transient transfections of the eosinophil-committed HL-60-C15 cell line (54). These sequences contain two consensus GATA binding sites, a purine-rich sequence representing potential binding sites for members of the *ets* family of transcription factors, as well as sequences described in other myeloid-specific promoters. This is the first demonstration of an eosinophil promoter that could serve as a model for characterizing the *cis* elements and *trans*-acting factors that regulate gene expression during differentiation of the eosinophil lineage.

III. EXPRESSION AND ENZYMATIC ACTIVITY OF RECOMBINANT CLC PROTEIN

We have examined the expression of the gene encoding CLC protein in transiently transfected monkey kidney (COS) cells and stably transfected Chinese hamster ovary (CHO) cell lines (28). Expression of recombinant CLC (rCLC) protein was evaluated both by Western blot and radioimmunoassay inhibition analyses of COS and CHO cell extracts and by indirect immunofluorescent staining and ultrastructural immunogold analyses of intact cells. The rCLC protein was immunochemically indistinguishable from native eosinophil-derived CLC protein, and each transfected CHO or COS cell expressed approximately 2.3 or 11 pg of rCLC protein, respectively, as measured by double-antibody radioimmunoassay. The rCLC protein was mainly cell-associated, with diffuse cytoplasmic and nuclear localizations in both COS and CHO cells as demonstrated by immunofluorescent light microscopy and ultrastructural immunogold analyses. By immunogold labeling, rCLC protein was present in the nucleus, perinuclear cisternae, endoplasmic reticulum, cytoplasm, and plasma membrane of transfected COS cells. In addition, lysates of transfected COS and CHO cells expressed considerable lysophospholipase activity, comparable to that of eosinophils. Furthermore, the rCLC protein expressed in both COS and CHO cells formed the distinctive intracytoplasmic and intranuclear hexagonal bipyramidal crystals that are characteristic of native eosinophil- and basophil-derived CLC protein. Transfected COS cells spontaneously formed CLCs in culture, and both transfected COS and CHO cells could be induced to form CLCs by exposure to hypotonic buffers. Thus, we have shown that the gene encoding human eosinophil CLC protein directs the expression of lysophospholipase activity in transfected cells and demonstrated that rCLC protein is immunochemically indistinguishable from and forms hexagonal bipyramidal crystals morphologically identical to the native eosinophil-derived enzyme. These results provide confirmation of the authenticity of the cloned CLC cDNA (45) and the expression of lysophospholipase activity (13) by this unique and prominent eosinophil (and basophil) constituent.

A. Transfections and Gene Expression

Transfection of the CLC cDNA into COS cells directed the expression of considerable lysophospholipase enzyme activity (Table 2) (28), confirming that the native eosinophil CLC protein functions as a lysophospholipase (12,13). A full-length, 598-bp cDNA insert containing the entire CLC coding region (45) was subcloned in sense (+) or antisense (−) orientations into the EcoRI cloning site of the expression vector, pMT2 (87), and transfected transiently into COS cells and stably into CHO cells. Transfection of the recombinant constructs into COS cells was performed by the DEAE-dextran technique (88). Transfection efficiency in COS cells averaged 33% ± 4 (n = 7) as determined by differential counts of COS cells expressing rCLC protein after staining by immunofluorescence. The CLC cDNA directed the synthesis of a prominent ~16-kDa band that was detectable by Western blotting in the lysates of the transfected COS cells, but not in COS cells transfected with the antisense (−) construct. Further, the rCLC protein remained primarily cell-associated and was not demonstrable in the neat culture super-

Table 2 Expression of Lysophospholipase Activity by Recombinant CLC Protein

Samples	Lysophospholipase activity (units)[a]
Exp. 1—Transiently transfected COS cells[d]	
Sense (+) construct transfected COS cell lysates	6.30[b]
Antisense (−) construct transfected COS cell lysates	0.46[b]
Nontransfected COS cell lysates	0.36[b]
Native eosinophil-derived CLC protein	8.00
Exp. 2—Stably transfected CHO cells	
pMT2-CLC (+) CHO cell lysate	2.70[c]
Nontransfected CHO cell lysate	0.17[c]
Native eosinophil-derived CLC protein	2.39

[a]One unit of lysophospholipase activity represents 1 μmol of fatty acid released per hour per mg CLC protein at 37°C and pH 7.5.
[b]Per 2.7 × 10^8 COS cells.
[c]Per 4.4 × 10^8 CHO cells.
[d]Modified and updated from Zhou et al. (28).

natants of transfected COS cells, but was detectable only after a 10-fold concentration. A kinetic analysis of rCLC protein production demonstrated that expression was maximal between 48 and 72 h posttransfection.

The subcellular localization of rCLC protein expressed in COS cells was evaluated by indirect immunofluorescence (28). Transfected COS cells expressing rCLC protein showed strong cytoplasmic, perinuclear, and nucleolar staining, with weaker nuclear staining; COS cells transfected with the antisense (−) construct showed no fluorescent staining for CLC protein. Likewise, in stably transfected CHO cell lines, the rCLC protein showed a diffuse cytoplasmic and nuclear distribution. The nuclear and cytoplasmic pattern of subcellular localization of rCLC protein in COS cells was similar to that observed in eosinophils and basophils. CLC protein in human eosinophils showed very intense staining in granule-free areas, especially beneath the plasma membrane, as well as in the perinuclear and nucleolar areas, but no visible staining of cytoplasmic granules; in human basophils, fluorescent staining for CLC protein was localized in the cytosol and nucleus, but with less intense nuclear staining than that observed in eosinophils. Previously, indirect immunofluorescent staining for CLC protein showed diffuse, cytoplasmic, plasma membrane, and distinct perinuclear localizations in the eosinophil (5,27) and cytoplasmic and perinuclear staining in the basophil (5,89). Ultrastructural analyses have shown that CLC protein was localized to a crystalloid-free primary granule population in mature human eosinophils (25), while it was a constituent of the large, particle-filled, basophil granule and their intragranular crystals (29). More recently, all these localizations of native CLC protein have been confirmed in activated eosinophils in vitro and in vivo using ultrastructural immunogold techniques (19,26).

B. Immunochemical Identity of Native and Recombinant CLC Proteins

The recombinantly expressed CLC protein was found to be immunochemically indistinguishable from native eosinophil-derived CLC protein as determined by radioimmunoassay inhibition analyses (28). These analyses showed that the rCLC protein expressed in COS and CHO cells was both immunochemically identical to and equally immunoreactive as the native eosinophil-derived protein. Transfected COS cells (positive for CLC protein by immunofluorescence) contained an average (± SD) of 11 ± 2.0 pg rCLC protein/cell while stable CHO lines expressed 2.3 ± 0.4 pg/rCLC protein cell.

C. Lysophospholipase Activity of Recombinant CLC Protein

Lysates of transiently transfected COS cells and stably transfected CHO cell lines producing rCLC protein expressed 14- to 17-fold more lysophospholipase activity

(6.3 and 2.7 μmol/h/mg CLC protein at 37°C and pH 7.5, respectively), with comparable activity to that of purified native eosinophil CLC protein (2–8 μmol/h/mg), than did equivalent numbers of antisense (−) construct transfected or nontransfected COS cells or nontransfected CHO cells (Table 2) (28). Lysates of both nontransfected and antisense (−) construct transfected COS cells and nontransfected CHO cells showed a low level of enzyme activity, likely representing activity of an endogenous lysophospholipase in these cells. Cleavage of the 1-carbon fatty acid from the ^{14}C-lysophosphatidylcholine substrate by rCLC protein in these assays was confirmed by thin-layer chromatography in comparison to the ^{14}C-palmitic acid cleavage products produced by native eosinophil lysophospholipase (CLC protein) or by a rat pancreatic lysophospholipase.

D. Formation of Hexagonal Bipyramidal Crystals by Recombinant CLC Protein

Transfected COS cells expressing rCLC protein formed CLC either spontaneously in culture or could be induced to form both intracellular and extracellular CLC under hypotonic conditions (Fig. 4) (28). Distinctive hexagonal bipyramidal crystals morphologically indistinguishable from eosinophil CLC (10–12,14) formed spontaneously in the cultures at 72 h in transiently transfected COS cells. Under hypotonic conditions, CLC were observed within the cytosol and the nucleus of swollen transfected COS and CHO cells as well as in transfected COS and CHO cell lysates. In contrast, the formation of crystals was never observed in the control antisense (−) construct transfected or nontransfected COS cells, in nontransfected CHO cells, or in their cell lysates.

The propensity for crystallization is perhaps the most interesting of the physicochemical properties of this polypeptide. The presence of these crystals at sites of eosinophil infiltration in tissues, and in body fluids and secretions, has been a hallmark of eosinophil involvement in parasitic, allergic, and inflammatory disorders (3,4). Although Charcot-Leyden crystals are uniquely formed by both eosinophils and basophils and not other leukocytes, the physiological role of CLC protein crystallization in the function of these granulocytes has not been elucidated; however, the crystallization of CLC protein inactivates its lysophospholipase activity (Zhou, Z, Clark, MA, Ackerman, SJ, unpublished observations). CLC protein and crystals have also been found in tissue macrophages in eosinophil-rich inflammatory reactions of the skin and in macrophages present in IL-5 containing cultures rich in eosinophils, suggesting a macrophage uptake and storage mechanism for CLC protein released from damaged and/or activated eosinophils (19,26); the effects of this lysophospholipase on macrophage function have not been determined. More recently, CLCs were localized by ultrastructural immunogold labeling in tumor cells of an epithelial neoplasm of the pancreas (90). These crystals were hypothesized to originate from adjacent tumor stroma eosino-

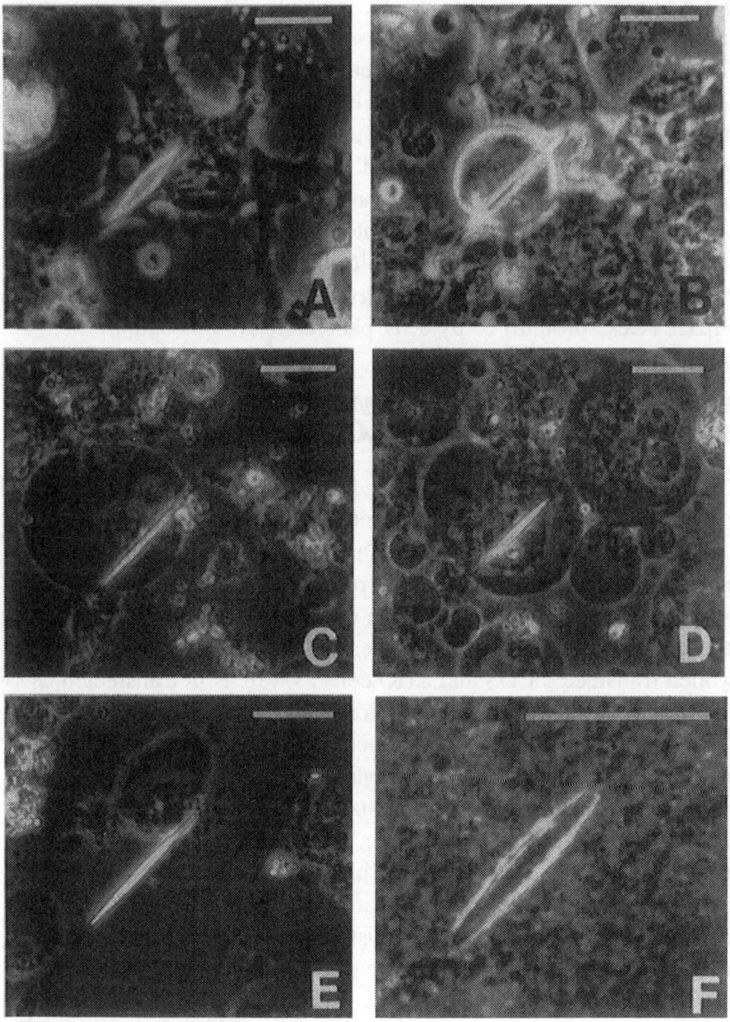

Figure 4 Spontaneous and induced formation of Charcot-Leyden crystals in transiently transfected COS cells expressing recombinant CLC protein. Spontaneously formed CLCs 72 h posttransfection in the culture medium (A) and intracellularly (B); (C) a single crystal formed within the cytosol of a hypotonically induced transfected COS cell; (D) a single crystal formed in the nucleus of a hypotonically induced transfected COS cell; (E) a crystal observed in hypotonic buffer containing transfected COS cells and (F) in the lysates from transfected COS cells. [From Zhou et al. (28).]

phils; tumors with stroma eosinophils are associated with improved prognosis in a wide variety of neoplasms (90).

The establishment of a eukaryotic expression system for obtaining recombinant CLC protein that possesses both lysophospholipase activity and the ability to undergo crystallization will facilitate using site-specific mutagenesis to study structure-function relationships for CLC protein to define the molecular basis for its lipolytic activity. Expression of both wild-type and mutant recombinant CLC protein will facilitate analyses of its marked hydrophobicity, unique propensity for crystallization, potential carbohydrate- or IgE-binding activities, and their relevance to the intracellular and extracellular functions of this protein in the biology of both eosinophils and basophils. Further, the transient or stable expression of rCLC protein in eukaryotic cell lines will facilitate the routine purification of this unique lysophospholipase for in vitro and animal model studies of its roles in eosinophil- and basophil-associated allergic inflammation (see below), as well as in eosinophil-parasite interactions.

IV. POTENTIAL FUNCTIONS OF EOSINOPHIL LYSOPHOSPHOLIPASE IN ASTHMA

Asthma is characterized by impaired respiratory function and airways inflammation. Lung pathology in many patients with moderate to severe asthma is marked by the infiltration and degranulation of eosinophils (91–93) as well as microatelectasis (localized regions of collapsed lung) (94). Eosinophils, through secretion of toxic granule proteins and lipid mediators, have been implicated as proinflammatory/cytotoxic effector cells in the pathogenesis of asthma (91). As noted above, lysophospholipase (CLC protein) is a major eosinophil constituent (11,13), and its lysolecithin substrate is a component of pulmonary surfactant; we have therefore studied the potential role of this lipolytic enzyme in the pathophysiology of human asthma (95). Pulmonary surfactant, a phospholipid-rich substance secreted by type II alveolar epithelial cells in the lung, is required for the reduction of surface tension in the terminal bronchioles and alveoli in order to prevent their collapse (for review, see 96). We tested the hypothesis that lysophosphatidylcholine is a critical component of surfactant activity and that eosinophil-derived lysophospholipase secreted into the airways degrades the lysophosphatidylcholine of surfactant, thereby altering its physical surface-active properties and compromising surfactant-mediated protection of the small airways against collapse (96). In support of this hypothesis, we found that the enzyme's preferred substrate, lysophosphatidylcholine, is a quantitatively minor but functionally significant constituent of pulmonary surfactant in that hydrolysis of this lysophosphatide by eosinophil lysophospholipase in vitro causes a marked increase in surfactant surface tension. Furthermore, BAL samples from asthmatics contained more CLC protein (lysophospholipase) than samples from normal individuals. Most importantly, this alteration was associated with asthma severity and suggests a novel biochemical mechanism for microatelectasis in asthmatic lung (94) and a partial explanation for compromised airways function in certain forms of asthma.

A. Tissue Localization of CLC Protein in Asthmatic Lung

CLC are routinely observed in the sputum of asthmatics (2), and markedly elevated levels of CLC protein are present in the sputum of asthmatics in association with both the acute and chronic disease, especially in patients hospitalized for acute asthmatic attacks (97). In our study of 116 sputum samples (97), CLC protein levels averaged 3.5 μg/ml, with values as high as 25 μg/ml in patients hospitalized for asthma. Thus, significant quantities of CLC protein are released from eosinophils in the lung and are present both in the bronchial lumen and possibly in the tissues. To determine whether eosinophils and eosinophil lysophospholipase are present in the surfactant-rich small airways of the lung, sections of postmortem lung biopsy specimens from two severe asthmatics were examined using immunochemical techniques and affinity-purified antipeptide antibodies specific for the amino terminus of eosinophil lysophospholipase. Numerous eosinophils and intense areas of both cell-associated and extracellular staining for CLC protein were observed in peribroncheolar regions, in and on the alveolar walls, and within the alveoli. The specificity of staining was confirmed by the lack of reaction product in serial sections of the same biopsies in which preimmunization serum was substituted for the primary antibody. Marked extracellular staining for CLC protein of both amorphous intra-alveolar material and the walls of the alveoli was observed, areas replete with surfactant. The specificity of the affinity-purified anti-CLC peptide antibody was confirmed by Western blot analyses; the antibody reacted with chromatographically purified eosinophil lysophospholipase and recognized only a single protein component of identical size in whole eosinophil lysates.

B. In Vitro Degradation of Pulmonary Surfactant: Effects on Surface Tension

The principal substrate of eosinophil lysophospholipase, lysophosphatidylcholine, is an important, albeit minor, component of pulmonary surfactant. Surfactant is primarily composed of phospholipids, the predominant one being 1,2-dipalmitoyl phosphatidylcholine (96). To determine the role of lysophosphatidylcholine in the maintenance of the surface-active properties of surfactant, various mixtures of synthetic 1,2-dipalmitoyl phosphatidylcholine and lysophosphatidylcholine were prepared and their surface tension characteristics measured using a pulsating bubble surfactometer (98). Mixtures deficient in lysophosphatidylcholine had very high surface tension (values nearly that of saline alone), whereas the addition of small amounts of lysophosphatidylcholine resulted in significant reductions in surface tension. Of particular interest, the minimal surface tension in these lipid mixtures was achieved when the content of lysophosphatidylcholine was between 4 and 5%, a concentration identical to that which we have measured in normal human pulmonary surfactant and similar to what other investigators have found in both canine and bovine surfactant (99,100). The effects of eosinophil lysophospholipase activity on the surface tensile properties of human surfactant lipids were also investigated. Results from these experiments showed that the enzyme produced both a time- and dose-dependent increase in the surface tension of human surfactant lipids from normal bronchoalveolar lavage (BAL) fluids, whereas the heat-inactivated enzyme had no effect. To demonstrate that the eosinophil lysophospholipase actually degraded the lysophosphatidylcholine component of the surfactant, Cs+ liquid secondary ion mass spectrometry (SIMS) was utilized to measure lysophosphatidylcholine and phosphatidylcholine concentrations (101); analysis of the spectra of the surfactant lipids indicated that the amount of lysophosphatidylcholine was significantly reduced by the lysophospholipase treatment, while the phospholipid concentration was unaffected. These findings demonstrated that eosinophil lysophospholipase is catalytically active in normal human pulmonary surfactant and capable of degrading the lysophosphatidylcholine component shown to be important for maintaining minimal surface tension.

C. In Vivo Associations of (Lysophospholipase) CLC Protein with Impaired Pulmonary Function in Asthma

Based on the marked effect of eosinophil lysophospholipase on the surface tensile properties of human surfactant in vitro, we have investigated whether the secretion of the enzyme in vivo is associated with impaired respiratory function in patients with asthma. This was addressed by studying BAL fluids available from a well-characterized group of nocturnal asthmatics (102) and normal control subjects with respect to the relationship between the lysophospholipase (as CLC protein) content of the BAL fluid and asthma severity as assessed by pulmonary function testing (% overnight drop in FEV_1). As previously reported for these patients, there was an overnight fall in FEV_1, which was related to the increased numbers of eosinophils obtained in the BAL (102). In association with the increased numbers of BAL eosinophils in these nocturnal asthmatics, there were also increased levels of eosinophil lysophospholipase (measured as CLC protein)

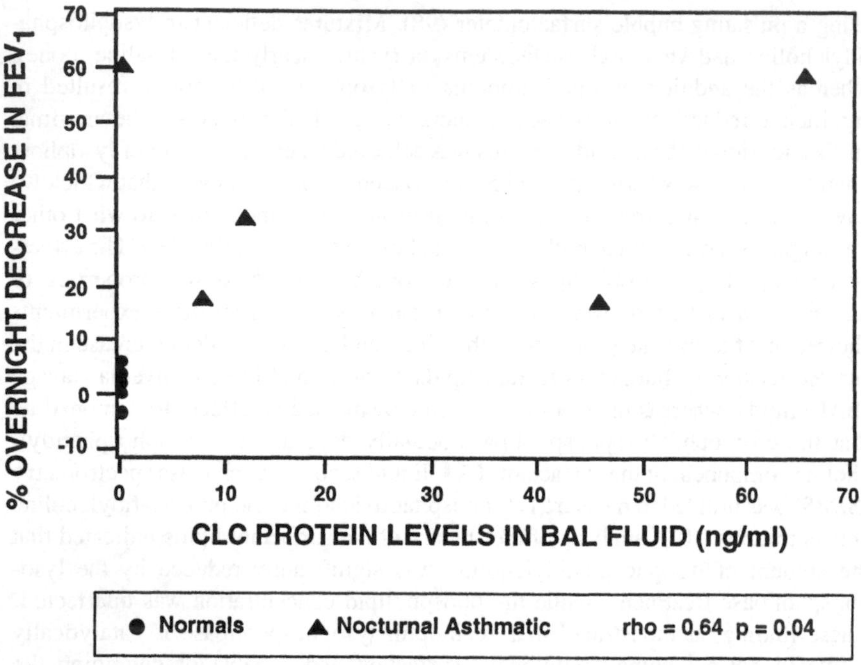

Figure 5 Relationship between pulmonary function and CLC protein levels in BAL fluid of normal and nocturnal asthmatic subjects. Pulmonary functions were assessed by the percent overnight decrease in FEV_1 from 22:00 h compared to 04:00 h. On the same night, bronchoalveolar lavage (BAL) was performed at 04:00 h and the concentration of eosinophil lysophospholipase in the BAL fluid (measured as CLC protein) was determined by radioimmunoassay as previously described (11,97). There was a statistically significant correlation (rho = 0.64, $p = 0.04$) between the levels of CLC protein in the BAL at 04:00 h and the severity of the overnight decrease in FEV_1; analysis was performed by Spearman's rank correlation.

in the BAL that were likewise associated with decreased pulmonary function in these individuals (Fig. 5). There was a significant positive relationship between CLC protein levels and asthma severity as assessed by percent overnight fall in FEV_1 (Fig. 5).

D. Potential Role of Surfactant Dysfunction in Asthma

The above results suggest the following potential pathological mechanism in asthma: (1) increased numbers of eosinophils, eosinophil degranulation/secretion, and release of CLC protein in the lungs of asthmatics results in increased levels of lysophospholipase; (2) the enzyme degrades the lysophosphatidylcholine component of surfactant, thereby increasing its surface tension properties; (3) the increased surface tension of surfactant in turn contributes both to microatelectasis (94,103,104) and possibly to overall impaired respiratory function of the asthmatic lung. The resulting inability of lysophosphatidylcholine-depleted surfactant to reduce alveolar surface tension might also contribute to the pathogenesis of airways obstruction by increasing air trapping (94,96) and possibly airways resistance, particularly during expiration (105,106). A surfactant defect could contribute to airway closure, also a feature of asthma; changes in surfactant surface tension may contribute to airway patency, as originally pointed out by Macklem and colleagues (107), and more recently by a number of other groups (106,108,109). Drazen and colleagues (108) have suggested that liquid in the airways and/or stimuli that increase the surface tension (γ_{min}) of surfactant might have substantially greater obstructive effects than do stimuli that only narrow airways. In this regard, surfactant dysfunction, potentially due to the action of eosinophil lysophospholipase, is more likely to contribute significantly to airways disease in severe asthma than in mild to moderate asthma, both of which are more fully reversible with β-agonists.

Increases in surfactant surface tension in relationship to disease severity have been reported for individuals suffering from allergic asthma (110,111), with improvement in surfactant activity noted after various types of asthma therapy (112), the only other studies of this relationship of which we are aware. Furthermore, increases in pulmonary surfactant surface tension have been demonstrated experimentally in actively sensitized guinea pigs upon aerosol exposure to antigen (113). The potential importance of surfactant in airways function in asthma is further supported by the recent finding that aerosol administration of surfactant to asthmatics can improve their respiratory function (114). It has been suggested that the microatelectasis seen in patients who died from asthma is due to postobstructive resorption of trapped gas; the biochemical mechanism suggested by the present studies would provide a novel explanation for the development of microatelectasis in severe asthma (94,103,104). We therefore suggest that eosinophil lysophospholipase may play a significant role in the development of the focal atelectasis seen in the lungs of patients who have died from asthma. Whether this also contributes importantly to the overall impairment of respiratory function in patients with asthma remains to be elucidated. Our recent cloning (45) and expression (28) of recombinant, enzymatically active CLC protein should facilitate the routine purification of sufficient enzyme for determination of the effects of intratracheal administration of lysophospholipase on pulmonary mechanics in animal models.

In addition to asthma, infiltration of the lungs with activated eosinophils is also observed to varying degrees in other diseases, including adult respiratory distress syndrome (115), interstitial pulmonary fibrosis (116), hypersensitivity pneumonitis

(117), and eosinophilic pneumonia (118); the above biochemical mechanism might also be relevant to surfactant and airways dysfunction in these diseases as well.

V. SUMMARY

Although Charcot-Leyden crystals have been recognized for well over 100 years, and the unique crystal-forming protein was shown to possess lysophospholipase activity more than 10 years ago, the intracellular and extracellular role(s) of this enzyme in eosinophil biology and inflammation remains unclear. The cloning of the cDNA encoding CLC protein and expression of the recombinant enzyme have verified the lipolytic activity of this major eosinophil constituent and enabled the characterization of the amino acid sequence. CLC protein bears no resemblance in primary sequence to other eukaryotic or prokaryotic lysophospholipases, but instead appears related to the superfamily of S-type β-galactoside-binding animal lectins, including certain IgE- and laminin-binding proteins. However, the significance of these sequence similarities and whether CLC protein is bifunctional, i.e., possesses any carbohydrate- or IgE-binding activities, remains to be determined. Structure-function analyses of CLC protein using site-specific mutagenesis should enable characterization of the enzyme's active site and possibly the regulatory mechanisms for intracellular or extracellular expression of lysophospholipase activity by the eosinophil, as well as address the unique propensity and biological role for crystallization. Finally, a potential pathophysiological role has been identified for this lysophospholipase in asthma involving the degradation of pulmonary surfactant lysophospholipids and alterations of the surface tensile properties of surfactant. Increased surfactant surface tension could contribute to the development of focal atelectasis, to airways closure, and possibly to overall impaired respiratory activity in the asthmatic lung, especially in severe asthma. The potential contribution of surfactant dysfunction to decreased airways patency in asthma requires further investigation. An in vivo assessment of the effects of purified eosinophil lysophospholipase on pulmonary mechanics in animal models should provide more direct evidence for a role for CLC protein in the pathogenesis of airways disease.

ACKNOWLEDGMENTS

This work was supported in part by NIH grants AI25230, AI22660, and AI33043 (to SJA), CA41456 (to DGT), and HL36577 (to CGI) and the American Heart Association of Colorado (to Y-PT). We thank the following individuals for their invaluable contributions as collaborators or consultants on various aspects of the work reviewed here. Cloning of CLC protein: Helene Rosenberg, Stephanie Corrette, Joshua Bennett, David Mastrianni, Anne Nicholson-Weller, Peter Weller, David Chin (Department of Medicine, Beth Israel Hospital and Harvard

Medical School, Boston, MA); ultrastructural localization and expression of CLC protein: Ann Dvorak (Department of Pathology, Beth Israel Hospital and Harvard Medical School, Boston, MA); eosinophil lysophospholipase in asthma: Diane Garsetti and Frederick Holtzburg (Allergy Division, Schering-Plough Research, Kennelworth, NJ); Guiseppi Pieatra and Sergio Villaschi (Department of Anatomic Pathology, College of Medicine, University of Pennsylvania, Philadelphia, PA); Jack Elias and John Rankin (Department of Medicine, School of Medicine, Yale University, New Haven CT); Richard Martin (Department of Medicine, National Jewish Center for Immunology and Respiratory Medicine, Denver, CO).

REFERENCES

1. Charcot JM, Robin C. Observation de Leocythemie. CR Mem Soc Biol 1853; 5:44.
2. Leyden E. Zur Kenntniss des Bronchial-asthma. Arch Pathol Anat 1872; 54:324.
3. Beeson PB, DA Bass. The eosinophil. In: Smith LH, ed. Major Problems in Internal Medicine. Vol. 14. Philadelphia: Saunders, 1977:39–42.
4. Ottesen EA, Cohen SG. The eosinophil, cosinophilia and eosinophil related disorders. In: Middleton E, Reed CE, Ellis EF, eds. Allergy: Principles and Practice. Vol. 2. St. Louis: Mosby, 1978:584.
5. Ackerman SJ, Weil GJ, Gleich GJ. Formation of Charcot-Leyden crystals by human basophils. J Exp Med 1982; 155:1597.
6. Archer TG, Blackwood A. Formation of Charcot-Leyden crystals in human eosinophils and basophils and study of the composition of the isolated crystal. J Exp Med 1965; 122:173.
7. Dvorak AM, Monahan RA. Crohn's disease. Ultrastructural studies showing basophil leukocyte granule changes and lymphocyte parallel tubular arrays in peripheral blood. Arch Pathol Lab Med 1982; 106:145.
8. Dvorak AM. Morphologic expressions of maturation and function can affect the ability to identify mast cells and basophils in man, guinea pig and mouse. In: Befus AD, Bienenstock J, Denburg JA, eds. Mast Cell Differentiation and Heterogeneity. New York: Raven Press, 1986:95.
9. Dvorak AM. Morphologic and immunologic characterization of human basophils (1879 to 1985). Riv Immunol Immunofarmacol 1988; 8:50.
10. Gleich GJ, Loegering DA, Mann KG, Maldonado JE. Comparative properties of the Charcot-Leyden crystal protein and the major basic protein from human eosinophils. J Clin Invest 1976; 57:633.
11. Ackerman SJ, Loegering DA, Gleich GJ. The human eosinophil Charcot-Leyden crystal protein: biochemical characteristics and measurement by radioimmunoassay. J Immunol 1980; 125:2118.
12. Weller PF, Goetzl EJ, Austen KF. Identification of human eosinophil lysophospholipase as the constituent of Charcot-Leyden crystals. Proc Natl Acad Sci USA 1980; 77:7440.
13. Weller PF, Bach DS, Austen KF. Biochemical characterization of human eosinophil Charcot-Leyden crystal protein (lysophospholipase). J Biol Chem 1984; 259:15100.

14. Weller PF, Bach D, Austen KF. Human eosinophil lysophospholipase: the sole protein component of Charcot-Leyden crystals. J Immunol 1982; 128:1346.
15. Adewusi K, Goven AJ. Enhancement of lysophospholipase activity with *Trichinella spiralis* antigen: evidence for cell cooperation. J Parasitol 1987; 72:716.
16. Ackerman SJ, Coburn EW, Weller PF. Biosynthesis of myristylated Charcot-Leyden crystal protein by HL-60 cells. Clin Res 1986; 34:276A.
17. Ackerman SJ, Weller PF. Charcot-Leyden crystal protein. Myristylation in HL-60 cells and eosinophils. In: Abstracts, 6th International Congress of Immunology, July 6–11, Toronto, Canada, 1986:653 (Abstract #5.28.19).
18. Dvorak AM, Ackerman SJ, Weller PF. Subcellular morphology and biochemistry of eosinophils. In: Harris JR, ed. Blood Cell Biochemistry: Megakaryocytes, Platelets, Macrophages and Eosinophils. Vol. 2. London: Plenum Press, 1990:237–344.
19. Dvorak AM, Weller PF, Monahan-Earley RA, Letourneau L, Ackerman SJ. Ultrastructural localization of Charcot-Leyden crystal protein (lysophospholipase) and peroxidase in macrophages, eosinophils, and extracellular matrix of the skin in the hypereosinophilic syndrome. Lab Invest 1990; 62:590.
20. Zucker-Franklin D. Eosinophil structure and maturation. In: Mahmoud AA, Austen KF, eds. The Eosinophil in Health and Disease. New York: Grune & Stratton, 1980:43.
21. Spry CJF. Eosinophils: A Comprehensive Review and Guide to the Scientific and Medical Literature. New York: Oxford University Press, 1988:66.
22. McDonald S, Shaw AF. Persistent eosinophilia and splenomegaly. Br Med J 1922; 2:966.
23. Sieker LC, Turley S, LeTrong I, Stenkamp RE, Weller PF, Ackerman SJ. Crystallographic characterization of human eosinophil Charcot-Leyden crystals. J Mol Biol 1988; 204:489.
24. Weller PF, Wasserman SI, Austen KF. Selected enzymes preferentially present in the eosinophil. In: Mahmoud AA, Austen KF, eds. The Eosinophil in Health and Disease. New York: Grune & Stratton, 1980:115.
25. Dvorak AM, Letourneau L, Login GR, Weller PF, Ackerman SJ. Ultrastructural localization of the Charcot-Leyden crystal protein (lysophospholipase) to a distinct crystalloid-free granule population in mature human eosinophils. Blood 1988, 72:150.
26. Dvorak AM, Furitsu T, Letourneau L, Ishizaka T, Ackerman SJ. Mature eosinophils stimulated to develop in human cord blood mononuclear cell cultures supplemented with recombinant human interleukin-5. Part 1. Piecemeal degranulation of specific granules and distribution of Charcot-Leyden Crystal protein. Am J Pathol 1991; 138:69.
27. Ackerman SJ, Kephart GM, Habermann TM, Greipp PR, Gleich GJ. Localization of eosinophil granule major basic protein in human basophils. J Exp Med 1983; 158:946.
28. Zhou Z-Q, Tenen DG, Dvorak AM, Ackerman SJ. The gene for human eosinophil Charcot-Leyden crystal protein directs the expression of lysophospholipase activity and crystallization in transfected COS cells. J Leuk Biol 1992; 52:588.
29. Dvorak AM, Ackerman SJ. Ultrastructural localization of the Charcot-Leyden

crystal protein (Lysophospholipase) to granules and intragranular crystals in mature human basophils. Lab Invest 1989; 60:557.

30. Silverman BA, Weller PF, Shin ML. Effect of erythrocyte membrane modulation by lysolecithin on complement-mediated lysis. J Immunol 1984; 132:386.
31. Tai PC, Spry CJF. The mechanisms which produce vacuolated and degranulated eosinophils. Br J Haematol 1981; 49:219.
32. Adewusi K, Goven AJ. Effect of anti-thymocyte serum on the eosinophil and lysophospholipase responses in mice infected with *Trichinella spiralis*. Parasitology 1987; 94:115.
33. Hirata F, Corcoran BA, Venkatasubramanian K, Schiffman E, Axelrod J. Chemoattractants stimulate degradation of methylated phospholipids and release of arachidonic acid in rabbit leukocytes. Proc Natl Acad Sci USA 1979; 76:2640.
34. Takenawa T, Homma Y, Nagai Y. Role of Ca^{2+} in phosphatidylinositol response and arachidonic acid release in formylated tripeptide- or Ca^{2+} ionophore A23187-stimulated guinea pig neutrophils. J Immunol 1983; 130:2849.
35. Ichikawa I, Yokoyama E. Effect of short-term exposure to ozone on the lecithin metabolism of rat lung. J Toxicol Environ Health 1982; 10:1005.
36. Keller R. Nachweis von lecithin bzw. lysolecithin in lipidextrakten normaler bzw. anaphylaktischer mastzellen. Helv Physiol Pharmacol Acta 1962; 20:c66.
37. Echetebu CO. Extracellular lipase and proteinase of *Basidiobolus haptosporus*: possible role in subcutaneous mycosis. Mycopathologia 1982; 80:171.
38. Williams AO, von Lichtenberg J, Smith JH. Ultrastructure of phycomycosis due to Entomophthora, basidiobolus, and associated "Splendore-Hoeppli" phenomenon. Arch Pathol 1969; 87:459.
39. Furlong ST, Caulfield JP. *Schistosoma mansoni*: synthesis and release of phospholipids, lysophospholipids, and neutral lipids by schistosomula. Exp Parasitol 1989; 69:65.
40. Golan DE, Furlong ST, Brown CS, Caulfield JP. Monopalmitoyl-phosphatidylcholine incorporation into human erythrocyte ghost membranes causes protein and lipid immobilization and cholesterol depletion. Biochemistry 1988; 27:2661.
41. Modolell M, Munder PG. The action of purified phospholipase B in inflammation and immunity. Int Arch Allergy Appl Immunol 1972; 43:724.
42. Abu-Ghazaleh RI, Gleich GJ, Prendergast FG. Interaction of eosinophil granule major basic protein with synthetic lipid bilayers: A mechanism for toxicity. J Membr Biol 1992; 128:153–164.
43. Weltzien HU. Cytolytic and membrane-perturbing properties of lysophosphatidylcholine. Biochim Biophys Acta 1979; 559:259.
44. Munder PG, Modolell M, Andreesen R, Weltzien HV, Westphal O. Lysophosphatidylcholine (lysolecithin) and its synthetic analogues. Immune modulating and other biologic effects. Semin Immunopathol 1979; 2:187.
45. Ackerman SJ, Corrette SE, Rosenberg HF, Bennett JC, Mastrianni DM, Nicholson-Weller A, Weller PF, Chin DT, Tenen DG. Molecular cloning and characterization of human eosinophil Charcot-Leyden crystal protein (lysophospholipase): similarities to IgE-binding proteins and the S-type animal lectin superfamily. J Immunol 1993; 150:456.

46. Rosenberg HF, Corrette SE, Tenen DG, Ackerman SJ. Rapid cDNA library screening using the polymerase chain reaction. Biotechniques 1991; 10:53.
47. Barondes SH. Soluble lectins: a new class of extracellular proteins. Science 1984; 221:1260.
48. Hirabayashi J, Ayaki H, Soma G, Kasai K. Cloning and nucleotide sequence of a full-length cDNA for human 14 kDa β-galactoside-binding lectin. Biochim Biophys Acta 1988; 1008:85.
49. Drickamer K. Two distinct classes of carbohydrate-recognition domains in animal lectins. J Biol Chem 1988; 263:9557.
50. Han JH, Stratowa C, Rutter WJ. Isolation of full-length putative rat lysophospholipase cDNA using improved methods for mRNA isolation and cDNA cloning. Biochemistry 1987; 26:1617.
51. Kyger EM, Wiegand RC, Lange LG. Cloning of the bovine pancreatic cholesterol esterase/lysophospholipase. Biochem Biophys Res Commun 1989; 164:1302.
52. Mastrianni DM, Eddy RL, Rosenberg HF, Corrette SE, Shows TB, Tenen DG, Ackerman SJ. Localization of the human eosinophil Charcot-Leyden crystal protein (lysophospholipase) gene to chromosome 19 and the human ribonuclease 2 (eosinophil-derived neurotoxin) and ribonuclease 3 (eosinophil cationic protein) genes to chromosome 14. Genomics 1992; 13:240–242.
53. Kozak M. Point mutations define a sequence flanking the AUG initiation codon that modulates translation by eukaryotic ribosomes. Cell 1986; 44:283.
54. Gomolin H, Yamaguchi Y, Paulpillai AV, Ackerman SJ, Tenen DG. Human eosinophil Charcot-Leyden crystal protein: characterization of a lysophospholipase gene promoter. Blood. Submitted for publication.
55. Wellner D, Panneerselvam C, Horecker BL. Sequencing of peptides and proteins with blocked N-terminal amino acids: *N*-acetylserine or *N*-acetylthreonine. Proc Natl Acad Sci USA 1990; 87:1947.
56. Hart GW, Brew K, Grant GA, Bradshaw RA, Lennarz WJ. Primary structural requirements for the enzymatic formation of the *N*-glycosidic bond in glycoproteins. Studies with natural and synthetic peptides. J Biol Chem 1979; 254:9747.
57. Kyte J, Doolittle RR. A simple method for displaying the hydropathic character of a protein. J Mol Biol 1982; 157:105.
58. von Heijne G. A new method for predicting signal sequence cleavage sites. Nucleic Acids Res 1986; 14:4683.
59. von Heijne G, Liljestrom P, Mikus P, Andersson H, Ny T. The efficiency of the uncleaved secretion signal in the plasminogen activator inhibitor type 2 protein can be enhanced by point mutations that increase its hydrophobicity. J Biol Chem 1991; 266:15240.
60. Tabe L, Krieg P, Strachan R, Jackson D, Wallis E, Colman A. Segregation of mutant ovalbumins and ovalbumin-globin fusion proteins in *Xenopus* oocytes. Identification of an ovalbumin signal sequence. J Mol Biol 1984; 180:645.
61. Chou PY, Fassman GD. Prediction of the secondary structure of proteins from their amino acid sequence. Adv Enz 1978; 47:45.
62. Tomonaga M, Gasson JC, Quan SG, Golde DW. Establishment of eosinophilic sub-

lines from human promyelocytic leukemia (HL-60) cells: demonstration of multipotentiality and single lineage commitment of HL-60 stem cells. Blood 1986; 67: 1433.

63. Liu F-T, Albrandt K, Mendel E, Kulczycki Jr A, Orida NK. Identification of an IgE-binding protein by molecular cloning. Proc Natl Acad Sci USA 1985; 82:4100.
64. Albrandt K, Orida NK, Liu F-T. An IgE-binding protein with a distinctive repetitive sequence and homology with an IgG receptor. Proc Natl Acad Sci USA 1987; 84:6859.
65. Cherayil BJ, Chaitovitz S, Wong C, Pillai S. Molecular cloning of a human macrophage lectin specific for galactose. Proc Natl Acad Sci USA 1990; 87:7324.
66. Robertson MW, Albrandt K, Keller D, Liu F-T. Human IgE-binding protein: a soluble lectin exhibiting a highly conserved interspecies sequence and differential recognition of IgE glycoforms. Biochemistry 1990; 29:8093.
67. Laing JG, Robertson MW, Gritzmacher CA, Wang JL, Liu F-T. Biochemical and immunological comparisons of carbohydrate-binding protein 35 and an IgE-binding protein. J Biol Chem 1989; 264:1907.
68. Laing JG, Wang JL. Identification of carbohydrate binding protein 35 in heterogeneous nuclear ribonucleoprotein complex. Biochemistry 1988; 27:5329.
69. Cherayil BJ, Weiner SJ, Pillai S. The Mac-2 antigen is a galactose-specific lectin that binds IgE. J Exp Med 1989; 170:1959.
70. Paroutaud P, Levi G, Tiechberg VI, Strosberg AD. Extensive amino acid sequence homologies between animal lectins. Proc Natl Acad Sci USA 1987; 84:6345.
71. Leffler H, Masiarz FR, Barondes SH. Soluble lactose-binding vertebrate lectins; a growing family. Biochemistry 1989; 28:9222.
72. Gitt MA, Barondes SH. Evidence that a human soluble β-galactoside-binding lectin is encoded by a family of genes. Proc Natl Acad Sci USA 1986; 83:7603.
73. Clerch LB, Whitney P, Hass M, Brew K, Miller T, Werner R, Massaro D. Sequence of a full-length cDNA for rat lung β-galactoside-binding protein: primary and secondary structure of the lectin. Biochemistry 1988; 27:692.
74. Leffler H, Barondes SH. Specificity of binding of three soluble rat lung lectins to substituted and unsubstituted mammalian β-galactosides. J Biol Chem 1986; 261: 10119.
75. Southan C, Aitken A, Childs RA, Abbott WM, Feizi T. Amino acid sequence of β-galactoside-binding bovine heart lectin. FEBS Lett 1987; 214:301.
76. Hirabayashi J, Kasai K. Complete amino acid sequence of a β-galactoside-binding lectin from human placenta. J Biochem 1988; 104:1.
77. Ohyama Y, Hirabayashi J, Oda Y, Oono S, Kawasaki H, Suzuki K, Kasai K. Nucleotide sequence of chick 14K β-galactoside-binding lectin mRNA. Biochem Biophys Res Commun 1986; 134:51.
78. Roff GF, Wang JL. Endogenous lectins from cultured cells. Isolation and characterization of carbohydrate-binding proteins from 3T3 fibroblasts. J Biol Chem 1983; 17:10657.
79. Gritzmacher CA, Robertson MW, Liu F-T. IgE-binding protein. Subcellular location and gene expression in many murine tissues and cells. J Immunol 1988; 141:2801.

80. Capron M, Spiegelberg HL, Prin L, Bennich H, Butterworth AE, Pierce RJ, Ouaissi MA, Capron A. Role of IgE receptors in effector function of human eosinophils. J Immunol 1984; 132:462.
81. Weller PF, Bach DS, Austen KF. Expression of lysophospholipase activity by intact human eosinophils and their Charcot-Leyden crystals. Trans Assoc Am Physicians 1981; 44:165.
82. Hinek A, Wrenn DS, Mecham RP, Barondes SH. The elastin receptor: a galactoside-binding protein. Science 1988; 239:1539.
83. Woo HJ, Shaw LM, Messier JM, Mercurio AM. The major non-integrin laminin binding protein of macrophages is identical to carbohydrate binding protein 35 (Mac-2). J Biol Chem 1990; 265:7097.
84. Butterfield JH, Ackerman SJ, Scott RE, Pierre RV, Gleich GJ. Evidence for secretion of human eosinophil granule major basic protein and Charcot-Leyden crystal protein during eosinophil maturation. Exp Hematol 1984; 12:163.
85. Barondes SH, Haywood-Reid PL. Externalization of an endogenous chicken muscle lectin with in vivo development. J Cell Biol 1981; 91:568.
86. Wells V, Mallucci L. Identification of an autocrine negative growth factor: mouse beta-galactoside-binding protein is a cytostatic factor and cell growth regulator. Cell 1991; 64:91.
87. Kaufman RJ, Davies MV, Pathak VK, Hershey JWB. The phosphorylation state of eucaryotic initiation factor 2 alters translational efficiency of specific mRNAs. Mol Cell Biol 1989; 9:946–958.
88. Seed B, Aruffo A. Molecular cloning of the CD2 antigen, the T-cell erythrocyte receptor, by a rapid immunoselection procedure. Proc Natl Acad Sci USA 1987; 84: 3365–3369.
89. Golightly LM, Thomas LL, Dvorak AM, Ackerman SJ. Charcot-Leyden crystal protein in the anaphylactic degranulation and recovery of activated basophils. J Leukoc Biol 1992; 51:386–392.
90. Dvorak AM, Letourneau L, Weller PF, Ackerman SJ. Ultrastructural localization of Charcot-Leyden crystal protein (lysophospholipase) to intracytoplasmic crystals in tumor cells of primary solid and papillary epithelial neoplasm of the pancreas. Lab Invest 1990; 62:608–615.
91. Gleich GJ. The eosinophil and bronchial asthma: current understanding. J Allergy Clin Immunol 1990; 85:422–436.
92. Bousquet J, Chanel P, LaCoste JY, Barneon G, Ghavania N, Enander I, Venge P, Ahlstedt S, Simony-Lafontain J, Godard P, Michel FB. Eosinophilic inflammation in asthma. N Engl J Med 1990; 323:1033–1039.
93. Dunnill MS. The pathology of asthma. In: Porter R, Birch J, eds. The Identification of Asthma. Ciba Foundation Symposium. London: Churchill Livingstone, 1971:35–56.
94. Macklem PT. Airway obstruction and collateral ventilation. Physiol Rev 1971; 51: 368–436.
95. Clark MA, Garsetti D, Holtzburg R, Das PR, Pieatra G, Villaschi S, Elias JA, Rankin JA, Irvin CG, Martin RJ, Tu Y-P, Zhou Z, Ackerman SJ. Catalysis of surfactant by eosinophil lysophospholipase: potential role in asthma. In preparation.

96. Van Golde LMG, Batenburg JJ, Robertson B. The pulmonary surfactant system: biochemical aspects and functional significance. Physiol Rev 1988; 68:374–455.
97. Dor PJ, Ackerman SJ, Gleich GJ. Charcot-Leyden crystal protein and eosinophil granule major basic protein in sputum of patients with respiratory diseases. Am Rev Respir Dis 1984; 130:1072–1077.
98. Robertson B, Enhorning G, Malmquist E. Quantitative determination of pulmonary surfactant with pulsating bubble. Scand J Clin Lab Invest 1972; 29:45–49.
99. Liau DF, Barrett R, Bell L, Ryan SF. Normal surface properties of phosphatidylglycerol-deficient surfactant from dog after acute long injury. J Lipid Res 1985; 26:1338–1344.
100. Holm BA, Keicher K, Liu M, Sokolowski J, Enhorning G. Inhibition of pulmonary surfactant function by phospholipases. J Appl Phyisol 1991; 71:317–321.
101. Pramanik BN, Zechman JM, Das PR, Bartner PL. Bacterial phospholipid analysis by fast atom bombardment mass spectrometry. Biomed Envrion Mass Spectrom 1990; 19:164–170.
102. Martin RJ, Cicutto LC, Smith HR, Ballard RD, Szefler SJ. Airways inflammation in nocturnal asthma. Am Rev Respir Dis 1991; 143:351–357.
103. Lecks HI, Wood DW, Kravis LP, Sutnick AI. Pulmonary surfactants, segmental atelectasis, and bronchial asthma. Clin Pediatr 1967; 6:270–276.
104. Lecks HI, Whitney T, Wood DW, Kravis LP. Newer concepts in occurrence of segmental atelectasis in acute bronchial asthma and status asthmaticus in children. J Asthma Res 1966; 4:65–74.
105. Liu M, Wang L, Enhorning G. 60 years of surfactant research. Presented at Floating Congress on the River Rhine, Nov 11–17, 1989.
106. Liu M, Wang L, Li E, Enhorning G. Pulmonary surfactant will secure free airflow through a narrow tube. J Appl Physiol 1991; 71:742–748.
107. Macklem PT, Proctor DF, Hogg JC. The stability of peripheral airways. Respir Physiol 1970; 8:191–201.
108. Yager D, Butler JP, Bastacky J, Israel E, Smith G, Drazen JM. Amplification of airway constriction due to liquid filling of airway interstices. J Appl Physiol 1989; 66:2873–2884.
109. Gaver DP, Samsel RW, Solway J. Effects of surface tension and viscosity on airway reopening. J Appl Physiol 1990; 69:74–85.
110. Mirrakhimov MM, Brimkulov NN, Liamtsev VT, Belov GV. Changes in the surface activity of bronchoalveolar washings and their cellular composition in bronchial asthma. Ter Arkh 1987; 59:31–36.
111. Ado AD, Volkova NV, Chervinskaia TA. Status of the surface-active properties of surfactant in bronchial asthma. Klin Med (Mosk) 1984; 62:41–46.
112. Ado Ad, Volkova NV, Chervinskaia TA. Effect of various types of bronchial asthma therapy on the surfactant content of the lungs. Klin Med (Mosk) 1985; 63:32–36.
113. Katoh H. A study of the pulmonary surface tension in experimentally induced asthma. With reference to alteration of the pulmonary surface tension in active sensitization and local desensitization. Nippon Ika Daigaku Zasshi 1981; 48: 286–296.
114. Kurashima K, Ogawa H, Ohka T, Fujimura M, Matsuda T, Kobayashi T. A pilot

study of surfactant inhalation in the treatment of asthmatic attack. Jpn J Allergol 1991; 40:160–163.
115. Hallgren R, Samuelsson T, Venge P, Modig J. Eosinophil activation in the lung is related to lung damage in adult respiratory distress syndrome. Am Rev Respir Dis 1987; 135:639–642.
116. Hallgren R, Bjermer L, Lundgren R, Venge P. The eosinophil component of the alveolitis in idiopathic pulmonary fibrosis. Signs of eosinophil activation in the lung are related to impaired lung function. Am Rev Respir Dis 1989; 139:373–377.
117. Rosenberg M, Patterson R, Mintzer R, Cooper BJ, Roberts M, Harris KE. Clinical and immunologic criteria for the diagnosis of allergic bronchopulmonary aspergillosis. Ann Intern Med 1977; 86:405–414.
118. Liebow AA, Carrington CB. The eosinophilic pneumonias. Medicine (Baltimore) 1969; 48:251–285.

DISCUSSION (Speaker: S. J. Ackerman)

M. Capron: In relation to the sequence homology between CLC and various IgE-binding proteins, did you look at whether the CLC-transfected cells could bind IgE?

Ackerman: We are in the process of investigating whether CLC protein possesses IgE-binding activity based on its similarities to the carboxy-terminal halves of the various IgE-binding proteins, human and mouse, Mac-2, rat IgE-binding protein, and mouse CBP35. Recombinant CLC protein expressed in COS or CHO cells, either transiently or in stably transfected lines, is not secreted by the cells. CLC protein is not detectable by immunostaining on the surface of eosinophils, although we have not yet attempted to localize it on the cell surface of transfected cells.

Denburg: When does CLC appear in sputum after allergen challenge in asthmatics?

Ackerman: Our studies on CLC protein levels in sputum utilized 106 consecutive sputa submitted to the microbiology laboratory for routine culture as well as sputum specimens from 10 patients hospitalized for asthma (1). In these samples, elevated levels of CLC protein were associated with both acute and chronic asthma. Studies of CLC protein in sputum (or BAL) after allergen challenge have not yet been conducted, but would be extremely important in view of the data we have presented today.

Gleich: Your findings of a biological function for CLC protein are exceedingly interesting and relevant to asthma. But could you comment on the possibility that CLC protein might exert a protective function in helminth infection?

Ackerman: Helminths such as the schistosomula stage of *S. mansoni* produce large amounts of the biologically active lysophosphatide monopalmitoyl-lysophosphatidyl choline that apparently mediates membrane fusion and lysis of cytotoxic effector cells. Since eosinophils, but not neutrophils, can selectively kill schistosomula in vitro, they may be resistant to the lytic and membrane-perturbing effects of parasite lysophosphatides because of their considerable content of this lysophospholipase. We plan to test this hypothesis.

Lee: Your hypothesis can possibly explain conditions such as hypersensitivity pneumonitis, but I am not so sure about asthma. If loss of surfactant in terminal bronchioles and alveolar spaces is a significant factor in asthma, one would expect much greater evidence physiologically of loss of volume and a restrictive defect—that is not the physiology of asthma, which is a reversible obstruction defect.

Ackerman: Whether or not surfactant dysfunction occurs in asthma is unclear; the available scientific evidence I alluded to is provocative, though not conclusive on this point. However, you are correct that if surfactant function was the only thing affected in asthma, one would observe solely a restrictive process. Rather we would point out that a surfactant defect in addition to other processes might explain some features of asthma, e.g., the microatelectasis observed in pathological specimens of patients who die of asthma. Additionally, surfactant surface tension may contribute to airway patency as originally pointed out by Macklem et al. (2) and more recently by Liu et al. (3), Gaver et al. (4), and Yager and colleagues (5). In this regard, a surfactant defect could explain airway closure and air trapping, which are also features of asthma. Further, we suggest that surfactant dysfunction, potentially due to the action of eosinophil lysophospholipase, is more likely to contribute significantly to airways disease in severe asthma than in mild to moderate asthma, which are more fully reversible with β-agonists. Your suggestion regarding surfactant function in conditions such as hypersensitivity pneumonitis is well taken; we too have envisioned a potential role for this enzyme in a variety of pathological conditions in which eosinophils may be involved, including adult respiratory distress syndrome, eosinophilic pneumonia, hypersensitivity pneumonitis, and interstitial pulmonary fibrosis.

Busse: In nocturnal asthma not only is there increased airway inflammation, but sleep also contributes to altered ventilation. One change is an increase in functional residual capacity. Do you feel that the contribution of CLC and altered surfactant activity is unique to nocturnal episodes of asthma or arises in other forms of asthma with positional or sleep-induced abnormalities in ventilation?

Ackerman: We would predict that eosinophil lysophospholipase-mediated increases in the surface tension of surfactant and surfactant function should not be restricted to nocturnal asthma, but should be detectable as well in other forms of asthma where significant eosinophil infiltration and active secretion of the enzyme occurs in the alveolar spaces and terminal bronchioles. We obviously would like to expand these observations to include other forms of asthma, and welcome collaborative interactions to obtain BAL for analysis.

Spry: What are the dynamics of CLC/surfactant interactions, and is the resistance of CLC proteins to proteolysis linked to your exciting new findings? Are there surfactants in the intestine where CLC are often found?

Ackerman: No information is available with regard to enzyme/ligand interactions for lysophospholipase-mediated degradation of the lysophosphatidylcholine component of pulmonary surfactant or, for that matter, pure substrate. I have no knowledge of gastrointestinal surfactants or of the specific source (eosinophil or basophil) of CLC that have been observed in various types of gastrointestinal inflammation.

Gleich: Have you measured the content of CLC protein in eosinophil nuclei? Your immunofluorescence data suggest that appreciable CLC protein is present in the nucleus.

Ackerman: We have not yet done subcellular fractionation and isolation of nuclei to quantitate the amount of CLC protein in the eosinophil nucleus. However, these studies may be difficult to interpret because of the extremely hydrophobic nature of CLC protein, which may associate nonspecifically with hydrophobic domains within the eosinophil once the cell is disrupted.

Townley: To answer the question and your hypothesis that surfactant is being destroyed by CLC, it would seem appropriate to administer surfactant to mild asthmatics with normal spirometry but abnormal small airway function, e.g., closing volume, and study whether surfactant alters the small airway function.

Ackerman: We are currently investigating the effects of intratracheal administration of active eosinophil lysophospholipase in rodent models of asthma in order to assess the enzyme's effect on pulmonary mechanics, surfactant surface tension, and pathological changes in airway structure, e.g., the ability of the enzyme to induce focal atelectasis. The suggested studies on mild asthmatics with normal spirometry, but abnormal small airways function, would be extremely interesting and could be utilized to assess the importance of pulmonary surfactant in maintaining small airways function in these patients.

Hansel: Would you expect mucolytic agents to overcome the effect of CLC protein on surfactant-related surface tension in asthma?

Ackerman: We have no data as yet addressing the effects of lysophospholipase-mediated increases in surfactant surface tension on clearance of mucus from the lung.

DISCUSSION REFERENCES

1. Dor PJ, Ackerman SJ, Gleich GJ. Charcot-Leyden crystal protein and eosinophil granule major basic protein in sputum of patients with respiratory diseases. Am Rev Respir Dis 1984; 130:1072–1077.
2. Macklem PT, Proctor DF, Hogg JC. The stability of peripheral airways. Respir Physiol 1970; 8:191–201.
3. Liu M, Wang L, Li E, Enhorning G. Pulmonary surfactant will secure free airflow through a narrow tube. J Appl Physiol 1991; 71:742–748.
4. Gaver DP, Samsel RW, Solway J. Effects of surface tension and viscosity on airway reopening. J Appl Physiol 1990; 69:74–85.
5. Yager D, Butler JP, Bastacky J, Israel E, Smith G, Drazen JM. Amplification of airway constriction due to liquid filling of airway interstices. J Appl Physiol 1989; 66:2873–2884.

3

G-Protein Regulation of Eosinophil Exocytosis

Oliver Cromwell
National Heart and Lung Institute, London, England

Bastien D. Gomperts
University College, London, England

Manfred Lindau
Max Planck Institute for Medical Research, Heidelberg, Germany

I. INTRODUCTION

The nature of the interactions between receptors, G-proteins, and effector systems is of central importance to the understanding of the intracellular signaling mechanisms that control cell function and that may be subject to modulation in disease states and by pharmacological intervention. Our present understanding of the mechanisms controlling the latter stages of the process of regulated secretion of granule proteins (exocytosis) has come about largely through the use of permeabilized secretory cells in which it is possible to obtain direct access to the cytosolic compartment. The initial regulatory steps imposed by surface receptors, their associated G-proteins, and effector enzymes can therefore be bypassed and the conditions pertaining in the cytosolic compartment become open to direct manipulation. Permeabilization can be achieved by a variety of methods that generate lesions of various sizes and tenure, and the techniques have been applied to a diverse range of cell types. In our studies of the eosinophil we have used two methods of permeabilization: (1) the bacterial cytolysin streptolysin-O (SL-O) and (2) patch clamping in the whole-cell configuration. SL-O renders the intra- and extracellular compartments continuous by generating persistent membrane lesions that allow free traffic of molecules up to 400 kDa in both directions, whereas the patch pipette involves a single lesion that permits dialysis of the cytosol while maintaining the separation of the two compartments.

Our studies of granule protein secretion with SL-O permeabilized guinea pig eosinophils reveal a dependence on both Ca^{2+} and a nonhydrolyzable GTP analog, guanosine-5′-*O*-(3-thiotriphosphate) (GTP-γ-S), suggesting roles for both calcium and GTP-binding proteins (1). These observations are in line with similar experiments with permeabilized mast cells and neutrophils (2,3) and also support the view that a notional G-protein, designated G_E, is involved in the exocytotic mechanisms at a point distal to the phospholipase C coupled receptor-associated G-protein, G_p (4). In addition, our investigation of the fine structure of the capacitance changes associated with degranulation in patch clamp experiments reveals sequential exocytosis of individual granules in response to GTP-γ-S (5,6) and occasionally a subsequent capacitance decrease consistent with endocytosis of small vesicles (6).

II. PERMEABILIZATION

The effectiveness of SL-O as a permeabilization agent on guinea pig eosinophils is demonstrated by its ability to release the cytosol enzyme lactate dehydrogenase (LDH). Leakage of the enzyme occurred at all concentrations of SL-O (0.05–0.4 U/ml) with Ca^{2+} buffered at pCa 7 [which approximates the resting cytosol concentration in these cells (7)], but as the concentration of SL-O increased, the rate leakage of LDH (MW 140 kDa) increased progressively to a point where it started within 1 min and was complete (96%) by 15 min. Under these conditions there was no measurable release of the granule-associated enzyme *N*-acetyl-β-D-glucosaminidase (hexosaminidase) indicating that the granule membranes are resistant to attack by SL-O, which, with a molecular weight of approximately 64 kDa, can be expected to penetrate the cells by passing through its own lesions. Electron micrographs of thin sections show that permeabilization of the cell membrane results in an increase in cell volume reflected in an increase in cell diameter and, in some instances, in swelling of the nucleus, which is accompanied by margination of the chromatin, possibly as a consequence of permeabilization of the nuclear membrane. The cytoplasm in the permeabilized cells appears less electron dense, an observation that can be attributed to the loss of diffusible cytoplasmic constituents.

III. Ca^{2+}- AND GTP-γ-S-DEPENDENT EXOCYTOSIS

When both Ca^{2+} and a guanine nucleotide (GTP-γ-S) are present at the time of permeabilization, hexosaminidase is released to the exterior. The process is dependent on the concentrations of both these effectors (Fig. 1) and can occur in the absence of Mg.ATP (1 mM). To ensure suppression of endogenous ATP, cells were treated with the metabolic inhibitors 2-deoxyglucose and antimycin-A prior

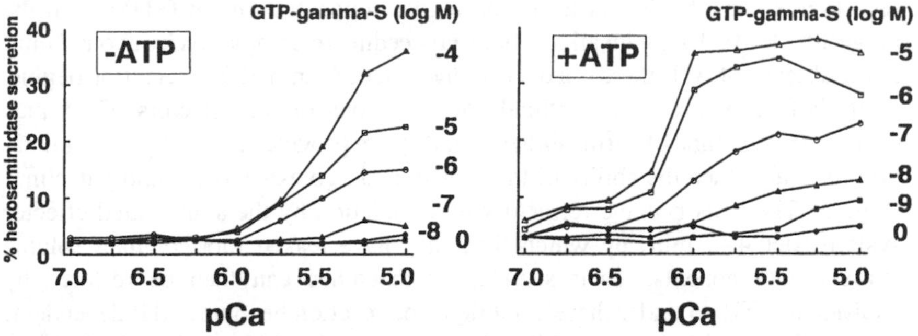

Figure 1 Ca^{2+}- and GTP-γ-S-dependent release of hexosaminidase from SL-O-permeabilized eosinophils in the absence and presence of ATP. Cells were preincubated with metabolic inhibitors for 5 min at 37°C prior to transfer to tubes containing SL-O (0.4 U/ml), calcium buffer, and GTP-γ-S, with or without ATP, and incubation (37°C, 20 min). [Reproduced with permission from *Journal of Immunology* (1).]

to permeabilization. The fact that exocytosis can be induced in the absence of exogenous ATP suggests that there is no obligatory role for a phosphorylation reaction in the latter stages of the exocytotic process in eosinophils. However, ATP may have an important function through its ability to enhance the effective affinities of both Ca^{2+} and GTP-γ in this process. In the intact cell the presence of ATP also ensures the presence of GTP through the nucleoside diphosphate kinase reaction. The inclusion of ATP in the presence of 10^{-5} M GTP-γ-S changes the EC_{50} [Ca^{2+}] from pCa 5.57 ± 0.04 (2.69 μM) to pCa 6.16 ± 0.03 (0.69 μM) ($n = 5$), equivalent to a fourfold increase in affinity for the effector. Similarly, lower concentrations of the guanine nucleotide are required to elicit secretion in cells provided with exogenous ATP. Comparison of electron micrographs of untreated cells with cells permeabilized with SL-O and cells permeabilized in the presence of optimal concentrations of Ca^{2+} and GTP-γ-S confirms that release of hexosaminidase from the permeabilized cells results in the loss of the majority of the granules from most of the cells, and this is accompanied by the formation of vacuoles of various sizes.

IV. LOSS OF EXOCYTOTIC RESPONSIVENESS WITH TIME FOLLOWING PERMEABILIZATION

ATP also influences the extent of secretion of hexosaminidase (Fig. 1), but the explanation for this may lie with the relatively slow time course of release (compared with mast cells), which requires that eosinophils maintain their secretory competence for a longer period after the initial permeabilization. The rate of secretion is considerably faster at the higher concentrations of GTP-γ-S in the range 10^{-9} to 10^{-4} M, with all reactions proceeding to completion by about 7 min. In the absence of ATP, there is a delay of the order of 1 min before secretion begins.

If, instead of providing optimal concentrations of the effectors (Ca^{2+} plus GTP-γ-S) at the time of permeabilization, they are provided after a "permeabilization interval," then the ability of the cells to undergo exocytosis rapidly declines (Fig. 2). The cells become refractory to stimulation by the simple dual effector system after 4–5 min, by which time they have leaked most of their soluble cytosolic components. Their secretory competence can then be restored by addition of ATP, so under these conditions the reaction becomes ATP dependent. If the cells are initially permeabilized in the presence of Mg.ATP, their ability to undergo late stimulation can be extended so that 50% of the control response can still be induced even after 4 min (Fig. 2). It seems likely that ATP is acting to maintain a phosphorylation state in which the exocytotic apparatus can continue to recognize the effectors. This view is supported by the observation that for cells pretreated with phorbol ester (PMA) prior to permeabilization, the period during which the cells retain responsiveness to Ca^{2+} plus GTP-γ-S, even in the absence of ATP, can be considerably extended (Fig. 2). The fact that the ATP-restored response is greater than that from cells permeabilized in the presence of the

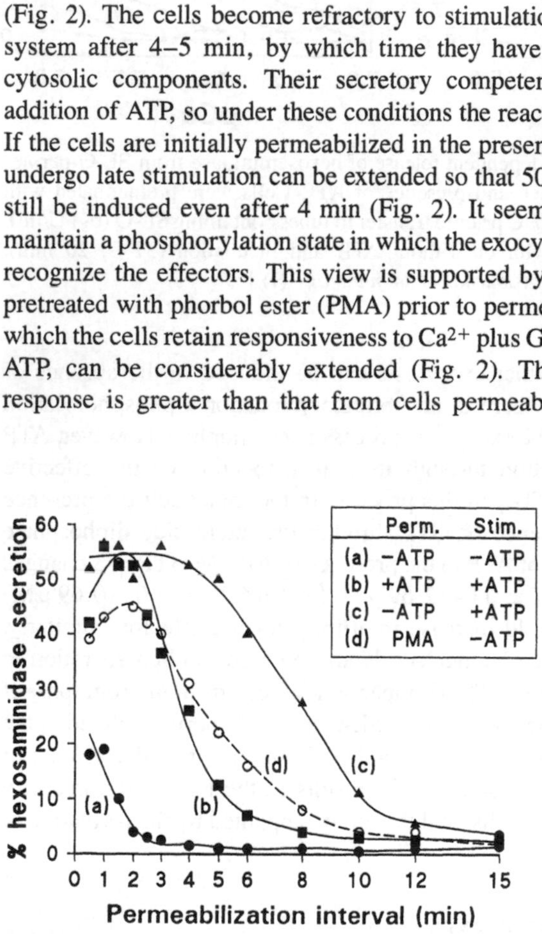

Figure 2 Decrease in exocytotic responsiveness with time after permeabilization and the ability of ATP or PMA pretreatment to restore and maintain responsiveness. Cells permeabilized (a) in the absence of ATP; (b) in the presence of ATP; (c) in the absence of ATP, with ATP added with the stimulus; (d) after 5 min pretreatment with PMA (100 nm) in the absence of ATP, and then stimulated to secrete by addition of Ca^{2+} (pCa 5.0) and GTP-γ-S (20 μM) after the permeabilization interval. [Reproduced with permission from *Journal of Immunology* (1).]

effectors (Fig. 2) may be attributed to the loss of inhibitory cytosolic factors, which in the case of mast cells are released rapidly via the SL-O-induced pores (8). Thus, cells that can be presumed to be in a hyperphosphorylated state prior to permeabilization can undergo exocytosis after an extended permeabilization interval, even under conditions in which no phosphorylation reaction can occur (i.e., no ATP). The experiment indicates, furthermore, that the maintenance reaction involves the phosphorylation of a substrate for protein kinase C, and indeed restoration of the secretory response by exogenous ATP can be partly inhibited by the protein kinase C pseudosubstrate peptide PKC-I (9) when it is added to the permeabilized cells 2 min before the stimulus. Again, this suggests that the action of ATP in enhancing both the effective affinities for Ca^{2+} and GTP-γ-S and the extent of exocytosis (Fig. 1) may be accounted for through its action as a phosphoryl donor in a protein-kinase-C-catalyzed phosphorylation.

V. CHANGES IN MEMBRANE CAPACITANCE MEASURED IN PATCH CLAMP EXPERIMENTS

Attachment of individual eosinophils to a glass patch pipette in the whole-cell configuration allows for direct regulation of the composition of the cytosol and the continuous measurement of electrical capacitance, which is directly proportional to membrane area. Introduction of Ca^{2+} and GTP-γ-S into the cell via the pipette induces an increase in capacitance that parallels exocytosis and directly registers the incorporation of granule membranes into the plasma membrane.

VI. TWO DISTINCT PHASES OF CAPACITANCE INCREASE

Resting eosinophils have a membrane capacitance of 2.70 ± 0.53 pF (SD, $n = 79$). Following stimulation with 20 μl GTP-γ-S at pCa 5.8 (1.5 μM), membrane capacitance increases in two distinct phases. An initial rapid increase proceeds immediately after stimulation of the cell and is comprised of capacitance steps of less than 4 fF, equivalent to granules or vesicles of less than 450 nm in diameter. The second phase commences after a variable delay of 2–10 min, possibly reflecting different primed states of the individual cells within the preparations. The overall slope is comparable with that of the first phase, of the order of 300 fF/min, but it differs in that it is characterized by a series of abrupt stepwise increments in capacitance between 7 and 15 fF (Fig. 3).

The stepwise phase was absolutely dependent on GTP-γ-S, but the presence of Ca^{2+} in the pipette was sufficient to induce a rapid graded increase, although with reduced amplitude. Together with the temporal separation of the two phases, this observation suggests that a different control mechanism may be involved. The

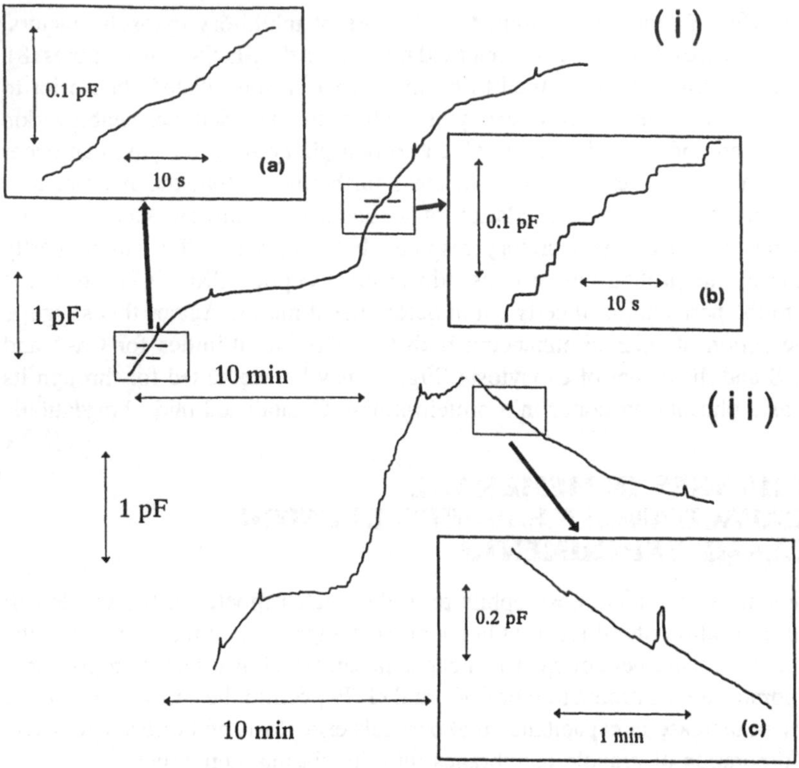

Figure 3 Membrane capacitance measurements in eosinophils stimulated with GTP-γ-S (20 μM) and Ca^{2+} (pCa 5.8/1.5 μM). (i) (a) Immediate and (b) delayed phases of exocytosis; (ii) immediate and delayed phases of exocytosis followed by (c) endocytosis.

increases in capacitance occurred in the absence of conductance changes, indicating stability of the seal, maintenance of distinct extra- and intracellular compartments, and a true exocytotic membrane fusion mechanism for both phases.

The stepwise phase of degranulation is, on average, composed of 210 steps, a figure that is in excellent agreement with the morphometric estimate of approximately 190 crystalloid granules per cell (6). The specific capacitance of biological membranes is in the range of 0.6–1.0 μF/cm² (10–13). Using this value, the capacitance step size distribution can be converted to a granule size distribution that is found to be in excellent agreement with the morphometrically obtained values (6). The formation of large exocytotic vacuoles by the fusion of several granules followed by fusion of the vacuole with the plasma membrane would generate a capacitance change with the same total amplitude, but the change would occur in a small number of much larger steps. Such changes were not observed; the few large capacitance steps could be equated to the observation of particularly large granules in the electron micrograph sections.

The early graded increase is not compatible with the fusion of crystalloid granules. A recent report showed that empty granules occur during the development of mature eosinophils in cord blood mononuclear cell cultures with interleukin-5 (14). In addition, eosinophil peroxidase was found in small vesicles clustered in perigranular and sub–plasma membrane regions, and this phenomenon was interpreted as reflecting "piecemeal degranulation" (PMD) involving transport of granular material by small vesicles to the plasma membrane. Secretion by such a mechanism would generate a graded increase in capacitance. However, the stepwise capacitance increase that we observe accounts for fusion of all the crystalloid granules and on the face of it does not support activation of PMD by GTP-γ-S. We cannot exclude absolutely the possibility that there may be partial loss of granule membrane and granule contents by PMD, but it is still not established whether PMD is related to stimulation of secretion or is a particular feature of development.

In rat peritoneal mast cells, high intracellular calcium induces continuous capacitance changes without steps and visible degranulation (15), and it was suggested that these may reflect calcium-dependent shifts of the equilibrium between constitutive exocytosis and endocytosis involving small granules (15). Small granules less than 500 nm in diameter have been identified and characterized in rat peritoneal eosinophils in terms of their unmasked acid phosphatase activity (16). Human eosinophils contain a population of small granules measuring 100–500 nm in diameter, which are moderately and uniformly dense when observed by electron microscopy and exhibit unmasked acid phosphatase activity and intense aryl sulfatase activity, unlike the larger crystalloid granules (17). Different granule classes have also been distinguished by the temporal difference in the release of aryl sulfatase (small granules) and markers contained in the large crystalloid containing granules (18), but the half-times are 4 min for small and 40 min for crystalloid granules. Human peripheral blood eosinophils contain between two and eight small granules per equatorial profile, and a few small granules can be seen in electron micrographs of guinea pig eosinophils (6). However, exocytosis of all small granules would generate a capacitance increase of less than 0.15 pF, which is three times less than the observed amplitude in the presence of GTP-γ-S.

A class of so-called specific microgranules is present in eosinophils from at least 20 species, including guinea pig, which measure 20–200 nm (19). These are more abundant during eosinophilia (20) and are preferentially located at the periphery of the cell, which suggests that they may form part of the cell's secretory apparatus. In electron micrographs the specific microgranules appear as spheres, double spheres, dumbbell, and cup-like structures. The cup-like and coated vesicles were, however, shown to be involved in endocytosis (16) (see below), but the dumbbell-like vesiculotubular structures, which are present in guinea pig eosinophils (19), are apparently not involved in endocytosis (16) and are thus possible candidates for the organelles generating the membrane area expansion underlying the rapid phase of the capacitance change.

VII. A CAPACITANCE DECREASE INDICATING ENDOCYTOSIS

In some cells a continuous capacitance decrease was observed, which was sometimes manifest between the two phases of the exocytotic process but was generally most evident after degranulation was complete. The continuous nature of the changes suggests that the process involves internalization of entities having membrane areas less than 0.3 μm^2 and diameters less than 300 nm. Such decreases are compatible with internalization of membrane by an endocytotic mechanism involving small vesicles or coated pits. Microendocytosis has been observed in rat peritoneal eosinophils after intraperitoneal infusion of saline or fetal calf serum (16) and is associated with the appearance of small membrane limited structures ($<$100 nm), some of which incorporate colloidal gold classifying them as endocytotic vesicles. Such a process would manifest itself in the capacitance change as a decrease consisting of very small steps as observed in our experiments. Electron micrographs of human eosinophils stimulated with the calcium ionophore A23187 revealed exocytotic fusion of crystalloid granules with the plasma membrane and increased numbers of small granules (21). It has been suggested that such structures may represent newly formed granules or the remains of crystalloid granules that have secreted most of their contents (22). The observed capacitance decrease indicates that they may instead be formed by an endocytotic mechanism of membrane reuptake.

VIII. SUMMARY

In cells permeabilized with SL-O in a simple NaCl-based electrolyte solution, secretion is dependent on the presence of both Ca^{2+} (at concentrations buffered in the micromolar range) and a guanine nucleotide. Secretion of *N*-acetyl-β-D-glucosaminidase correlates well with the extent of degranulation as assessed by morphological inspection of thin sections. There is no absolute requirement for ATP, but in its presence the affinities for both Ca^{2+} and GTP-γ-S are enhanced by a protein-kinase-dependent mechanism. In all these respects, the mechanism closely resembles that of other myeloid cell types, although there are differences, mainly relating to the more extended time over which exocytosis occurs. Patch clamp experiments indicate that exocytosis involves fusion of discrete granules with the plasma membrane in two temporally dissociated phases. The first phase involves small granules or vesicles and is at least partly Ca^{2+} dependent. The second phase reflects sequential fusion of all the crystalloid granules as individual units with the plasma membrane and shows an absolute requirement for GTP-γ-S. Membrane reuptake then occurs in the form of small vesicles or coated pits.

REFERENCES

1. Cromwell O, Bennett JP, Hide I, Kay AB, Gomperts BD. Mechanisms of granule enzyme secretion from permeabilized guinea pig eosinophils: dependence on Ca^{2+} and guanine nucleotides. J Immunol 1991; 147:1905–1911.
2. Gomperts BD, Barrowman MM, Cockcroft S. Dual role for guanine nucleotides in stimulus-secretion coupling: an investigation of mast cells and neutrophils. Fed Proc 1986; 45:2156–2161.
3. Gomperts BD. G_E: a GTP-binding protein mediating exocytosis. Annu Rev Physiol 1990; 52:591–606.
4. Gomperts BD, Churcher Y, Koffer A, Kramer IM, Lillie T, Tatham PER. The role and mechanism of the GTP-binding protein G_E in the control of regulated exocytosis. Biochem Soc Symp 1990; 56:85–101.
5. Nüsse O, Lindau M, Cromwell O, Kay AB, Gomperts BD. Intracellular application of guanosine-5′-O-(3-thiotriphosphate) induces exocytotic granule fusion in guinea pig eosinophils. J Exp Med 1990; 171:775–786.
6. Lindau M, Nüsse O, Bennett J, Cromwell O. The fine structure of the capacitance changes in degranulating guinea pig eosinophils. J. Cell Sci. 1993; 104:203–209.
7. Henderson WR, Chi EY, Jorg A, Klebanoff SJ. Horse eosinophil degranulation induced by the ionophore A23187. Am J Pathol 1983; 111:341–349.
8. Koffer A, Gomperts BD. Soluble proteins as modulators of the exocytotic reaction of permeabilised rat mast cells. J Cell Sci 1989; 94:585–591.
9. House C, Kemp BE. Protein kinase C contains a pseudosubstrate prototype in its regulatory domain. Science 1987; 238:1726–1728.
10. Fernandez JM, Neher E, Gomperts BD. Capacitance measurements reveal stepwise fusion events in degranulating mast cells. Nature 1984; 312:453–455.
11. Breckenridge LJ, Almers W. Final steps in exocytosis observed in a cell with giant secretory granules. Proc Natl Acad Sci USA 1987; 84:1945–1949.
12. Zimmerberg J, Curran M, Cohen FS, Brodwick M. Simultaneous electrical and optical measurements show that membrane fusion precedes secretory granule swelling during exocytosis of beige moust mast cells. Proc Natl Acad Sci USA 1987; 84:1585–1589.
13. Nüsse O, Lindau M. The dynamics of exocytosis in human neutrophils. J Cell Biol 1988; 107:2117–2124.
14. Dvorak AM, Furitsu T, Letourneau L, Ishizaka T, Ackerman SJ. Mature eosinophils stimulated to develop in human cord blood mononuclear cell cultures supplemented with recombinant human interleukin-5. Part 1. Piecemeal degranulation of specific granules and distribution of Charcot-Leyden crystal protein. Am J Pathol 1991; 138: 69–82.
15. Almers W, Neher E. Gradual and stepwise changes in the membrane capacitance of rat peritoneal mast cells. J Physiol 1987; 386:205–218.

16. Komiyama A, Spicer SS. Microendocytosis in eosinophilic leukocytes. J Cell Biol 1975; 64:622–635.
17. Parmley RT, Spicer SS. Cytochemical and ultrastructural identification of a small type granule in human late eosinophils. Lab Invest 1977; 30:557–567.
18. Kroegel C, Yukawa T, Dent G, Venge P, Chung KF, Barnes PJ. Stimulation of degranulation from human eosinophils by platelet-activating factor. J Immunol 1989; 142:3518–3526.
19. Schaefer HE, Hubner G, Fischer R. Spezifische Mikrogranula in Eosinophilen. Acta Haematol (Basel) 1973; 50:92–104.
20. Zucker-Franklin D. Eosinophil structure and maturation. In: Mahmoud AA, Austen KF, Simon AS, eds. The Eosinophil in Health and Disease. New York: Grune & Stratton, 1980:43–59.
21. Henderson WR, Chi Y. Ultrastructural characterization and morphometric analysis of human eosinophil degranulation. J Cell Sci 1985; 73:33–48.
22. Spry CJF. Eosinophils: A Comprehensive Review and Guide to the Scientific and Medical Literature. Oxford: Oxford University Press, 1988.

DISCUSSION (Speaker: O. Cromwell)

Konig: Did you use radiolabelled GTP-γ-S to identify the G (exocytosis) protein?

Cromwell: We are obviously very keen to characterize G_E, but as you can appreciate, there are numerous GTP-binding proteins in the cell and binding of radiolabeled GTP-γ-S alone will not provide evidence of identity. Much of the evidence to support the concept of G_E, a GTP-binding protein involved in the latter stages of the signal transduction events leading to exocytosis, has come from studies with the mast cell. These have been reviewed in detail by Lindau and Gomperts (1), who at the same time argue the case for G_E being one of the heterotrimeric G-proteins with some resemblance to the group exemplified by G_i. All experiments conducted so far suggest that the fundamental elements of the exocytotic machinery in mast cells and eosinophils are very similar.

Kita: In our experience of eosinophil degranulation induced by sIgA or FMLP, release of crystalloid granule proteins plateaus within 2 h after stimulation. In your system with Ca^{2+} and GTP-γ-S stimulation, hexosaminidase plateaus by 5 min. How can you explain the difference of time course of your system of degranulation?

Cromwell: In your system you are using intact cells and employing ligation of surface receptors to initiate signal transduction processes that culminate in the release of granule proteins. In our system with permeabilized cells we are bypassing the surface receptors and their associated G-proteins and any influence they may have on the time course of secretion. In addition, the success of immunoglobulin-coated beads as a stimulus for secretion is based on their resemblance to large parasitic targets onto which the cells are able to spread. The time required for spreading will obviously contribute to the time course of events, but the process appears to facilitate secretion and enhances the cytotoxic potential of the granule proteins by focusing them onto the target.

Egan: Although the exocytosis may be slow and prolonged, that does not necessarily rule out the involvement of G-proteins that are also involved in the initial stages of eosinophil activation. For example, we have demonstrated that phospholipase D is present in the human eosinophil, and based on comparison with neutrophils, a G protein is essential to PLD activity and the activity is quite prolonged. Is there evidence that the G protein you describe in exocytosis is unique to the exocytosis event?

Cromwell: First, it is important to stress that G_E exists as a concept on the basis of experiments in cell physiology. Identification and characterization have yet to be achieved. Although I am suggesting that the activity of G_E is fundamental to the exocytosis process, I have no grounds for claiming that its activities are exclusive to exocytosis. As you rightly say, phospholipase D activity has long been recognized in eosinophils and one would reasonably expect that its activity is G-protein-dependent by analogy with other cells of myeloid origin [for example, Geny and Cockcroft (2)].

Spry: Peter Kernen also has evidence for a G_i protein in the response of human eosinophils to agonists within a few seconds. There are also very slow secretory events that take several hours or days. It would be interesting to relate this to your work. As micropinocytosis and granule expulsion into vacuoles or externally in human eosinophils are seldom seen, do you think there will be species differences in the patch clamp work?

Cromwell: Unfortunately, I am not familiar with Kernen's work. As you have said, exocytotic events are rarely captured on electron micrographs and consequently it is difficult to appreciate the subtleties of the process using this technique. Our data with guinea pig eosinophils show excellent agreement between the morphometric estimation of granule numbers and the specific conductance steps measured in the patch clamp experiments. Susanne Scopek, working in Manfred Lindau's laboratory, has obtained similar results with both horse eosinophils and eosinophils derived from human cord blood mononuclear cells cultured in the presence of interleukin-5.

Roos: The time course of exocytosis depends on the assay. If exocytosis is measured by following the expression of proteins in the granule membrane on the plasma membrane of the cells, the kinetics of exocytosis—at least in neutrophils—are much faster than was the case in your experiments. Is this due to your experimental setup with permeabilized cells?

Cromwell: The principal eosinophil granule proteins are highly cationic, a property that seems to render them particularly sticky. Several reports in the literature have noted this fact. Our attempts to measure eosinophil peroxidase activity in the permeabilized cell supernatants were fraught with difficulties, and for this reason we opted to measure *N*-acetyl-D-glucosaminidase (hexosaminidase) activity. I think it is reasonable to expect that the cationic proteins, which predominate, may have some influence on the release of other proteins, but at the same time I would expect to be able to appreciate the expression of granule membrane markers on the plasma membrane before granule proteins had time to diffuse into the extracellular milieu.

Gleich: Dr. Butterworth, in your studies of eosinophil killing of schistosomula of *S. mansoni*, killing took 24–48 h to become evident. Were you able to detect eosinophil degranulation at earlier time points?

Butterworth: Yes, we assay killing of schistosomula after 48 h, simply because that is a convenient time for detecting death of these large multicellular organisms. However, the membrane damage that leads eventually to death occurs much earlier, and eosinophil degranulation is in fact mediated within a matter of minutes. The major limitation on timing is the physical constraints of the assay, in that eosinophils accumulate progressively on the surface of the antibody-coated schistosomulum.

Cromwell: This point again highlights the different questions we are seeking to answer. Experiments with intact cells give very little indication as to what is going on within the cell during the process of exocytosis. In the case of schistosomula killing, for example, the time course of events will be influenced by various factors such as cell spreading on the target and adhesion.

Galli: Ilan Hammel and David Lagunorf used an ultrastructural morphometric approach to demonstrate that the cytoplasmic granules of rat peritoneal mast cells occur in a multimodal size distribution in which the modes are integral multiples of the volume of the smallest size class. Recently, Dr. Hammel reported the same finding in rat eosinophils. However, the multimodal size distributions of these granules is appreciated when the granules are analyzed according to volume, not area. In a patch clamp study of stimulated secretion in rat peritoneal mast cells, Fernandez et al. found stepwise changes in mast cell membrane capacitance, which they interpreted as reflecting the fusion with the plasma membrane of individual cytoplasmic granules that represent members of a population with a multimodal volume distribution. I formed the impression that in the patch clamp studies of the second phase of secretion in guinea pig eosinophils, the calculated areas of the granules fusing with the plasma membranes did not exhibit a multimodal distribution. It is possible that you would obtain evidence for a multimodal size distribution of the eosinophil granules if you converted your granule area data to granule volumes. Another possibility is that the granules in unstimulated eosinophils have a multimodal volume distribution but that this was obscured by the effects of the piecemeal degranulation that you have suggested may occur during the first phase of secretion in your model. Theoretically, the loss of granule membrane and content that is thought to occur during piecemeal degranulation could change the granules' volume sufficiently to eliminate a preexisting multimodal volume distribution.

Cromwell: These are possibilities that we cannot exclude at this time and it is important that we continue with the detailed analysis of these events.

Durham: Could you comment on the influence of drugs, particularly corticosteroids, on eosinophil degranulation?

Cromwell: Drs. Kita and Gleich have investigated the effects of various glucocorticoids on immunoglobulin-induced eosinophil degranulation and interleukin-5 enhancement of degranulation. I believe that I am correct in saying that they found no evidence that glucocorticoids had any direct effect on these events.

Gleich: That's right. These were human peripheral blood eosinophils from normal volunteers and one patient with episodic angioedema (3).

DISCUSSION REFERENCES

1. Lindau M, Gomperts BD. Techniques and concepts in exocytosis: focus on mast cells. Biochim Biophys Acta 1991; 1071:429–471.
2. Geny B, Cockcroft S. Synergistic activation of phospholipase D by protein kinase C- and G-protein-mediated pathways in streptolysin-O-permeabilized HL-60 cells. Biochem J 1992; 284:531–538.
3. Kita H, Abu-Ghazaleh R, Sanderson CJ, Gleich GJ. Effects of steroids on immunoglobulin-induced eosinophil degranulation. J Allergy Clin Immunol 1991; 87:70–77.

4

Regulation of Eosinophil Function by P-Selectin

Mathew A. Vadas, C. M. Lucas, J. R. Gamble, and Angel F. Lopez
Hanson Centre for Cancer Research, Adelaide, South Australia, Australia

M. P. Skinner and M. C. Berndt
Baker Medical Research Institute, Prahran, Victoria, Australia

I. INTRODUCTION

The accumulation of eosinophils in tissues is dependent on their removal from the axial stream to roll along blood vessels and then to migrate through endothelial junctions into tissues. The capture of leukocytes from the axial stream is mediated by selectins (1). P-selectin (previously known as GMP-140) in particular has been shown to mediate the shear resistant adhesion and rolling of neutrophils and to allow the more stable integrin ICAM-1-dependent adhesion reactions to take place that appear necessary for transmigration (2).

The adhesion of neutrophils to P-selectin is not associated with spreading (3) or with the generation of O_2^- anions typically seen after neutrophils adhere to extracellular matrix proteins via their integrin receptors (3). These findings suggested that P-selectin serves not only to capture leukocytes from the axial stream, but also to inhibit certain aspects of their function presumably to prevent damage to EC during the process of transmigration.

We now show that P-selectin is a powerful adhesion molecule for eosinophils, that the eosinophils fail to spread or produce O_2^- while adherent to P-selectin, and furthermore are also prevented from degranulating in response to PMA—suggesting that soluble or surface-bound P-selectin is a strong inhibitor of eosinophil function. By contrast, P-selectin does not decrease degranulation in response to Ig Sepharose or antibody-dependent, cell-mediated cytotoxicity. As adhesion to fibrinogen and spreading are functions that are also inhibited by antibodies to $\alpha_2\beta_1$, we suggest that P-selectin is a signaling molecule that changes $\alpha_2\beta_1$ integrin function (4). Furthermore, as P-selectin is a normal constituent of plasma (5), we suggest that it tonically acts to inhibit certain aspects of eosinophil function and may be responsible for the lack of vessel damage during most forms of eosinophilia.

II. MATERIALS AND METHODS

Eosinophils were purified from normal human blood by dextran sedimentation and discontinuous, slightly hypertonic metrizamide gradients as described (6). Preparations were >90% pure and >95% viable as assessed by Giemsa stain and trypan blue exclusion, respectively. Neutrophils were purified by dextran sedimentation and Ficoll/Hypaque gradient centrifugation as described (3). All assays were performed in RPMI + 2.5% fetal calf serum.

P-selectin was extracted from human platelets, purified by affinity chromatography, and then for assay, either immobilized onto microtiter wells, or used in the fluid phase after Triton X-100 removal by Extractigel-D (Pierce) (7).

The CHO cell line stably expressing P-selectin was obtained as follows. CHO cells were suspended in phosphate-buffered saline and 5×10^5 cells were electroporated at 1300 volts/25 μFd with the neomycin-resistant plasmid SV_2Neo and MTGMP plasmid containing the cDNA for P-selectin. Cells were selected with neomycin (35 μg/ml), and Facstar Plus (Becton Dickinson) was used to obtain a clone positive for P-selectin expression as detected by the polyclonal antibody to P-selectin.

A. Adhesion

Eosinophils were labeled with ^{51}Cr, washed, and 5×10^5 allowed to adhere to protein-coated microtiter wells or to confluent monolayers of CHO cells at 37°C for 30 min. Nonadherent cells were then washed away and adherent cells solubilized and counted. Percent adherence was calculated from the total number of counts added. The graph shown is representative of five separate experiments.

B. Superoxide Anion Release

O_2^- release by eosinophils was measured by reduction of cytochrome C at 550 nm in microtiter plates as described (3). The graph shown is representative of four separate experiments.

C. Degranulation

Eosinophil cationic protein (ECP) release was determined as a measure of eosinophil degranulation. Cells were primed with GM-CSF (30 ng/ml) or medium for 15 min at 37°C and then stimulated in microtiter wells with PMA (10 ng/ml) or Ig-coated sepharose 4B beads (1:10) at 37°C in the presence of soluble P-selectin at 10 μg/ml or control buffer. The supernatant was collected for ECP estimation using the Pharmacia ECP double-antibody radioimmunoassay kit. The data shown are the results from one preliminary experiment.

D. Antibody-Dependent Cell Cytotoxicity (ADCC) Assay

ADCC was performed as previously described (8). Briefly, ^{51}Cr-labeled and trinitrophenyl (TNP)-coupled P815 BALB mastocytoma cells were incubated in V-bottom microtiter plates for 2 h at 37°C with antibody to TNP and eosinophils or neutrophils as effector cells. Percent cytotoxicity was calculated from the amount of ^{51}Cr release from the "killed" P815 cells compared to total counts added. The graph shown is representative of three experiments.

III. RESULTS

A. Eosinophils Adhere to P-Selectin

Purified human eosinophils adhered weakly to microtiter plates coated with fetal calf serum. There was a stronger adhesion when the microtiter plates were coated with P-selectin; significantly increased adhesion was evident even at P-selectin concentrations as low as 0.1 μg/ml (Fig. 1). This concentration is approximately five- to 10-fold lower than those needed to show neutrophil adhesion. Antibody to P-selectin totally inhibited adhesion to dishes coated with 10 μg/ml P-selectin (data not shown). It is noteworthy that control dishes showed higher eosinophil adhesion than dishes coated with P-selectin in the presence of anti-P-selectin antibodies, suggesting that P-selectin totally coats the dishes used. Strong adhesion to P-selectin-transfected CHO cells was also observed (Fig. 2).

The shape of P-selectin eosinophils was round, and microscopically they remained refractile. This was observed even in the presence of GM-CSF, a stimulator of eosinophil spreading and polarization (9) (Figs. 3 and 4).

B. Inhibition of Eosinophil O_2^- Production and Degranulation by P-Selectin

Adherent eosinophils generate O_2^- anion, and this process is enhanced by the eosinophilopoietic growth factors including GM-CSF. The O_2^- generation of eosinophils adherent to various surfaces was examined. Figure 5 shows that whereas strong O_2^- generation was seen on serum-coated plastic or fibrinogen-coated plastic, little O_2^- was produced on P-selectin. This inhibition was seen even in the presence of GM-CSF.

Eosinophil degranulation was quantitated by measuring ECP release. In a

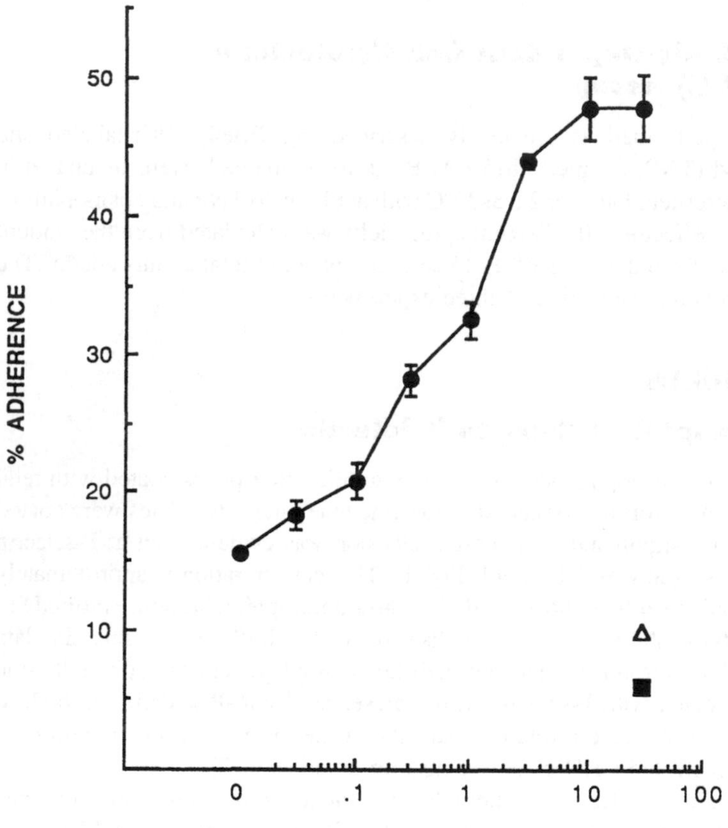

Figure 1 Binding of purified human eosinophils to graded doses of purified P-selectin coated onto plastic microtiter plates (●). The binding to fibrinogen (30 μg/ml) (△) and to fibronectin (■) (30 μg/ml) is also shown. Adherence to 0.1 μg/ml P-selectin was significantly increased compared to adherence to fibrinogen ($p < 0.001$). Each point is the arithmetic mean of triplicate determinations ± SEM.

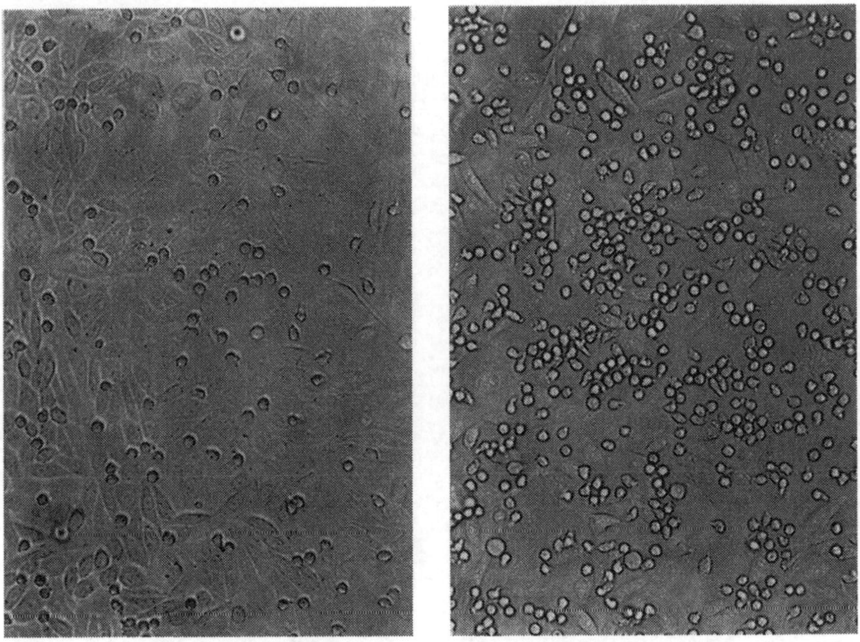

Figure 2 Binding of purified human eosinophils to CHO cells (left) or CHO cells transfected with P-selectin cDNA (right). The more intense binding to the CHO transfectants is evident.

preliminary experiment, P-selectin decreased basal release as well as release in response to PMA. However, P-selectin did not decrease degranulation in response to Ig Sepharose (Fig. 6).

C. Functional Effects of P-Selectin Are Selective

It was notable that cells incubated with solid or fluid-phase P-selectin did not have altered viability over a 48-h period (data not shown) nor did they lose their capacity to kill antibody-coated tumor targets (Fig. 7). This observation was in keeping with the P-selectin not altering degranulation in response to Ig sepharose (Fig. 6).

IV. DISCUSSION

P-selectin appears to be a physiologically relevant molecule that is expressed on endothelial cell surfaces after treatment with thrombin or histamine (10) and on platelets after activation (11). P-selectin is also found in the plasma of normal individuals (5).

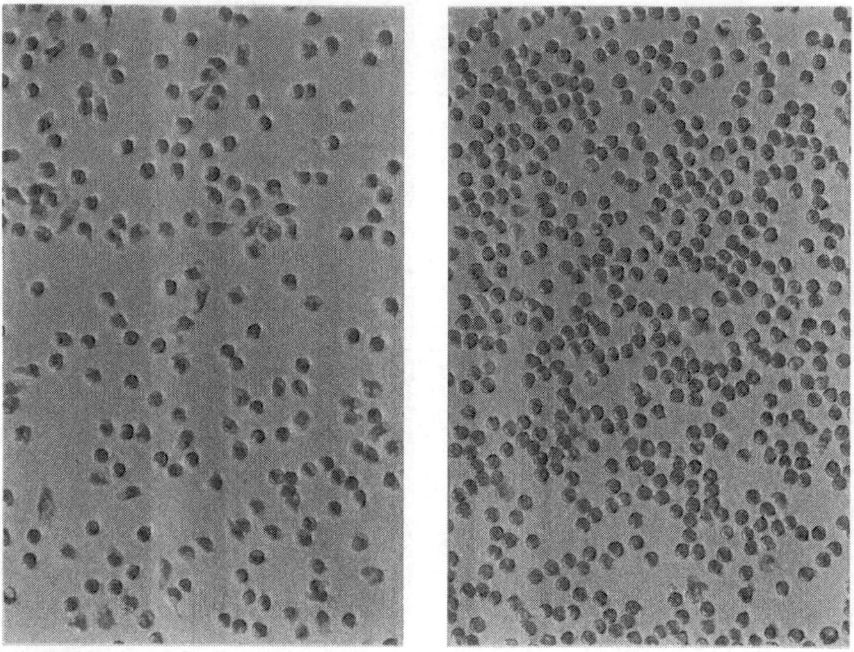

Figure 3 Phase contrast microscopy (×200) of eosinophils binding to fibrinogen or P-selectin. The more intense binding and lack of polarization of binding of cells to P-selectin is evident.

We found that eosinophils adhere strongly to purified P-selectin (Fig. 1) or CHO cells expressing recombinant P-selectin (Fig. 2) and that this adhesion is not associated with spreading or polarization (Figs. 3 and 4) normally associated with eosinophil activation. The adhesion was inhibitable by a polyclonal antibody against P-selectin (data not shown). In comparison to cells adherent to extracellular matrix proteins, eosinophils adherent to P-selectin were deficient in O_2^- generation. Fluid-phase P-selectin also inhibited eosinophil activation even after adhesion to fibrinogen (data not shown). This suggests that P-selectin signals cells and interrupts the normal activation process. Similar inhibitory effect of fluid-phase P-selectin was observed on degranulation in response to phorbol myristate acetate but not Ig Sepharose (Fig. 6), suggesting that P-selectin does not interrupt all activation pathways. This was supported by the lack of effect of P-selectin on antibody-dependent cytotoxicity (Fig. 7).

The structure on eosinophils that binds P-selectin has not been defined, although on neutrophils the carbohydrate moieties sialyl Lewis X (12) and sulfatide (13) appear to play significant roles. It thus appears that occupation of

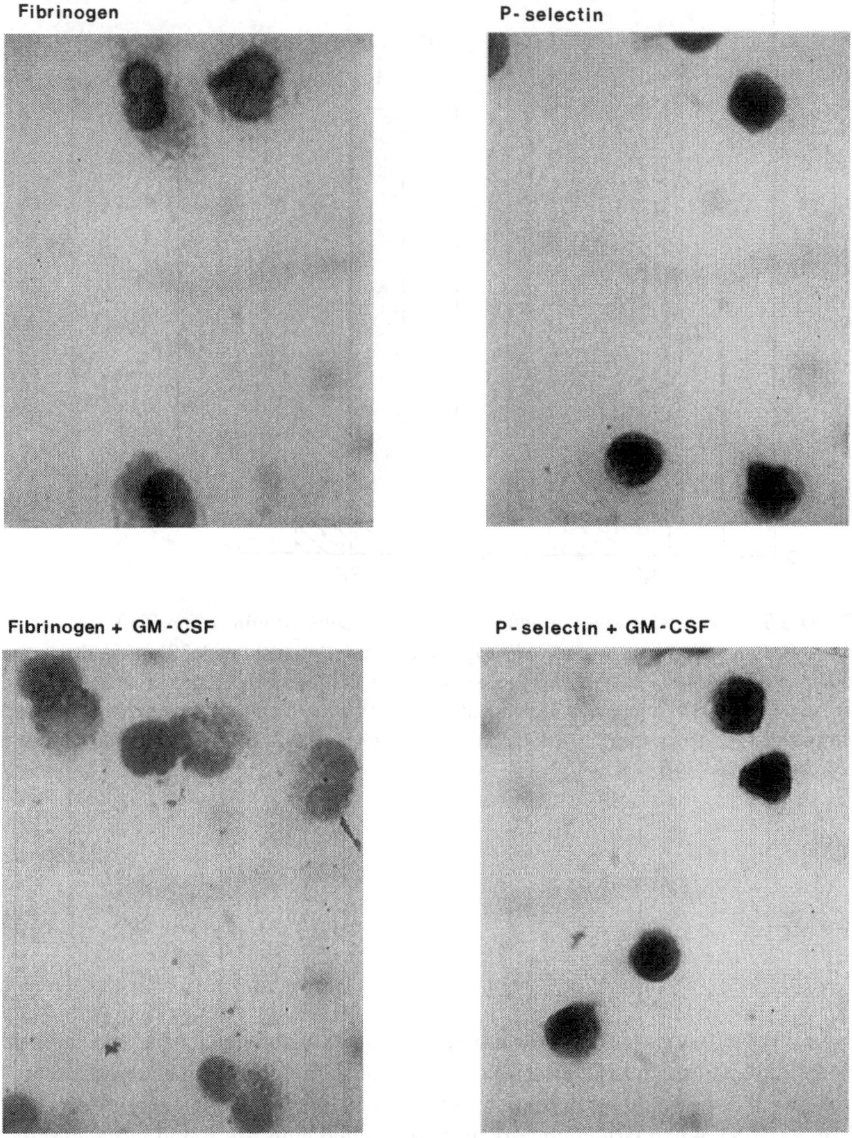

Figure 4 Hematoxylin-eosin stain of eosinophils bound to fibrinogen or P-selectin in the presence or absence of 30 ng/ml GM-CSF. The spreading of the cells and readily defined granules are evident in the absence of P-selectin.

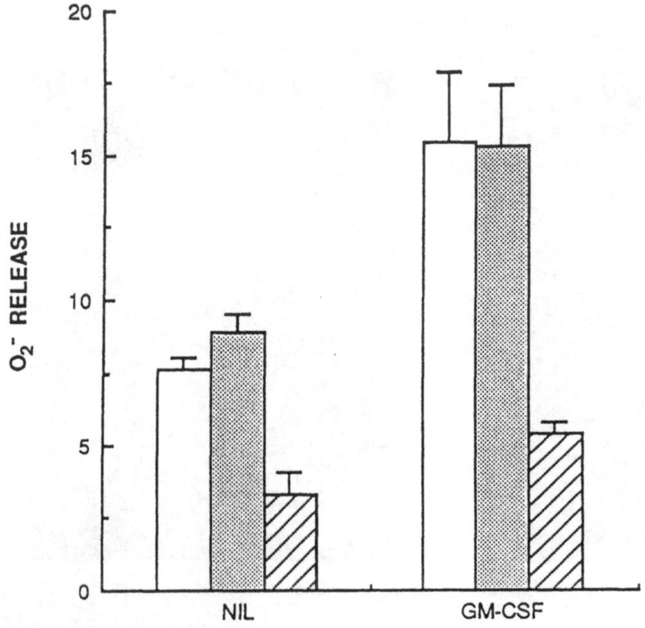

Figure 5 Superoxide anion production (nmol/10^6 cells/60 min) of purified eosinophils over a 60-min period in uncoated wells (open bars) or wells coated with 30 μg/ml fibrinogen (dots) or 10 μg/ml P-selectin (diagonal slashes) in the absence (left) or presence (right) of 30 ng/ml GM-CSF. There was significant ($p < 0.05$) reduction of superoxide dismutase inhibitable O_2^- generation on P-selectin. Each point is the arithmetic mean of triplicate determinations ± SEM.

these receptor(s) confers onto eosinophils the unactivated phenotype that is found in the circulation. Indeed eosinophil activation is not seen under most circumstances of eosinophilia, the major exception being the hypereosinophilic syndrome (HES), when large numbers of degranulated cells are found in the circulation and there is evidence of damage of the lining of blood vessels. It could thus be hypothesized that P-selectin may prevent this type of activation and that tissue damage in HES is in part due to a deficiency of activation-inhibiting protein of which one example may be P- selectin.

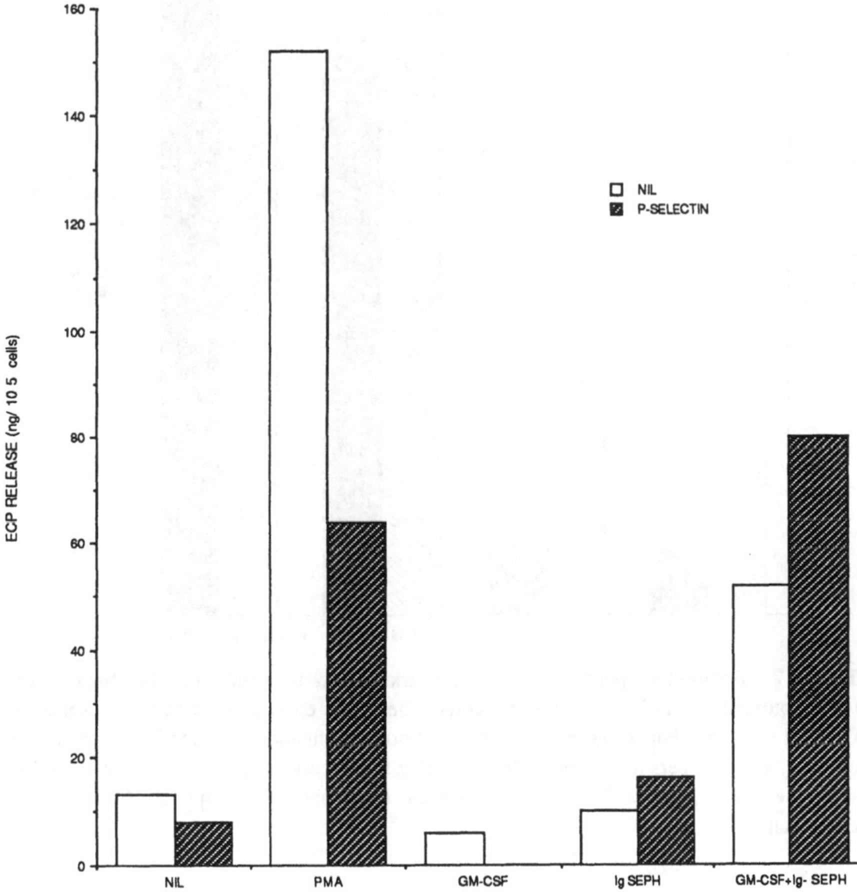

Figure 6 Eosinophil cationic protein (ECP) release of purified eosinophils in the presence of buffer (open bars) or P-selectin (slashed bars). There was a decrease in ECP release over 30 min in PMA and GM-CSF columns. Each point is a single determination.

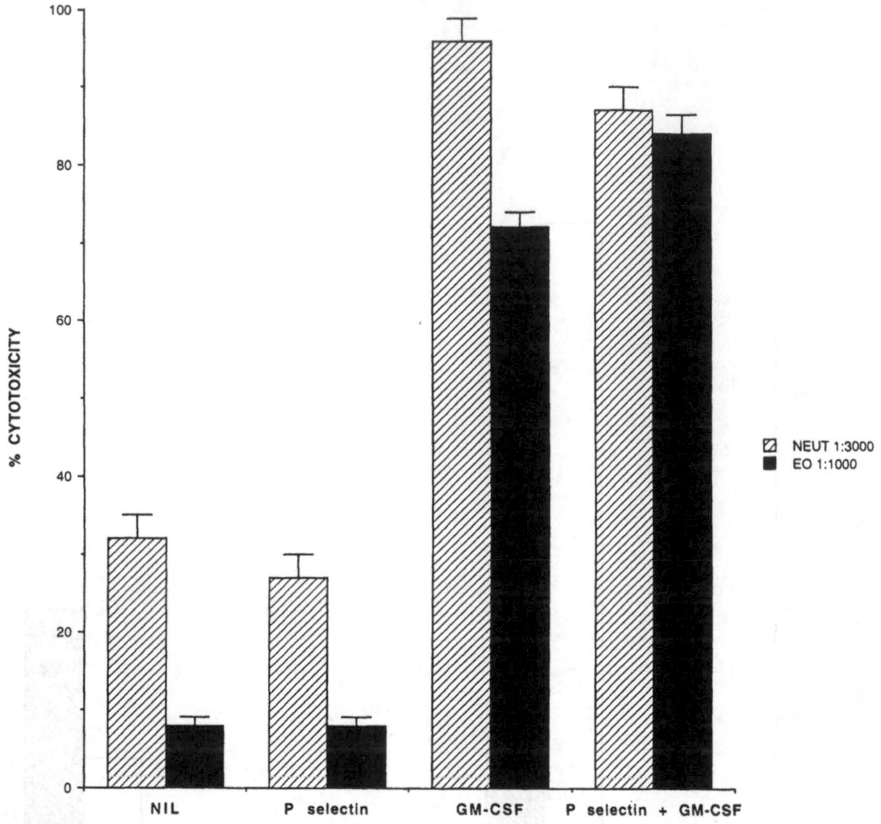

Figure 7 Antibody-dependent eosinophil (dark bars) or neutrophil (slashed bars) mediated cytotoxicity of P815 mastocytoma cells. There was enhancement in the presence of 10 ng/ml GM-CSF, but 10 μg/ml P-selectin did not significantly alter the killing by either cell type. The numbers next to the cell types refer to the concentration of anti-TNP used in each case. Each point is the mean of triplicate (neutrophils) or duplicate (eosinophils) determinations ± SD.

REFERENCES

1. Butcher EC. Leukocyte-endothelial cell recognition: three (or more) steps to specificity and diversity. Cell 1991; 67:1033–1036.
2. Lawrence MB, Springer TA. Leukocytes roll on a selectin at physiologic flow rates: distinction from and prerequisite for adhesion through integrins. Cell 1991; 65:859–873.
3. Wong CS, Gamble JR, Skinner MP, Lucas CM, Berndt MC, Vadas MA. Adhesion protein GMP140 inhibits superoxide anion release by human neutrophils. Proc Natl Acad Sci USA 1991; 88:2397–2401.

4. Gamble JR, Skinner MP, Berndt MC, Vadas MA. Prevention of activated neutrophil adhesion to endothelium by soluble adhesion protein GMP140. Science 1990; 249: 414–417.
5. Dunlop LC, Skinner MP, Bendall LJ, Favaloro EJ, Castaldi PA, Gorman JJ, Gamble JR, Vadas MA, Berndt MC. Characterisation of GMP-140 (selectin) as a circulating plasma protein. J Exp Med 1992; 175:1147–1150.
6. Vadas MA, David JR, Butterworth AE, Pisani NT, Siongok TA. A new method for the purification of human eosinophils and neutrophils and a comparison of the ability of these cells to damage schistosomula of *Schistosoma mansoni*. J Immunol 1979; 122: 1228–1236.
7. Skinner MP, Fournier DJ, Andrews RK, Gorman JJ, Chesterman CN, Berndt MC. Characterization of human platelet GMP-140 as a heparin-binding protein. BBRC 1989; 164:1373–1379.
8. Vadas MA, Nicola N, Metcalf D. Activation of antibody dependent cell-mediated cytotoxicity of human neutrophils and eosinophils by separate colony stimulating factors. J Immunol 1983; 130:795–799.
9. Lopez AF, Williamson DJ, Gamble JR, Begley CG, Harlan JM, Klebanoff SJ, Waltersdorph A, Wong G, Clark SC, Vadas MA. Recombinant human granulocyte-macrophage colony-stimulating factor (rH GM-CSF) stimulates in vitro mature human neutrophil and eosinophil function, surface receptor expression and survival. J Clin Invest 1986; 78:1220–1228.
10. Hattori R, Hamilton KK, Fugote RD, McEver RP, Sims PJ. Stimulated secretion of endothelial von Willebrand Factor is accompanied by rapid redistribution to the cell surface of the intracellular granule membrane protein GMP-140. J Biol Chem 1989; 264:7768–7771.
11. Berman CL, Yeo EL, Wencel-Drake JD, Furie BC, Ginsberg MH, Furie B. A platelet alpha granule membrane protein that is associated with the plasma membrane after activation. J Clin Invest 1986; 78:130–137.
12. Larsen E, Palabrica T, Sajer S, Gilbert G, Wagner D, Furie BC, Furie B. PADGEM-dependent adhesion of platelets to monocytes and neutrophils is mediated by a lineage-specific carbohydrate, LNF111 (CD15). Cell 1990; 63:467–474.
13. Aruffo A, Kolanus W, Walz G, Fredman P, Seed B. CD62/P-selectin recognition of myeloid and tumor cell sulfatides. Cell 1991; 67:35–44.

DISCUSSION (Speaker: M. Vadas)

Gleich: Do not cytokines themselves (in the absence of other stimuli) mediate signal transduction in eosinophils?

Vadas: I am aware of the experiments to which you refer and would like to point out that the control of spreading and so forth without P-selectin was not done (indeed, as flow conditions were being examined, could not be done), and I suspect that the degree of activation observed in static assays would be greater without P-selectin.

Roos: You showed generation of O_2^- at a rate of 10–20 nmol per 10^6 eosinophils per hour, which is 1–2% of the total capacity of these cells. So, although GMP-140 may inhibit this basal rate of O_2^- production, I am not surprised that you did not see inhibition with GMP-140 in an ADCC assay. In ADCC, the oxidase is probably activated to a much higher extent.

Vadas: I have two responses. First, our ADCC assay does not appear to involve O_2^-. Second, the level of O_2^- generation with GM-CSF and adherence is approximately a third of that we see with PHA. However, I agree it is less than maximal capacity of the cell.

Bochner: Is the monoclonal antibody that blocks eosinophil adhesion to P-selectin directed against the P-selectin or the eosinophil? Do you know what the counterreceptor for P-selectin is on eosinophils?

Vadas: The antibody is a rabbit polyclonal against P-selectin. No, we don't know the counterreceptor. Several possibilities exist in the literature—Lewis X, sialyl Lewis X, and sulfatide. In addition, sugars such as heparin may be involved.

Schleimer: I agree with Dr. Ackerman's earlier comment that the context of adhesive interactions is quite important. This is best illustrated by the beautiful studies of Lawrence and Springer showing that under flow conditions neutrophils rolled on GMP-140 in an artificial membrane, adhered, but did not spread and form tight adhesion unless ICAM-1 was also present. With ICAM-1 alone no adherence was observed. This illustrates the sequential cooperative adhesion responses that probably also occur with eosinophils.

Vadas: I would also like to point out that the effect of P-selectin is rapidly reversible—indeed it seems to be associated with a rapid on-and-off rate of binding (a necessity if the ligand is to mediate rolling). Thus to understand functional effects the number of binding molecules will have to be quantified.

Konig: Does P-selectin affect the binding of zymosan? Is it known that zymosan interacts with an α-mannan receptor? Is there an interaction of P-selectin with the receptor?

Vadas: Unfortunately, we do not yet know. I only mentioned the data as they shed a different light on the interpretation of our experiments, and I did not want you to draw a conclusion in its absence.

Thomas: Is the adhesion of eosinophils to V-selectin occurring in a calcium-dependent manner?

Vadas: The binding of P-selectin to cells is Ca^{2+} dependent. However, the binding of heparin or sulfatide to P-selectin is not.

5

Structural Basis for the GM-CSF- and IL-3-Mediated Stimulation of Eosinophil Production and Function in Humans

Angel F. Lopez, J. Woodcock, T. Hercus, S. Barry, P. Dyson, P. Simmons, M. F. Shannon, and Mathew A. Vadas
Hanson Centre for Cancer Research, Adelaide, South Australia, Australia

I. INTRODUCTION

Human granulocyte-macrophage colony-stimulating factor (GM-CSF) and interleukin-3 (IL-3) are glycoproteins produced by activated T lymphocytes that control the proliferation, differentiation, functional activation, and survival of human eosinophils in vitro (1–3). Infusion of GM-CSF and IL-3 in monkeys and humans indicates that at least some of these eosinophil stimulatory properties are also observed in vivo (4,5).

The effects of GM-CSF and IL-3 are exerted throughout the differentiation pathway of the eosinophil lineage (6) and occur following the interaction of GM-CSF and IL-3 with their receptors on the eosinophil surface (7). The receptors for these cytokines have been characterized and shown to exhibit a unique pattern of cross-competition on eosinophils, indicating the presence of a common component (8). The recent cloning of these receptors has shown that each comprises two chains: a ligand-specific α chain and a β chain, which is common to both GM-CSF and IL-3 receptors as well as to the IL-5 receptor (9–12). Importantly, while GM-CSF and IL-3 bind their respective α chains with low affinity, their binding to the common β chain alone is undetectable. Instead, the β chain confers high-affinity binding when coexpressed with the α chain.

Although GM-CSF and IL-3 stimulate multiple eosinophil functions, the underlying mechanisms have remained elusive. In particular, little is known about the structural basis for the interaction of GM-CSF and IL-3 with their receptors on eosinophils. This study is essential to (1) identify regions and residues in the GM-CSF and IL-3 molecules involved in binding and function, (2) determine which component/s of the GM-CSF and IL-3 receptors mediate the enhancement of eosinophil production, and (3) devise strategies for the generation of novel GM-CSF and IL-3 molecules with more selective or antagonistic eosinophil-activating properties. We show here that a glutamic acid at position 21 in GM-CSF is critical for GM-CSF stimulation of eosinophil function and for binding to the β chain of the receptor, implicating the β chain in eosinophil signaling. In addition, mutations in the helix D of IL-3 led to a mutant with increased binding to the IL-3 receptor α chain and biological activity. These results identify residues of GM-CSF and IL-3 involved in binding to the α and β chains of their receptors and suggest that both the α and β chains of these receptors participate in signal transduction in activated eosinophils.

II. MATERIALS AND METHODS

A. Human Eosinophils

Eosinophils were purified from the blood of slightly eosinophilic individuals by dextran sedimentation followed by centrifugation on a hypertonic gradient of metrizamide as described (13).

B. Production of GM-CSF and IL-3 Mutants

Site-directed mutagenesis of a human GM-CSF cDNA clone provided by Dr. S. Clark (Genetics Institute, Cambridge, MA), was performed in phage M13 as described (14). Mutant plaques were screened by the 3M tetramethyl ammonium chloride procedure (15). The presence of the correct mutation was confirmed by chain termination sequencing (16) using a Sequenase Kit (United States Biochemical, Cleveland, OH). The mutant GM-CSF cDNAs were then excised from the M13 clones RF DNA and subcloned into the transient mammalian expression vector pJL4 (17). All plasmid constructs were sequenced at the site of the mutation prior to transfection (18).

Human IL-3 mutants were constructed by either site-directed mutagenesis (14) or polymerase chain reaction (PCR) mutagenesis (19). The template for site-directed mutagenesis was a single-stranded M13 vector containing a synthetic hIL-3 cDNA (20). The IL-3 protein produced from this synthetic cDNA has been shown to have the same properties as the recombinant hIL-3 (20). A two-part polymerase chain reaction using 3 primers (19) was used to create mutants in the IL-3 expression vector pJLA+IL-3 (20). Two primers were external to the IL-3 gene and the third was the mutagenic oligonucleotide. In the first step the external primer (which binds to the antisense strand) was used with the mutagenic oligonucleotide (which binds to the sense strand). Twenty-five cycles of PCR with these primers resulted in amplification of a portion of the gene. This portion contained the mutant sequence and was used as a primer together with the other external primer (which binds to the sense strand) for the second PCR reaction.

IL-3 analogs created by site-directed mutagenesis were digested with BamHI and SacI and cloned together with a SacI/EcoRI DNA fragment containing SV40 polyadenylation signal into BamHI/EcoRI sites of pJL4 (20).

GM-CSF and IL-3 mutants in the pJL4 expression vector were transiently transfected in COS cells and quantitated by radioimmunoassay as previously described (21,22).

C. Eosinophil Antibody-Dependent Cytotoxicity (ADCC)

^{51}Cr-labeled, trinitrophenyl-coupled P815 cells (4×10^3 cells in 40 μl) were incubated with 24 μl of rabbit antitrinitrophenyl antibody (Miles-Yeda, Rehovot, Israel), 80 μl of purified human eosinophils (1.3×10^5 cells) as effector cells, and 16 μl of cytokine or medium for 2.5 h at 37°C in V-bottom microtiter plates. The percent cytotoxicity was calculated as described (2).

D. Bone Marrow Colony Assay

Light-density, nonadherent human bone marrow cells prepared as previously described (3) were incubated in 0.3% agar in the presence of different cytokines for 14 days at 37°C. The agar discs were then fixed with 3% (vol/vol) glutaraldehyde and transferred onto individual 5 × 8 cm glass slides. The discs were dried at room temperature and stained with luxol fast blue and a combined specific and nonspecific esterase stain. In some cases, the number of cells present in each clone was enumerated to establish the clone size.

E. Binding Assays

Yeast-derived human GM-CSF or IL-3 (gift from Dr. L. Park, Immunex Corporation, Seattle, WA) were radioiodinated by the ICR method (23). Iodinated protein was separated from free ^{125}I by chromatography on a Sephadex G-25 PD 10 column (Pharmacia, Uppsala, Sweden) equilibrated in PBS containing 0.02% Tween 20, and stored at 4°C for up to 4 weeks. Before use, the iodinated protein was purified from Tween and non-protein-associated radioactivity by cation exchange chromatography on a 0.3-ml CM-Sepharose CL-6B column (Pharmacia) and stored at 4°C for up to 5 days. The radiolabeled GM-CSF and IL-3 retained >90% biological activity as judged from titration curves using noniodinated GM-CSF as controls.

Eosinophils were suspended in binding medium consisting of RPMI-1640 supplemented with 20 mmol/L HEPES, 0.5% bovine serum albumin (BSA), and 0.1% sodium azide. Typically, equal volumes (50 μl) of 4 × 10^6 eosinophils, 70 pM iodinated GM-CSF or IL-3, and different concentrations of cytokines were mixed in siliconized glass tubes for 3 h at 4°C and then centrifuged over a cushion of FCS. Competition for binding to low-affinity GM-CSF receptors used CHO cells permanently transfected with the GM-CSF receptor α chain expressing 80,00 binding sites per cell. Specific counts were determined by first subtracting the counts obtained in the presence of excess wild-type (WT) GM-CSF or IL-3. In each case cell suspensions were overlaid on 0.2 ml FCS at 4°C, centrifuged in a Beckman Microfuge 12, and the tip of each tube containing the visible cell pellet cut off and counted in a gamma counter.

III. RESULTS

A. Importance of Glu^{21} in the Eosinophil-Stimulating Activity of GM-CSF

Initial structure-function studies of GM-CSF using chemically synthesized peptides of different lengths revealed that a region between residues 14 and 24 was required for the stimulation of eosinophils and other cell types (24). We have now performed site-directed mutagenesis in this region, in particular, in the hydrophilic peak comprising residues Gln^{20} and Glu^{21}. We found that these two residues were structurally required to stimulate eosinophil function. Deletion mutagenesis of $Gln^{20}Glu^{21}$ resulted in a GM-CSF analog incapable of stimulating the formation of eosinophil colonies (Table 1).

Substitution mutagenesis at positions 20 and 21 revealed that Glu^{21} but not Gln^{20} was required for optimal stimulation of human eosinophils. The GM-CSF mutant Ala^{20} was as effective as WT GM-CSF at stimulating eosinophil colonies; however, the mutant GM-CSF Ala^{21} showed impaired stimulatory activity (Table 1).

The difference between GM-CSF mutants was reflected in the size of the eosinophil colonies they stimulated. Thus, GM-CSF Ala^{20} stimulated clones of similar size to those stimulated by WT GM-CSF, while GM-CSF Ala^{21} stimulated smaller eosinophil clones none of which exceeded 60 cells per clone (Fig. 1). Deletion of both residues $Gln^{20}Glu^{21}$ resulted in an inert GM-CSF molecule unable to stimulate eosinophil clones above control levels (mock-transfected COS cells).

To examine the role of Gln^{20} and Glu^{21} in the stimulation of mature eosinophil function, the GM-CSF mutants GM-CSF Ala^{20}, GM-CSF Ala^{21}, and GM-CSF deletion 20,21 were tested for their ability to stimulate eosinophils to kill antibody-coated target cells. The results showed that GM-CSF deletion 20,21 was inactive, while GM-CSF Ala^{20} and GM-CSF Ala^{21} stimulated eosinophils in a dose-dependent manner. Importantly, however, while GM-CSF Ala^{20} was equipotent to

Table 1 Stimulation of Eosinophil Colonies from Human Bone Marrow by GM-CSF Mutants

	Dilution of COS cell supernatant						
GM-CSF	1/10	1/20	1/40	1/80	1/160	1/320	0
Wild type	10.3 ± 8.7[a]	13 ± 8.2	12 ± 3.5	12.3 ± 7.6	12 ± 7	6.3 ± 2.3	0
Gln20Ala	15.6 ± 3.5	14.3 ± 6	16.7 ± 4.1	13.3 ± 5.8	7.7 ± 3.8	7.3 ± 3	0
Glu21Ala	10.3 ± 8.7	3 ± 1	0	0	0	0	0
Deletion Gln20, Glu21	0	0	0	0	0	0	0

[a]Total number of eosinophil colonies/10^5 cells. 4×10^4 nonadherent, T-cell-depleted bone marrow cells were incubated with different GM-CSF–containing COS cell supernatants for 14 days at 37°C. Plates were scored for the presence of eosinophil colonies (>40 cells/clone).

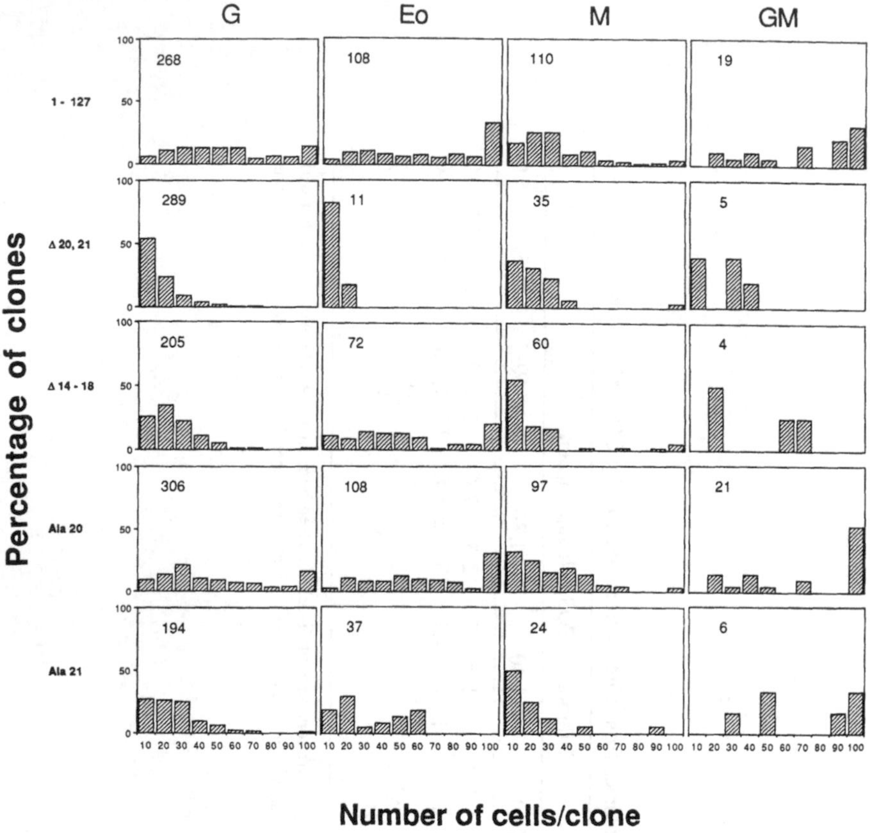

Figure 1 Size of human bone-marrow-derived clones stimulated by different GM-CSF mutants at 30 ng/ml. 4×10^4 nonadherent, T-cell-depleted bone marrow cells were incubated for 14 days at 37°C. The plates were fixed and stained with tristain to reveal colony morphology. The number of cells present in each clone is shown, with the numbers in each box representing the total number of clones examined.

WT GM-CSF, GM-CSF Ala^{21} was about 10-fold less active at stimulating eosinophil function (Fig. 2).

B. Interaction of Glu^{21} with the GM-CSF Receptor of Eosinophils

After Glu^{21} was identified as critical for both eosinophil production and eosinophil activation, mutagenesis at this position was carried out to introduce different residues. The substitution of Glu^{21} for Arg^{21} or Lys^{21} was found to cause the most significant effects in biological activity, reducing GM-CSF potency by 300-fold (21).

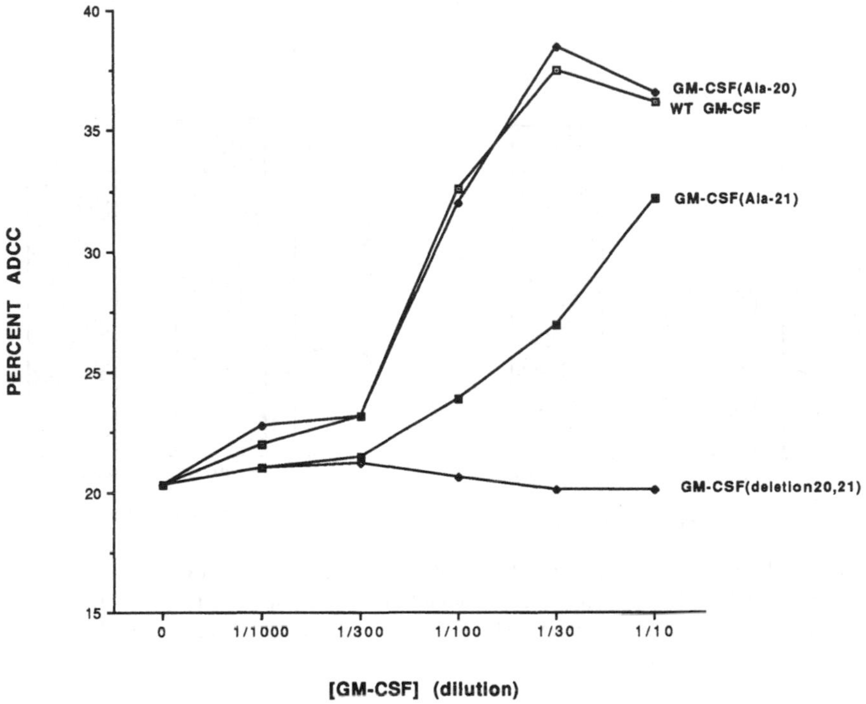

Figure 2 Stimulation of antibody-dependent eosinophil-mediated cytotoxicity (ADCC) of tumor cells by WT GM-CSF and GM-CSF mutants. GM-CSF was tested in the form of dilution of supernatants of COS cells transfected with WT GM-CSF cDNA or different mutant GM-CSF cDNAs.

The GM-CSF receptor is composed of two chains, an α chain, which binds GM-CSF with low affinity (9), and a β chain, which does not show detectable binding on its own but which confers high-affinity binding when coexpressed with the α chain (10). Since GM-CSF Arg^{21} was deficient at stimulating eosinophil bioactivity (21), it was important to determine the effects of this mutation in binding to the receptor α and β chains. To this end, we examined the ability of GM-CSF Arg^{21} to compete with ^{125}I-GM-CSF for binding to (1) human eosinophils that express the GM-CSF high-affinity receptor (αβ chain complex) only (7), and (2) CHO fibroblasts expressing only the receptor α chain. A titration showed that GM-CSF (Arg^{21}) exhibited a 300-fold decrease in ability to bind to the high-affinity receptor (αβ complex) of eosinophils but was equipotent to WT GM-CSF at binding the low-affinity (α chain) receptor. This difference was most evident at around the 50% binding values where GM-CSF (Arg^{21}) could be seen to be selectively deficient at interacting with the αβ receptor complex (Fig. 3).

These experiments strongly suggested that Glu^{21} was important for binding to

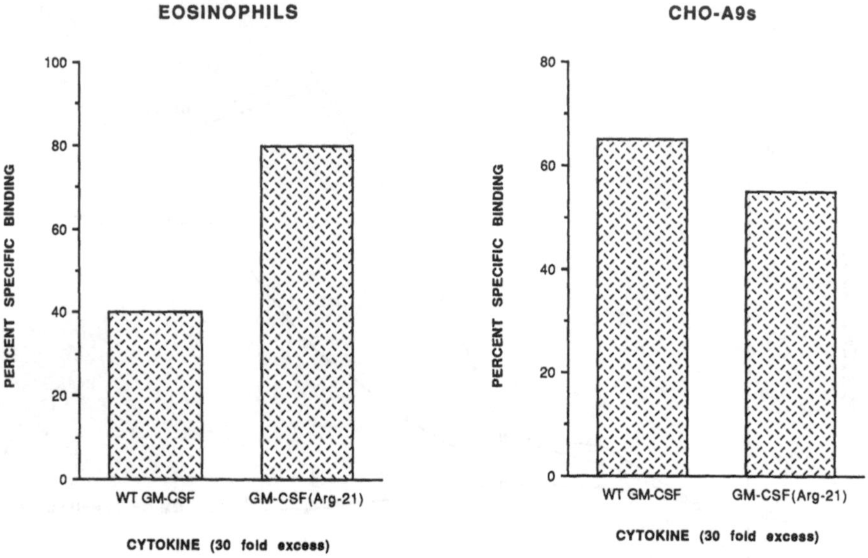

Figure 3 GM-CSF (Arg^{21}) is deficient at competing for ^{125}I-GM-CSF binding to the high-affinity GM-CSF receptor of eosinophils but is equipotent to WT GM-CSF at competing for binding to the cloned low-affinity (α chain) of the GM-CSF receptor expressed in COS cells.

the eosinophil GM-CSF receptor β chain but not to the α chain. To confirm this finding, we took advantage of the cross-competition observed between GM-CSF, IL-3, and IL-5 for binding to eosinophils (8), a phenomenon likely to be the result of these receptors bearing the common β chain. Using pM amounts of labeled cytokines to detect mainly high-affinity (αβ) receptors, we compared the level of cross-competition achieved with WT GM-CSF and GM-CSF (Arg^{21}). The results showed that GM-CSF (Arg^{21}) was much less effective than WT GM-CSF at cross-competing for IL-3 and IL-5 binding to human eosinophils (Fig. 4). These results further emphasize the importance of Glu^{21} in recognizing the β chain and identifies in eosinophils the β chain as the common component of the GM-CSF, IL-3, and IL-5 receptors.

C. Is There an Analogous Glutamic Acid in IL-3 and IL-5?

Since the β chain appears to be shared by the GM-CSF, IL-3, and IL-5 receptors of eosinophils, it is likely that IL-3 and IL-5 also bind to the β chain; however, the regions or residues involved are unknown. Clues as to the regions involved have been missing, particularly since there is very little primary sequence homology

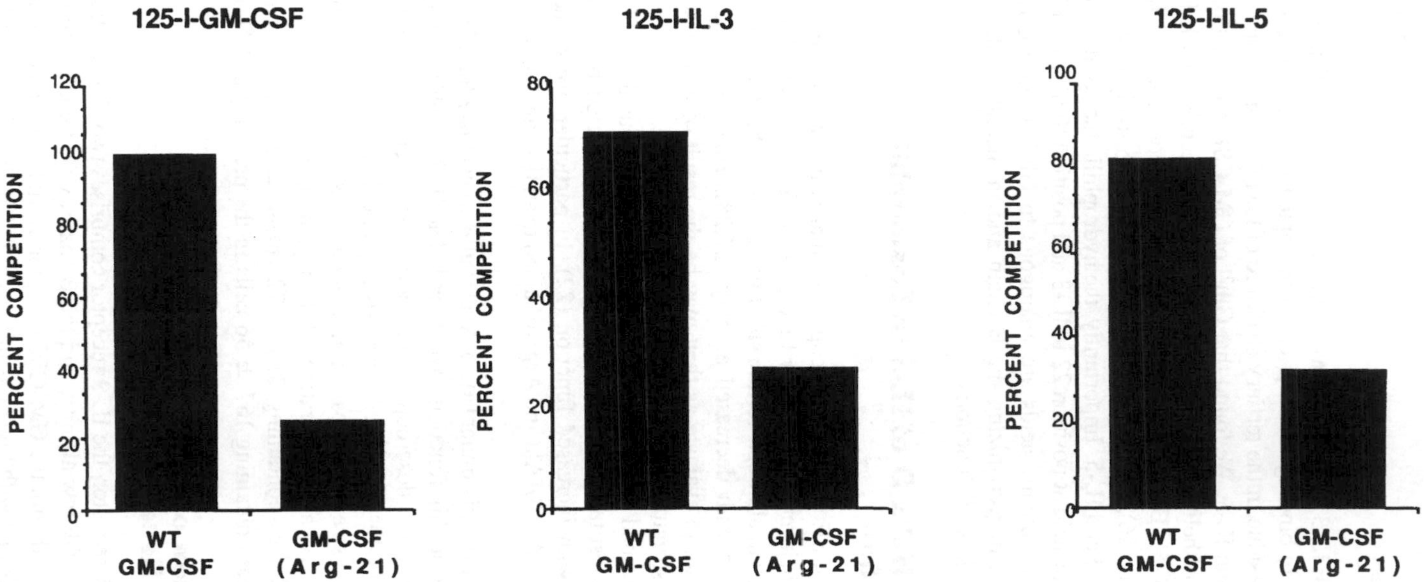

Figure 4 Decreased ability of GM-CSF (Arg^{21}) to cross-compete for ^{125}I-IL-3 and ^{125}I-IL-5 binding to human eosinophils. Radioligand was used at 100 pM and WT GM-CSF and mutant GM-CSF (Arg^{21}) were used at 30-fold excess.

between GM-CSF, IL-3, and IL-5. Nevertheless, the observation that Glu^{21} of GM-CSF is involved in binding to the β chain prompted us to examine closely the position of this residue in the tertiary structure of GM-CSF and in the analogous regions of IL-3 and IL-5. We found that Glu^{21} of GM-CSF was positioned in the hydrophilic face of helix A in the GM-CSF structure recently revealed by X-ray crystallography (25). By performing predictive studies on IL-3 and IL-5 using several algorithms (26), we found that this region is also likely to be an amphipathic helix in IL-3 and IL-5. Importantly, the hydrophilic face of these cytokines contains a glutamic acid at position 22 in IL-3 and position 10 in IL-5 (Fig. 5). We predict that these glutamic acids are analogous to Glu^{21} in GM-CSF and are critically involved in recognizing the β chain and in mediating IL-3 and IL-5 stimulation of eosinophil function.

D. Role of Helix D of IL-3 in Eosinophil Stimulation

Similarly to GM-CSF, IL-3 is a glycoprotein that stimulates the production of eosinophils and their multiple effector functions (3). The structural basis for IL-3 activity is not known, although certain substitutions throughout the molecule have resulted in analogs with decreased activity (27,28). However, these losses of activity are difficult to evaluate as they may be the result of modifying binding sites or of gross structural alterations of the IL-3 molecule.

We have performed site-directed mutagenesis throughout the IL-3 molecule and found that substitutions in, or adjacent to, the predicted helix D resulted in IL-3 analogs with increased function (22). In particular, we found that the combined mutation Asp101Ala and Arg116Val led to an IL-3 analog with enhanced potency (22).

We have now used this mutant IL-3 $Ala^{101}Val^{116}$ to evaluate the effect of these mutations on eosinophil function. We found that IL-3 $Ala^{101}Val^{116}$ stimulated eosinophil colonies in a dose-dependent manner, and that at each concentration tested, IL-3 $Ala^{101}Val^{116}$ stimulated more eosinophil colonies than WT IL-3 (Table 2). In addition, the size of the eosinophil colonies was greater in the presence of IL-3 $Ala^{101}Val^{116}$. At a concentration of 10 ng/ml IL-3, $Ala^{101}Val^{116}$ gave rise to eosinophil colonies containing 316 ± 72 (mean ± SD) cells compared to eosinophil colonies containing 167 ± 56 cells in the presence of WT IL-3.

E. Interaction of Helix D of IL-3 with the IL-3 Receptor

Like the GM-CSF receptor, the IL-3 receptor comprises two chains, the α chain, which binds IL-3 with low affinity, and the β chain, which confers high-affinity binding and is shared with the GM-CSF receptor. Initial experiments showed that IL-3 $Ala^{101}Val^{116}$ bound with higher affinity than WT IL-3 to the high-affinity IL-3

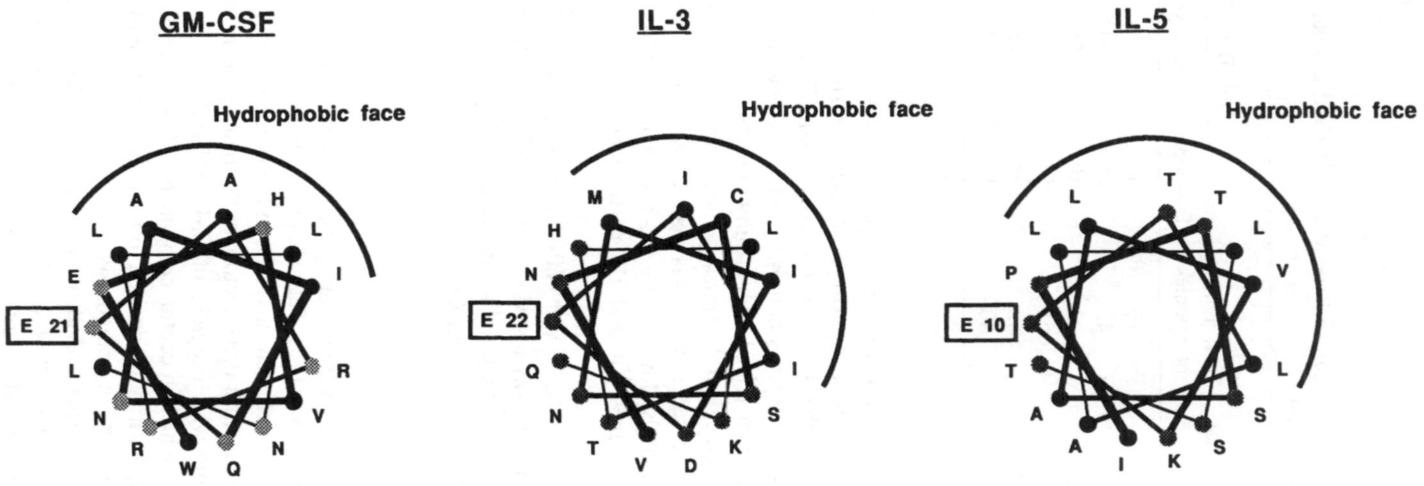

Figure 5 Comparison of the amphipathic helix A of GM-CSF and the predicted helices A of IL-3 and IL-5. The conserved glutamic acid (E) in the exposed face of GM-CSF (25) is likely to be exposed also in helix A of IL-3 and IL-5 and to contact the common β chain of the GM-CSF, IL-3, and IL-5 receptors.

Table 2 Increased Stimulation of Human Bone-Marrow-Derived Eosinophil Colonies by the IL-3 Mutant IL-3 $Ala^{101}Val^{116}$

Stimulus	Concentration (ng/ml)	Total number of colonies	Colony type (%)		
			Eosinophil	Monocyte	Neutrophils
IL-3 WT	1	16	44	56	0
IL-3 $Ala^{101}Val^{116}$	1	42	66	34	0
IL-3 WT	3	22	49	51	0
IL-3 $Ala^{101}Val^{116}$	3	61	55	45	0
IL-3 WT	30	42	34	65	1
L-3 $Ala^{101}Val^{116}$	30	60	53	40	7

receptor of eosinophils. However, since eosinophils express only the high-affinity (αβ complex) IL-3 receptor, these experiments could not distinguish whether IL-3 $Ala^{101}Val^{116}$ bound better to the IL-3 receptor α or β chain. To determine this, we used human monocytes that express both high (αβ)- and low (α)-affinity receptors. The results showed that under high-affinity conditions (100 pM ^{125}I-IL-3), IL-3 $Ala^{101}Val^{116}$ exhibited about 15 times greater affinity than WT IL-3 and, importantly, that this increased affinity was maintained under conditions (2 nM ^{125}I-IL-3) that detect mainly low-affinity IL-3 receptors (22). These results strongly suggest that residues 101 and 116 participate in binding to the IL-3 receptor α chain.

To examine the ability of IL-3 $Ala^{101}Val^{116}$ to cross-compete for binding on eosinophil IL-5 receptors, it was tested in parallel with WT IL-3. The results showed that IL-3 $Ala^{101}Val^{116}$ exhibited an enhanced ability to cross-compete for ^{125}I-IL-5 binding on eosinophils (Fig. 6).

IV. DISCUSSION

Human GM-CSF and IL-3 regulate the production and function of human eosinophils in vitro and in vivo (1–5). They are produced by activated T cells, and increased mRNA has been detected in T cells in the bronchi and skin of allergic individuals (29,30), implicating these factors in the pathogenesis of asthma and skin disease. By defining which regions of GM-CSF and IL-3 are required for eosinophil stimulation, which receptor components they recognize, and their role in signaling, strategies may be devised to alter the structure of these factors to generate selective agonists or antagonists.

The results shown here provide evidence that Glu^{21} of GM-CSF is critical for the eosinophil-stimulating activity of GM-CSF. Receptor studies on eosinophils and transfected cell lines strongly suggest that this residue is involved in binding

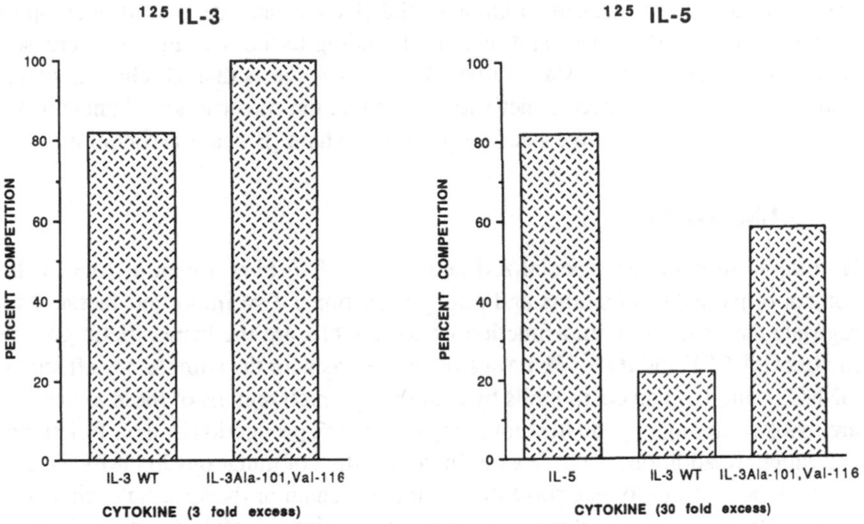

Figure 6 Increased ability of the IL-3 $Ala^{101}Val^{116}$ to compete for ^{125}I-IL-3 binding and to cross-compete for ^{125}I-IL-5 binding to purified human eosinophils. Radioligands were used at 200 pM, and 100% competition represents the competition obtained in the presence of 100-fold excess unlabeled homologous ligand.

to the β chain. Given the common nature of the β chain and the presence of a glutamic acid in analogous positions in IL-3 and IL-5, it is postulated that substituting this residue will also achieve substantial decreases in eosinophil activation by IL-3 and IL-5.

The significance of this finding is that high-affinity binding and eosinophil stimulation can be greatly reduced without necessarily affecting binding to the α chain of the GM-CSF receptors. This implies two distinct binding sites in these molecules and suggests that the construction of antagonists may be possible by careful and selective mutagenesis in these molecules.

While the results with GM-CSF implicate Glu^{21} in interacting with the β chain, the results with IL-3 indicate that helix D is important for recognition of the receptor α chain. In particular, the results showed that binding to the α chain can be improved and thus more potent eosinophil activators can be constructed. This finding may also bring the construction of antagonists of eosinophil activators nearer by offering the possibility of combining mutations that increase binding to the α chain (IL-3 $Ala^{101}Val^{116}$-like) with mutations that decrease binding to the β chain (GM-CSF Arg^{21}-like).

Finally, in terms of receptor signaling, the results indicate that both α and β chains of these receptors are involved in eosinophil activation. This is shown by the correlation of decreased binding to the β chain and decreased eosinophil activation (GM-CSF Arg^{21}), and increased binding to the α chain and increased eosinophil function (IL-3 $Ala^{101}Val^{116}$). This also suggests that αβ chain association is likely to be required for activation, and that the construction of antagonists should take into account the need to prevent α chain–β chain association.

V. SUMMARY

Human eosinophils are specialized cells believed to play a central role in the control of parasitic infections and allergic reactions. Two important factors that regulate the production and function of eosinophils are the hemopoietic growth factors GM-CSF and IL-3. We have studied the basis for the stimulatory effects of GM-CSF and IL-3 on eosinophils by examining which regions of these molecules are involved in binding and in stimulating eosinophil production or their functional activation. In addition, we have examined the effect of mutations in GM-CSF and IL-3 on their ability to recognize the common β chain or the specific α chains of their receptors, seeking clues as to the contribution of each receptor chain in receptor activation and signaling in eosinophils.

The results show that a glutamic acid in the first helix of GM-CSF is crucial for optimal stimulation of eosinophil production and function. Binding studies revealed that this residue selectively interacts with the β chain of the GM-CSF receptor, thus implicating this chain in signaling. Furthermore, given that the β chain of the GM-CSF receptor is also part of the IL-3 and IL-5 receptors, and that a glutamic acid exists in an analogous position in IL-3 and IL-5, these results predict that similar dissociation of binding will be observed in the other two eosinophil activators, IL-3 and IL-5.

Mutagenesis of IL-3 revealed two residues involved in function, substitution of which led to an IL-3 analog with increased capacity to stimulate eosinophil production and function. Binding experiments indicate that these residues are involved in binding to the α chain of the IL-3 receptor.

Taken together, these results identify crucial yet distinct regions in GM-CSF and IL-3 involved in eosinophil stimulation and point to both the α and β chains of their receptors as important for receptor activation and signaling in eosinophils. The identification of these two regions offers, in addition, the possibility of constructing new mutants with more defined functional specificity and with the potential to behave as selective agonists or antagonists of eosinophil function.

REFERENCES

1. Metcalf D, Begley CG, Johnson GR, Nicola NA, Vadas MA, Lopez AF, Williamson DJ, Wong GG, Clark SC, Wang EA. Biological properties in vitro of a recombinant human granulocyte-macrophage colony-stimulating-factor. Blood 1986; 67:37–45.

2. Lopez AF, Williamson DJ, Gamble JR, Begley CG, Harlan JM, Klebanoff SJ, Waltersdorph A, Wong G, Clark SC, Vadas MA. Recombinant human granulocyte-macrophage colony-stimulating factor (rH GM-CSF) stimulates in vitro mature human neutrophil and eosinophil function, surface receptor expression and survival. J Clin Invest 1986; 78:1220–1228.
3. Lopez AF, Dyson P, Elliott M, Milton S, To LB, Juttner C, Russell J, Yang Y-C, Clark SC, Vadas MA. Recombinant interleukin-3 stimulation of hemopoiesis in man. Loss of responsiveness of neutrophilic granulocytes with differentiation. Blood 1988; 72: 1797–1804.
4. Donahue RE, Seehra J, Metzger M, Lefebvre D, Rock B, Carbone S, Nathan DG, Garnick M, Sehgal PK, Laston D et al. Human IL-3 and GM-CSF act synergistically in stimulating hematopoiesis in primates. Science 1988; 241:1820–1823.
5. Ganser A, Lindemann A, Seipelt G, Ottmann OG, Herrmann F, Eder M, Frisch J, Schulz G, Mertelsmann R, Hoelzer D. Effects of recombinant human IL-3 in patients with normal hematopoiesis and in patients with bone marrow failure. Blood 1990; 76: 666–676.
6. Clutterbuck EJ, Sanderson CJ. Regulation of human eosinophil precursor production by cytokines: comparison of recombinant human interleukin-1 (rhIL-1), rhIL-3, rhIL-5, rhIL-6, and rh granulocyte macrophage colony-stimulating factor. Blood 1990; 75:1774–1779.
7. Lopez AF, Eglinton JM, Gillis D, Park LS, Clark S, Vadas MA. Reciprocal inhibition of binding between interleukin 3 and granulocyte-macrophage colony-stimulating factor to human eosinophils. Proc Natl Acad Sci USA 1989; 86:7022–7026.
8. Lopez AF, Vadas MA, Woodcock J, Milton SE, Lewis A, Elliott MJ, Gillis D, Ireland R, Olwell E, Park LS. Selective interaction of the human eosinophil interleukin-5 receptor with interleukin-3 and granulocyte-macrophage colony-stimulating factor. J Biol Chem 1991; 266:24741–24747.
9. Gearing DP, King JA, Gough NM, Grail D, Dunn AR. Expression cloning of a receptor for human granulocyte-macrophage colony-stimulating factor. EMBO J 1985; 8:3667–3676.
10. Hayashida K, Kitamura T, Gorman DM, Arai K, Yokota T, Miyajima A. Molecular cloning of a second subunit of the receptor for human GM-CSF. Reconstitution of a high affinity GM-CSF receptor. Proc Natl Acad Sci USA 1989; 86:1213–1217.
11. Kitamura T, Sato N, Arai K-I, Miyajima A. Receptor cloning of the human IL-3 receptor cDNA reveals a shared beta subunit for the human IL-3 and GM-CSF receptors. Cell 1991; 66:1165–1174.
12. Tavernier J, Devos R, Cornelis S, Tuypens T, Van der Heyden J, Fiers W, Plaetinck G. A human high affinity interleukin-5 receptor (IL-5R) is composed of an IL-5-specific α chain and a β chain shared with the receptor for GM-CSF. Cell 1991; 66:1175–1184.
13. Vadas MA, David JR, Butterworth AE, Pisani NT, Siongok TA. A new method for the purification of human eosinophils and neutrophils and a comparison of the ability of these cells to damage schistosomula of *Schistosoma mansoni*. J Immunol 1979; 122: 1228–1236.
14. Zoller MJ, Smith M. Labaratory Methods. Oligonucleotide-directed mutagenesis: a

simple method using two oligonucleotide primers and a single-stranded DNA template. DNA 1984; 3:479–488.
15. Wood WI, Gitschier J, Lasky LA, Lawn RM. Base composition-independent hybridization in tetramethylammonium chloride: a method for oligonucleotide screening of highly complex gene libraries. Proc Natl Acad Sci USA 1985; 82:1585–1588.
16. Sanger F, Nicklen S, Coulson AR. DNA sequencing with chain-terminating inhibitors. Proc Natl Acad Sci USA 1977; 74:5463–5467.
17. Gough NM, Metcalf D, Gough J, Grail D, Dunn AR. Structure and expression of the mRNA for murine granulocyte-macrophage colony stimulating factor. EMBO J 1985; 4:645–654.
18. Chen EY, Seeburg PH. Laboratory Methods. Supercoil sequencing: a fast and simple method for sequencing plasmid DNA. DNA 1985; 4:165–170.
19. Kammann M, Laufs J, Shell J, Gronenborn B. Rapid insentional mutagenesis of DNA by polymerase chain reaction (PCR). Nucleic Acids Res 1989; 17:5404–5410.
20. Phillips JA, Lopez AF, Milton SE, Vadas MA, Shannon MF. Synthesis and expression of the gene for human interleukin-3. Gene 1990; 84:501–507.
21. Lopez AF, Shannon MF, Hercus T, Nicola NA, Cambareri B, Dottore M, Layton MJ, Eglinton L, Vadas MA. Residue 21 of human granulocyte-macrophage colony-stimulating factor is critical for biological activity and for high but not low affinity binding. EMBO J 1992; 11:909–916.
22. Lopez AF, Shannon MF, Barry S, Phillips JA, Cambareri B, Dottore M, Simmons P, Vadas MA. Structure-function studies of human interleukin-3 identify an analog with increased biological and binding properties. Proc Natl Acad Sci USA 1992; 89: 11842–11846.
23. Contreras MA, Bale WF, Spar IL. Iodine monochloride (ICR) iodination techniques. Methods Enzymol 1983; 92:277–292.
24. Clark-Lewis I, Lopez AF, To LB, Vadas MA, Schrader JW, Hood LE, Kent SBH. Structure-function studies of human GM-CSF. Identification of residues required for activity. J Immunol 1988; 141:881–889.
25. Diederichs K, Boone T, Karplus PA. Novel fold and putative receptor binding site of granulocyte-macrophage colony-stimulating factor. Science 1991; 254:1779–1782.
26. Goodall G, Bagley C, Vadas MA, Lopez AF. The structure of the GM-CSF, IL-3 and IL-5 receptors: motif conservation and its implication for structure-function studies. Growth Factors 1993; 8:87–97.
27. Lokker NA, Movva NR, Strittmatter U, Fagg B, Zenke G. Structure activity relationship study for human IL-3. Identification of residues required for biological activity in site-directed mutagenesis. J Biol Chem 1991; 266:10624–10631.
28. Lokker NA, Zenke G, Strittmatter U, Fagg B, Rao Movva N. Structure-activity relationship study of human IL-3: role of the C-terminal region for biological activity. EMBO J 1991; 10:2125–2131.
29. Kay AB, Ying S, Varney V, Gaga M, Durham SR, Moqbel R, Wardlaw AJ, Hamid Q. Messenger RNA expression of the cytokine gene cluster, Interleukin 3 (IL-3), IL-4, IL-5 and granulocyte/macrophage colony-stimulating factor, in allergen-induced late-phase cutaneous reactions in atopic subjects. J Exp Med 1991; 173:775–778.
30. Robinson DS, Hamid Q, Ying S, Tsicopoulos A, Barkans J, Bentley AM, Corrigan C,

Durham SR, Kay AB. Predominant T_{H2}-like bronchoalveolar T-lymphocyte population in atopic asthma. N Engl J Med 1992; 326:298–304.

DISCUSSION (A. Lopez/M. Vadas)

Moqbel: Do you have transgenic mouse models to test your hypothesis?

Vadas: No, but the lack of cross-reactivity of human and mouse GM-CSF may make interpretation difficult.

Abrams: Since IL-5 appears to be a disulfide-bonded dimer, how do you rationalise the fact that monomer GM-CSF is still more potent than IL-5 in most of the bioassays reported, given your proposed model involving dimerization and increased affinity?

Vadas: As IL-5 is a head-to-tail dimer, it may only present to the receptor a single site binding to the α and β chains of the receptor.

Takatsu: Did you say that IL-3Rα chain dimer with high affinity can transduce IL-3 signals? What do you think about the roles of β chain in signal transduction? I would like to make a comment. IL-5 makes a dimer in an antiparallel manner. IL-5 monomer does not bind to high-affinity IL-5R. This may be interpreted that N-terminal α-helix region IL-5 binds to β chain and C-terminal region binds to IL-5Rα.

Lopez: No, we do not have any evidence that the IL-3R α chain by itself transduces signal. We do now know (unpublished) that the $Ala^{101}Val^{116}$ mutant of IL-3 binds better than wild type to the cloned α chain, consistent with the C-terminus binding to the IL-3Rα chain and the N-terminus binding to the common β chain.

6

The Role of Adhesion Molecules and Cytokines in Eosinophil Recruitment

Robert P. Schleimer, Motohiro Ebisawa, Steve N. Georas, and Bruce S. Bochner
Johns Hopkins Asthma and Allergy Center, Baltimore, Maryland

I. INTRODUCTION

It is well established that large numbers of eosinophils can be found at allergic reaction sites and in allergic diseases. These cells are now believed to contribute to the pathological destruction of airway tissues and to airways hyperreactivity in asthma. In allergic individuals, whole lung or segmental lung antigen challenge results in large numbers of eosinophils in the lavage fluid 18–24 h later (1–4). In fact, as many as 60% or more of the infiltrating cells can be eosinophils. The question we would like to address here is what are the mechanisms by which allergens elicit an eosinophil-rich inflammatory cell infiltrate? While a number of important events occur in leukocyte recruitment, this discussion will be confined to the molecular basis of the adhesive interaction between eosinophils and endothelial cells and the influence of cytokines on this interaction.

The studies to be discussed here have largely been performed in vitro with isolated cell types. We have primarily used eosinophils purified from peripheral blood with a combination of a discontinuous Percoll purification method and CD 16-positive (i.e., neutrophil) cell depletion with immunomagnetic beads (5). Human umbilical vein endothelial cells cultured on tissue culture plates have been used to study the interaction between the endothelial cells and purified leukocytes. For some adhesion assays, the endothelial cells are stimulated with cytokines, leukocytes are added and nonadherent cells are washed away after a 10-min incubation. In some cases, the stimulus is present when the leukocytes are in the 10-min adhesion assay with the endothelial cells. This distinction is made because factors that induce adherence can be divided into two main groups: those that activate endothelium and those that activate leukocytes (some factors can do both) (6). In general, endothelial activation requires several hours whereas leukocyte activation is almost instantaneous, as one might imagine since a leukocyte moving through a capillary bed must respond very rapidly in order to extravasate locally.

II. MECHANISMS OF EOSINOPHIL ADHESION TO CYTOKINE-ACTIVATED ENDOTHELIUM

One of the important observations made in this system many years ago was that exposure of endothelial cells to interleukin-1 (IL-1), tumor necrosis factor (TNF), phorbol esters, or bacterial lipopolysaccharide (LPS) caused them to express adhesive properties for neutrophils (7–9). Subsequent studies revealed that the same basic results occur with eosinophils and basophils as with neutrophils (10,11). Studies with protein and RNA synthesis inhibitors revealed that induction of de novo expression of adhesion molecules on endothelial cells increases the adherence of all three cell types. To date, three major cytokine-induced endothelial adherence molecules have been characterized and identified which are felt to be important in leukocyte adhesion. They include E-selectin (formerly ELAM-1, endothelial leukocyte adhesion molecule-1), ICAM-1 (intercellular adhesion molecule-1, CD54), and VCAM-1 (vascular cell adhesion molecule-1) (12). Unstimulated cells do not express either E-selectin or VCAM-1 whereas ICAM-1 is expressed basally on resting cells; the level of expression for each of these molecules can be further induced by activation. Induction of the expression of these three adhesion molecules by IL-1, as detected by flow cytometry, is shown in Figure 1. Thus, in a few short years our concept of leukocyte recruitment mechanisms has been expanded to include an active role for endothelium mediated, in part, by the de novo expression of adhesion molecules. Although the number of identified adhesion molecules has rapidly expanded, within the context of eosinophil recruitment, we first asked whether there was any selectivity among these three adhesion molecules for eosinophils.

A. Eosinophil Adhesion to IL-1 Activated Endothelium

In collaboration with Drs. Michael Gimbrone, Jr., F. William Luscinskas, and Walter Newman, we tested the ability of specific anti-ICAM-1, anti-E-selectin, and anti-VCAM-1 antibodies to inhibit adherence of eosinophils, neutrophils, and basophils to IL-1-stimulated endothelial cell monolayers (13). Specific $F(ab')_2$ preparations of antibodies to ICAM-1 and E-selectin inhibited adherence of all

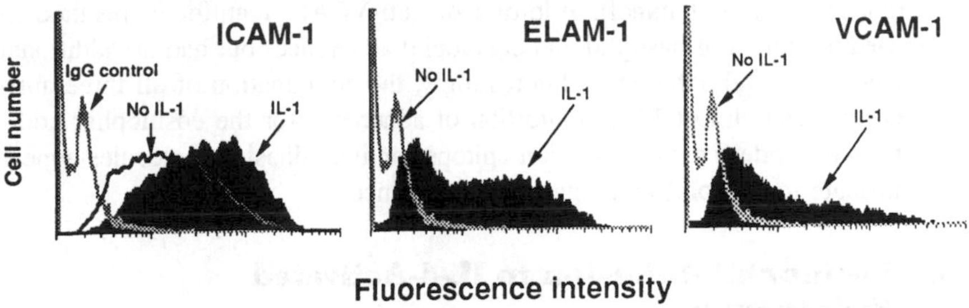

Figure 1 Treatment of human umbilical vein endothelial cells with IL-1 (5 ng/ml, 4 h) induces expression of ICAM-1, E-selectin (ELAM-1), and VCAM-1. After stimulation, single cell suspensions were generated and labeled for analysis by indirect immunofluorescence using specific monoclonal antibodies (Hu5/3, H18/7, and 2G7, respectively).

three leukocyte types by approximately 20 or 30%; i.e., there was no apparent selectivity among leukocytes. Specific anti-VCAM-1 antibody was the most effective in inhibiting eosinophil adherence (see Fig. 2, left). Anti-VCAM-1 also significantly inhibited basophil adherence, although to a modest degree, but had no effect on IL-1-induced adherence of neutrophils. Combinations of the antibody types revealed an extension of the same observation. Antibodies to ICAM-1 and E-selectin together produced roughly additive inhibitory effects (Fig. 2, right), which did not discriminate among these three cell types (note the scale is different from the left panel). Addition of anti-VCAM-1 antibody resulted in increased inhibition of basophil and eosinophil adherence, but had no additional

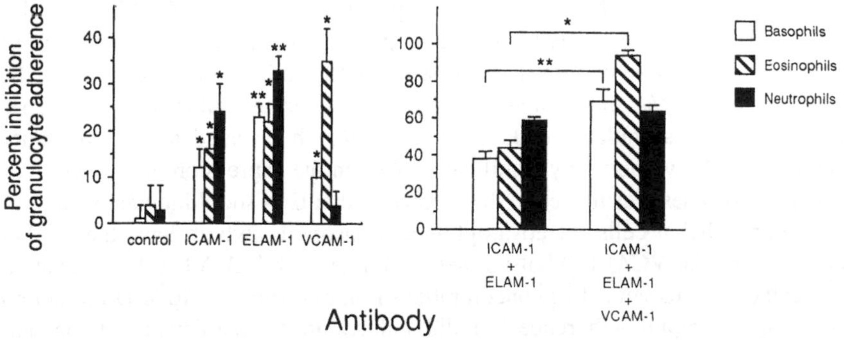

Figure 2 Effect of $F(ab')_2$ preparations of anti-ICAM-1 (Hu5/3), anti-E-selectin (H18/7), anti-VCAM-1 (2G7), or control mAb (W6/32, anti-HLA class I) on adherence of basophils, eosinophils, and neutrophils to IL-1-treated (4 h, 5 ng/ml) umbilical vein endothelial cells. $*p < 0.05$; $**p < 0.01$. (From Ref. 13.)

different from the left panel). Addition of anti-VCAM-1 antibody resulted in increased inhibition of basophil and eosinophil adherence, but had no additional effect on neutrophil adherence. Interestingly, the combination of all three antibodies produced almost 100% inhibition of adherence for the eosinophil; additional, as-yet-undetermined adhesion epitopes and/or adhesion molecules appear to contribute to basophil and neutrophil adherence.

B. Eosinophil Adhesion to IL-4-Activated Endothelium

It has been established that a counterligand for VCAM-1 on lymphocytes is VLA-4, one of the β1 integrin adhesion molecules also referred to as CD49d/CD29 (14). Results of the analysis of the expression of VLA-4 on eosinophils, basophils, and neutrophils fit surprisingly well with the adherence data: neutrophils, which did not show any anti-VCAM-1-inhibitable adherence, also did not express VLA-4, whereas eosinophils and basophils both expressed VLA-4 (13,15–17). Indeed, antibody to VLA-4 inhibited eosinophil adhesion to IL-1-stimulated endothelium (15,16). The tentative conclusion from these studies therefore is that VCAM-1 on endothelial cells can act as a substrate for VLA-4-mediated adherence of eosinophils and basophils. To independently confirm that eosinophils and basophils can adhere to VCAM-1, the ability of these cells to adhere to a truncated recombinant form of VCAM-1, which has been immobilized to plastic plates, was tested. Eosinophils and basophils (unlike neutrophils) adhered more avidly to this surface compared to control plates (35 and 18% adherence vs. 11 and 8%, respectively), and the adhesion was inhibited using specific anti-VCAM-1 and anti-VLA-4 antibodies (18,19) (see Fig. 3). For eosinophils, these results are consistent with other recently published data employing a similar in vitro adhesion system (17).

These results raised the possibility that specific induction of VCAM-1 on endothelial cells would selectively induce eosinophil and basophil (but not neutrophil) adherence. It had been previously suggested that IL-4 might selectively lead to the induction of VCAM-1 expression (20). This was confirmed using flow cytometric analysis of IL-4-stimulated endothelial cells, which revealed a marked induction of VCAM-1 without any significant effect on the expression of E-selectin or ICAM-1 (18). These results led to the prediction that IL-4 should induce adhesiveness in endothelial cells for eosinophils and basophils (which have the VLA-4 counterligand for VCAM-1) but not neutrophils (which lack VLA-4). Incubation of endothelial cells with IL-4 concentrations ranging from 1 to 1000 U/ml had no effect on neutrophil adherence but did induce, in a dose-dependent manner, eosinophil and basophil adherence (Fig. 4) (18). Antibody-blocking studies showed that greater than 70% of this adherence can be inhibited with either anti-VCAM-1 or anti-VLA-4 antibodies in the case of both the eosinophil and basophil

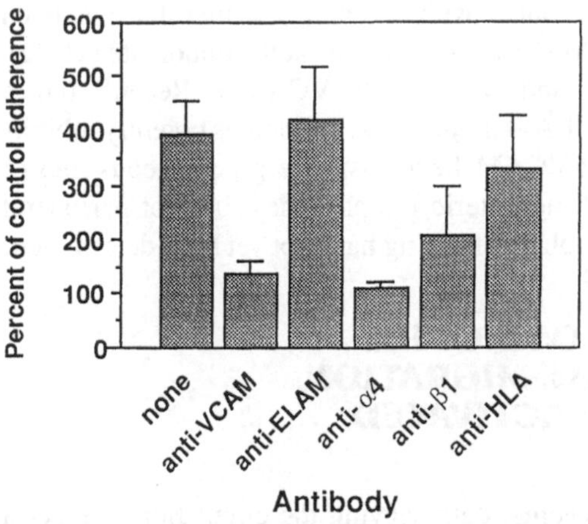

Figure 3 Adhesion of eosinophils to plates coated with a soluble form of VCAM-1 (50 ng/well) and blocked with 1% BSA or wells coated with BSA alone (control) was tested in the presence or absence of specific monoclonal antibodies as indicated. Values represent mean ± SEM, for percent of control adherence, which was 11 ± 2%; $n \geqslant 3$.

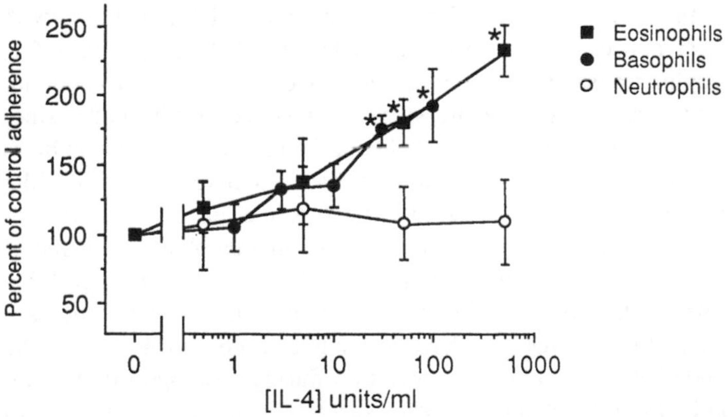

Figure 4 IL-4 induces adhesiveness in endothelial cells for eosinophils and basophils but not neutrophils. Endothelial cells were stimulated with the indicated concentrations of IL-4 for 24 h. Monolayers were rinsed, and eosinophils, basophils, or neutrophils were added for a 10-min incubation after which unbound cells were rinsed away. $*p < 0.05$. (From Ref. 18.)

with the observation that mice inoculated with a tumor cell line that has been transfected with the IL-4 gene develop local eosinophilia at the tumor site (21,22). Lymphocytes also express VLA-4 and can bind to VCAM-1. Recent studies suggest that administration of anti-VLA-4 monoclonal antibodies potently inhibits migration of lymphocytes to sites of VCAM-1 expression [e.g., cutaneous sites of inflammation, Peyer's patches, and mesenteric lymph nodes, but not peripheral lymph nodes (23)]; effects on eosinophil trafficking have not yet been determined.

III. MECHANISMS OF EOSINOPHIL TRANSENDOTHELIAL MIGRATION ACROSS CYTOKINE-ACTIVATED ENDOTHELIUM

In addition to the initial adhesion events, cells leaving the circulation to enter a local inflammatory site must undergo transendothelial migration. Previous in vitro studies have revealed that transendothelial migration of neutrophils across IL-1- or TNF-activated endothelial cells requires, or is facilitated by, endothelial-derived IL-8, a chemoattractant that presumably activates elements of the locomotive process necessary for penetration of the endothelial barrier (24). It is not yet known whether the same is true for eosinophils and/or basophils or whether there are other factors derived from endothelial cells or other tissue cells that assume this role. However, we have begun to explore the molecular mechanisms regulating eosinophil transendothelial migration in vitro (25,25a). For these experiments, umbilical vein endothelial cells were grown to confluence on collagen-coated Transwell culture inserts with a 5-μm pore size. Treatment of endothelial monolayers with IL-1 or TNF, which induces the expression of E-selectin, ICAM-1, and VCAM-1, increased eosinophil transendothelial migration. Treatment with IL-4 alone, which selectively induces VCAM-1 expression, also resulted in enhanced eosinophil transmigration. Interestingly, treatment of endothelial cells with the combination of IL-1 and IL-4 or TNF and IL-4 (conditions that are synergistic for VCAM-1 expression, as mentioned above) resulted in a synergistic effect on transendothelial migration. These preliminary results demonstrate that cytokine activation of endothelial cells promotes eosinophil transendothelial migration and suggest that adhesion molecules on the endothelial cell actively participate in this process.

Based on these preliminary data, we have begun to explore the roles of specific leukocyte and endothelial cell adhesion molecules during eosinophil transendothelial migration by using blocking antibodies specific for these molecules (25a) (see Fig. 5). Transendothelial migration through IL-1-treated endothelium was inhibited by antibodies to CD18 (β_2 integrin) on the leukocyte, while CD29 (β_1 integrin) antibody had little or no effect. Likewise, antibodies directed against ICAM-1 but not VCAM-1 were effective in inhibiting transmigration, consistent with the previously reported roles of these molecules in mediating transmigration of other leukocytes (26). These data support the hypothesis that transendothe-

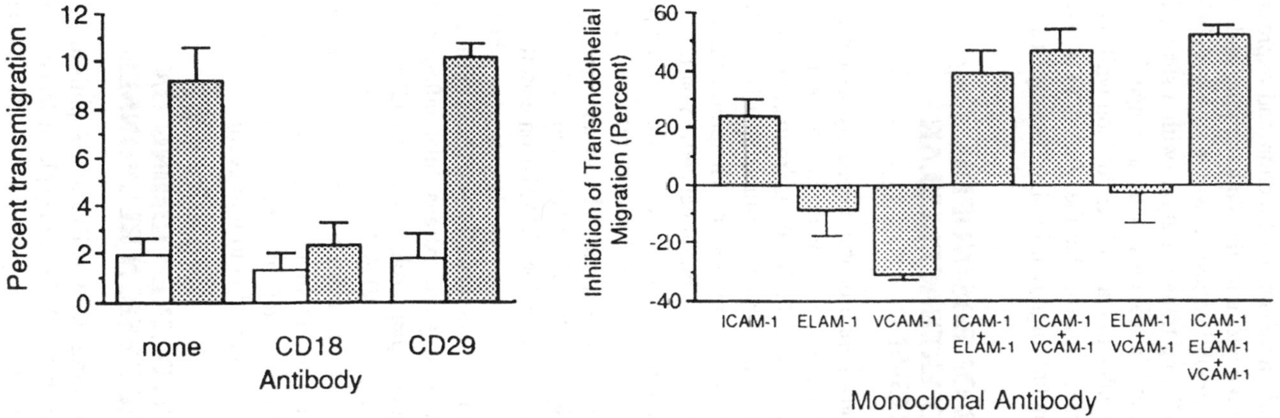

Figure 5 (Left) Effect of anti-CD18 (H52) and anti-CD29 (18D3) monoclonal antibodies on eosinophil transendothelial migration across unstimulated (open bars) or IL-1 treated (4 h, 5 ng/ml, closed bars) endothelium ($n = 3$). (Right) Effect of $F(ab')_2$ anti-ICAM-1 (Hu5/3), anti-VCAM-1 (2G7), and anti-E-selectin (H18/7) monoclonal antibodies on eosinophil transendothelial migration across IL-1 treated (4 h, 5 ng/ml) endothelium ($n = 3$). (From Ref. 25a.)

lial migration of eosinophils involves the function and expression of adhesion molecules on both the leukocyte and the endothelium and suggest that the mechanisms regulating eosinophil adhesion (with a strong VCAM-1 component) may differ from those mediating transmigration (with a strong ICAM-1 component). The observation that, in patients with a congenital deficiency of β_2 integrins, neutrophils are unable to leave the circulation while eosinophils can be found in extravascular tissues suggests that the VLA-4/VCAM-1 system may be able to sustain transendothelial migration in vivo, however (27).

IV. MECHANISMS OF EOSINOPHIL ADHESION TO EXTRACELLULAR MATRIX PROTEINS

Selective endothelial responses are probably not solely responsible for the preferential recruitment and migration of eosinophils and basophils to allergic reaction sites. Once a cell enters a tissue, the local environmental milieu consisting of various proteins that comprise the extracellular matrix likely interacts with leukocytes during chemotaxis and migration. Therefore, the ability of a given tissue to retain or trap leukocytes may be influenced, in part, by the ability of the leukocyte to bind to extracellular matrix proteins. Most of the β_1 integrins function as receptors for these molecules, and it is therefore of interest that among the β_1 integrins, eosinophils also express VLA-6 and are capable of binding laminin, one of its known ligands (28). The VLA-4 molecule, which as mentioned above functions as a counterreceptor for VCAM-1 on endothelial cells, can also function as an alternative receptor on lymphocytes for fibronectin (14); this appears to be the case for eosinophils as well (29). Thus, the ability of eosinophils to remain and localize within a given extravascular inflammatory site may be enhanced in tissues expressing higher levels of laminin and/or fibronectin. However, studies examining the expression of VLA-4 on human eosinophils isolated from sites of allergic inflammation (e.g., from bronchoalveolar lavage following antigen challenge) have failed to find any increases in expression (4), unlike observations made with lymphocytes isolated from rheumatoid synovium where increased levels of VLA-4 and enhanced adhesion to fibronectin have been observed (30).

V. EFFECTS OF CYTOKINE PRIMING ON EOSINOPHIL ADHESION RESPONSES

The responsiveness of circulating eosinophils and basophils to a given stimulus is another important parameter that we believe may play a decisive role in the cellular recruitment pattern seen following allergen challenge (31). In particular, there is good reason to believe that circulating eosinophils and perhaps basophils have been "primed" by cytokines such as IL-3, IL-5, and GM-CSF both in asthmatic subjects with ongoing disease as well as following antigen challenge of asymptomatic individuals in vivo (1,2,32). This priming of eosinophils can lead to enhanced adhesive and migratory behavior.

Studies were initiated to determine whether GM-CSF will induce changes in the adhesive properties or expression of adhesion molecules in human eosinophils (33). Challenge of fresh eosinophils with fmet-peptide or PAF induced very little increase in the expression of the CD11b adherence molecule. After priming for up to 4 days with GM-CSF, the ability of these cells to express CD11b was potentiated, and the adherence of primed eosinophils to endothelial cells and gelatin-coated plastic plates was increased up to twofold. Therefore, exposure to cytokines in vivo could potentially increase the adhesive and perhaps migratory responses of eosinophils.

These in vitro studies suggested that upregulation of CD11b by fmet-peptide or PAF might be a useful marker of priming in vivo. This parameter was, therefore, monitored in vivo in a patient with episodic hypereosinophilia and myalgia who had high numbers of circulating eosinophils (34). Analysis of the eosinophil responses showed that they had this "priming" for CD11b upregulation. When this individual was treated with prednisone, the number of eosinophils dropped precipitously, the percentage of hypodense eosinophils fell, and the increased CD11b up-regulation response returned to the normal level. As the steroid treatment was tapered, the eosinophil number began to climb, the proportion of hypodense cells increased, and the priming of the CD11b response returned. Unlike normal sera, analysis of this patient's sera for cytokine bioactivity using the M-07e cell line, which proliferates in response to GM-CSF and IL-3, revealed elevated activity during the symptomatic period. This signal was, for the most part, neutralized with anti-GM-CSF antibodies, but there was also some remaining activity. As one would expect from in vitro results from a number of laboratories, prednisone treatment reduced the serum levels of GM-CSF in this individual. Such treatment reduced serum cytokine levels with a time course similar to that seen for the decreases in eosinophil number, low-density eosinophil number, and priming for the CD11b response. Interestingly, elevations in serum cytokine levels returned following tapering of the steroid dosage and were associated with a symptomatic relapse. These results suggest that cytokines can potentiate eosinophil responses in vivo in a way that should augment their propensity to infiltrate tissues at sites of antigen exposure and they further support the suggestion that inhibition of cytokine production is an important action of glucocorticoids (31,35–37).

Cytokine priming of eosinophils probably also occurs in allergic reactions in the airways. Several groups, including our own, have obtained evidence for GM-CSF production in vivo in humans following antigen exposure (38,39). Studies in collaboration with Dr. Mark Liu have revealed that the amount of GM-CSF in bronchoalveolar lavage fluid is 5- to 30-fold higher after antigen challenge than after saline challenge of allergic individuals (38a). Other groups have detected IL-3 or IL-5 in analogous studies (2,40). The levels of GM-CSF detected are well within the range in which priming and enhanced survival of eosinophils occurs

(33,41). Thus, eosinophil-priming cytokines can be detected in vivo during allergic reactions, providing circumstantial evidence to suggest that eosinophils may be primed and their survival within tissues prolonged by cytokines released locally as a result of antigen stimulation in vivo. Many cell types, including T lymphocytes, macrophages, mast cells, endothelial cells, and perhaps even eosinophils themselves, have the capability of synthesizing some or all of the abovementioned cytokines, although the specific cells responsible for their production in vivo remain to be determined.

VI. EVIDENCE OF CYTOKINE-INDUCED ENDOTHELIAL ACTIVATION IN VIVO DURING ALLERGIC INFLAMMATION

A number of laboratories have established that endothelial activation does indeed occur following challenge of allergic individuals with allergen as measured by the expression of E-selectin, ICAM-1, and VCAM-1 (42–44). The fact that both E-selectin and VCAM-1 are expressed in vivo suggests that the endothelial cells are activated by a cytokine such as IL-1 or TNF. This is further supported by the observations that IL-1 is released during cutaneous allergic reactions (45), that the human skin mast cell can release TNF, which is capable of activating endothelial cell E-selectin expression (46), and that antigen-induced E-selectin expression can be inhibited in skin biopsies if cultured with a combination of antibodies to IL-1 and TNF (43). Thus, based only on expression of endothelial cell adhesion molecules, one would expect little selectivity among granulocytes. Analysis of cytokine mRNA in human cutaneous allergen challenge reactions has established that IL-4 mRNA is produced (47). Furthermore, in vitro studies have established that IL-4 can synergize with IL-1 and/or TNF in promoting expression of VCAM-1 (48,49). Thus, even in the context of nonselective activation of endothelial cells by IL-1 or TNF, the concurrent presence of IL-4 could contribute to favor VCAM-1-mediated recruitment of eosinophils in vivo.

VII. CONCLUSIONS

Several lines of evidence suggest that adhesion molecules and cytokines may play an important role in the selective migration of eosinophils to allergic reaction sites. The model displayed in Figure 6 summarizes several of these proposed mechanisms. At the level of the endothelium, either selective or prominent expression of VCAM-1 along with the other endothelial adhesion molecules may assist in promoting eosinophil binding to endothelium and subsequent infiltration at sites of exposure to allergen. Previous exposure of circulating eosinophils to priming cytokines may also potentiate enrichment of eosinophils at a given tissue site by increasing their adhesivity and migratory responses. Finally, prolongation of eosinophil survival by cytokines at a local tissue site may, by virtue of attrition, enrich the infiltrating cell populations for eosinophils.

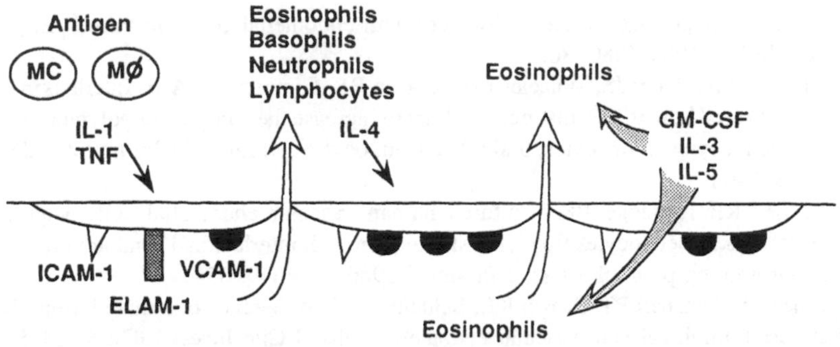

Figure 6 Proposed sequence of cytokine production and endothelial activation in eosinophil recruitment during allergic inflammation. Initial antigen exposure leads to the local release of IL-1 and TNF from mast cells (MC) and macrophages (MØ), resulting in endothelial expression of ICAM-1, E-selectin (ELAM-1), and VCAM-1. This pattern of endothelial cell activation may result in the adhesion and transendothelial migration of many different leukocyte types, although the subsequent production of IL-4 by T-cells and perhaps mast cells, either alone or in synergy with IL-1 and TNF, may lead to preferential VCAM-1 expression and eosinophil recruitment. In addition, exposure of eosinophils to cytokines such as IL-3, IL-5, and/or GM-CSF may result in "priming" of various functions including adhesion, migration, and survival.

REFERENCES

1. Frick WE, Sedgwick JB, Busse WW. The appearance of hypodense eosinophils in antigen-dependent late phase asthma. Am Rev Respir Dis 1989; 139:1401–1406.
2. Sedgwick JB, Calhoun WJ, Gleich GJ, Kita H, Abrams JS, Schwartz LB, Volvitz B, Ben-Yaakov M, Busse WW. Immediate and late airway response of allergic rhinitis patients to segmental antigen challenge. Am Rev Respir Dis 1991; 144:1274–1281.
3. Liu MC, Hubbard WC, Proud D, Stealey B, Galli S, Kagey-Sobotka A, Bleecker ER, Lichtenstein LM. Immediate and late inflammatory responses to ragweed antigen challenge of the peripheral airways in asthmatics: cellular, mediator, and permeability changes. Am Rev Respir Dis 1991; 144:51–58.
4. Georas SN, Liu MC, Newman W, Beall WD, Stealey BA, Bochner BS. Altered adhesion molecule expression and endothelial activation accompanies the recruitment of human granulocytes to the lung following segmental antigen challenge. Am J Respir Cell Mol Biol 1992; 7:261–269.
5. Hansel TT, Pound JD, Pilling D, Kitas GD, Salmon M, Gentle TA, Lee SS, Thompson RA. Purification of human eosinophils by negative selection using immunomagnetic beads. J Immunol Methods 1989; 122:97–103.
6. Bochner BS, Lamas AM, Benanati SV, Schleimer RP. On the central role of vascular endothelium in allergic reactions. In: Dorsch R, ed. Late Phase Allergic Reactions. Boca Raton, FL: CRC Press, 1990:221–235.
7. Dunn CJ, Fleming WE. Increased adhesion of polymorphonuclear leukocytes to vascular endothelium by specific interaction of endogenous (interleukin-1) and

exogenous (lipopolysaccharide) substances with endothelial cells "in vitro." Eur J Rheum Inflam 1984; 7:80–86.
8. Bevilacqua MP, Pober JS, Wheeler ME, Cotran RS, Gimbrone MA Jr. Interleukin 1 acts on cultured human vascular endothelium to increase the adhesion of polymorphonuclear leukocytes, monocytes, and related leukocytic cell lines. J Clin Invest 1985; 76:2003–2011.
9. Schleimer RP, Rutledge BK. Cultured human vascular endothelial cells acquire adhesiveness for leukocytes following stimulation with interleukin-1, endotoxin, and tumor-promoting phorbol esters. J Immunol 1986; 136:649–654.
10. Bochner BS, Peachell PT, Brown KE, Schleimer RP. Adherence of human basophils to cultured umbilical vein vascular endothelial cells. J Clin Invest 1988; 81:1355–1360.
11. Lamas AM, Mulroney CR, Schleimer RP. Studies on the adhesive interaction between human eosinophils and cultured vascular endothelial cells. J Immunol 1988; 140: 1500–1505.
12. Springer TA. Adhesion receptors of the immune system. Nature 1990; 346:425–434.
13. Bochner BS, Luscinskas FW, Gimbrone MA Jr., Newman W, Sterbinsky SA, Derse-Anthony C, Klunk D, Schleimer RP. Adhesion of human basophils, eosinophils, and neutrophils to IL-1 activated human vascular endothelial cells: contributions of endothelial cell adhesion molecules. J Exp Med 1991; 173:1553–1557.
14. Elices MJ, Osborn L, Takada Y, Crouse C, Luhowskyj S, Hemler ME, Lobb RR. VCAM-1 on activated endothelium interacts with the leukocyte integrin VLA-4 at a site distinct from the VLA-4/fibronectin binding site. Cell 1990; 60:577–584.
15. Walsh GM, Mermod J, Hartnell A, Kay AB, Wardlaw AJ. Human eosinophil, but not neutrophil, adherence to IL-1-stimulated human umbilical vascular endothelial cells is α4β1 (very late antigen-4) dependent. J Immunol 1991; 146:3419–3423.
16. Dobrina A, Menegazzi R, Carlos TM, Nardon E, Cramer R, Zacchi T, Harlan JM, Patriarca P. Mechanisms of eosinophil adherence to cultured vascular endothelial cells: eosinophils bind to the cytokine-induced endothelial ligand vascular cell adhesion molecule-1 via the very late activation antigen-4 integrin receptor. J Clin Invest 1991; 88:20–26.
17. Weller PF, Rand TH, Goelz SE, Chi-Rosso G, Lobb RR. Human eosinophil adherence to vascular endothelium mediated by binding to vascular cell adhesion molecule 1 and endothelial leukocyte adhesion molecule 1. Proc Natl Acad Sci USA 1991; 88: 7430–7433.
18. Schleimer RP, Sterbinsky SA, Kaiser J, Bickel CA, Klunk DA, Tomioka K, Newman W, Luscinskas FW, Gimbrone MA Jr., McIntyre BW, Bochner BS. Interleukin-4 induces adherence of human eosinophils and basophils but not neutrophils to endothelium: association with expression of VCAM-1. J Immunol 1992; 148:1086–1092.
19. Bochner BS, Newman W, McIntyre BW, Yamada T, MacGlashan DW Jr., Schleimer RP. Counter-receptors for endothelial cell adhesion molecules on human eosinophils and basophils. FASEB J 1992; 6:A1435 (Abstr.).
20. Thornhill MH, Kyan-Aung U, Haskard DO. IL-4 increases human endothelial cell adhesiveness for T cells but not for neutrophils. J Immunol 1990; 144:3060–3065.
21. Tepper RI, Pattengale PK, Leder P. Murine interleukin-4 displays potent anti-tumor activity in vivo. Cell 1989; 57:503–512.

22. Golumbek PT, Lazenby AJ, Levitsky HI, Jaffee LM, Karasuyama H, Baker M, Pardoll DM. Treatment of established renal cancer by tumor cells engineered to secrete interleukin-4. Science 1991; 254:713–716.
23. Issekutz TB. Inhibition of in vivo lymphocyte migration to inflammation and homing to lymphoid tissues by the TA-2 monoclonal antibody—a likely role for VLA-4 in vivo. J Immunol 1991; 147:4178–4184.
24. Huber AR, Kunkel SL, Todd RF III, Weiss SJ. Regulation of transendothelial neutrophil migration by endogenous interleukin-8. Science 1991; 254:99–102.
25. Ebisawa M, Bochner BS, Lichtenstein LM, Bickel C, Klunk D, Schleimer RP. Effect of cytokines on eosinophil transendothelial migration (abstr.). J Allergy Clin Immunol 1992; 89:296.
25a. Ebisawa M, Bochner BS, Georas SN, Schleimer RP. Eosinophil transendothelial migration induced by cytokines. I. Role of endothelial and eosinophil adhesion molecules in IL-1B-induced transendothelial migration. J Immunol 1992; 149:4021–4028.
26. Smith CW. Transendothelial migration. In: Harlan JM, Liu DY, eds. Adhesion: Its Role in Inflammatory Disease. New York: WH Freeman, 1992:83–115.
27. Anderson DC, Schmalstieg FC, Finegold MJ, Hughes BJ, Rothlein R, Miller LJ, Kohl S, Tosi MF, Jacobs RL, Waldrop TC, Goldman AS, Shearer WT, Springer TA. The severe and moderate phenotypes of heritable Mac-1, LFA-1, p150,95 deficiency: their quantitative definition and relation to leukocyte dysfunction and clinical features. J Infect Dis 1985; 152:668–689.
28. Georas SN, Sterbinsky SA, McIntyre BW, Ebisawa M, Schleimer RP, Bochner BS. Expression and function of β1 integrins on human eosinophils. FASEB J 1992; 6: A1402.
29. Anwar ARE, Walsh GM, Kay AB, Wardlaw AJ. Adhesion of human eosinophils to fibronectin. Am Rev Respir Dis 1992; 145:A667 (Abstr.).
30. Laffon A, Garciavicuna R, Humbria A, Postigo AA, Corbi AL, Delandazuri MO, Sanchez-Madrid F. Upregulated expression and function of VLA-4 fibronectin receptors on human activated T-cells in rheumatoid arthritis. J Clin Invest 1991; 88:546–552.
31. Schleimer RP, Benenati SV, Friedman B, Bochner BS. Do cytokines play a role in leukocyte recruitment and activation in the lungs? Am Rev Respir Dis 1991; 143: 1169–1174.
32. Sedgwick JB, Calhoun WJ, Busse WW. Functional and density comparison of blood and airway eosinophils following segmental antigen challenge. Am Rev Respir Dis 1991; 143:A42 (Abstr.).
33. Tomioka K, Bochner BS, Derse-Anthony C, Lichtenstein LM, Schleimer RP. GM-CSF enhances PAF- and FMLP-induced eosinophil expression of adherence molecules (CD11b) and increases FMLP-induced adherence to endothelial cells. J Allergy Clin Immunol 1991; 87:303 (Abstr.).
34. Bochner BS, Friedman B, Krishnaswami G, Schleimer RP, Lichtenstein LM, Kroegel C. Episodic eosinophilia-myalgia-like syndrome in a patient without L-tryptophan use: association with eosinophil activation and increased serum levels of granulocyte-macrophage colony-stimulating factor. J Allergy Clin Immunol 1991; 88:629–636.
35. Schleimer RP. Effects of glucocorticoids on inflammatory cells relevant to their therapeutic applications in asthma. Am Rev Respir Dis 1990; 141:S59–S69.

36. Lamas AM, Leon OG, Schleimer RP. Glucocorticoids inhibit eosinophil responses to granulocyte-macrophage colony-stimulating factor. J Immunol 1991; 147:254–259.
37. Wallen N, Kita H, Weiler D, Gleich GJ. Glucocorticoids inhibit cytokine-mediated eosinophil survival. J Immunol 1991; 147:3490–3495.
38. Massey W, Friedman B, Kato M, Cooper P, Kagey-Sobotka A, Lichtenstein LM, Schleimer R. Appearance of granulocyte-macrophage colony-stimulating factor activity at allergen-challenged cutaneous late-phase reaction sites. J Immunol 1993; 150:1084–1092.
38a. Massey W, Friedman B, Kato M, Lichtenstein LM, Liu M, Schleimer R. Allergen-induced late-phase reaction sites contain IL-3 and GM-CSF activity (Abstr). J Allergy Clin Immunol 1991; 87:207.
39. Broide DH, Firestein GS. Endobronchial allergen challenge in asthma—demonstration of cellular source of granulocyte macrophage colony-stimulating factor by in situ hybridization. J Clin Invest 1991; 88:1048–1053.
40. Ohnishi T, Kita H, Mayeno A, Sur S, Gleich GJ, Broide DH. Eosinophil-active cytokines and an inhibitor of cytokine activity in the bronchoalveolar lavage fluids (BALF) of symptomatic patients with asthma. J Allergy Clin Immunol 1992; 89:214 (Abstr.).
41. Owen WF Jr., Rothenberg ME, Silberstein DS, Gasson JC, Stevens RL, Austen KF. Regulation of human eosinophil viability, density, and function by granulocyte/macrophage colony-stimulating factor in the presence of 3T3 fibroblasts. J Exp Med 1987; 166:129–141.
42. Kyan-Aung U, Haskard DO, Poston RN, Thornhill MH, Lee TH. Endothelial leukocyte adhesion molecule-1 and intercellular adhesion molecule-1 mediate the adhesion of eosinophils to endothelial cells in vitro and are expressed by endothelium in allergic cutaneous inflammation in vivo. J Immunol 1991; 146:521–528.
43. Leung DYM, Pober JS, Cotran RS. Expression of endothelia-leukocyte adhesion molecule-1 in elicited late phase allergic reactions. J Clin Invest 1991; 87:1805–1809.
44. Benenati S, Bochner B, Horn T, Farmer E, Schleimer R. Endothelial-leukocyte adhesion molecule-1 (ELAM-1) expression following cutaneous allergen challenge (abstr.). J Allergy Clin Immunol 1991; 87:304.
45. Bochner BS, Charlesworth EN, Lichtenstein LM, Gillis S, Dinarello CA, Derse CP, Schleimer RP. Interleukin-1 is released at sites of human cutaneous allergic reactions. J Allergy Clin Immunol 1990; 86:830–839.
46. Walsh LJ, Trinchieri G, Waldorf HA, Whitaker D, Murphy GF. Human dermal mast cells contain and release tumor necrosis factor α, which induces endothelial leukocyte adhesion molecule 1. Proc Natl Acad Sci USA 1991; 88:4220–4224.
47. Kay AB, Ying S, Varney SR, Gaga M, Durham SR, Moqbel R, Wardlaw AJ, Hamid Q. Messenger RNA expression of the cytokine gene cluster, interleukin-3 (IL-3), IL-4, IL-5, and granulocyte/macrophage colony-stimulating factor, in allergen-induced late-phase cutaneous reactions in atopic subjects. J Exp Med 1991; 173: 775–778.
48. Thornhill MH, Haskard DO. IL-4 regulates endothelial cell activation by IL-1, tumor necrosis factor, or IFN-γ. J Immunol 1990; 145:865–872.
49. Masinovsky B, Urdal D, Gallatin WM. IL-4 acts synergistically with IL-1β to promote lymphocyte adhesion to microvascular endothelium by induction of vascular cell adhesion molecule-1. J Immunol 1990; 145:2886–2895.

DISCUSSION (Speakers: R. Schleimer and B. Bochner)

Konig: Did functional parameters change after adhesion events? For instance, did you look for LTC_4 release during adherence or subsequent transendothelial migration? Are the eosinophils more susceptible to subsequent stimulation?

Schleimer: This has been studied by Dr. Toshimitsu Yamada. We have not observed any induction or potentiation of LTC_4 release in eosinophils adherent to insolubilized VCAM-1 or E-selectin. We have not studied the function of transmigrated cells.

Roos: You showed that adherence of neutrophils and eosinophils to endothelial cells was partly inhibited by antibodies against L-selectin. However, when we strip L-selectin from neutrophils by incubation with fMLP, which induces shedding of L-selectin, this hardly affects adherence to endothelial cells. It only slows down the process slightly but it does not affect the extent of adherence. What might be the reason for the discrepancy with your results?

Bochner: Dr. Edward Knol, working in our laboratory, has adapted the method of Spertini et al. (1) to study adhesion of eosinophils under rotational conditions. The data I discussed, relating to the role of L-selectin under this in vitro "flow" condition, involve experiments performed at 4°C and at 60 rpm on a horizontal rotator platform (2). These conditions appear to dramatically reduce the contribution of integrins during adhesion and likely explain the differences you mentioned.

Coffman: Anti-IL-4 antibodies fail to inhibit eosinophil accumulation in granulomas induced by *Nippostrongylus* or *Schistosoma* in mice. What do you think this tells us about the in vivo role of IL-4-induced eosinophil adhesion?

Schleimer: This actually emphasizes our concept quite well. We believe that *either* eosinophil-activating (IL-5, IL-3, GM-CSF) *or* endothelial-activating cytokines (IL-1, TNF, but especially IL-4) can activate eosinophil accumulation. Thus, in the model you describe, we speculate that IL-5, IL-3, and/or GM-CSF may figure prominently. Ultimately, cell-specific accumulation probably occurs by the combined actions of relatively selective (but not specific) cytokines, adhesion molecules, and chemoattractants.

Busse: Are you sure the bronchoalveolar lavage (BAL) eosinophils had enhanced transendothelial migration, since they have already migrated? You showed that BAL eosinophils also had increased transendothelial migration. Since these cells had already undergone migration into the lungs, it is possible that they may be "desensitized" as a consequence of this response. In this regard, have you had a chance to determine whether eosinophils prepared by your in vitro migration method had altered migration activity?

Schleimer: Again, we have not done functional studies on in vitro transmigrated cells. There is no a priori reason to believe that priming ceases after an eosinophil transmigrates in vivo. However, certainly most of the other events in priming, i.e., increased survival, killing, mediator release, and so forth, seem likely to be most useful *after* the cell has migrated into a given tissue.

Spry: I have been interested for several years in the mechanisms that regulate eosinophil migration from the circulation following the observation (3) that T-cell lymphoma cells can cause the retention of eosinophils in the blood of rats for very long periods. In nonmalignant disease, the higher the blood eosinophil count, the longer their half-life in the blood. It would be interesting to examine these results again using the reagents you describe. Could T cells affect the expression of binding molecules on eosinophils or endothelial cells to *reduce* their mutual binding?

Bochner: Studies performed by Jutila et al. (4) have shown that when neutrophils that have shed L-selectin (either as a result of activation in vitro by chemoattractants or during recruitment in vivo) are then infused intravenously, they have lost the ability to migrate out of the circulation, despite enhanced CD11b expression. One possible explanation for your observation might therefore be loss of L-selectin from circulating eosinophils, but this is only speculation.

Vadas: Did you look at the effect of GM-CSF on movement across membranes without endothelium? Do you think the induction of IL-8R on eosinophils by GM-CSF is a possible mechanism?

Schleimer: We have not tested the response in the absence of an endothelial monolayer. Although it is possible that IL-8R induction plays a role, this is unlikely since: the response is quite rapid; it is not inhibited by an anti-IL-8 antibody; and we have not observed IL-8 to be a chemoattractant for eosinophils.

Denburg: Does the VLA-4/VCAM-1 interaction occur selectively in asthma?

Schleimer: Probably not. However, several recent studies indicate that VCAM-1 is expressed in vivo in humans after antigen challenge of skin or airways. Drs. Bentley, Kay, and Durham have recently shown what appears to be selective up-regulation of VCAM-1 24 hr after challenge in the lungs, and that the VCAM-1 expression correlates with tissue infiltration by eosinophils.

DISCUSSION REFERENCES

1. Spertini O, Luscinskas FW, Kansas GS, Munro JM, Griffin JD, Gimbrone MA, Jr, Tedder TF. Leukocyte adhesion molecule-1 (LAM-1, L-selectin) interacts with an inducible endothelial cell ligand to support leukocyte adhesion. J Immunol 1991; 147: 2565–2573.
2. Knol EF, Kansas GS, Tedder TF, Schleimer RP, Bochner BS. Human eosinophils use L-selectin to bind to endothelial cells under non-static conditions (Abstr). J Allergy Clin Immunol 1993; 91:334.
3. Spry CJ. Mechanism of eosinophilia. VII. Eosinophilia in rats with lymphoma. Br J Haematol 1972; 22:407–413.
4. Jutila MA, Rott L, Berg EL, Butcher E. Function and regulation of the neutrophil MEL-14 antigen in vivo: comparison with LFA-1 and Mac-1. J Immunol 1989; 143: 3318–3324.

7

The Role of Adhesion in Eosinophil Accumulation and Activation in Asthma

Andrew J. Wardlaw, Garry M. Walsh, A. R. E. Anwar, Adele Hartnell, Andrew M. Bentley, and A. Barry Kay
National Heart and Lung Institute, London, England

I. INTRODUCTION

A. The Role of the Eosinophil in Asthma

Eosinophils are associated with a number of diseases, most prominently asthma, allergic disease, and infection with helminthic parasites. The association between asthma and eosinophils has been well documented. In recent years the debate about the precise role of eosinophils in the pathogenesis of asthma has been brought into sharp focus by advances in our understanding of the pathology of asthma. This has been brought about primarily by use of the fiberoptic bronchoscope to obtain bronchoalveolar lavage cells and endobronchial biopsies from asthmatic airways (1,2). These studies have consistently shown an increase in the number of eosinophils in the airways of symptomatic asthmatics, with a broad association between the severity of asthma and the number of eosinophils (3). Although most studies have investigated young atopic asthmatics, similar findings have been demonstrated in intrinsic and occupational asthma (4,5). Eosinophils in the airways of asthmatics are activated as determined by positive staining with the anti-ECP monoclonal antibody EG2 and are found in association with activated, CD25-positive, CD4-positive T lymphocytes (6). Paralleling these studies, there has been an increased awareness that eosinophils can generate large amounts of mediators relevant to the pathological and physiological features of asthma (7).

Eosinophils can synthesize and release leukotriene C_4 (LTC_4) and platelet-activating factor (PAF), which can cause mucus hypersecretion, bronchoconstriction, and leukocyte migration. The eosinophil basic granule proteins have been shown to be cytotoxic to the bronchial epithelium in vitro and more recently, eosinophils have been shown to be cytokine-secreting cells (8,9). These observations have led to the persuasive hypothesis that eosinophil-derived mediators are responsible for the tissue damage characteristic of asthma.

Two important questions regarding the role of eosinophils in asthma are (1) the mechanism by which eosinophils selectively (i.e., without neutrophils) accumulate in the airways and (2) the mechanisms by which eosinophils are triggered to release their mediators. Inhibition of either of these steps may be an effective approach to the treatment of asthma as well as offering clues to the pathogenesis of this common and debilitating disease.

B. Potential Role of Adhesion in Eosinophil Accumulation

Intensive study has focused on the possibility that a specific eosinophil chemoattractant is responsible for the eosinophil accumulation in asthma. However, to date, although a number of eosinophil chemoattractants have been defined, including PAF, LTB_4, C5a, and the cytokines IL-5, IL-3, GM-CSF, and IL-8. Only IL-5 and IL-3 are not active on neutrophils. IL-5 and IL-3 are only moderately effective in vitro and then only on eosinophils from normal individuals (10).

There are at least two other possible mechanisms for selective eosinophil accumulation: (1) a specific eosinophil adhesion pathway and (2) prolonged survival under the influence of cytokines such as IL-3, IL-5, and GM-CSF. A third mechanism is in situ differentiation from eosinophil precursors resident in the airways. Although this is an attractive possibility, at the moment, there is limited evidence that this occurs.

Migration of eosinophils into the bronchial mucosa involves first adhesion and transmigration through the venular endothelium of the bronchial circulation, and second interaction with the extracellular matrix, which is composed of a number of proteins such as fibronectin, laminin, and collagen. Adhesion to endothelium involves an initial transient stage, which occurs under the high-sheer-stress conditions present under normal blood flow (Fig. 1). This initial adhesion is not firm and results in leukocytes rolling along the endothelial surface. The next stage is flattening of the leukocyte against the vessel wall followed by transmigration (11). These adhesion steps are mediated by a number of receptors, which fall into three main superfamilies, defined by structural similarities most clearly characterized at the gene level (12). These gene superfamilies are (1) selectins, (2) integrins, and (3) membranes of the immunoglobulin gene superfamily (Table 1).

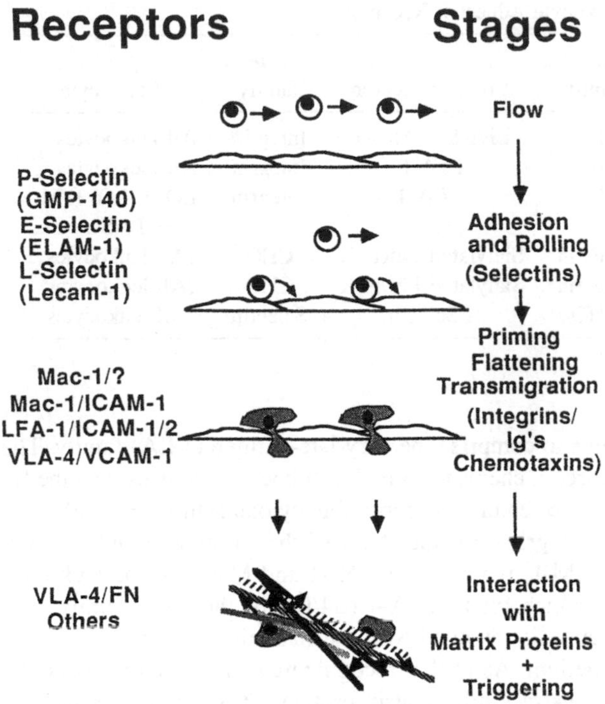

Figure 1 Schematic representation of the stages and adhesion receptors involved in eosinophil migration from the vascular space into tissue.

Selectins are transmembrane proteins characterized by an N-terminal C-lectin domain connected to a single epithelial growth factor (EGF)-like domain and a variable number of domains related to the complement-binding proteins, such as decay-accelerating factor (DAF). The three members of the selectin family so far described are E-selectin (ELAM-1), expressed on endothelium, P-selectin (GMP-140), expressed on endothelium and platelets, and L-selectin (LECAM-1), expressed on leukocytes. The counterreceptors or ligands for the selectins have yet to be fully defined but consist in part of a carbohydrate moiety containing the sialylated Lewis X motif (13). Selectins appear to be involved in the early, high-sheer-stress stage of leukocyte adhesion to endothelium. Integrins are α- and β-chain heterodimers, noncovalently bound on the cell surface (14). So far, eight β chains and 14 α chains have been characterized. Although there is a degree of promiscuity, three main subfamilies of integrins can be defined according to common β chain combining with a number of α chains. Thus the β1 chain

Table 1 Endothelial/Leukocyte Adhesion Receptors

Endothelium	Gene family	Counterreceptor	Gene family	Leukocyte
ICAM-1	Ig	LFA-1 + Mac-1	Integrin	All leukocytes
ICAM-2	Ig	LFA-1	Integrin	All leukocytes
VCAM-1	Ig	VLA-4	Integrin	EO + LO + MO + BASO
E-Selectin	Selectin	Sialylated Lewis X	CHO	?All leukocytes
P-Selectin	Selectin	Sialylated Lewis X	CHO	?All leukocytes
Sialylated Lewis X?	CHO	L-Selectin	Selectin	?All leukocytes

combines with six α chains to comprise the very-late-antigen (VLA) family. The β1 chain combines with three α chains to form the leukocyte integrins, and the β3 chain combines with two α chains to form the cytoadhesins (Fig. 2). The counterreceptors for the integrins are members of the immunoglobulin superfamily. These include ICAM-1, which binds LFA-1 and Mac-1 of the leukocyte integrins, and VCAM-1, which binds VLA-4 (α4/β1) of the VLA family. The integrin-Ig family interaction is involved with the flattening and transmigration step of adhesion to endothelium. As well as recognizing membrane receptors, the integrins are receptors for extracellular matrix proteins. There is a considerable degree of overlap between integrins and their matrix protein ligands. This is due, in part, to a common recognition sequence in the matrix proteins consisting of the oligopeptide arginine-glycine-aspartic acid (RGD). The stages involved in the migration of eosinophils into tissue and the adhesion receptors that are potentially involved are summarized in Figure 1.

II. EOSINOPHIL ADHESION RECEPTORS

A. Selectins

Indirect evidence suggests that eosinophils express the counterreceptors for E-selectin and P-selectin. Thus, monoclonal antibodies against E-selectin block adhesion to IL-1-stimulated human umbilical vein endothelial cells (HUVEC) (15) (which express E-selectin) and eosinophils form rosettes with COS cells transfected with the E-selectin cDNA. In addition, eosinophils can bind to purified P-selectin coated on plastic wells and rosette with COS cells transfected with a P-selectin cDNA (Chapter 4, this volume). Eosinophils also express L-selectin, although to a lesser extent than neutrophils. Eosinophils, therefore, appear to be able to interact with all three members of the selectin family and their counterreceptors in adhesion interactions.

Alpha Chain \ Beta Chain	β1	β2	β3	β4	β5	β6	β7	β8
α1	VLA-1 (LN,Coll)							
α2	VLA-2 (FN,LN,Coll)							
α3	VLA-3 (LN,Coll)							
α4	VLA-4 (FN,VCAM-1)						α4/β7 (FN, VCAM-1)	
α5	VLA-5 (FN)							
α6	VLA-6 (LN)			α6/β4 (LN?)				
α7	α7/β1 (LN)							
α8	α8/β1 (?)							
α_{IEL}							α_{IEL}/β7 (?)	
αL		LFA-1 (ICAM-1,2,3)						
αM		Mac-1 (ICAM-1,C3bI,FB)						
αX		p150,95 (FB, ? C3bI)						
αIIb			gpIIb/IIIa (FN,FB)					
αV	α_V/β1 (VN)		α_V/β3 (VN,FN)		α_V\β5 (VN)	α_V/β6 (FN)		α_V/β8 (?)

FN: Fibronectin. FB: Fibrinogen. VN: Vitronectin
LN: Laminin. Coll: Collagen

Figure 2 Representation of the integrin superfamily, showing heterodimer pairs and some of the ligands (in parentheses) for the various receptors.

B. Integrins

Eosinophils express a number of integrin receptors. LFA-1 and Mac-1 of the β2 leukocyte integrins are well expressed, although the third member of this family, p150-95, is less well expressed (16). Mac-1 appears to be the major receptor involved in eosinophil adhesion to unstimulated HUVEC. This adhesion interaction is up-regulated by IL-3 and IL-5 (17). These cytokines have no effect on neutrophil adhesion. Eosinophils, unlike neutrophils, express VLA-4. As a result,

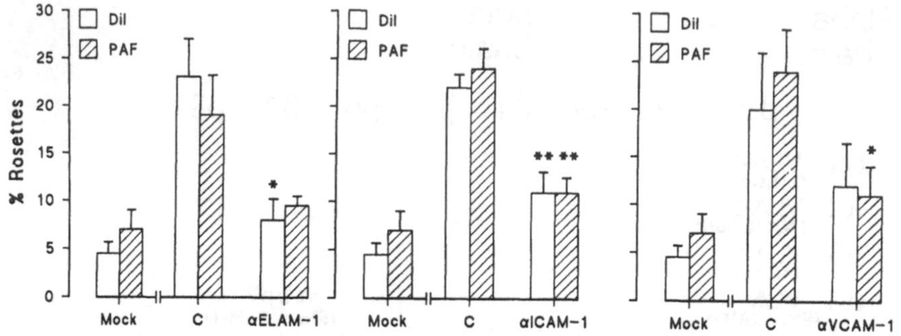

Figure 3 Eosinophils rosetting to COS cells transfected with E-selectin, ICAM-1, and VCAM-1, demonstrating comparable increases in rosetting with no effect of eosinophil activation with PAF and inhibition with specific antibodies.

they can bind VCAM-1, as shown by a number of groups including ourselves (18–21). Thus, eosinophil adhesion to IL-1-stimulated HUVEC was inhibited by both anti-VCAM-1 and anti-VLA4 monoclonal antibodies (MAb). Eosinophils specifically rosetted with VCAM-1-transfected COS cells. In addition, VCAM-1 was selectively induced on HUVEC in vitro by IL-4, and IL-4 up-regulated eosinophil, but not neutrophil, adhesion to HUVEC (22,23). As IL-4 appears to be generated in the airways in asthma, this seems an attractive mechanism by which eosinophils can be recruited into the airways, at least in atopic asthmatics. However, as detailed below, expression of VCAM-1 in the bronchial mucosa does not appear to be increased in asthma.

Peripheral blood eosinophils from subjects with a mild eosinophilia appear to express gpIIb/IIIa ($\alpha_{IIb}/\beta3$), but not $\alpha v\beta3$. The apparent expression of this platelet-specific receptor is due to platelets tightly adhering to eosinophils, presumably at the time of venesection. The receptors involved in this interaction are unclear, but P-selectin has been shown to be involved in platelet adherence to neutrophils and macrophages. This very strong adhesive reaction may be important in vivo in "carrying" platelets into sites of allergic inflammation.

C. In Vivo Expression of Adhesion Molecules in Asthma

Eosinophils have the potential to use a number of endothelial adhesion pathways. Of these, only VLA-4/VCAM-1 is not shared by neutrophils. In vitro expression of adhesion receptors on HUVEC is subtly modulated, with different inflammatory mediators and different kinetics giving distinct patterns of receptor expression. Thus, while unstimulated HUVEC express only low amounts of ICAM-1 and no E-selectin or VCAM-1, stimulation of HUVEC with cytokines such as IL-1 results

Table 2 ICAM-1 Expression on Bronchial Epithelium

	Chronic asthma		Allergen challenge	
Staining	Asthma*	Nonasthmatic	Allergen*	Diluent
Positive	13	2	7	2
Negative	5	7	6	11

*$p < 0.05$.

in the induction of E-selectin and VCAM-1 expression after about 4 h and a marked increase in ICAM-1 expression, which is optimal at 24 h. However, the extent to which these changes occur in vivo is not clear. We found that ICAM-1 and E-selectin were well expressed in both the upper and lower airways of normal individuals, with no evidence for increased expression in subjects with atopic asthma or allergic rhinitis, although there was an increase in both the intensity and extent of expression in intrinsic asthmatics. VCAM-1 was weakly expressed in normal subjects and both groups of asthmatics, with a tendency toward down-regulation of expression in atopic asthmatics. After allergen challenge, there was a trend for an increase in VCAM-1 expression, and expression of VCAM-1 after allergen challenge correlated with the number of eosinophils in the bronchial mucosa. A striking difference between asthmatic and normal individual subjects was a significant increase in the expression of ICAM-1 on the basal epithelium of asthmatic subjects both in steady-state asthma and after allergen challenge (Table 2). This could play a part in the vulnerability of the epithelium to damage in asthma. These results therefore suggest that the pattern of expression of adhesion receptors in vitro is of little relevance to their in vivo expression in the lung. Although in vivo expression suggests that ICAM-1 and VCAM-1 may be involved in eosinophil accumulation after allergen challenge, these results provide little evidence that modulation of known endothelial adhesion receptors is the mechanism for differential cell accumulation in asthma.

III. EOSINOPHIL FIBRONECTIN INTERACTIONS

A. Eosinophil Adhesion to Fibronectin

After migration through the endothelium, eosinophils come into contact with a number of matrix proteins including fibronectin. Fibronectin has two receptor recognition sites, an RGD site contained in a 120-kDa chymotrypsin fragment and an alternative site at the C-terminal end, which is contained in a 40-kDa heparin-binding fragment and is non-RGD-dependent (24). A number of integrin receptors appear to bind fibronectin through its RGD site, including the classical fibronectin receptor VLA-5 as well as VLA-3, $\alpha v\beta 3$, and gpIIb/IIIa. VLA-4 recognizes the non-RGD site that is defined by the peptide region termed CS-1 (Fig. 4). Eosinophils adhere to fibronectin-coated plates in a time- and dose-dependent fashion with an optimal concentration of 100 μg/ml and a time of 1 h. Adhesion was enhanced by stimulation of eosinophils with PAF and inhibited both by an antifibronectin antiserum and by MAbs against both the α and β chains of VLA-4. In contrast to mononuclear cells, RGDS peptide was ineffective at inhibiting eosinophil adhesion to fibronectin. Adherence to fibronectin was effectively inhibited by preincubation of eosinophils with the 40-kDa non-RGD-dependent fragment, but only weakly blocked by the 120-kDa fragment. Similarly, eosinophils bound strongly to the 40-kDa fragment coated on 96 well plates but only weakly to the 120-kDa fragment. These data further support the important role of VLA-4 as a major eosinophil fibronectin receptor.

B. Fibronectin Promotes Eosinophil Survival

Adhesion to fibronectin primed eosinophils for mediator release. Eosinophils adhering to fibronectin-coated wells for 1 h generated twice as much LTC_4 after stimulation with calcium inophore compared to the same number of eosinophils in BSA-coated wells. This enhanced mediator generation was inhibited by anti-VLA-4 MAb but not by an anti-CD45 MAb used as a control. Fibronectin had even more profound effects on eosinophils in long-term culture. Culture of eosinophils in fibronectin-coated wells resulted in prolonged survival compared to eosinophils cultured on BSA or plastic (Fig. 5). Survival was inhibited by anti-VLA-4 MAbs as well as by polyclonal antibodies against IL-3 and GM-CSF, but not antiserum against IL-5. Using commercially available ELISA kits, we also found that eosinophils cultured on fibronectin released substantial quantities of GM-CSF and IL-3. Cytokine release was inhibited by antifibronectin and anti-VLA-4 monoclonal antibodies.

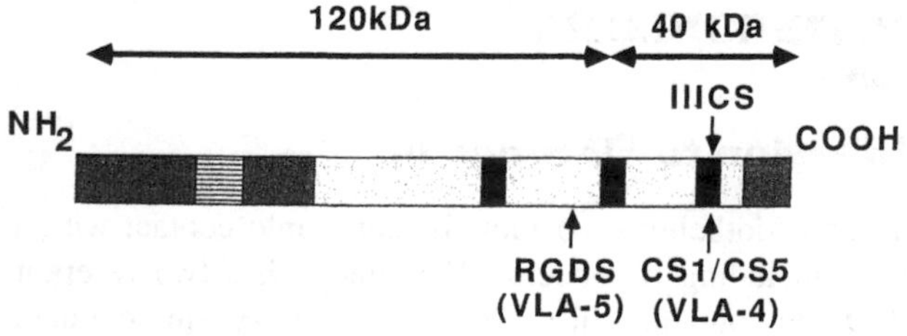

Figure 4 Schematic representation of binding sites for cell adhesion receptors on fibronectin.

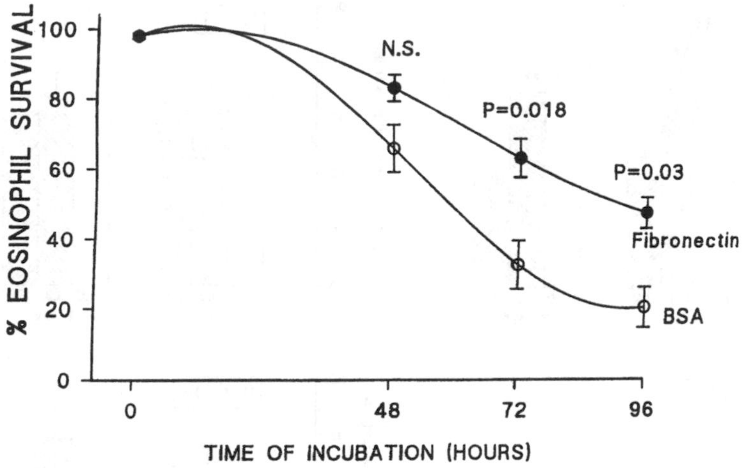

Figure 5 Survival of eosinophils, as determined by trypan blue exclusion, following culture on fibronectin and BSA.

IV. CYTOKINES AND EOSINOPHIL RECEPTORS

A. IL-3 and IFN-γ Have Differential Effects on Eosinophil Receptor Expression

In addition to prolonging eosinophil survival, IL-3, GM-CSF, and other cytokines have a number of effects on mature eosinophils. One of the most interesting is modulation of eosinophil receptor expression.

Culture of eosinophils with IL-3 for 24 h caused a marked increase in expression of Mac-1(CR3) in a dose-dependent fashion (25). This increased expression was inhibited by cycloheximide, suggesting it was dependent on protein synthesis rather than simply due to recruitment from intracellular stores. In addition, the effect of IL-3 on Mac-1 was inhibited by culturing the eosinophils in dexamethasone with an IC50 of 10^{-8} M. Different cytokines affect the expression of different receptors. Thus IL-3 increased or induced the expression of Mac-1, HLA-DR, and to a lesser extent ICAM-1, but had no effect on LFA-1 expression, whereas IFN-γ was effective in inducing expression of ICAM-1 and HLA-DR but had no effect on the expression of either Mac-1 or LFA-1 (Fig. 6).

INF-γ, but not IL-3, also induced Fcγ receptor expression on peripheral blood eosinophils. After culturing eosinophils for 1–2 days in IFN-γ, we observed both prolonged eosinophil survival and induction of expression of FcγRI (CD64) and FcγRIII (CD16) (26). Expression was induced in a time- and dose-dependent

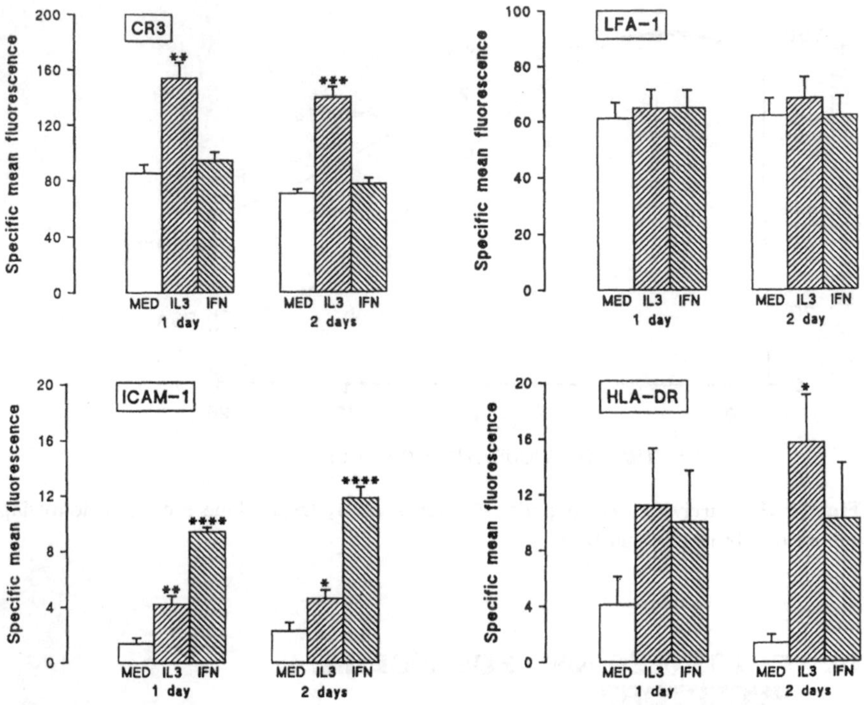

Figure 6 Specific mean fluorescence of Mac-1 (CR3), LFA-1, ICAM-1, and HLA-DR on peripheral blood eosinophils after 1–2 days of culture in medium, IL-3 (300 units/ml) or IFN-γ (500 units/ml), $n = 4$. $*p < 0.05$, $**p < 0.01$, $***p < 0.001$, $****p < 0.0001$ compared with the medium control at each time point (unpaired *t*-test).

manner. Expression of CD16 was protein synthesis dependent and was associated with increased expression of mRNA for CD16. Expression was inhibited by dexamethasone. CD16 on eosinophils was functional as perturbation with an anti-CD16 monoclonal antibody resulted in membrane depolarization and LTC_4 release.

B. Eosinophils Express CD69

Modulation of eosinophil phenotype by cytokines suggests that the profile of leukocyte receptor expression on tissue eosinophils may be very different from that on peripheral blood eosinophils. The receptors involved in triggering eosinophil mediator release at sites of allergic inflammation are still unclear. One possibility is that a receptor whose expression was induced as the eosinophils migrated into the extravascular tissue may be involved. CD69 is a receptor that is

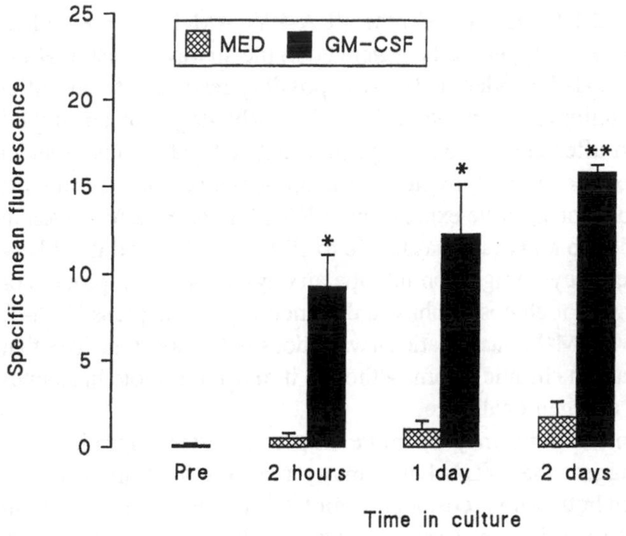

Figure 7 Time course of expression of CD69 by eosinophils cultured with GM-CSF (10 nM). Expression was determined by indirect immunofluorescence using a specific anti-CD69 MAb (Leu 23, Becton Dickinson) and flow cytometry.

induced on T lymphocytes within 2–3 h of activation (27). Perturbation of CD69 with monoclonal antibodies in concert with PMA induced peripheral blood T cells to proliferate. In addition, CD69 is constitutively expressed on platelets, where it acts as a receptor for platelet aggregation and degranulation (28). We found that although peripheral blood eosinophils did not express CD69, within 1–2 h of culture with GM-CSF, CD69 expression on eosinophils was detected. Expression was maximal by 1 day and sustained for at least 2 days (Fig. 7). Expression was dose dependent and induced by GM-CSF but not by PAF. Expression was not inhibited by dexamethasone but was partly inhibited by cyclohexamide. Neutrophils did not express CD69 after culture with GM-CSF. Importantly, BAL eosinophils from asthmatics expressed similar amounts of CD69 to in vitro GM-CSF-stimulated peripheral blood eosinophils, confirming the in vivo relevance of this observation.

V. CONCLUSIONS

We have investigated the role of adhesion in eosinophil migration and activation particularly in relation to the hypothesis that eosinophil-derived mediators cause asthma. Two adhesion pathways have been defined that are available to eosinophils but not neutrophils: (1) IL-5- and IL-3-mediated enhancement of eosinophil adhesion to unstimulated HUVEC and (2) the VLA-4/VCAM-1-mediated adhesion. Although IL-5 and IL-3 appear to be generated in the airways of asthmatics, the relevance of this Mac-1-dependent pathway is possibly restricted to an initial insult to a previously uninflamed tissue such as the early stages of eosinophil migration into the skin after antigen challenge in sensitized individuals. Our in vivo studies of adhesion receptor expression in the bronchial submucosa suggest that there is considerable constitutive expression of ICAM-1 and ELAM-1 even in normal individuals. These pathways are available to all leukocytes, and it is likely that there is constant leukocyte migration into the airways as part of a process of immune surveillance. Although eosinophils, unlike neutrophils, express VLA-4, the low expression of VCAM-1 in asthmatic airways does not support the idea that this pathway is important in chronic asthma although it may have a role in eosinophil recruitment after allergen challenge.

VLA-4-mediated interactions may be more important after migration into tissue. Our studies suggest that VLA-4 is a major receptor for fibronectin on eosinophils. Fibronectin both primes eosinophils for LTC_4 release in the short term and prolongs eosinophil survival over several days in culture through autocrine generation of IL-3 and GM-CSF. Fibronectin-induced eosinophil survival could therefore play an important role in eosinophil accumulation in the bronchial mucosa in asthma.

In addition to prolonging eosinophil survival, IL-3 and GM-CSF modulate eosinophil receptor expression. The receptors involved in triggering eosinophil degranulation in vivo are not clearly defined. One of the curious observations regarding in vitro Fc-triggered degranulation is the very slow kinetics, with optimal release occurring only after several hours (29). In contrast, degranulation triggered through direct protein G stimulation occurs within minutes (30). This phenomenon could be explained by the requirement for a coreceptor for degranulation whose expression was induced by cell activation. We have made the interesting observation that eosinophils express the early activation antigen CD69, which in T cells and platelets is involved in activation and mediator release. CD69 was expressed both in vitro after stimulation with GM-CSF and in vivo on BAL eosinophils from asthmatics. It remains to be determined whether CD69 is involved in eosinophil degranulation.

VI. SUMMARY

Two possible mechanisms whereby eosinophils preferentially accumulate in tissue in diseases such as asthma and allergic disease are (1) eosinophil-specific adhesion pathways and (2) prolonged survival under the influence of cytokines such as IL-3, GM-CSF, and IL-5. We have examined mechanisms of adhesion in vivo and in vitro in eosinophil migration and activation in relation to the pathogenesis of asthma. Eosinophil adhesion in vitro to unstimulated HUVEC is selectively enhanced by IL-5 and IL-3. This pathway appears to be Mac-1 dependent. At sites of chronic inflammation, an array of adhesion molecules are likely to be induced on venular endothelium. Eosinophils, like other leukocytes, can potentially utilize all three selectin adhesion receptors as well as the immunoglobulin family member ICAM-1. In addition, unlike neutrophils, eosinophils express VLA-4 and can use the VLA-4/VCAM-1 pathway. However, although in vitro IL-4 selectively up-regulates VCAM-1 on HUVEC, VCAM-1 expression was not increased in the airways of symptomatic extrinsic asthmatics compared with normal subjects.

We have found the VLA-4 on eosinophils is a major receptor for fibronectin. Adhesion to fibronectin primed eosinophils for increased LTC_4 release and prolonged eosinophil survival by inducing autocrine generation of IL-3 and GM-CSF by eosinophils.

In addition, IL-3, GM-CSF, IL-5 as well as IFN-γ can modulate receptor expression, inducing membrane expression of Fc receptors and activation and adhesion antigens such as HLA-DR and ICAM-1. We have found that GM-CSF induced expression of the early activation antigen CD69 in vitro and that this receptor was expressed in vivo on BAL eosinophils from asthmatics. This receptor may play a role in triggering eosinophil degranulation in the airways in asthma.

ACKNOWLEDGMENTS

We thank Lindsay Needham, British Biotechnology, Oxford, and Paul Hellewell, Department of Applied Pharmacology, NHLI, London, for their collaboration with the COS cell transfectants. We thank Dr. Redwan Moqbel and Dr. Oliver Cromwell for their advice and assistance with the fibronectin/eosinophil adhesion interactions and Dr. Peter Jeffery for the high resolution scanning electron micrographs demonstrating platelet adhesion to eosinophils.

REFERENCES

1. Gleich GJ. The eosinophil and bronchial asthma: current understanding. J Allergy Clin Immunol 1990; 85:422–436.
2. Wardlaw AJ, Kay AB. The role of the eosinophil in the pathogenesis of asthma. Allergy 1987; 42:321–335.
3. Bousquet J, Chanez P, Lacoste JY, Barneon G, Gavanian N, Enander I, Venge P, Ahlstedt S, Simony-Lafontaine J, Godard P, Michel F-B. Eosinophilic inflammation in asthma. N Engl J Med 1990; 323:1033–1039.
4. Bentley AM, Menz G, Storz C, Robinson DS, Bradley B, Jeffery PK, Durham SR, Kay AB. Identification of T lymphocytes, macrophages and activated eosinophils in the bronchial mucosa in intrinsic asthma: relationship to symptoms and bronchial responsiveness. Am Rev Respir Dis. In press.

5. Bentley AM, Maestrelli P, Saetta M, Fabbri LM, Robinson DS, Bradley B, Jeffery PK, Durham SR, Kay AB. Activated T lymphocytes and eosinophils in the bronchial mucosa in isocyanate induced asthma. J Allergy Clin Immunol. In press.
6. Azzawi M, Bradley B, Jeffery PK, Frew AJ, Wardlaw AJ, Knowles G, Assoufi B, Collins JV, Durham S, Kay AB. Identification of activated T lymphocytes and eosinophils in bronchial biopsies in stable atopic asthma. Am Rev Respir Dis 1990; 142:1407–1413.
7. Weller PF. The immunobiology of eosinophils. N Engl J Med 1991; 324:1110–1118.
8. Moqbel R, Hamid Q, Sun Ying, Barkans J, Hartnell A, Tsicopoulos A, Wardlaw AJ, Kay AB. Expression of mRNA and immunoreactivity for the granulocyte/macrophage colony-stimulating factor in activated human eosinophils. J Exp Med 1991; 174: 749–752.
9. Rita H, Ohnishi T, Oqubo DY, Weiler D, Abrams JS, Gleich GJ. GM-CSF and IL-3 release from peripheral blood eosinophils and neutrophils. J Exp Med 1991; 174:45–49.
10. Sehmi R, Wardlaw AJ, Cromwell O, Kurihara K, Waltmann P, Kay AB. Interleukin-5 selectively enhances the chemotactic response of eosinophils obtained from normal but not eosinophilic subjects. Blood 1992; 79:2952–2959.
11. Hynes RO, Lander AD. Contact and adhesive specificities in the associations, migrations and targeting of cells and axons. Cell 1992; 68:303–322.
12. Springer TA. Adhesion receptors of the immune system. Nature 1990; 346:425–434.
13. Springer TA, Lasky LA. Sticky sugars for selectins. Nature 1991; 349:196–197.
14. Hynes RO. Integrins: versatility, modulation and signaling in cell adhesion. Cell 1992; 69:11–25.
15. Kyan-Aung U, Haskard DO, Poston RN, Thornhill MH, Lee TH. Endothelial leukocyte adhesion molecule-1 and intercellular adhesion molecule-1 mediated the adhesion of eosinophils to endothelial cells in vitro and are expressed by endothelium in allergic cutaneous inflammation in vivo. J Immunol 1991; 146:521–528.
16. Hartnell A, Moqbel R, Walsh GM, Bradley B, Kay AB. Fc gamma and CD11/18 receptor expression on normal density and low density human eosinophils. Immunology 1990; 69:264–270.
17. Walsh GM, Hartnell A, Wardlaw AJ, Kurihara K, Sanderson CJ, Kay AB. IL-5 enhances the in vitro adhesion of human eosinophils, but not neutrophils, in a leukocyte integrin (CD11/18)-dependent manner. Immunology 1990; 71:258–265.
18. Walsh GM, Hartnell A, Mermod J-J, Kay AB, Wardlaw AJ. Human eosinophil, but not neutrophil adherence to IL-1 stimulated HUVEC is a4b1 (VLA-4) dependent. J Immunol 1991; 146:3419–3423.
19. Bochner BS, Luscinskaas FW, Gimbrone MA, Newmann W, Sterbinsky SA, Derse-Anthony C, Klunk D, Schleimer RP. Adhesion of human basophils and neutrophils to IL-1 activated human vascular endothelial cells: contribution of endothelial cell adhesion molecules. J Exp Mod 1991; 173:1553.
20. Dobrina A, Menegazzi R, Carlos TM, Nardon E, Cramer R, Zacchi T, Harlan JM, Patriarca P. Mechanisms of eosinophil adherence to cultured vascular endothelial cells: eosinophils bind to the cytokine-induced endothelial ligand vascular cell adhesion molecule-1 via the very late antigen-4 integrin receptor. J Clin Invest 1991; 88:20.

21. Weller PF, Rand TH, Goelz SE, Chi-Rosso G, Lobb RR. Human eosinophil adherence to vascular endothelium mediated by binding to vascular cell adhesion molecule-1 and endothelial leukocyte adhesion molecule 1. Proc Natl Acad Sci USA 1991; 88:7430.
22. Thornhill MH, Kyan-Aung U, Haskard DO. IL-4 increases human endothelial cell adhesiveness for T cells but not neutrophils. 1990; 144:3060.
23. Schleimer RP, Sterbinsky SA, Kaiser J, Bickel CA, Klunk DA, Tomioka K, Newman W, Luscinskas FW, Gimbrone MA, McIntyre BW, Bochner B. IL-4 induces adherence of human eosinophils and basophils but not neutrophils to endothelium. J Immunol 1992; 148:1086–1092.
24. Humphries MJ, Akiyama SK, Komoriya A, Olden K, Yamada KM. Identification of an alternatively spliced site in human plasma fibronectin that mediates cell type-specific adhesion. J Cell Biol 1986; 103:2637–2647.
25. Hartnell A, Kay AB, Wardlaw AJ. IL-3 induced upregulation of CR3 expression on human eosinophils is inhibited by dexamethasone. Immunology. In press.
26. Hartnell A, Kay AB, Wardlaw AJ. IFN-g induces expression of FcgRIII (CD16) on human eosinophils. J Immunol 1991; 148:1471–1478.
27. Hara T, Jung LKL, Bjorndahl JM, Shu Man Fu. Human T cell activation III. Rapid induction of a phosphorylated 28/32 kDa di-sulphide-linked early activation antigen (EA-1) by 12-O-tetradecanoyl phorbol-13-acetate, mitogens and antigens. J Exp Med 1986; 164:1988–2005.
28. Testi R, Pulcinella F, Frati L, Gazzaniga PP, Santoni A. CD69 is expressed on platelets and mediates platelet activation and aggregation. J Exp Med 1990; 172:701–707.
29. Fujisawa T, Abu-Ghazaleh R, Kita H, Sanderson CJ, Gleich GJ. Regulatory effects of cytokines on eosinophil degranulation. J Immunol 1990; 144:642–646.
30. Nusse O, Lindau M, Cromwell O, Kay AB, Gomperts BD. Intracellular application of guanosine-5′-O-(3-thiotriphosphate) induces exocytic granule fusion in guinea pig eosinophils. J Exp Med 1990; 171:775–786.

DISCUSSION (Speaker: A. J. Wardlaw)

Gleich: Dr. Gundel has told me that MBP stimulates expression of ICAM-1 on respiratory epithelial cells, an observation of potential importance in our understanding of the role of adhesion molecules in asthma.

Wardlaw: I agree. ICAM-1 expression by bronchial epithelium in asthmatics could explain the exacerbating effect of upper respiratory tract viral infections in asthma by allowing the bronchial epithelium to become infected with rhinovirus (ICAM-1 is a receptor for rhinovirus). In addition, the eosinophil-mediated epithelial damage characteristic of asthma may require eosinophil adhesion to bronchial epithelium through receptors such as ICAM-1.

Galli: Have you looked at the effect of adherence to laminin on eosinophil survival and cytokine generation?

Wardlaw: No.

Ackerman: Do peripheral blood eosinophils, either normodense or hypodense, express any detectable CD16 (patient vs. normal eosinophils)? Does the negative selection technique potentially deplete CD16 positive eosinophils in the process of depleting neutrophils?

Wardlaw: Unstimulated peripheral blood eosinophils, whether from normal or eosinophilic individuals, do not express CD16.

Ackerman: Do tissue eosinophils express CD16?

Wardlaw: BAL eosinophils from asthmatics do not express CD16. We have not looked at CD16 expression on tissue eosinophils in diseases where there is evidence of IFN-γ production such as Crohn's disease.

Lee: In collaboration with Stephen Holgate, Peter Howarth, and Dorian Haskard, we have also shown that there is increased expression of ICAM and ELAM in bronchial biopsies from both asthmatics and normal subjects. Furthermore, nasal mucosa of patients with seasonal rhinitis demonstrates increased expression of ICAM, ELAM, and VCAM during the pollen season. There may be a difference between enhanced expression caused by allergen challenge and the chronic situation where patients with chronic asthma seem to be similar to normal individuals in adhesion molecule expression.

Wardlaw: I agree. We have found that there is a trend toward increased VCAM-1 expression in bronchial biopsies after allergen challenge and eosinophil counts in the bronchial mucosa after allergen correlated with expression of ICAM-1 and VCAM-1. In the nose we found marked constitutive expression of ICAM-1, E-selectin, and, to a lesser extent, VCAM-1 in normals, with no difference between normals and seasonal rhinitics in season. We have not looked at expression in perennial rhinitis.

Schleimer: I think it is premature to dismiss VCAM-1 from having a role in asthma or other allergic diseases. Although VCAM-1 was not elevated in the asthmatics that you studied, we can speculate that it has been expressed by virtue of the generalized endothelial activation that appears to have occurred. Furthermore, recent studies by Drs. Bentley, Kay, and Durham have clearly shown antigen-induced VCAM-1 expression in the bronchial mucosa in correlation with eosinophil infiltration.

Wardlaw: I agree. It is possible that the initiating event in the development of asthma involves eosinophil migration via VCAM-1 and that eosinophils then remain in the tissue through mechanisms such as prolonged survival. Nevertheless, it is at variance with the in vitro data that we found such weak expression of VCAM-1 in atopic asthmatics where we know IL-4 is being generated.

Silberstein: Respiratory epithelial cells secrete copious amounts of the mucin known as episialin or DF3 antigen. This molecule is large (300–400 kDa) with an extended and rigid structure, such that it would extend beyond adhesion molecules. It is known that episialin has antiadhesion properties and, for example, will inhibit eosinophil adhesion to schistosomula. On the epithelial cells, this mucin is restricted to the apical surfaces. In your biopsies of the asthmatic patients, could you see any adhesion of eosinophils to the apical surfaces? What was the distribution of the ICAM on the epithelial cells?

Wardlaw: In our biopsies the epithelium was generally sloughed, leaving only the basal layer where ICAM-1 expression could be clearly seen. Eosinophils are approaching the epithelium from the basal surface, and their effects may be due to detachment of the epithelium from its basal layer under the influence of eosinophil mediators. Attachment of eosinophils to the apical surface may therefore not be required for eosinophil-mediated damage.

Sanderson: On the question of the specific localization of eosinophils and the mechanisms controlling this; some years ago we made films of eosinophils and neutrophils phagocytosing antibody-coated red blood cells. Eosinophils move differently from neutrophils. For example, eosinophils appear to have a very active membrane. They phagocytose and regurgitate particles very rapidly, and move rapidly. As shown by Butterworth, eosinophils are uniquely adapted to flatten out onto large parasites. Thus it seems possible that eosinophils have a different cytoskeletal anatomy than neutrophils. My point is that we may need to look further than just the adhesion molecules to explain specific localization of eosinophils. In addition to differences in adhesion molecules that eosinophils utilize, there may be differences in the way eosinophils use the information they receive from their surface molecules.

Wardlaw: A very interesting comment.

Butterworth: Is the induction of FcRI and FcRIII mediated only by IFN-γ, or do you get the same effect with other cytokines? Are the effects of IFN-γ qualitatively different, or simply slower?

Wardlaw: Of the limited range of cytokines we have tested, only IFN-γ appears to affect Fc-γ receptor expression. There do appear to be both qualitative and quantitative differences in the effects of cytokines such as IFN-γ versus IL-3, IL-5, and GM-CSF in their ability to induce expression of membrane receptors.

Repp: We found that ADCC of eosinophils after 24 h of incubation with IFN-γ is still dependent on Fc-γ RII. Do you have direct evidence that Fc-γ RI or Fc-γ RIII induced by IFN-γ is involved in ADCC after 24 h of incubation with IFN-γ?

Wardlaw: We did not examine the function of Fc-γ RIII or Fc-γ RI in an ADCC assay. We demonstrated that perturbation of Fc-γ RIII by anti-CD16 monoclonal antibodies resulted in the generation of LTC_4 and membrane depolarization, showing that it was functional on eosinophils.

Vadas: I would like to comment on the concept that ICAM induction is an inevitable association of induction of other adhesion molecules. Endothelial cells exist in association with smooth muscle cells or pinocytes, and the combination of these cells produces *active* TGFβ. TGF-β has no effect on ICAM expression but suppresses that of E-selectin.

Wardlaw: The down-regulation of adhesion molecules by cytokines such as TGF-β is very interesting. Such a mechanism may explain the low expression of VCAM-1 in our atopic asthmatics.

Schleimer: The increase in survival due to fibronectin binding was completely inhibited by α-IL-3 and half inhibited by α-GM-CSF. How can this be explained?

Wardlaw: I agree this was odd and I have no satisfactory explanation. We have been able to detect both IL-3 and GM-CSF in supernatants from eosinophils adhering to fibronectin, and it is unexpected that αIL-3 should therefore completely abrogate survival. We are currently undertaking in situ hybridization to help confirm fibronectin-stimulated eosinophil autocrine production.

Bochner: We have been unable to find a preparation of fibronectin that contains significant amounts of the CS-1 domain. I therefore was curious as to the data presented in which anti-VLA-4 monoclonal antibody was as effective as anti-CR3 MAb in inhibiting adhesion to fibronectin. Can you comment on this?

Wardlaw: CR3 is not regarded as a receptor for fibronectin, although the literature is inconclusive. It is possible that αCR3 inhibition of eosinophil binding is due to CR3 being a receptor for fibronectin. An alternative explanation is that eosinophils are aggregating and therefore CR3 is having an indirect role, although there is little evidence of aggregation on direct visual inspection.

Spry: What happens to VLA-4 expression in patients treated with cytokines?

Wardlaw: VLA-4 expression appears to be unaffected by both in vivo and in vitro activation by cytokines or PAF.

Venge: Did you investigate whether IL-5 might have an autocrine effect in adhesion by eosinophils to fibronectin?

Wardlaw: An antibody against IL-5 did not inhibit fibronectin-induced eosinophil survival.

8

Eosinophil-Active Cytokines in Human Disease: Development and Use of Monoclonal Antibodies to IL-3, IL-5, and GM-CSF

John S. Abrams and Jon E. Silver
DNAX Research Institute of Molecular and Cellular Biology, Palo Alto, California*

Robert E. Van Dyke
Arris Pharmaceutical Corporation, South San Francisco, California

Gerald J. Gleich
Mayo Clinic and Mayo Foundation, Rochester, Minnesota

I. INTRODUCTION

There is a growing body of evidence that the colony-stimulating factors GM-CSF, IL-3, and IL-5 each, or in combination, can have an effect on eosinophil proliferation and differentiation from early lineage committed precursor cells (reviewed in 1). In addition, these factors can interact with the mature, end-stage differentiated eosinophil to prolong its survival in vitro and to regulate several of its effector functions (reviewed in 2). Within this factor triad, IL-5 is considered to be the most late-acting in the lineage differentiation sequence, and this cytokine's spectrum of biological activities appears to be more narrowly confined (at least in the human) to the eosinophil and basophil lineages (3–5). In comparison, GM-CSF and IL-3 stimulate the development of a broader spectrum of leukocyte cell types.

The ability to detect and quantify these eosinophil-active cytokines in clinical and biological samples is facilitated by the use of specific and sensitive immuno-

*The DNAX Research Institute is entirely supported by the Schering-Plough Corporation.

assays. Cytokine detection with antibody-based techniques is particularly important because this series of eosinophil-active factors demonstrate overlapping or parallel bioactivities in many in vitro systems, and it is extremely difficult to distinguish and measure them in complex mixtures on the basis of biological activity alone. Furthermore, additional factors such as IL-2 (6), IFN-γ (7–15), TGF-β (10,11,16), PDGF (17), or glucocorticoids (12), which may also be present in samples containing IL-3, IL-5, or GM-CSF, have been reported to modulate the eosinophil-active effects of these colony-stimulating factors. Therefore, both neutralizing antibodies in bioassay systems and cytokine immunoassays of complex samples are required.

This report describes the preparation and properties of a series of rat antihuman monoclonal antibodies (MAb) against each of the three eosinophil-active cytokines: IL-3, IL-5, and GM-CSF. Pairs of MAbs recognizing spatially distinct epitopes on each of these molecules have been identified, and sensitive immunoenzymetric assays have been configured for each of these cytokines. Each of the antibodies in the immunoenzymetric pair is capable of neutralizing the colony-stimulating factor biological activity. This panel of immunoassays has been used to investigate the presence of eosinophil-active cytokines in various clinical samples obtained from patients with diverse eosinophilias. Results from these studies are summarized in the second part of this chapter.

II. DEVELOPMENT OF NEUTRALIZING ANTIBODIES TO EOSINOPHIL-ACTIVE CYTOKINES

A. Anti-GM-CSF

1. *Anti-GM-CSF Monoclonal Antibodies*

Rat antihuman GM-CSF monoclonal antibodies were prepared from splenocytes of male Lewis rats immunized with *Escherichia coli*–expressed recombinant GM-CSF. The rat cells were fused with the nonsecreting mouse myeloma fusion partner P3X63-AG8.653 to yield heterohybridomas secreting rat immunoglobulin. The animals were immunized as follows: day 0, 50 μg rhGM-CSF (provided by Schering-Research, Bloomfield, NJ) IP in complete Freund's adjuvant (CFA); day 19, 50 μg rhGM-CSF IP in incomplete Freund's adjuvant (IFA); day 158, combined immunizations with 50 μg IP + 50 μg IV of rhGM-CSF in PBS; day 162 splenocyte fusion. A testbleed was obtained at day 29.

A major goal was the selection of several anti-hGM-CSF antibodies that neutralized and were capable of recognizing spatially distinct GM-CSF epitopes for immunoenzymetric assay development of both natural and recombinant material. Two different initial immunochemical screening assays were employed in parallel. Day 10 postfusion hybridoma supernatants were tested by indirect

ELISA on *E. coli*–expressed rhGM-CSF-coated microtiter plates. Supernatants were also tested in an immunoenzymetric assay format employing plates coated with protein A–purified rabbit anti-GM-CSF IgG as the capture antibody and a mouse antirat Ig immunoperoxidase conjugate to detect ligand-bound hybridoma supernatant rat Ig anti-GM-CSF.

A total of 116/3520 hybridoma supernatants were identified. These were further screened in parallel biological activity and immunoprecipitation assays. The series of 116 hybridoma supernatants were examined for their ability to inhibit GM-CSF-induced increased proliferation of the KG-1 myelomonocytic cell line. Relative functional blocking avidity was ascertained by serial titration analysis. This series was also tested in parallel for their ability to immunoprecipitate ^{125}I-radiolabeled recombinant GM-CSF. A subset consisting of eight of the more potent neutralizing hybridoma supernatants (titers of 1:1000 or greater), were retested by serial titration immunoprecipitation analysis. Also, a selected subset of 13 hybridoma supernatants (obtained from expanded cultures grown in 48 well plates) were tested for their ability to inhibit GM-CSF-induced colony formation of human cord blood precursors after 14 days in methyl cellulose culture. Three of the hybridoma supernatants were identified as containing potent anti-GM-CSF antibodies that were active in all five assay systems, and these were subsequently subcloned by limiting dilution.

The three hybridomas identified were BVD2-5A2 (a rat IgG1), BVD2-21C11 (a rat IgG2a), and BVD2-23B6 (a rat IgG2a). An example of their relative blocking avidities in the KG-1C (hGM-CSF induced) proliferation bioassay is shown in Figure 1a. Each of these antibodies is a potent antagonist of GM-CSF in this assay; the 23B6 antibody appeared to have a 5- to 10-fold greater neutralizing potency. For this antibody, an antibody:antigen molar ratio of approximately 2:1 provided half-maximal inhibition of GM-CSF in vitro bioactivity. Also, it has been determined that Fab fragments of each of these antibodies blocked GM-CSF function with roughly the same avidity as the parent IgG.

2. *GM-CSF Immunoenzymetric Assay*

The three rat monoclonal anti-hGM-CSF IgG's were purified and derivatized with the hapten, nitroiodophenyl (NIP) acetate (18). They were tested pairwise against each other for ability to detect hGM-CSF in an immunoenzymetric assay format utilizing the J4 anti-NIP immunoperoxidase conjugate (18) as part of the detecting system. It was observed that the 23B6 MAb bound to a determinant that was spatially distinct from that recognized by the 5A2 or 21C11 MAbs; these latter two antibodies bound to overlapping determinants. The most sensitive immunoenzymetric assay format was provided by the 23B6 IgG as coating MAb and the 21C11-NIP IgG as the detecting MAb. A representative standard curve is shown in Figure 1b. This immunoassay format has been routinely used for the quantification of GM-CSF in patient samples.

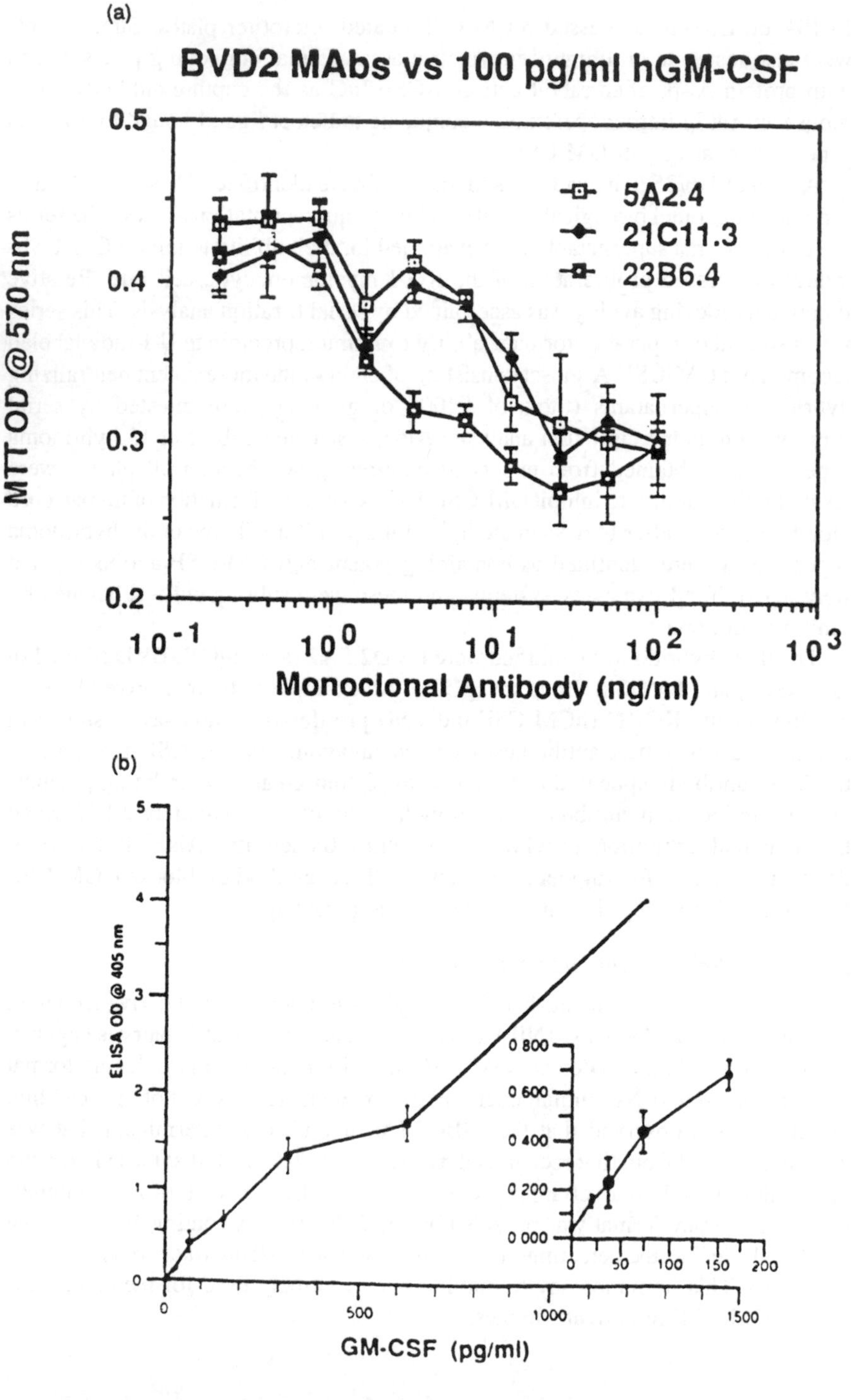
(a)
BVD2 MAbs vs 100 pg/ml hGM-CSF
MTT OD @ 570 nm
0.5
0.4
0.3
0.2
5A2.4
21C11.3
23B6.4
10^-1
10^0
10^1
10^2
10^3
Monoclonal Antibody (ng/ml)
(b)
ELISA OD @ 405 nm
0
1
2
3
4
5
0
500
1000
1500
GM-CSF (pg/ml)
0 000
0 200
0 400
0 600
0 800
0
50
100
150
200

B. Anti-IL-5

1. Anti-IL-5 Monoclonal Antibodies

A pair of rat anti-hIL-5 MAbs that neutralize the biological activity of the factor as well as provide a sensitive immunoenzymetric assay have been produced. These were part of a large hybridoma library of anti-hIL-5 heterohybridomas produced from the splenocytes of a rat immunized with partially purified, mammalian (COS7)-expressed hIL-5. The rat was immunized with crude material emulsified in FCA, IP on day 0, and with the same material emulsified in IFA on days 14 and 47. On days 97 and 196, 75 μg of semipurified hIL-5 in IFA, IP was administered, and on day 351 (4 days prior to fusion), the animal was immunized with a combination of 75 μg IV and 25 μg IP. Test bleeds were obtained at days 26, 57, 106, and 204 during the course of the immunization to monitor serum anti-IL-5 antibody response.

Hybridomas producing anti-IL-5 antibody were identified in two parallel screening assays. Antibodies recognizing microtiter plate-immobilized (noncovalent) hIL-5 were identified by indirect ELISA. Antibodies that neutralized the biological activity of hIL-5 were identified by their ability to neutralize the IL-5-induced proliferation of the TF-1 erythroleukemic, multifactor-dependent cell line (19). A scatter plot depicting the value of each hybridoma in the two screening assays is presented in Figure 2. It can be seen that the majority of hybridomas obtained produced neutralizing anti-IL-5 antibody with relatively low ELISA signal. Antibodies that were strongly positive in both assays grouped in the upper left quadrant, exemplified by the 5A10 and 39D10 MAbs.

A serial titration analysis of a selected panel of neutralizing anti-IL-5 hybridoma supernatants using the TF-1 cell line bioassay is shown in Figure 3. Supernatants presented in the upper left panel are the most potent at neutralizing IL-5 bioactivity in this series, and the 39D10 MAb furnished complete inhibition at supernatant dilutions considerably greater than 1:1000.

Serial dilutions of these same four hybridoma supernatants, tested in the histamine content induction bioassay using alkaline-passaged HL-60 cells stimulated with rhIL-5 in the presence of sodium butyrate (5), demonstrated the same rank order of antibody potency. A comparison of the potency of the JES1-39D10 MAb for IL-5 bioactivity neutralization in the TF-1 proliferation and alkaline-

---

Figure 1 Neutralization of GM-CSF functional activity and immunoenzymetric assay using monoclonal antibodies to hGM-CSF. BVD2-5A2, BVD2-21C11, and BVD2-23B6 MAbs were compared (a) for their ability to neutralize hGM-CSF-induced proliferation of the KG1C cell line. Proliferation (MTT OD 570 nm) was determined using a colorimetric signal first described in Ref. 69. A representative standard curve (b) in the hGM-CSF immunoenzymetric assay is shown using BVD2-23B6 as coating antibody and BVD2-21C11-NIP as the detecting antibody. (From U.S. Patent 5,070,013.)

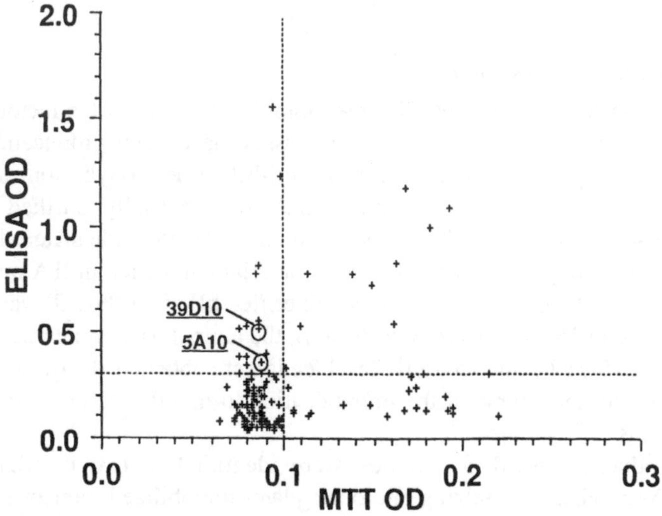

Figure 2 Anti-hIL-5 hybridoma supernatants compared in parallel by indirect ELISA and bioactivity neutralization. rhIL-5 coated microtiter plates were used to assess IL-5 immunoreactivity (ELISA OD) and inhibition of the IL-5-induced proliferation of the TF-1 cell line (MTT OD) was carried out to identify function blocking antibodies. The same colorimetric assay used in Figure 1 was performed. Neutralizing antibodies group to the left of the vertical line.

passaged HL-60 histamine content induction bioassay is shown in Figure 4. This antibody was roughly equipotent in neutralizing the same quantity of IL-5 in these two different assay systems. An approximate 10:1 molar ratio of MAb:IL-5 provided half-maximal inhibition of 500 pg/ml IL-5 in these two assay systems. This is in accord with the observed K_d of the 39D10 MAb for radiolabeled rhIL-5, approximately 150 pM, obtained by an antibody-capture Scatchard analysis procedure (Chou, C-C, personal communication).

2. *Immunoenzymetric Assay*

The utility of the preexisting TRFK4 and TRFK5 rat antimouse IL-5 (hIL-5 cross-reactive) MAbs for immunoenzymetric assay has been reported (20). Based on the relatively lower sensitivity of this assay format for hIL-5, we attempted to improve upon it using the new series of anti-hIL-5 MAbs. Making use of a screening strategy that relies on isotype differences between the coating and detecting antibodies, depicted in Figure 5, we identified the neutralizing and ELISA positive rat IgG2a 5A10 MAb (also see Fig. 2) to be a useful detecting antibody paired with the rat IgG1 TRFK5 MAb coat. Subsequently, when the NIP-derivatized 5A10 MAb was tried in combination with the 39D10 MAb as coating antibody, a very sensitive and robust hIL-5 immunoenzymetric assay (Fig. 6a) was obtained

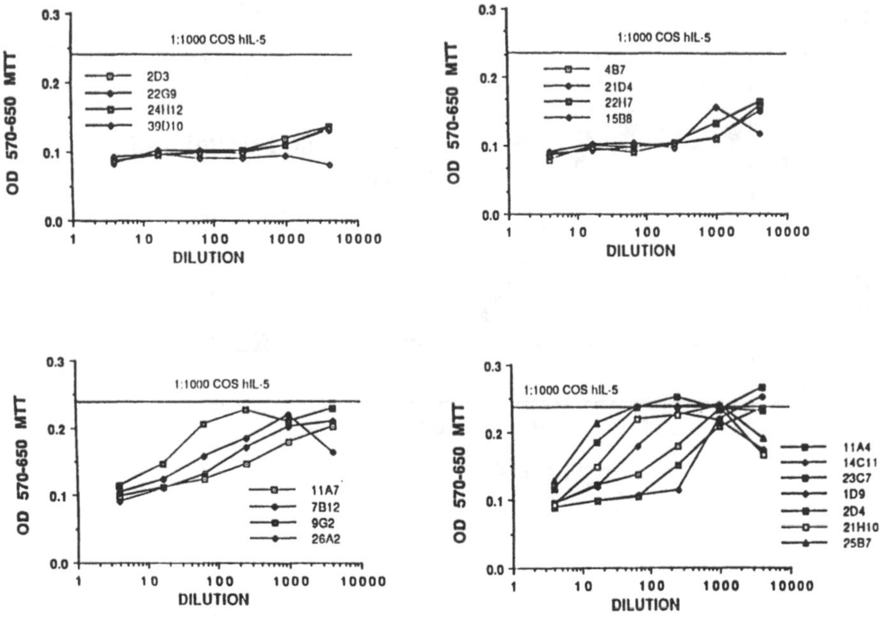

Figure 3 Blocking of rhIL-5 in the TF-1 bioassay by selected anti-IL-5 hybridoma supernatants. Serial titration analyses were carried out to establish rank order and identify the highest-avidity blocking antibody.

with a working sensitivity approaching 20 pg/ml. This assay format has been used for subsequent IL-5 determinations in patient samples.

C. Anti-IL-3

The development of the anti-IL-3 neutralizing antibody pair of BVD3-1F9 as coating antibody and BVD8-3G11 as detecting antibody has been described in a recently submitted manuscript (18). An epitope analysis of the regions where these and other anti-IL-3 monoclonal antibodies bind to hIL-3 has been carried out, and a schematic of this result is shown in Figure 7. Like the other anti-GM-CSF and IL-5 MAbs described above, Fab fragments of the 1F9 and 3G11 antibodies also blocked factor bioactivity, like the parent IgG in the TF-1 proliferation bioassay (Fig. 8). The BVD8-3G11 MAb demonstrated the superior blocking potency.

D. Immunoassay Versus Bioassay

A comparison of relative sensitivities of the IL-5 immunoenzymetric assay compared with the TF-1 and HL-60 bioassays is presented in Figure 6a and 6b. The working range of these assays is not radically different. In contrast, the

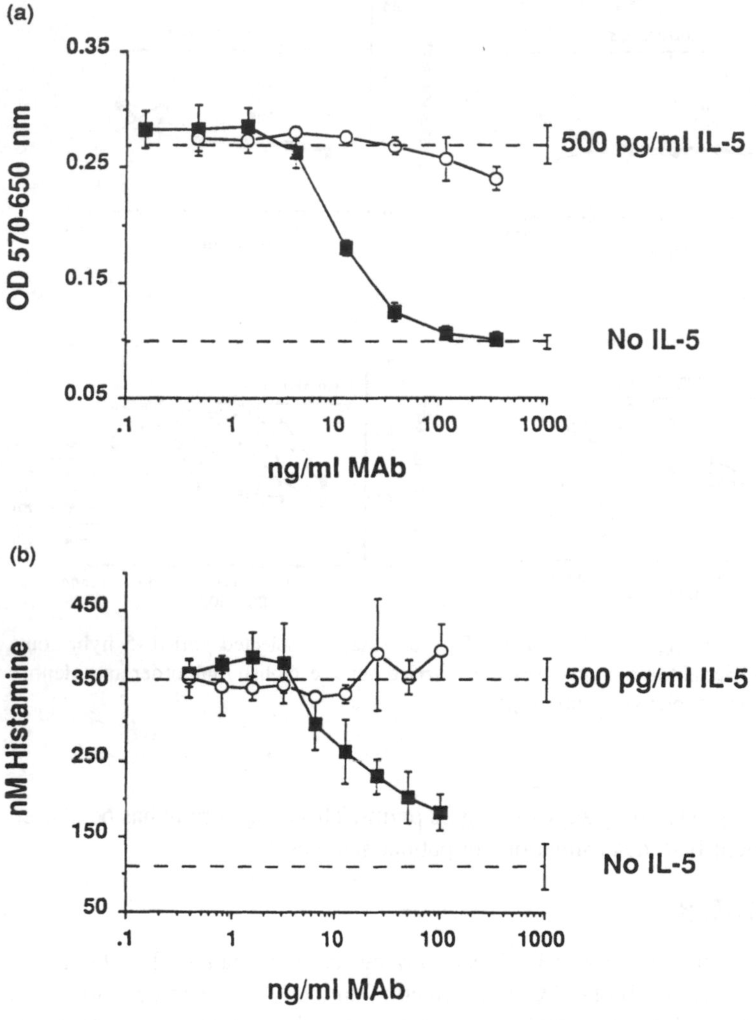

Figure 4 Antibody inhibition of IL-5 bioactivity. Rat anti-hIL-5 MAb JES1-39D10 was tested for the ability to neutralize IL-5-induced (a) proliferation of TF-1 cells and (b) histamine content induction in alkaline-passaged HL-60 cells. Varying concentrations of purified MAbs (39D10 closed symbol) or GL117 (anti–*E. coli* b-galactosidase) rat IgG2a isotype control (open symbol) were tested with a constant amount of purified L-cell-expressed IL-5. (From Ref. 5.)

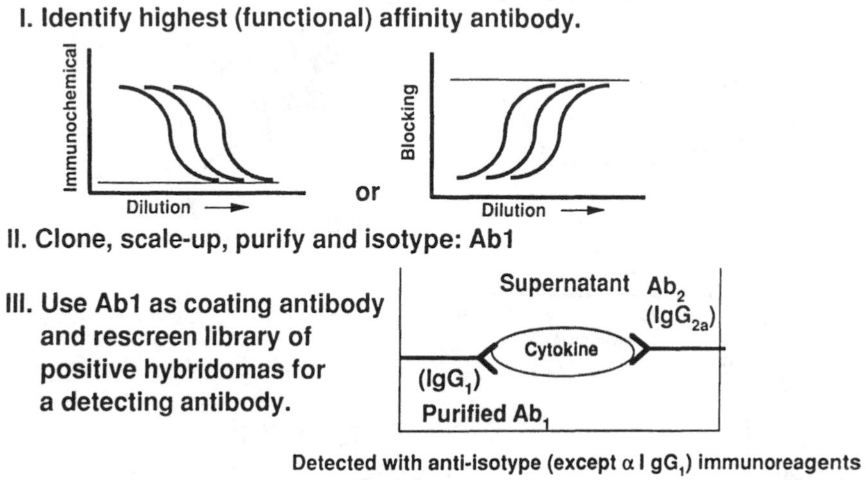
I. Identify highest (functional) affinity antibody.
Immunochemical
Dilution
or
Blocking
Dilution
II. Clone, scale-up, purify and isotype: Ab1
III. Use Ab1 as coating antibody and rescreen library of positive hybridomas for a detecting antibody.
Supernatant Ab2
(IgG2a)
Cytokine
(IgG1)
Purified Ab1
Detected with anti-isotype (except α I gG1) immunoreagents

Figure 5 Screening assay schematic. Isotype differences between the coating and detecting antibodies were exploited to identify IgG2a antibodies recognizing epitopes on the IL-5 molecule that were spatially distinct from that recognized by the TRFK5 rat IgG1.

eosinophil survival bioassay (Fig. 6c), which measures the ability of factors such as IL-3, IL-5, and GM-CSF to prolong the survival of purified eosinophils in cell culture after 4 days, is at least an order of magnitude more sensitive for each of these factors. It is noteworthy that the same rank order of potency of these factors in the eosinophil survival bioassay was reflected in the activities of these same factors in the TF-1 proliferation bioassay (data not shown). Perhaps cytokine-mediated eosinophil survival can be triggered with fewer receptor-mediated "events" than hematopoietic target cell lines carried in culture. It is not clear whether this represents triggering of fewer receptors on eosinophils at similar fractions of receptor occupancy between the different cell types, or whether this represents eosinophil triggering at a lower receptor occupancy. The eosinophil survival bioassay has proven to be an important approach to the detection of low levels of eosinophil-active cytokines in patient samples that are below the threshold of detection of the immunoassays (21). The inclusion of anticytokine MAbs in this assay is required to render it monospecific.

III. EOSINOPHIL-ACTIVE CYTOKINES IN CLINICAL SAMPLES

Interleukin-5 was first discovered as a murine and human B-cell differentiation factor (22–24). Its biological activities include eosinophil-differentiating activity (25,26), in vitro eosinophil survival and activation (21,27,28), and parasite-

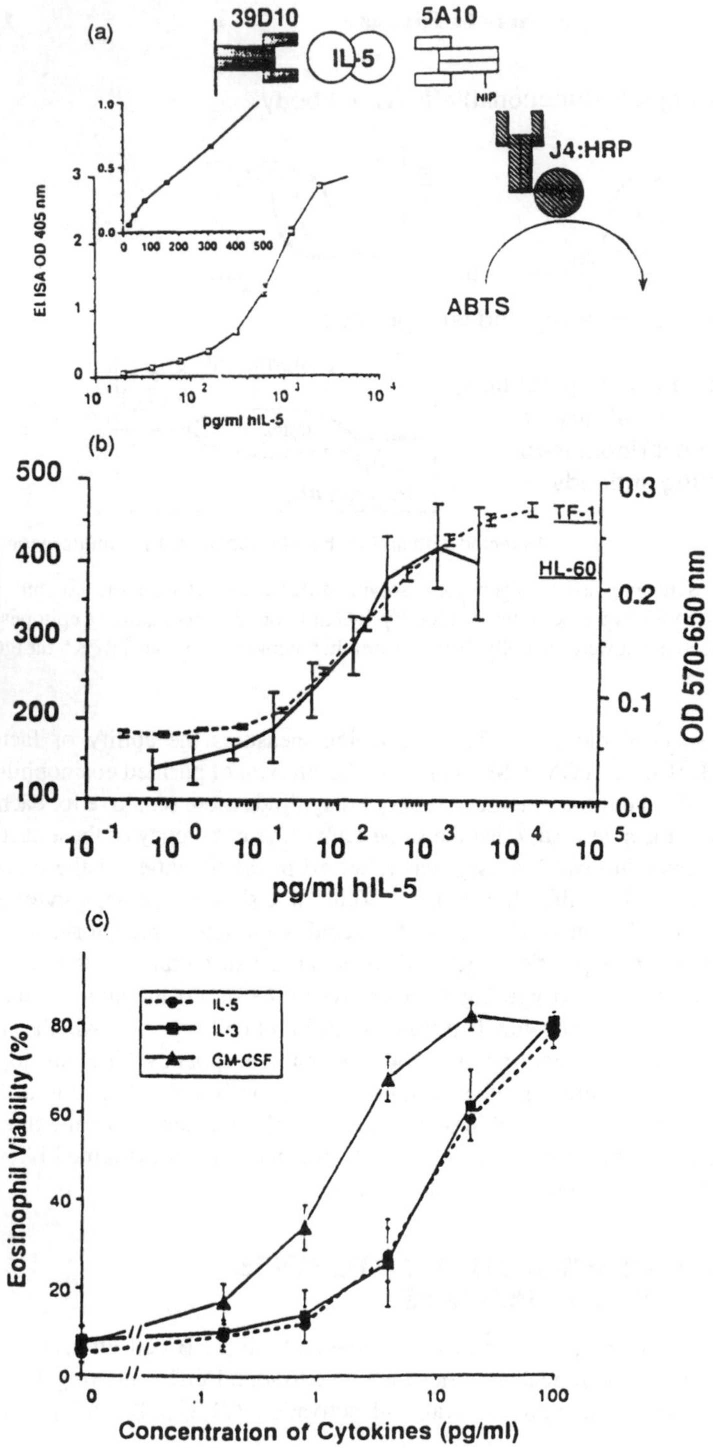

(a)
39D10
IL-5
5A10
NIP
J4:HRP
ABTS
ELISA OD 405 nm
pg/ml hIL-5
(b)
nM Histamine
OD 570-650 nm
TF-1
HL-60
pg/ml hIL-5
(c)
IL-5
IL-3
GM-CSF
Eosinophil Viability (%)
Concentration of Cytokines (pg/ml)

induced eosinophilia in a mouse model (29). It has been shown that IL-5 also has basophilopoietic activity, since it can induce histamine content in HL-60 cells and basophilic differentiation in peripheral blood basophil-eosinophil progenitors (5) and influence basophil function (4). IL-5 is a cytokine with a reasonably restricted profile of producer cell types and a relatively narrow spectrum of biological activities. T cells and mast cell lines (30) have been reported to make IL-5 protein; Reed-Sternberg cells of Hodgkin's disease (31) and eosinophils themselves (32) have been reported to make IL-5 mRNA by in situ hybridization. Eosinophils have also been reported to express other cytokines after various activations, such as: GM-CSF (13,33), IL-3 (33), IL-6 (14), and TGF-α (34–37). We have undertaken a series of studies to examine the presence of IL-5, as well as GM-CSF and IL-3, in patient samples from disease situations involving eosinophilia. Aspects of this work have also been reviewed elsewhere (38,39).

A. Microfilaria-Infected Patients

We have investigated whether the levels of the eosinophil-active cytokines GM-CSF, IL-3, and IL-5 were selectively elevated in helminth-infected patients by examining their PBMC stimulated ex vivo. Production of these cytokines, as determined by immunoassay, was compared between a group of patients with helminth-induced eosinophilia (loaiasis) and a noneosinophilic control group (40). The median level of IL-5 was approximately 20-fold higher in the eosinophilic patient PBMC supernatants induced by mitogen (50 ng/ml PMA and 1 μg/ml ionomycin), implicating IL-5 as an important mediator in inducing the eosinophilia present in filarial infections. In the initial set of six patients and 11 normals, there was no significant difference in the amount of GM-CSF, and there was a nonsignificant trend for the eosinophilic group to have slightly higher IL-3. In a subsequent study (41,42) this twofold difference in IL-3 became significant ($p < 0.02$) when the control group was expanded to 18 individuals. The patient group also showed a significantly higher median IL-4 level, around fivefold higher than the control group, in this antigen-nonspecific mode of activation. Like GM-CSF, the IFN-γ levels did not differ significantly between the two groups. While the IL-4 and IL-5 data are consistent with the hypothesis that a particular cytokine profile is produced in response to parasite challenge through the selective activation of a

---

Figure 6 Comparison of sensitivity ranges and potencies of IL-5 in various systems. The immunoenzymetric assay format (a) makes use of the antibody pair JES1-39D10 as coating antibody and JES1-5A10:NIP as part of the detecting system in conjunction with the J4 anti-NIP immunoperoxidase conjugate. In (b) the dose-response curves for IL-5-induced TF-1 proliferation and IL-5-induced histamine content in alkaline-passaged HL-60 cells are compared. (From Ref. 5.) In (c) the greater potency of these factors for inducing eosinophil survival is shown. (From Ref. 56.)

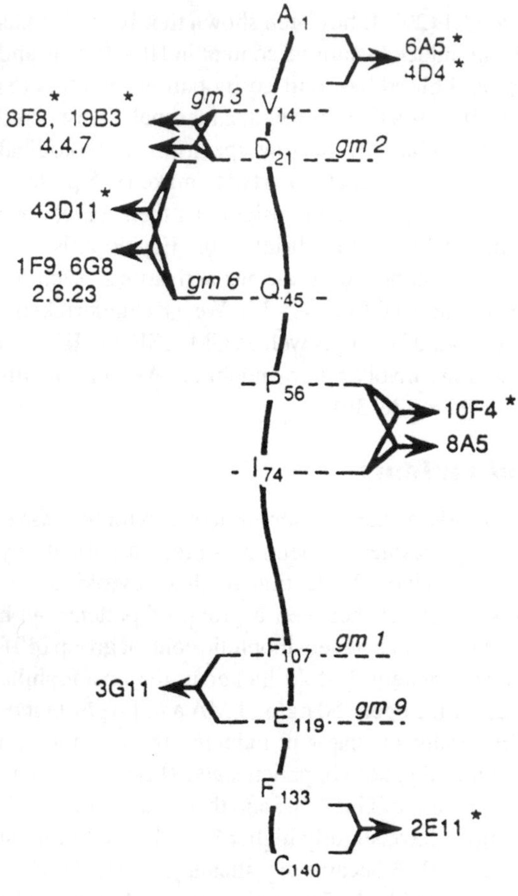

Figure 7 Schematic of the epitopes on hIL-3 recognized by different anti-IL-3 monoclonal antibodies. The 1F9 and 3G11 MAbs are seen to bind to different regions of the linear protein sequence, which must be spatially distinct in the tertiary structure as well. (From Ref. 18.)

particular T-cell helper subset, the slight increase in IL-3 is not easily explained by this paradigm. There was a direct correlation between IL-4 production and IL-5 production at the level of both protein production and frequency of T cells capable of producing these cytokines as determined by ELISPOT methodology. It is believed that this linkage may be responsible for the higher serum IgE levels observed in these patients.

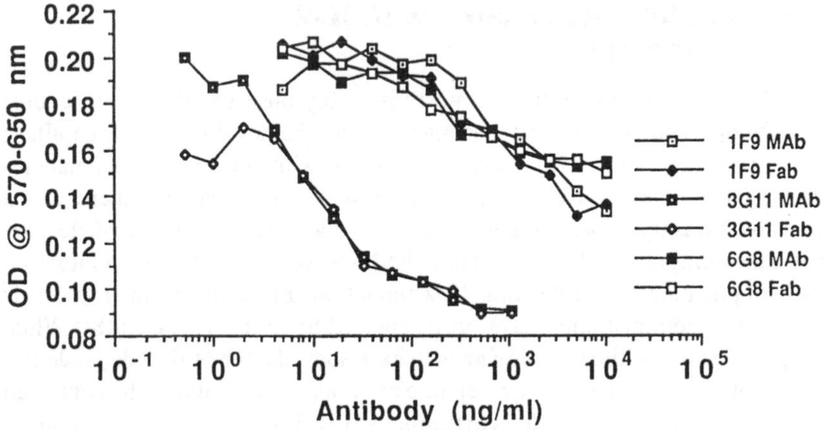

Figure 8 Neutralization of IL-3 functional activity using monoclonal antibodies (IgG and Fab) to IL-3. Monoclonal antibody IgG and Fab fragments were compared for their ability to neutralize hIL-3-induced proliferation of the TF-1 cell line. Proliferation (OD at 570–650 nm) was determined as in Figures 1–5.

B. Helminth Infection and Posttreatment Eosinophilia

Eosinophilia has been commonly reported following treatment of infections with *Schistosoma mansoni* and most of the filariae, including *Wuchereria bancrofti*, *Onchocerca volvulus*, *Loa loa*, and *Mansonella ozzardi*. We examined both the posttreatment eosinophilic response (known as the Mazzotti reaction) in a group of microfilaria-positive patients with onchocerciasis ($n = 10$) before and after treatment with diethylcarbamazine and the serum levels of eosinophil-active cytokines measured in serial samples obtained at various times following treatment (43). Serum levels of IL-5 rose sharply from pretreatment levels to a peak (70.5 ± 11 pg/ml) by 24 h after treatment. They remained elevated over the next 2–3 days, and declined toward baseline approximately 6 days after treatment, at which time the eosinophil levels were steadily increasing. The rise in serum IL-5 prior to the posttreatment eosinophilia seen in this group of patients indicates a temporal relationship between IL-5 and the subsequent development of eosinophilia. The other cytokines examined, IL-3 and GM-CSF, were not detectable in the serum at any time before or after treatment. The mechanism by which IL-5 is selectively induced to levels high enough to be detected in the serum by immunoassay is not clear. It may be a function of antigenic presentation to T cells of dead and dying parasite antigens, which are somehow seen by the immune system de novo and skew the cytokine response toward high IL-5.

C. Eosinophilia Associated with IL-2 Chemotherapy

Because peripheral eosinophilia is almost invariably observed during the course of IL-2 therapy and is frequently accompanied by the development of capillary leak syndrome, we tested the sera of patients with advanced malignancies undergoing IL-2 therapy for the presence of IL-5. In all patients there was a temporal relationship between the infusion of IL-2 and the appearance of elevated plasma concentrations of IL-5 (44). These findings were consistent with literature (45–47) implicating IL-5 following IL-2 induction. Elevated serum IL-5 levels following IL-2 administration have been reported by others as well (48). While IL-5 expression may ultimately be shown to have some degree of IL-2 dependency, the ability of IL-5 to induce IL-2 receptor expression (in the human) has not been demonstrated (6,49,50), in contrast to the mouse (51,52). Aspects of the morbidity associated with this therapy, namely the capillary leak syndrome, may be a direct result of the release of eosinophil products following degranulation induced by IL-5. In this study, the presence of serum IL-5 was confirmed by the eosinophil survival bioassay for a series of samples from one of the patients who demonstrated a false-positive immunoreactivity. Serum GM-CSF was not found in any of the three patients who were examined for this cytokine, and no IL-3 immunoassay determinations were carried out at that time.

D. Episodic Angioedema

A syndrome of cyclical weight gain, edema, and marked eosinophilia, termed episodic angioedema associated with eosinophilia, has been described (53). This condition is similar to the capillary leak syndrome seen during IL-2 therapy. We tested sera obtained longitudinally from a series of four patients for the presence of IL-5. Elevated serum IL-5, which peaked prior to flares of clinical symptoms and eosinophilia, was observed in these patients (54). The kinetics of appearance of eosinophilia following the peak IL-5 levels were reminiscent of that observed with the diethylcarbamazine-treated onchocerciasis patients described above and may represent activation of relatively late-stage eosinophil progenitors in both these cases. Serum GM-CSF was not detected in this patient series, and serum IL-3 determinations were not done.

E. Allergic Respiratory Disease

There is increasing evidence that eosinophils are important in the pathogenesis of allergic respiratory disease. We have utilized a technique of antigen challenge by segmental bronchoprovocation in allergic volunteers to characterize aspects of the immediate and late allergic airway response to allergen challenge (55). We believe this model demonstrates important parallels with asthma in terms of mimicking events occurring upon antigen challenge in that disease. Bronchoalveolar lavage (BAL) fluid was obtained from allergic rhinitis patients 12 min after segmental challenge with a ragweed antigen instillation, to obtain samples during the immediate response. A repeat segmental lavage 48 h later was carried out to obtain samples from the late response. Eosinophils and eosinophil products were determined as well as BAL IL-5 levels. In initial studies, the late lavage samples showed a significant increase in IL-5 concentration, which correlated with the presence of eosinophils and eosinophil granular proteins (55). The BAL fluids were also analyzed in the eosinophil survival bioassay using a panel of anticytokine antibodies (IL-3, IL-5, and GM-CSF) to render the assay monospecific. The eosinophil survival bioactivity present in the late-phase (48 h) BAL fluids following segmental bronchoprovocation with allergen was found to be primarily IL-5, since 5/7 samples were completely neutralized by anti-IL-5 MAb. The other 2/7, which were partially neutralized by anti-IL-5 alone, were completely blocked by the combination of anti-IL-5 and anti-GM-CSF. In an additional 10 patients, the eosinophil survival bioactivity was partially (75%) neutralized by anti-IL-5 and completely neutralized by anti-IL-5 and anti-GM-CSF in combination (56). These data suggest that eosinophils are attracted to the airway during the late-phase allergic reaction, and that local IL-5, which is the predominant eosinophil-active cytokine present in BAL fluid during allergen-induced late-phase inflammation, may produce changes in airway eosinophil density and promote release of granular proteins to cause airway injury.

F. Asthmatics

We have not been able to detect convincing IL-5 immunoreactivity by immunoenzymetric assay in BAL samples obtained from asthmatics from two different centers (Broide, D, San Diego; Bousquet, J, Montpellier), and we have not carried out extensive serum cytokine determinations on these patients. Recently, other workers have reported the presence of eosinophil-active cytokines in asthmatic patient sera (57,58). In this context, we were unable to detect serum IL-5 by immunoassay in the ragweed-challenged allergic patients described above. However, eosinophil survival bioactive cytokines were detected in the lung lavage fluids from a series of symptomatic asthmatics (59). Based on the antibody neutralization profile, it was concluded that the activity was primarily due to IL-5, with lesser contributions from GM-CSF and IL-3. This is consistent with published in situ mRNA studies carried out on mucosal bronchial biopsies from asthmatic patients (60). It is possible that high local concentrations of this mediator are produced within the lung microenvironment, but the sampling methodology results in a significant dilution. Future studies aimed at discovering eosinophil-active cytokine involvement in pulmonary disease may benefit from approaches utilizing direct cytokine localization techniques (61) in cells obtained from affected patients obtained with relatively noninvasive procedures (62).

G. Hypereosinophilic Syndrome

Idiopathic hypereosinophilic syndrome (HES) is another disease that has been investigated for IL-5 involvement. Although, we have been unable to demonstrate any circulating serum IL-5 by immunoassay, more sensitive bioassay detection methodology has revealed an association with IL-5 and HES (28,63). In our recent study, we have been able to detect eosinophil survival bioactivity in HES patient serum that could be neutralized to a great extent with anti-IL-5 (but not with anti-GM-CSF or anti-IL-3), although it should be noted that one of the patients demonstrated weak inhibition with each of the anticytokine antibodies tested. In vitro PBMC activation with PHA induced significantly more (4.5×) IL-5 compared to normal donors (but not increased GM-CSF). This latter finding is consistent with literature describing IL-5 induction in vitro following IL-2 treatment of T cells obtained from these types of patients (45).

H. B-Cell Acute Lymphocyte Leukemia with Translocation and Eosinophilia

A condition associated with eosinophilia, where the implicated cytokine was IL-3, has been described (64). This patient had a distinct subtype of acute leukemia characterized by the triad of B-lineage immunophenotype, eosinophilia, and the t(5;14)(q31;q32) translocation. We were able to detect circulating serum IL-3 in this patient, but no IL-5 or GM-CSF. This subtype of leukemia may arise in part due to the chromosome translocation that activates the IL-3 gene.

I. Eosinophilia-Myalgia Syndrome

The eosinophilia-myalgia syndrome (EMS) was recognized in late 1989 as an illness associated with consumption of L-tryptophan produced by a single manufacturer. EMS (65) is characterized by peripheral blood and tissue eosinophilia, myalgias, arthralgias, and cutaneous manifestations, including eosinophilic fasciitis [associated with elevated expression of the genes for TGFβ1 and type VI collagen (66)] and polyneuropathy. We examined sera samples from EMS patients for the presence of IL-5, but were unable to detect any by immunoassay. Supernatants of patient PBMC stimulated with implicated lots of tryptophan have been analyzed for eosinophil-active cytokines, both by eosinophil survival bioassay with monoclonal anticytokine antibodies and by immunoassay. While neither IL-3 or IL-5 was found in these analyses, the principal eosinophil-active cytokine detected was GM-CSF; IL-6 was also elevated in these supernatants (67). Further detailed chemical characterization of the cytokine-inducing entity in the contaminated lots of L-tryptophan indicated the material to be endotoxin. We do not have an explanation for our inability to find IL-5 in this condition, either in patient sera or in ex vivo stimulated PBMC supernatant, in view of a recent report describing an EMS serum:IL-5 association (68).

IV. SUMMARY AND CONCLUSIONS

Clinical conditions associated with eosinophilia and the presence of eosinophil-active cytokines, primarily IL-5, have been identified. Eosinophilia can result from a diverse set of etiologies with differing underlying mechanisms. IL-5 can be consistently observed in a variety of eosinophilic conditions such as IL-2-associated eosinophilia, the Mazzotti reaction, and episodic angioedema, which represent the high-end levels of circulating IL-5. Hypereosinophilic syndrome and certain types of asthmas may represent the low end, where barely detectable levels of circulating IL-5 are present, yet these levels are sufficient to maintain the hypereosinophilic state. One principal issue is whether low levels of IL-5 or other eosinophil-active cytokines are present in a particular compartment and, although not in general circulation, are potentially locally active. To answer this question, detection methodologies must shift from analyses of bulk fluids to the specific detection of cytokines within microenvironments or single cells in tissue. We are currently developing immunohistochemical methodology to carry out these analyses. The availability of such methods will be a key step in furthering understanding of the role of cytokines in disease.

In view of the clear implication of IL-5 in many pathological eosinophilias, the clinical application of IL-5 antagonists may offer some real benefit. In the near term, humanized anti-IL-5 therapeutic monoclonal antibodies may be ideal.

Finally, it may be important to consider that while many of the above-described pathologies involving eosinophilia appear to have a demonstrable association with IL-5, there are other eosinophil-active cytokines and new ones may ultimately be described. Since factors such as GM-CSF can be more potent activators on this cell type, it is conceivable that certain eosinophilic pathologies or patient subsets may have factors other than IL-5 involved in their disease. It may also turn out that there is an autocrine loop in certain eosinophilias if initial claims describing eosinophil production of self-active factors can be substantiated. Chronic eosinophilic diseases may therefore come to share certain characteristics of autoimmune conditions in which some mechanism keeps the cycle going through autostimulation.

REFERENCES

1. Weller PF. Cytokine regulation of eosinophil function. Clin Immunol Immunopathol 1992; 1:2.
2. Weller PF. The immunobiology of eosinophils. N Engl J Med 1991; 324:1110–1118.

3. Dvorak AM, Saito H, Estrella P, Kissell S, Arai N, Ishizaka T. Ultrastructure of eosinophils and basophils stimulated to develop in human cord blood mononuclear cell cultures containing recombinant human interleukin-5 or interleukin-3. Lab Invest 1989; 61:116–132.
4. Bischoff SC, Brunner T, De Weck A, Dahinden CA. Interleukin-5 modifies histamine release and leukotriene generation by human basophils in response to diverse agonists. J Exp Med 1990; 172:1577–1582.
5. Denburg JA, Silver JE, Abrams JS. Interleukin-5 is a human basophilopoietin: induction of histamine content and basophilic differentiation of HL-60 cells and of peripheral blood basophil-eosinophil progenitors. Blood 1991; 77:1462–1468.
6. Rand TH, Silberstein DS, Kornfeld H, Weller PF. Human eosinophils express functional interleukin 2 receptors. J Clin Invest 1991; 88:825–832.
7. Fujisawa T, Abu GR, Kita H, Sanderson CJ, Gleich GJ. Regulatory effect of cytokines on eosinophil degranulation. J Immunol 1990; 144:642–646.
8. Thorne KJ, Richardson BA, Mazza G, Butterworth AE. A new method for measuring eosinophil activating factors, based on the increased expression of CR3 alpha chain (CD11b) on the surface of activated eosinophils. J Immunol Methods 1990; 133: 47–54.
9. Valerius T, Repp R, Kalden JR, Platzer E. Effects of IFN on human eosinophils in comparison with other cytokines. A novel class of eosinophil activators with delayed onset of action. J Immunol 1990; 145:2950–2958.
10. Watanabe H, Kawabe T, Yodoi J, Tanaka M, Kim KM, Nambu M, Tsuruta S, Morita M, Yorifuji T, Mayumi M, et al. Transforming growth factor beta and dexamethasone suppress the expression of Fc epsilon receptor 2 (CD23) on a human eosinophilic cell line EoL-3. Immunol Lett 1990; 25:313–318.
11. Kawabe T, Maeda Y, Maekawa N, Tanaka M, Mayumi M, Mikawa H, Yodoi J. Role of PAF and cytokines in the modulation of Fc epsilon RII/CD23 expression on human eosinophils. Adv Prostaglandin Thromboxane Leukotriene Res 1991.
12. Wallen N, Kita H, Weiler D, Gleich GJ. Glucocorticoids inhibit cytokine-mediated eosinophil survival. J Immunol 1991; 147:3490–3495.
13. Moqbel R, Hamid Q, Ying S, Barkans J, Hartnell A, Taicopoulos A, Wardlaw AJ, Kay AB. Expression of mRNA and immunoreactivity for the granulocyte/macrophage-colony stimulating factor (GM-CSF) in activated human eosinophils. J Exp Med 1991; 174:749.
14. Hamid Q, Moqbel R, Ying S, Barkans J, Abrams JS, Kay AB. Human eosinophils can synthesize and secrete interleukin-6 in vitro. Blood 1992; 80:496–501.
15. Hartnell A, Kay AB, Wardlaw AJ. IFN-γ induces expression of FcγRIII (CD16) on human eosinophils. J Immunol 1992; 148:1471–1478.
16. Tanaka M, Lee KC, Yodoi J, Saito H, Iwai Y, Kim KM, Morita M, Mayumi M, Mikawa H. Regulation of Fc epsilon receptor 2 (CD23) expression on a human eosinophilic cell line EoL3 and a human monocytic cell line U937 by transforming growth factor beta. Cell Immunol 1989; 122:96–107.
17. Bach MK, Brashler JR, Stout BK, Johnson HG, Sanders ME. Platelet-derived growth factor can activate purified primate, phorbol myristate acetate-primed eosinophils. Int Arch Allergy Appl Immunol 1991; 94:167–168.

18. Kaushansky K, Shoemaker SG, Broudy VC, Lin NL, Matous JV, Alderman E, Aghajanian JD, Szklut PJ, Van Dyke RE, Pearce MK, Abrams JS. The structure-function relationships of interleukin-3: an analysis based on the function and binding characteristics of a series of interspecies chimera of gibbon and murine IL-3. J Clin Invest. In press.
19. Kitamura T, Takaku F, Miyajima A. IL-1 up-regulates the expression of cytokine receptors on a factor-dependent human hemopoietic cell line, TF-1. Int Immunol 1991; 3:571–577.
20. Schumacher JH, O'Garra A, Shrader B, Van Kimmenade A, Bond MW, Mosmann TR. The characterization of four monoclonal antibodies specific for mouse IL-5 and development of mouse and human IL-5 enzyme-linked immunosorbent assay. J Immunol 1988; 141:1576.
21. Okubo Y, Kita H, Weiler DA, Abrams JS, Gleich GJ. A sensitive method for the measurement of IL-3, IL-5, and GM-CSF using an eosinophil survival assay. In preparation.
22. O'Garra A, Umland S, de France T, Christiansen J. A review—"B-cell factors" are pleiotropic. Immunol Today 1988; 9:45–54.
23. Coffman RL, Seymour BWP, Lebman DA, Hiraki DD, Christiansen JA, Shrader B, Cherwinski HM, Savelkoul HFJ, Finkelman FD, Bond MW, Mosmann TR. The role of helper T cell products in mouse B cell differentiation and isotype regulation. Immunol Rev 1988; 102:5–28.
24. Yolota T, Arai N, de Vries J, Spits H, Banchereau J, Zlotnik A, Rennick D, Howard M, Takebe Y, Miyatake S, Lee F, Arai K. Molecular biology of interleukin 4 and interleukin 5 genes and biology of their products that stimulate B cells, T cells and hemopoietic cells. Immunol Rev 1988; 102:137–187.
25. Campbell HD, Tucker WQ, Hort Y, Martinson ME, Mayo G, Clutterbuck EJ, Sanderson CJ, Young IG. Molecular cloning, nucleotide sequence, and expression of the gene encoding human eosinophil differentiation factor (interleukin 5). Proc Natl Acad Sci USA 1987; 84:6629–6633.
26. Yokota T, Coffman RL, Hagiwara H, Rennick DM, Takebe Y, Yokota K, Gemmell L, Shrader B, Yang G, Meyerson P, et al. Isolation and characterization of lymphokine cDNA clones encoding mouse and human IgA-enhancing factor and eosinophil colony-stimulating factor activities: relationship to interleukin 5. Proc Natl Acad Sci USA 1987; 84:7388–7392.
27. Lopez AF, Sanderson CJ, Gamble JR, Campbell HD, Young IG, Vadas MA. Recombinant human interleukin 5 is a selective activator of human eosinophil function. J Exp Med 1988; 167:219–224.
28. Owen WF, Rothenberg ME, Petersen J, Weller PF, Silberstein D, Sheffer AL, Stevens RL, Soberman RJ, Austen KF. Interleukin 5 and phenotypically altered eosinophils in the blood of patients with the idiopathic hypereosinophilic syndrome. J Exp Med 1989; 170:343–348.
29. Coffman RL, Seymour BW, Hudak S, Jackson J, Rennick D. Antibody to interleukin-5 inhibits helminth-induced eosinophilia in mice. Science 1989; 245: 308–310.
30. Plaut M, Pierce JH, Watson CJ, Hanley-Hyde J, Nordan RP, Paul WE. Mast cell lines

produce lymphokines in response to cross-linkage of FcERi or to calcium ionophores. Nature 1989; 339:64–67.
31. Samoszuk M, Nansen L. Detection of Interleukin-5 messenger RNA in Reed-Sternberg cells of Hodgkin's disease with eosinophilia. Blood 1990; 75:13–16.
32. Desreumaux P, Janin A, Colombel JF, Prin L, Plumas J, Emilie D, Torpier G, Capron A, Capron M. Interleukin 5 messenger RNA expression by eosinophils in the intestinal muscosa of patients with coeliac disease. J Exp Med 1992; 175:293–296.
33. Kita H, Ohnishi T, Okubo Y, Weiler D, Abrams JS, Gleich GJ. GM-CSF and interleukin-3 released by human peripheral blood eosinophils and neutrophils. J Exp Med 1991; 174:745–748.
34. Elovic A, Galli SJ, Weller PF, Chang AL, Chiang T, Chou MY, Donoff RB, Gallagher GT, Matossian K, McBride J, et al. Production of transforming growth factor alpha by hamster eosinophils. Am J Pathol 1990; 137:1425–1434.
35. Wong DT, Weller PF, Galli SJ, Elovic A, Rand TH, Gallagher GT, Chiang T, Chou MY, Matossian K, McBride J, et al. Human eosinophils express transforming growth factor alpha. J Exp Med 1990; 172:673–681.
36. Todd R, Donoff BR, Chiang T, Chou MY, Elovic A, Gallagher GT, Wong DT. The eosinophil as a cellular source of transforming growth factor alpha in healing cutaneous wounds. Am J Pathol 1991; 138:1307–1313.
37. Ghiabi M, Gallagher GT, Wong DT. Eosinophils, tissue eosinophilia, and eosinophil-derived transforming growth factor alpha in hamster oral carcinogenesis. Cancer Res 1992; 52:389–393.
38. Abrams JS, Roncarolo M-G, Yssel H, Andersson U, Gleich GJ, Silver J. Strategies of anti-cytokine monoclonal antibody development: Immunoassay of IL-10 and IL-5 in clinical samples. Immunol Rev 1992; 127:5–24.
39. Gleich GJ, Butterfield JH, Leiferman KM, Kita H, Abrams J. Eosinophils, allergic diseases and cytokines. Proc of the XIV Int Cong on Allergology and Clinical Immunology. In press.
40. Limaye AP, Abrams JS, Silver JE, Ottesen EA, Nutman TB. Regulation of parasite-induced eosinophilia: selectively increased interleukin-5 production in helminth-infected patients. J Exp Med 1990; 172:399–402.
41. Mahanty S, Abrams JS, Limaye AP, Nutman TB. Linkage of interleukin-4 and interleukin-5 production in human helminth infections. Trans Assoc Am Physicians 1991; CIV:296–303.
42. Mahanty S, Abrams JS, King CL, Limaye AP, Nutman TB. Parallel regulation of IL-4 and IL-5 in human helminth infections. J Immunol. In press.
43. Limaye AP, Abrams JS, Awadzi K, Francis HF, Silver JE, Ottesen EA, Nutman TB. Interleukin-5 and the post-treatment eosinophilia in patients with onchocerciasis. J Clin Invest 1991; 88:1418–1421.
44. van Haelst Pisani C, Kovach JS, Kita H, Leiferman KM, Gleich GJ, Silver JE, Abrams JS. Administration of interleukin-2 (IL-2) results in increased plasma concentrations of IL-5 and eosinophilia in patients with cancer. Blood 1991; 78:1538–1544.
45. Enokihara H, Furusawa S, Nakakubo H, Kajitani H, Nagashima S, Saito K, Shishido H, Hitoshi Y, Takatsu K, Noma T, et al. T cells from eosinophilic patients produce interleukin-5 with interleukin-2 stimulation. Blood 1989; 73:1809–1813.

46. Yamaguchi Y, Suda T, Shiozaki H, Miura Y, Hitoshi Y, Tominaga A, Takatsu K, Kasahara T. Role of IL-5 in IL-2 induced eosinophilia. In vivo and in vitro expression of IL-5 mRNA by IL-2. J Immunol 1990; 145:873.
47. Fishel RS, Farnen JP, Hanson CA, Silver SM, Emerson SG. Role of IL-5 in IL-2 induced eosinophilia: in vivo and in vitro expression of IL-5 mRNA by IL-2. J Immunol 1990; 145:873–877.
48. MacDonald D, Gordon AA, Kajitani H, Enokihara H, Barrett AJ. Interleukin-2 treatment associated eosinophilia is mediated by interleukin-5 production. Br J Haematol 1990; 76:168.
49. Nagasawa M, Ohshiba A, Yata J. Effect of recombinant interleukin 5 on the generation of cytotoxic T cells (CTL). Cell Immunol 1991; 133:317–326.
50. Zola H, Weedon H, Thompson GR, Fung MC, Ingley E, Hapel AJ. Expression of IL-2 receptor p55 and p75 chains by human B lymphocytes: effects of activation and differentiation. Immunology 1991; 72:167–173.
51. Yoshimoto T, Nakanishi K, Matsui K, Hirose S, Hiroishi K, Tanaka T, Hada T, Hamaoka T, Higashino K. IL-5 up-regulates but IL-4 down-regulates IL-2R expression on a cloned B lymphoma line. J Immunol 1990; 144:183–190.
52. Nakanishi K, Matsui K, Hirose S, Yoshimoto T, Hiroishi K, Kono T, Hada T, Hamaoka T, Higashino K. Lymphokine-regulated differential expression of mRNA for p75kDa-IL-2R and p55kDa-IL-2R in a cloned B lymphoma line (BLC1-CL-3 cells). J Immunol 1990; 145:1423–1429.
53. Gleich GJ, Schroeter AL, Marcoux JP, Sachs MI, O'Connell EJ, Kohler PF. Episodic angioedema associated with eosinophilia. N Engl J Med 1984; 310:1621–1626.
54. Butterfield JH, Leiferman KM, Abrams JS, Silver JE, Bower J, Gonchoroff N, Gleich GJ. Elevated serum levels of interleukin-5 in patients with the syndrome of episodic angioedema and eosinophilia. Blood 1992; 79:688–692.
55. Sedgwick JB, Calhoun WJ, Gleich GJ, Kita H, Abrams JS, Schwartz LB, Volovitz B, Ben-Yaakov M, Busse WW. Immediate and late allergic airway response of allergic rhinitis patients to segmental antigen challenge: characterization of eosinophil and mast cell mediators. Am Rev Respir Dis 1991; 144:1274–1281.
56. Ohnishi T, Kita H, Weiler D, Sedgwick JB, Calhoun WJ, Busse WW, Abrams JS, Gleich GJ. IL-5 is the predominant eosinophil-active cytokine in the antigen-induced pulmonary late phase reaction. Am Rev Respir Dis. Submitted.
57. Walker C, Virchow JJ, Bruijnzeel PL, Blaser K. T cells and asthma. II. Regulation of the eosinophilia of asthma by T cell cytokines. Int Arch Allergy Appl Immunol 1991; 94:248–250.
58. Corrigan CJ, Haczku A, Gemou-Engesaeth V, Doi S, Kikuchi Y, Takatsu K, Durham SR, Kay AB. CD4 T-lymphocyte activation in asthma accompanied by increased serum concentrations of interleukin-5: effect of glucocorticoid therapy. Am Rev Respir Dis 1993; 147:540–547.
59. Ohnishi T, Kita H, Mayeno A, Sur S, Gleich GJ, Broide DH. Eosinophil-active cytokines and an inhibitor of cytokine activity in the bronchoalveolar lavage fluids (BALF) of symptomatic patients with asthma. American Academy of Allergy and Immunology, 1992, meeting abstract.
60. Hamid Q, Azzawi M, Ying S, Moqbel R, Wardlaw AJ, Corrigan CJ, Bradley B,

Durham SR, Collins JV, Jeffery PK, Quint DJ, Kay AB. Expression of mRNA for interleukin-5 in mucosal bronchial biopsies from asthma. J Clin Invest 1991; 87: 1541–1546.
61. Sander B, Hoiden I, Andersson U, Moller E, Abrams JS. Similar frequencies and kinetics of cytokine producing cells in murine peripheral blood and spleen: cytokine detection by immunoassay and intracellular immunostaining. J Immunol Methods. Submitted.
62. Gibson PG, Dolovich J, Denburg JA, Girgis GA, Hargreave FE. Sputum cell counts in airway disease: a useful sampling technique. Agents Actions 1990; 30 (Suppl): 161–172.
63. Kita H, Weiler DA, Ide M, Abrams JS, Gleich GJ. Eosinophil characteristics and cytokines involved in 12 patients with idiopathic hypereosinophilic syndrome. In preparation.
64. Meeker TC, Hardy D, William C, Hogan T, Abrams J. Activation of the interleukin-3 gene by chromosome translocation in acute lymphocyte leukemia with eosinophilia. Blood 1990; 76:285–289.
65. Hertzman PA, Blevins WL, Meyer J, Greenfield B, Ting BM, Gleich GJ. Association of the eosinophilia-myalgia syndrome with the ingestion of tryptophan. N Engl J Med 1990; 322:869–873.
66. Peltonen J, Varga J, Sollberg S, Uitto J, Jimenez SA. Elevated expression of the genes for transforming growth factor-beta 1 and type VI collagen in diffuse fasciitis associated with the eosinophilia-myalgia syndrome. J Invest Dermatol 1991; 96: 20–25.
67. Kita H, Weyand CM, Goronzy JJ, Weiler DA, Lundy SK, Mayeno AN, Abrams JS, Gleich GJ. Eosinophil-active cytokine from mononuclear cells cultured with L-tryptophan products: An in vitro method for investigation of the eosinophilia-myalgia syndrome. J Immunol. Submitted.
68. Owen WFJ, Petersen J, Sheff DM, Folkerth RD, Anderson JM, Corson JM, Sheffer AL, Austen KF. Hypodense eosinophils and interleukin 5 activity in the blood of a patient with the eosinophilia-myalgia syndrome. Proc Natl Acad Sci USA 1990; 87: 8647–8651.
69. Mosmann TR. Rapid colorimetric assay for cellular growth and survival: Application to proliferation and cytotoxicity assays. J Immunol Methods 1983; 65:55.

DISCUSSION (Speaker: J. Abrams)

Lopez: Did the antihuman IL-3 antibodies discussed that inhibited IL-3 activity recognize the N-terminus and C-terminus regions that Lopez and Vadas found to be critical for activity, and did they block binding of IL-3 to the α or β chain of the IL-3 receptor?

Abrams: The apparent antibody-binding epitopes of a series of antihuman IL-3-neutralizing antibodies are depicted in Figure 7. Based on collaborative work with Ken Kaushansky (1) using a series of interspecies chimeric IL-3 molecules, we have been able to identify a series of blocking antibodies that map toward the N-terminal portion and one neutralizing antibody that maps toward the C-terminal portion. At present, it is not clear which receptor subunit is inhibited from interacting with the cytokine ligand by each of these antibodies, and further work is required. I speculate that at least the 3G11 antibody, binding somewhere within residues 107 (F)–119 (E), would inhibit binding to the α subunit of the receptor.

Ackerman: Why are kits not yet available commercially for measurement of IL-5? Are there problems inherent in the assays that we have not heard about? Please comment on the bioassay versus ELISA—do they detect the same thing? Are there active and inactive forms of IL-5? What about the effects of soluble IL-5 receptor on these assays?

Abrams: It is not clear to me why there are no commercial cytokine ELISA kits on the market for hIL-5 at present. I do not know of any reason why IL-5 quantification by immunometric assay should present any inherent difficulties relative to the other cytokines. Perhaps part of the problem can be traced to the difficulty of producing high-specific-activity recombinant IL-5 in *E. coli*, which many groups favor as a source of recombinant cytokine immunogen. Regarding the bioassay versus immunoassay, we have observed (2) a situation where a patient material (serum) gave a false-positive reading relative to what was measured by subsequent confirmative bioassay (eosinophil survival). It is known that this immunoassay format can be affected by the presence of rheumatoid factor or heterophilic antibody, so care needs to be taken in sample handling. I do not have any information on how the presence of soluble receptor would affect the assay, but I would point out that the immunoassay is comprised of two function-blocking antibodies that presumably would compete for the same sites on IL-5 that bind to the receptor subunit(s). It is theoretically possible that the presence of soluble receptor would dampen the signal in both the immunoassay and the bioassay, but direct experiments need to be performed to clarify this.

Weller: With regard to the studies with Tom Nutman, I wonder what explanation there may be for the heightened IL-5 release elicitable by an ionophoric stimulus from T cells of filariasis patients. Does this increased formation of IL-5 reflect increases in Th2 lymphocytes in the filarial patients?

Abrams: We believe the increased IL-5 levels in the filariasis patient PBMC supernatants after mitogen stimulation to be caused by the clonal expansion of IL-5-producing cells. PMA/ionomycin stimulation is a good way to elicit relatively high levels of cytokine, presumably from cells that are already committed to produce them if they were triggered with the appropriate antigen. Also, in this system, the most probable source of IL-5 appears to be the CD4+ cell (3,4).

M. Capron: Do you know whether the results of IL-5 titration might be influenced by the binding of IL-5 to soluble IL-5 receptor α chain? Can the binding of IL-5 to its receptor explain the differences between ELISA results and biological activity?

Abrams: Again, I do not have any evidence that soluble IL-5 receptor α subunit is either present in any of the samples or interferes in the immunoassay. Probably the only way to answer this is going to be to assay for the presence of IL-5 receptor directly using antibodies that bind to the receptor but do not block biological function. Regarding perceived differences between immunoassay and biological activity, it may be that if mouse target cells (like BCL-1) are used, then the apparent activity of human IL-5, which has been observed to be 100- to 1000-fold less potent than the mouse factor on mouse target cells (5), may result in a relatively insensitive assay. The amounts present in the sample may therefore be below detection in the bioassay.

Devos: The affinity of the hIL-5R complex on eosinophils and differentiated HL60 cells is quite high (20 pM = K_d). What is the affinity of the "humanized" anti-hIL-5 monoclonal antibody?

Abrams: The apparent affinity (K_d) of the parent rat IgG2a anti-hIL-5 antibody is in the range of 150 pM, which I understand to be comparable to the affinity of IL-5 for its receptor on TF-1 cells and eosinophils.

Butterworth: In the collaborative study on filariasis that you carried out with Tom Nutman, did you also look for antigen-driven cytokine production? (We have preliminary evidence for IL-5 production in response to adult worm antigens by peripheral blood mononuclear cells from patients with schistosomiasis, before and after treatment.)

Abrams: Yes, we have tried to detect antigen-driven IL-5 production in these types of samples, but have not been successful in obtaining results that are as clear as those with PMA/ionomycin activation. I do not at present have an explanation why this later activation has worked better than antigen preparations. Ultimately, studies that can clearly demonstrate antigen-driven cytokine expression in association with a particular disease are the most satisfying.

Moqbel: Do you have any evidence for in situ release of IL-5 during the Mazzotti reaction, bearing in mind that this reaction appears to be associated with in vivo degranulation of eosinophils?

Abrams: No, unfortunately the only samples we have examined to date involve cytokine levels in the bulk fluid environment like serum/plasma. It is not clear from this approach what is the actual cell source or sources of the factor. As the various localization techniques (both immunohistochemistry and in situ RNA detection) improve, it would certainly be interesting to bring these methodologies to bear on this condition. I understand, however, that this may be difficult, given that DEC treatment is no longer considered to be the standard of care.

Spry: Are serum assays of cytokines at the limits of current technology worthwhile? They may be quantitating proteins of short half-life, and other factors may also make these assays difficult to interpret.

Abrams: This is a very important point. While we have made progress understanding some of the underlying aspects of cytokine association with disease by measuring levels in bulk fluids, ultimately we must turn our attention to what is actually happening within the tissue (and blood) microenvironment. That is why a good portion of my own lab's present effort is devoted to cytokine immunohistochemistry methods development for tissue. I foresee major advances in our understanding as these methods become more reliable and generally applied.

Vadas: Do you have any information about IL-5 release in vivo where there is massive eosinophil death?

Abrams: No, I don't know of any data relevant to this point.

Gleich: Treatment of patients with the syndrome of episodic angioedema and marked peripheral blood eosinophilia with high doses of glucocorticoids causes a gradual reduction in eosinophil counts and an increase of neutrophils along with a fall in serum IL-5 levels to baseline (undetectable). In two cases glucocorticoid treatment caused the above changes, but the retained fluid was not excreted and accumulated in the pleural spaces. In these patients treatment with diuretics caused a prompt secretion of fluid.

DISCUSSION REFERENCES

1. Kaushansky K, Shoemaker SG, Broudy VC, Lin NL, Matous JV, Alderman E, Aghajanian JD, Szklut PJ, Van Dyke RE, Pearce MK, Abrams JS. The structure-function relationships of interleukin-3: an analysis based on the function and binding characteristics of a series of interspecies chimera of gibbon and murine IL-3. J Clin Invest. In press.
2. van Haelst Pisani C, Kovach JS, Kita H, Leiferman KM, Gleich GJ, Silver JE, Abrams JS. Administration of interleukin-2 (IL-2) results in increased plasma concentrations of IL-5 and eosinophilia in patients with cancer. Blood 1991; 78:1538–1544.
3. Mahanty S, Abrams JS, Limaye AP, Nutman TB. Linkage of interleukin-4 and interleukin-5 production in human helminth infections. Trans Assoc Am Physicians 1992; CIV:296–303.
4. Mahanty S, Abrams JS, King CL, Limaye AP, Nutman TB. Parallel regulation of IL-4 and IL-5 in human helminth infections. J Immunol. In press.
5. O'Garra A, Barbis D, Wu J, Hodgkin P, Abrams J, Howard M. Conversion of the BCL-1 bioassay to a monospecific assay for IL-5 or GM-CSF. Cell Immunol 1989; 123:189.

9

Ultrastructural Studies on Mechanisms of Human Eosinophil Activation and Secretion

Ann M. Dvorak
Harvard Medical School and Beth Israel Hospital, Boston, Massachusetts

I. INTRODUCTION

Ultrastructural studies have advanced our understanding of human eosinophil activation and secretion (reviewed in 1). These studies include general observations of bone marrow, peripheral blood, and body cavity fluid and tissue eosinophils in human samples. More recently, rhIL-5-developed human eosinophils have provided an excellent source for ultrastructural studies of eosinophil differentiation, maturation, and function (2–5). Additional ultrastructural tools that have improved our understanding of human eosinophil cell biology include cytochemistry, immunocytochemistry, and ultrastructural autoradiography. In this review, we will briefly outline the ultrastructural morphology of mature human eosinophils and their immature myelocyte precursors, as well as the major contents of their granules and lipid bodies—two important organelles that respond to activation and secretion stimuli. Activation morphology of eosinophils in vivo and in vitro involves changes in the number, size, and contents of granules and lipid bodies. These morphological changes may reflect biochemical changes, as determined by a large number of studies implicating eosinophils in proinflammatory events. Ultrastructural secretion and release morphology will be presented in three general categories: (1) injury (including necrosis, apoptosis, and the morphological continuum of apoptosis → necrosis), (2) regulated secretion (morphologically analogous to anaphylactic degranulation of mast cells and basophils

(6) and characterized by extrusion of secretory granules in toto), and (3) piecemeal degranulation (PMD). Finally, we will conclude with ultrastructural studies that document vesicular transport of granule matrix peroxidase as a mechanism for effecting piecemeal degranulation of mature (5) and immature (7) human eosinophilic leukocytes.

II. NORMAL EOSINOPHIL MORPHOLOGY

Mature peripheral blood eosinophils (Fig. 1) are granulocytes with polylobed nuclei within which chromatin is generally condensed. Nucleoli are rarely present.

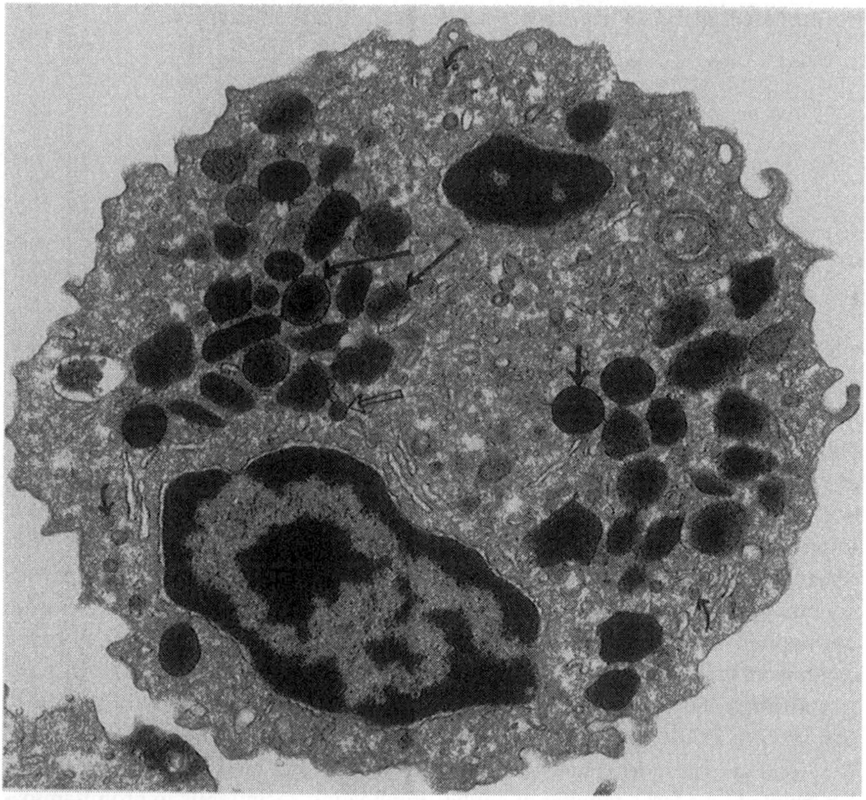

Figure 1 Mature human eosinophil from peripheral blood of a patient with HES has a bilobed nucleus with condensed chromatin, bicompartmental secondary granules (long closed arrows), homogeneously dense, core-free primary granules (short closed arrow), dense, round small granules (open arrow), and curved, elongated, and circular vesiculotubular structures (curved arrows). ×13,650.

Irregular, broad surface projections and, rarely, uropods characterize their surface profiles. The synthetic structures (Golgi, endoplasmic reticulum) are markedly reduced over their immature precursors. Cytoplasmic secretory granules of at least three types, in addition to nonmembrane-bound lipid bodies and vesiculotubular structures, characterize the remaining cytoplasmic organelles (1).

Eosinophilic myelocytes (Fig. 2), by contrast, are much larger cells which contain a single nucleus with dispersed chromatin. Ample Golgi structures, large numbers of dilated cisterns of rough endoplasmic reticulum, and increased numbers of large mitochondria characterize the general cytoplasmic organization. Interspersed among these organelles are large numbers of immature granules and small numbers of mature granules. Lipid bodies are exceedingly rare. Small granules and vesiculotubular structures also do not generally characterize these immature cells (1).

Eosinophil secretory granules provide a major means for the ultrastructural identification of eosinophils. Thus, the emergence of crystalloid core-containing secondary granules in small numbers in immature precursors allows the assignment of such cells to the eosinophil lineage (2). The granule compartment in mature eosinophils contains mostly bicompartmental secondary (specific) granules. Eosinophilic myelocytes represent a developmental continuum (1,8), ranging from cells with few immature granules and cytoplasmic vacuoles to cells filled with large, loosely textured granules that do not contain crystalline cores and are best considered to be immature primary granules (Dvorak AM, unpublished data).

Condensation of these granules produces spherical primary granules that persist in small numbers in mature eosinophils and have been shown to be the major site of eosinophil Charcot-Leyden crystal (CLC) protein by immunogold stains (1,4,9,10). Clearly, many primary granules do not survive the myelocyte stage, since their reduction is greater than one would expect merely related to cell division (with granule dilution) in the absence of further production. While this certainly would cause some diminution in their numbers, we have seen primary granule extrusion in normal bone marrow samples (Fig. 3) (1) and Charcot-Leyden crystals in the bone marrow interstitium (Fig. 4) of samples obtained from leukemic patients (1) (see later). Eosinophilic myelocytes also contain large numbers of perfectly round, homogeneously dense granules, some of which also contain small vesicles beneath the granule membrane (Figs. 2 and 5). These are immature secondary (specific) granules (2). For example, in eosinophilic myelocytes that have progressed farther along the maturation continuum (Fig. 5), one can find central, irregular crystalline cores condensing within these immature granules, thereby creating the familiar spherical, bicompartmental, core-containing specific eosinophil granule (2). Small, homogeneously dense granules (Fig. 6) and vesiculotubular structures that appear electron-lucent (Fig. 7) are also found in mature eosinophils, structures that increase in the activated cell (1,10–19). Lipid bodies are large, osmiophilic, nonmembrane-bound repositories of lipid (Fig. 8)

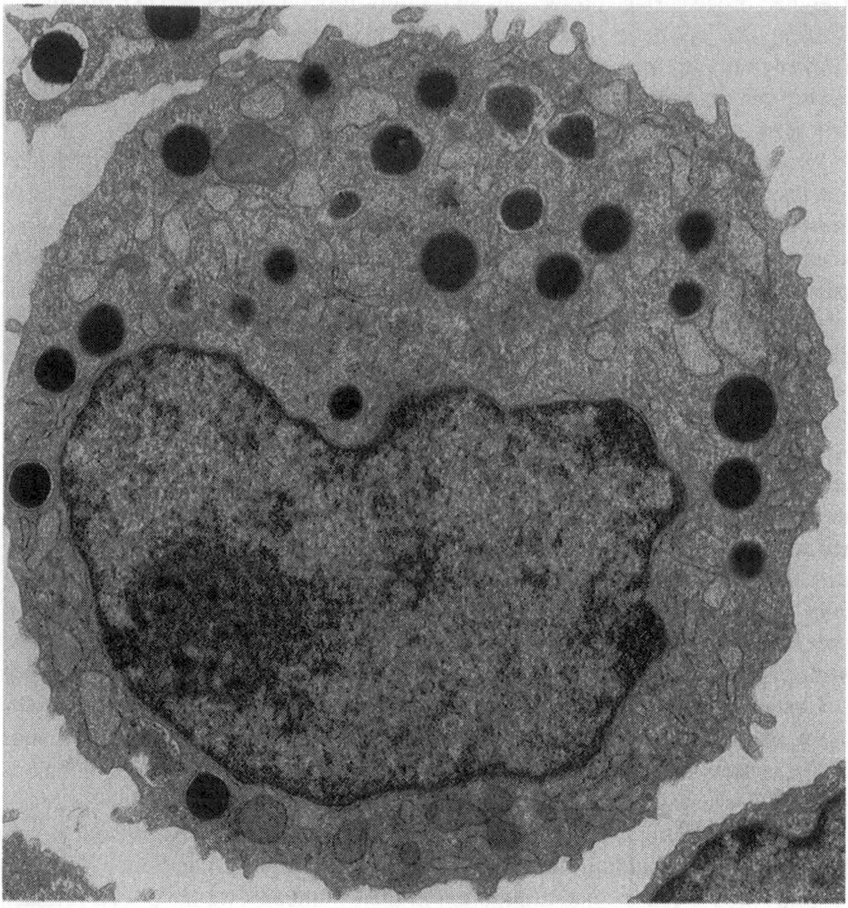

Figure 2 Human eosinophilic myelocyte in 3-week rhIL-5-containing culture of cord blood mononuclear cells shows large, eccentric, single-lobed nucleus with a large nucleolus and dispersed chromatin, numerous distended cisterns of rough endoplasmic reticulum, large mitochondria, and irregular short surface processes. Homogeneously dense, round, immature secondary granules and several immature granules, which contain intragranular vesicles and are less condensed, are present. ×11,360. (With permission from Ref. 3.)

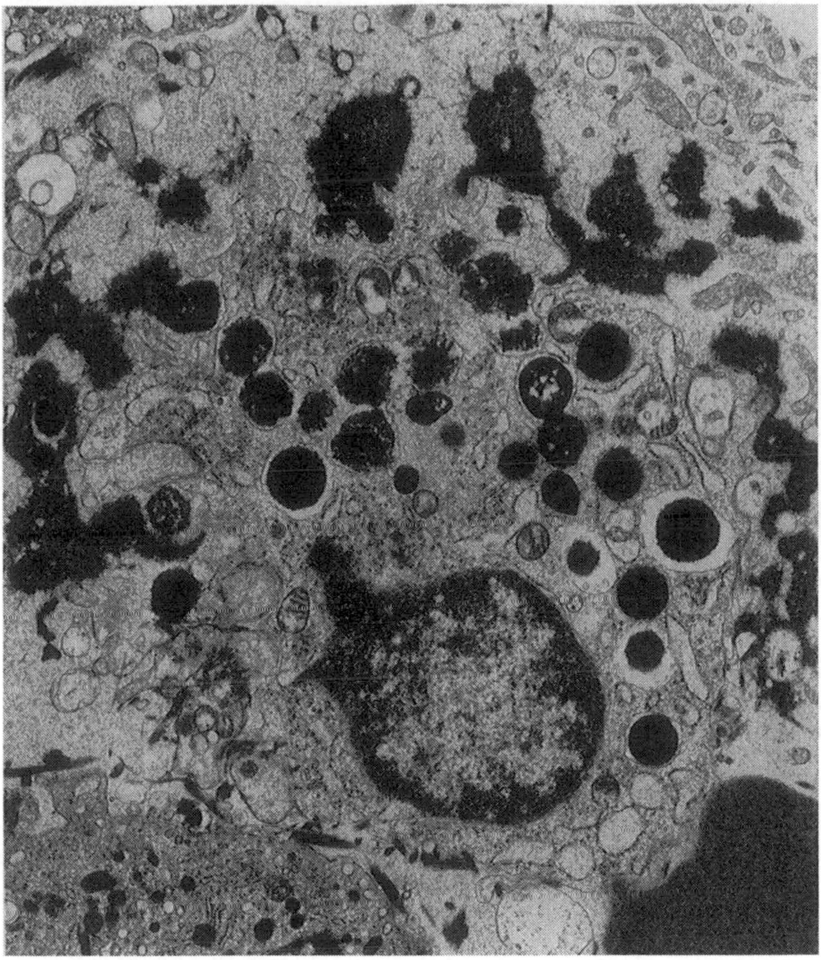

Figure 3 Eosinophilic myelocyte from a human bone marrow sample showing massive extrusion of primary granule contents from fused intracytoplasmic degranulation chambers in a viable cell. ×10,000. (With permission from Ref. 1.)

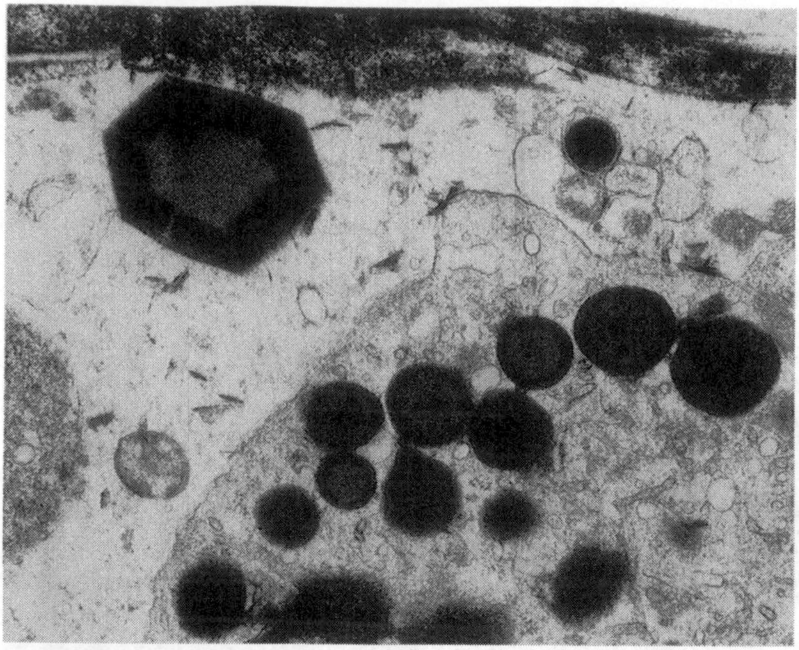

Figure 4 Bone marrow specimen, from a patient with myeloid leukemia, showing a portion of an eosinophil adjacent to the interstitial hexagonal CLC. ×20,000. (With permission from Ref. 1.)

that are present in small numbers in mature eosinophils and are more abundant in activated cells (1,20,21).

III. IMPORTANT MEDIATORS OF INFLAMMATION (OR THEIR RATE-LIMITING ENZYMES AND SUBSTRATES) HAVE BEEN LOCALIZED TO INDIVIDUAL EOSINOPHIL ORGANELLES (TABLE 1)

A. Secondary Granules

Bicompartmental, membrane-bound, spherical secondary granules have a centrally located crystalline core compartment and an outer matrix compartment (Fig. 9A) (1). Certain proinflammatory eosinophil granule products have been localized to these compartments in secondary granules using immunogold stain-

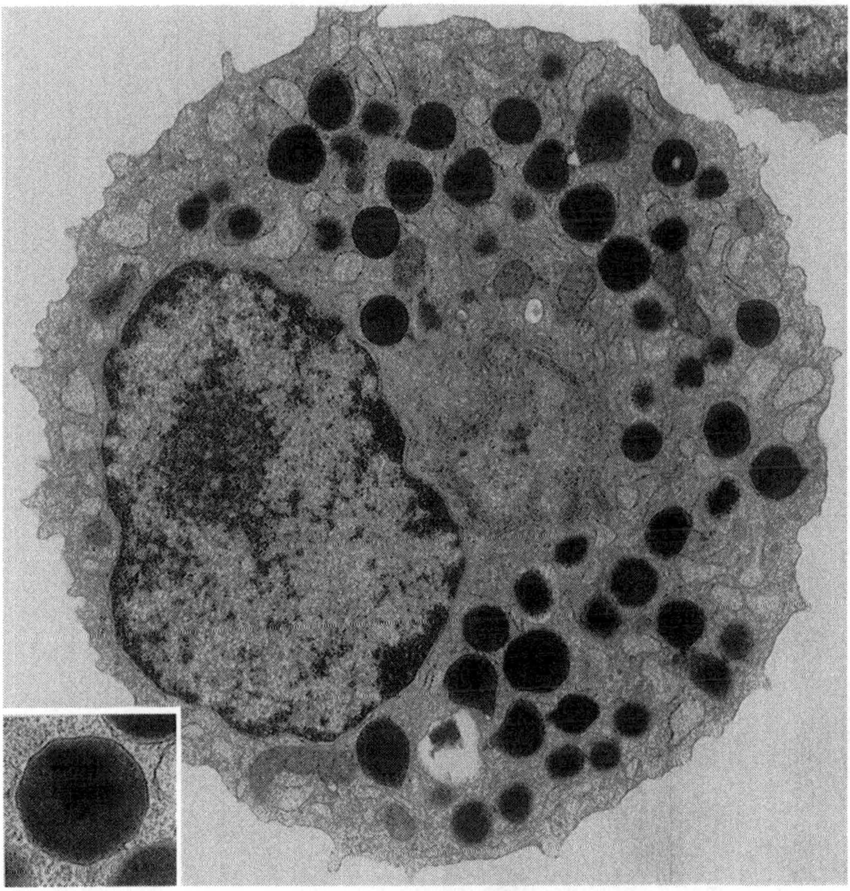

Figure 5 Eosinophilic myelocyte in a 3-week rhIL-3-containing culture of human cord blood mononuclear cells shows an active Golgi area, large mitochondria, dilated cisterns of rough endoplasmic reticulum, and an extensive population of immature and maturing secondary granules with central dense cores (inset, higher magnification of one specific granule with core) and less dense matrix compartments. ×10,560; inset ×29,040. (With permission from Ref. 1.)

ing. Major basic protein (MBP)—a cyotoxic protein known to damage a wide variety of mammalian cells (22,23) and parasites (24–27)—is present in the core compartment (28); eosinophil-derived neurotoxin (EDN) and eosinophil cationic protein (ECP)—neurotoxic (29,30) and membrane lytic (31) proteins—reside in the matrix compartment (28); eosinophil peroxidase (EPO), which amplifies tumor and parasite killing (32–36), is also present only in the matrix compartment (1,5). Secondary granules do not store CLC protein (9).

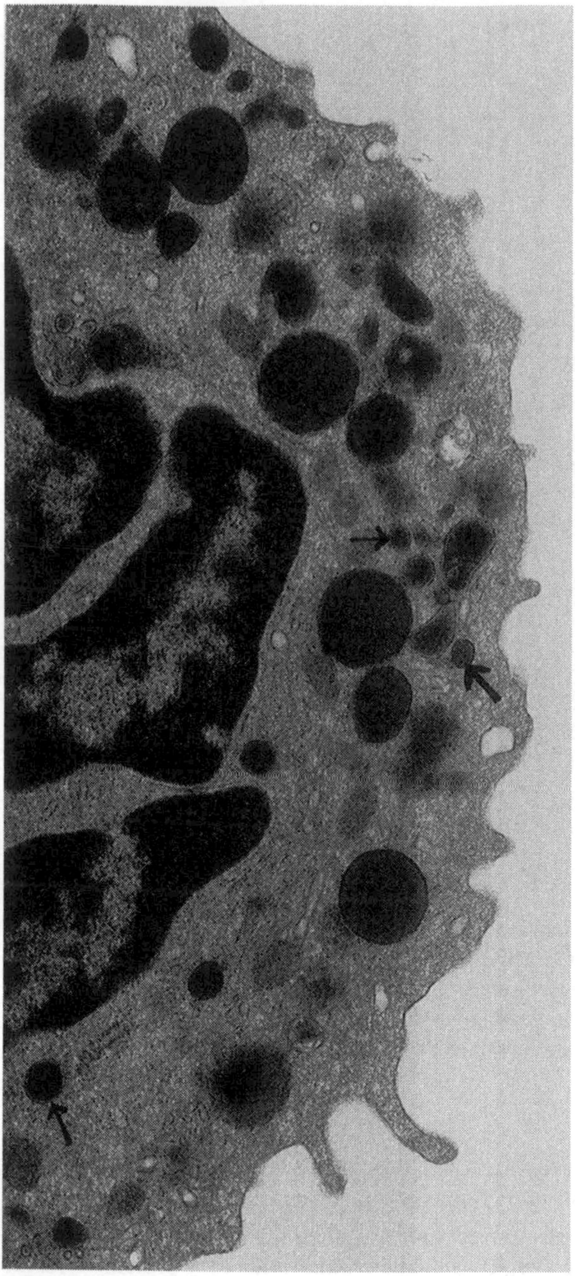

Figure 6 Mature eosinophil in a 5-week rhIL-5-containing culture of human cord blood mononuclear cells shows increased numbers of small dense granules (arrows). ×26,500. (With permission from Ref. 3.)

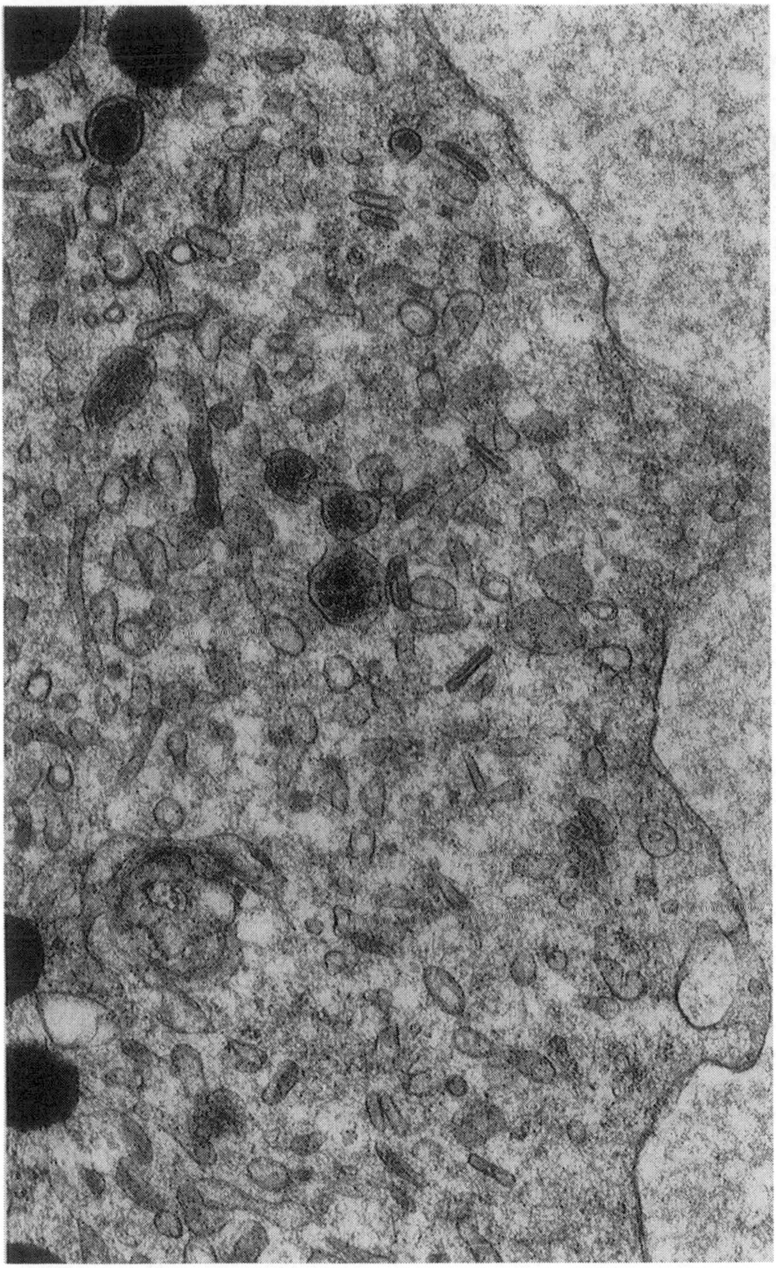

Figure 7 Peripheral blood eosinophil, from a patient with HES, showing cytoplasmic area devoid of secondary granules. This area is filled with vesiculotubular organelles. ×33,000. (With permission from Ref. 1.)

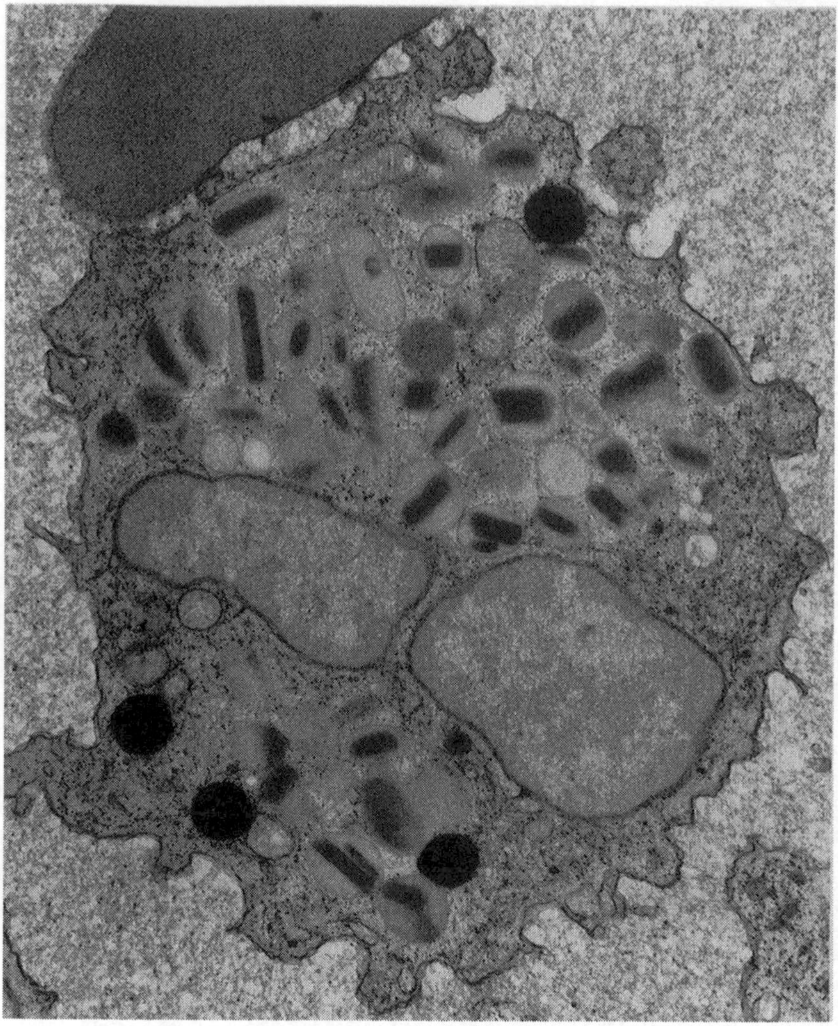

Figure 8 Mature peripheral blood eosinophil, from a patient with HES, shows a mixture of bicompartmental secondary granules, several primary granules that do not contain dense central cores, and four osmiophilic lipid bodies. ×14,000. (With permission from Ref. 1.)

Table 1 Subcellular Sites of Eosinophil Products Determined by Ultrastructural Methods

Product	Site	Ref.
Major basic protein	Secondary granule core compartment	28
Eosinophil-derived neurotoxin	Secondary granule matrix compartment	28
Eosinophil cationic protein	Secondary granule matrix compartment	28
Eosinophil peroxidase	Secondary granule matrix compartment	1,5
Eosinophil peroxidase	Immature granules, Golgi, cisterns of RER, and perinuclear area	1,2,7,68
Charcot-Leyden crystal protein	Primary granules	1,4,9
Charcot-Leyden crystal protein	Cytoplasm, nucleus	4,5,10,15
Charcot-Leyden crystal protein	CLC	1
Acid phosphatase	Small granules	17
Arylsulfatase	Small and secondary granules	17
Catalase	Small granules	37
Reactive oxygen products	Vesiculotubular structures	38
Arachidonic acid	Lipid bodies	1, 20, 21
Prostaglandin endoperoxide synthase	Lipid bodies	Dvorak AM, Weller PF, unpublished data

B. Primary Granules

This recently documented granule population (Fig. 9B) (9) is the sole granule repository of CLC protein, as determined by immunogold staining (Fig. 10) (9). The function(s) of CLC protein is currently not known.

C. Small Granules

These small structures (Fig. 6) (<0.5 μm) have been shown to contain hydrolytic enzymes (acid phosphatase, arylsulfatase) by cytochemical studies (17), and one immunogold analysis localized catalase to them (37). Thus, their functions may be analogous to that of lysosomes and peroxisomes (microbodies) in other cells.

D. Vesiculotubular Structures

These structures (Fig. 7) have recently been implicated in reactive oxygen product formation by immunogold staining of cytochrome b_{558} [a component of the

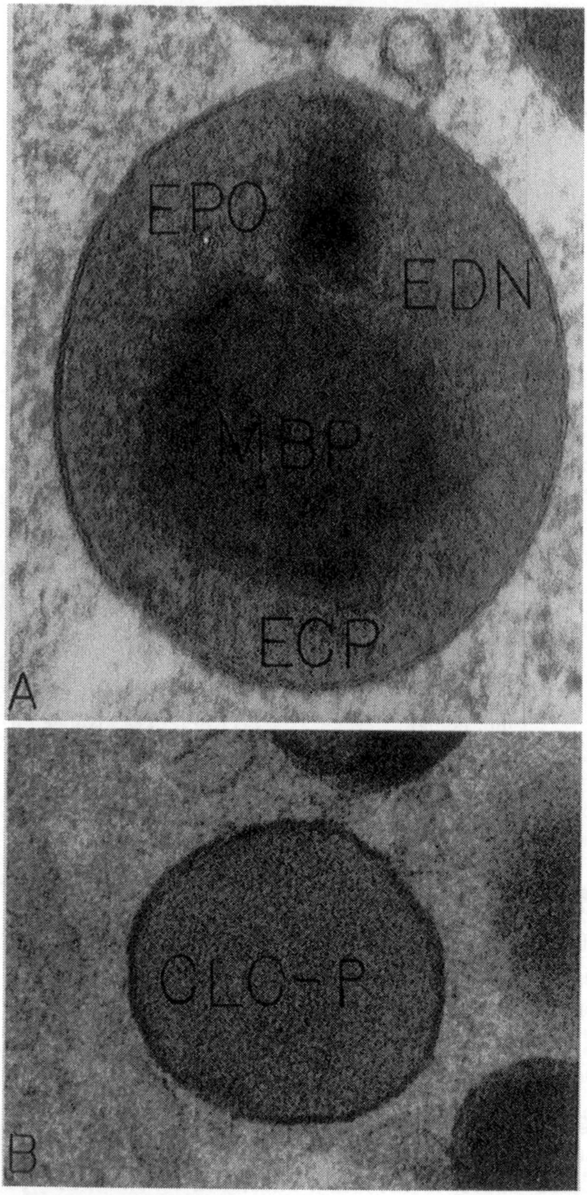

Figure 9 Two granule populations found in mature eosinophils have different contents. The secondary (specific) granule (A) contains MBP in the dense crystalline core and EPO, ECP, and EDN in the less dense matrix. The primary granule (B) does not contain a crystalline core and is the storage location of CLC protein (CLC-P). (A) ×152,000; (B) ×88,000.

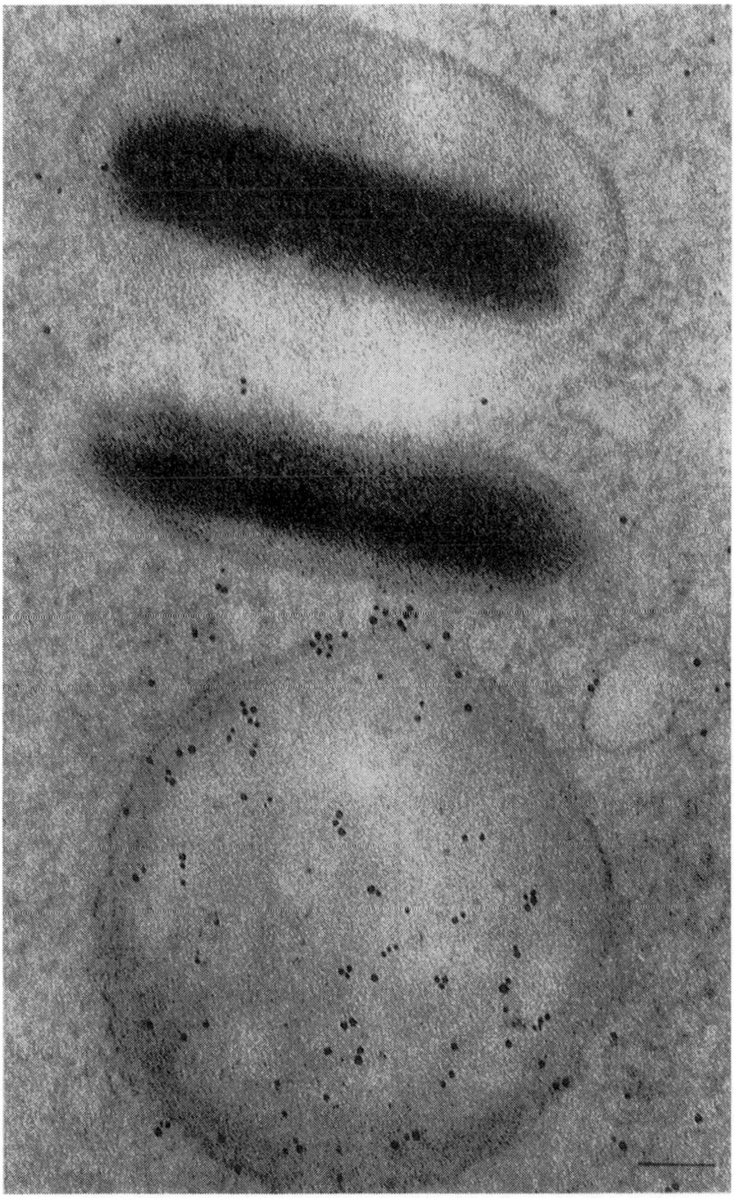

Figure 10 Eosinophil granules in peripheral blood eosinophil of a patient with HES that is stained for CLC protein with immunogold. The large, round, coreless primary granule contains many 5-nm gold particles; the large, core-containing specific granules are not stained. ×116,000. (With permission from Ref. 9.)

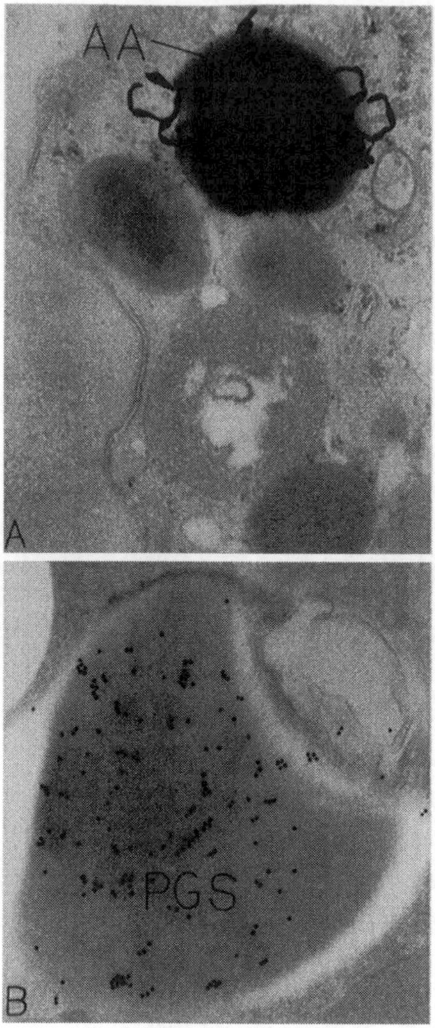

Figure 11 Preparations of eosinophils using autoradiography (A) and immunogold (B) show lipid bodies labeled with silver grains, indicating the presence of [^{3}H]-arachidonic acid (AA), the substrate for eicosanoid synthesis (A), and labeled with 20-nm gold, indicating the presence of prostaglandin endoperoxide synthase (PGS), the rate-limiting enzyme for oxidation of arachidonic acid to produce prostaglandins (B). (A) ×30,000 (with permission from Ref. 1); (B) ×46,000 (before reduction).

NADPH-oxide system (38)]. Superoxide O_2^- is essential for cytotoxicity to catalase-positive bacteria (39).

E. Lipid Bodies

Eosinophils participate in inflammation by generating proinflammatory mediators, derived from the oxidative metabolism of arachidonic acid (reviewed in 40). While the dogma (supported by little experimental data) is that eicosanoid pathways of production occur in the plasma membrane, we considered the possibility that a cytoplasmic organelle [the lipid body (Fig. 8) (41–44)] might have an important role in the generation of products of arachidonic acid metabolism. To this end, we performed combined ultrastructural-biochemical studies to determine the role(s) of lipid bodies in eosinophil metabolism leading to eicosanoid synthesis. Initially, we showed by ultrastructural autoradiography (20) that eosinophil lipid bodies (but not plasma membrane) incorporated [^{3}H]-arachidonic acid rapidly and in quantity (Fig. 11A). In this same study, we determined that the majority of the [^{3}H] label was incorporated into cellular phospholipids (20)—the predominant precursor material for arachidonic acid release. Subcellular fractions of eosinophils labeled with [^{3}H]-arachidonate were next prepared (21). These studies showed that, in purified lipid bodies, the [^{3}H]-arachidonate was esterified almost totally into glycerolipids, predominantly in classes of phospholipids including phosphatidylinositol and phosphatidylcholine (21). Thus, it seems clear that the substrate (phospholipids → arachidonic acid) for eicosanoid production is generally present in eosinophil lipid bodies. More recently, we have used combined immunomorphological methods to demonstrate the rate-limiting enzyme for prostaglandin synthesis in eosinophil lipid bodies (8,40; Dvorak AM et al., unpublished data). For example, using several specific primary antibodies and reporter systems by light and electron microscopy, it is clear that prostaglandin endoperoxide synthase (PG synthase) is contained within eosinophil lipid bodies (Fig. 11B) and is not present in plasma membranes (8,40; Dvorak AM et al., unpublished data). Thus, eosinophil lipid bodies [like lipid bodies in human mast cells, alveolar macrophages, type II alveolar pneumocytes, neutrophils (45), and 3T3 fibroblasts (40; Dvorak AM et al., unpublished data)] have the substrate and enzyme necessary to generate prostaglandins.

IV. ACTIVATED EOSINOPHIL MORPHOLOGY

Activation morphology of human mature eosinophils includes changes in number, size, and content of cytoplasmic organelles (reviewed in 1). For example, changes that generally accompany the activation state include increased number and size of lipid bodies (Fig. 12), structures that may also develop internal circumscribed

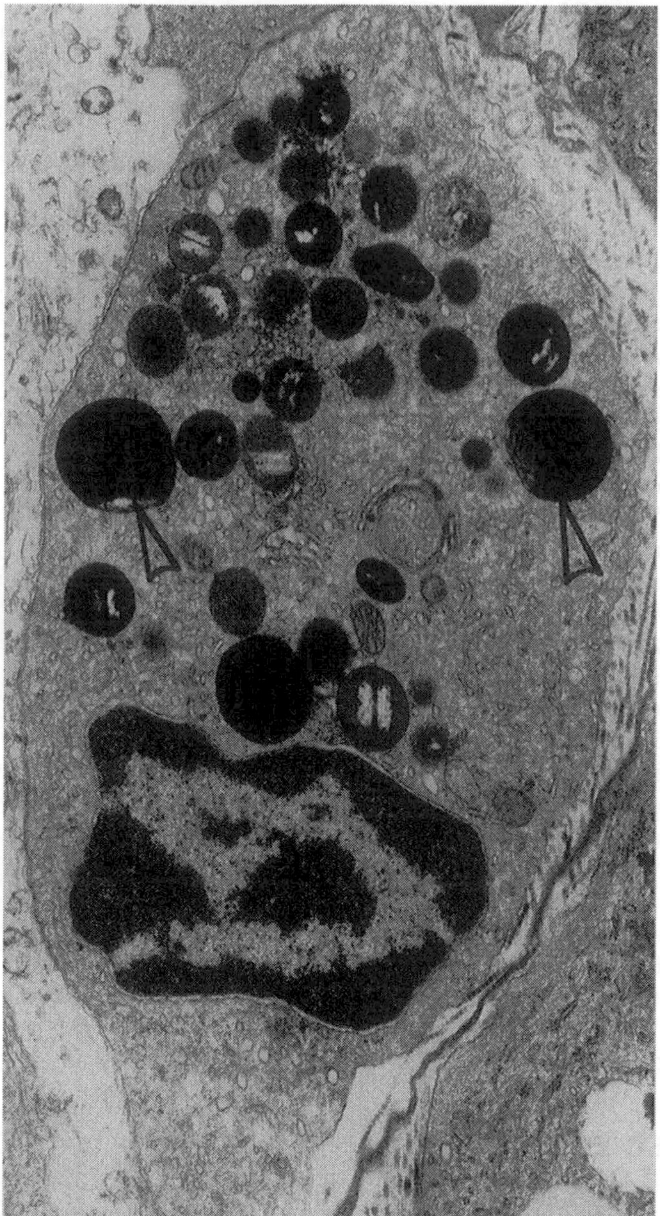

Figure 12 Tissue eosinophil in the skin of a patient with HES and a rash shows activation morphology. Lipid bodies are increased in size and number (arrowheads), and many secondary granule core compartments are focally lucent or empty. ×17,000.

lucencies and dense particles. Other organelles that increase in number include primary granules (Fig. 13), small granules (Fig. 14), and vesiculotubular structures (Fig. 7). Larger, smooth membrane-bound cytoplasmic vesicles and tubules of smooth endoplasmic reticulum, structures that are rarely present in pristine, circulating peripheral blood eosinophils, also make their appearance in activated eosinophils. Non-membrane-bound cytoplasmic CLCs may appear in activated cells (Fig. 15) (1). Specific crystalloid core-containing (secondary) granules are the only regularly present cytoplasmic organelles to be decreased in activated eosinophils (Figs. 16 and 17) (1). Immunogold studies of CLC protein localization

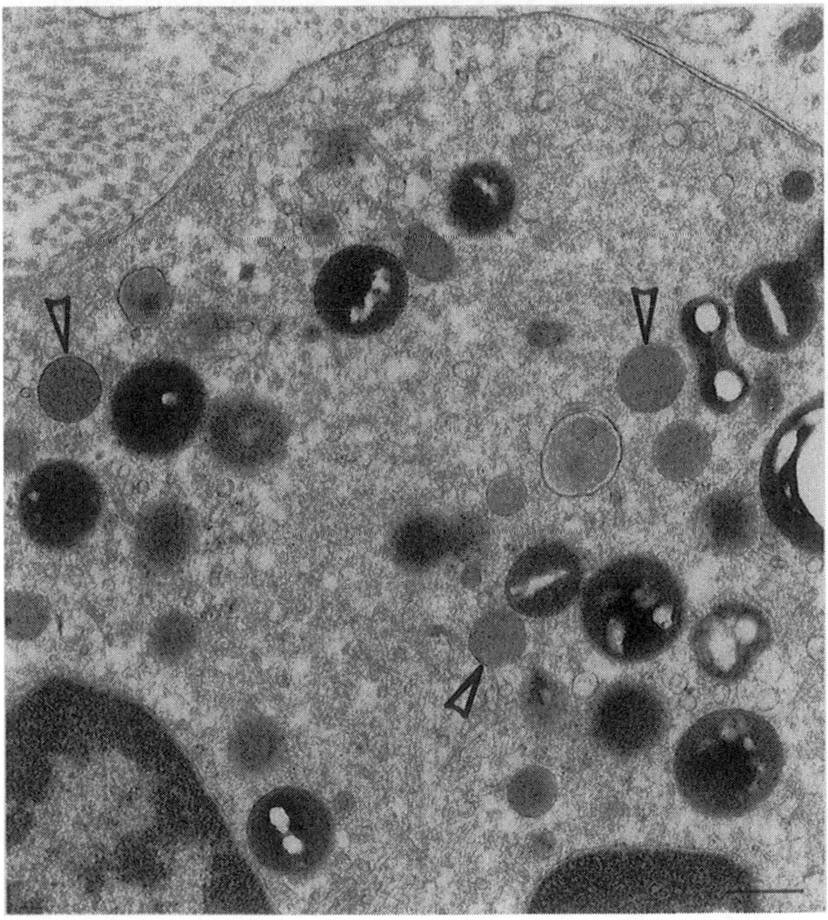

Figure 13 Tissue eosinophil (skin, HES) with increased numbers of core-free primary granules (arrowheads). Bar = 0.4 μm. (With permission, from Ref. 1.)

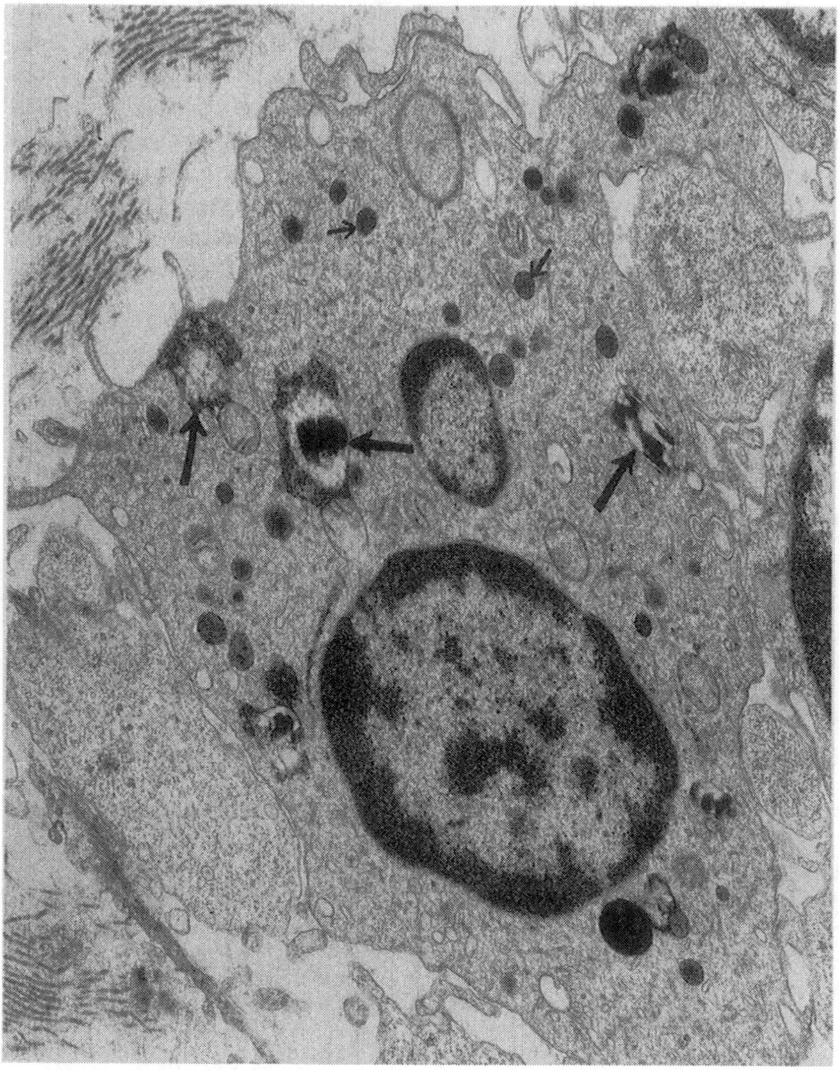

Figure 14 Tissue eosinophil (ileum, Crohn's disease) with increased small granules (short arrows), several residual secondary granules with cores (long arrows), and focal clouds of dense material bound to the cytoplasm adjacent to a secondary granule and the plasma membrane. ×10,000.

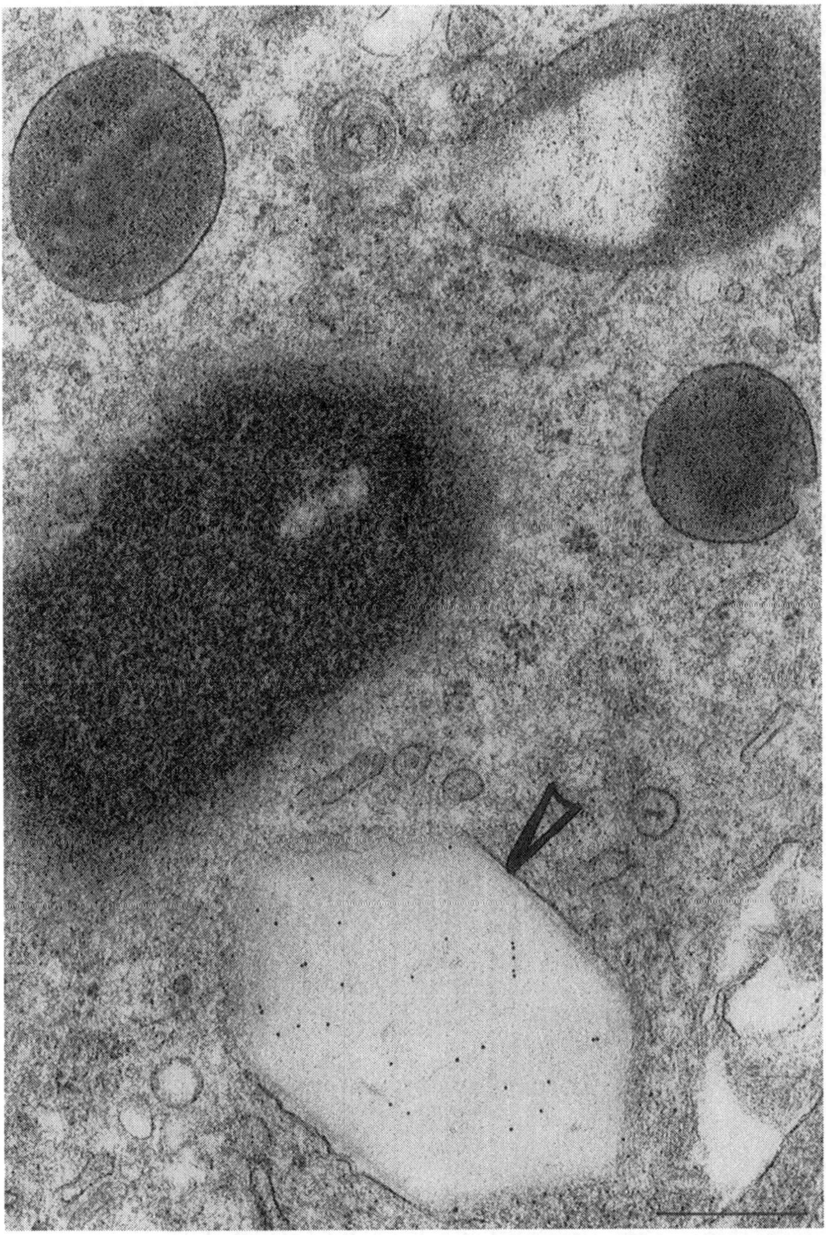

Figure 15 Tissue eosinophil (ileum, bacteria-infected) stained with immunogold shows the presence of CLC protein in the cytoplasmic hexagonal CLC (arrowhead). Bar = 0.4 μm. (With permission from Ref. 1.)

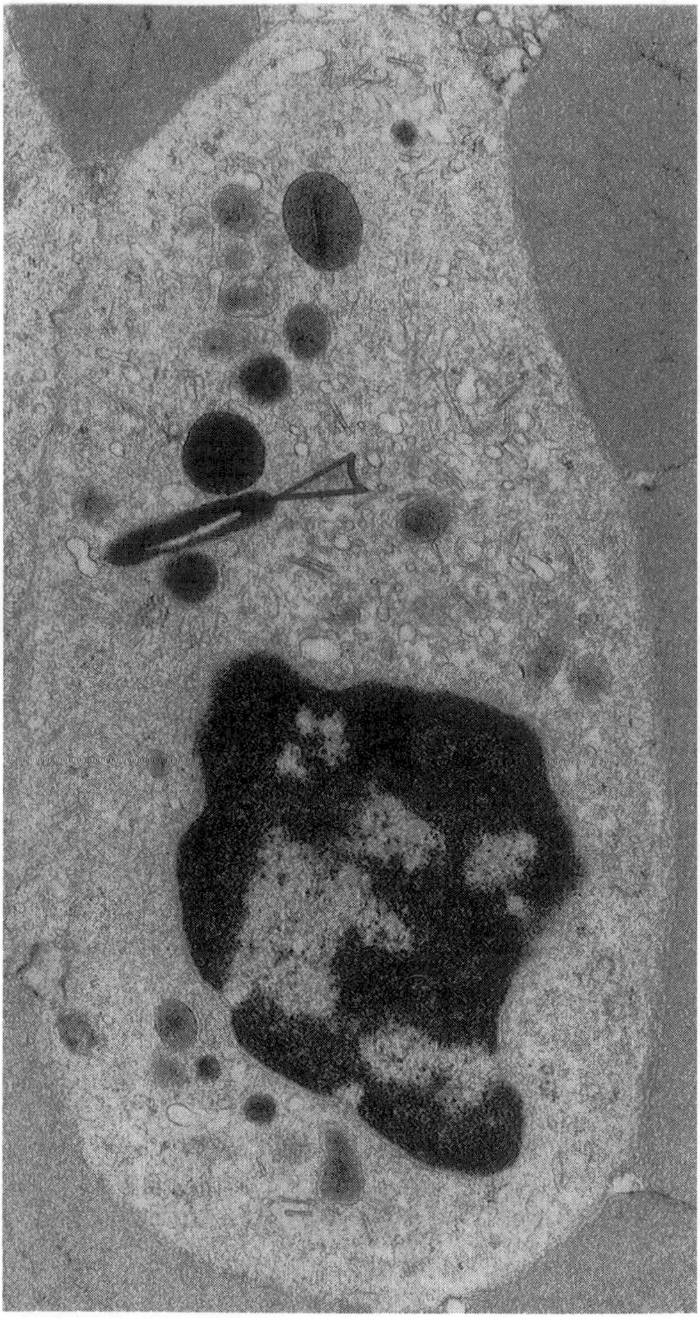

Figure 16 Peripheral blood eosinophil (HES) shows marked reduction in the specific granules (arrowhead). ×24,000.

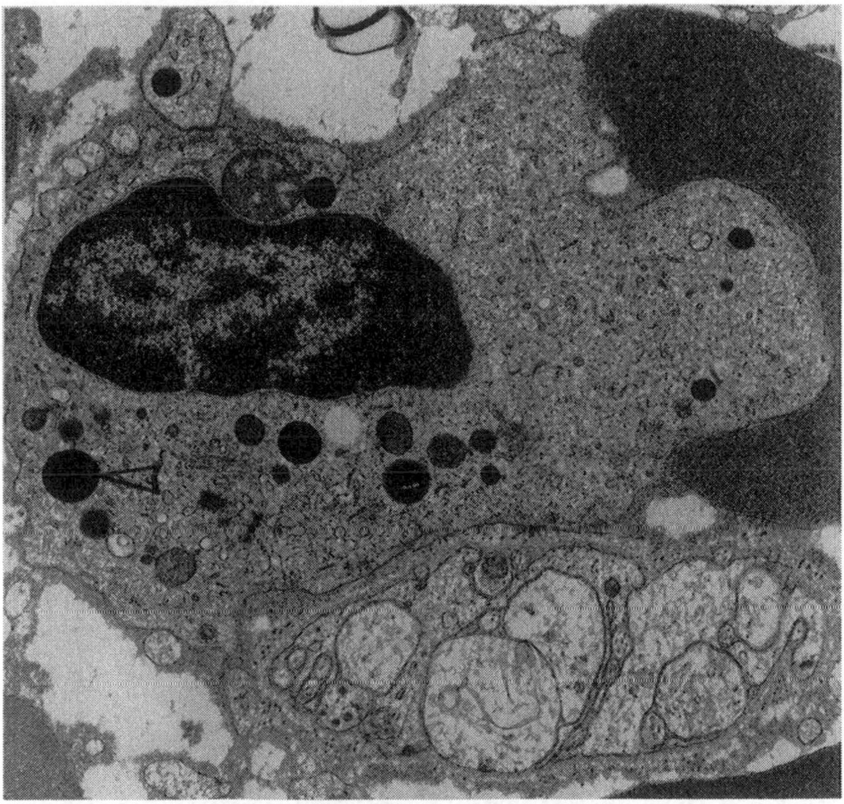

Figure 17 Tissue eosinophil (ileum, Crohn's disease) shows marked reduction in specific granules (arrowhead). Note the fascicle of swollen damaged axons beneath the hypogranular eosinophil. ×12,870.

in activated eosinophils reveal extensive, diffuse cytoplasmic and nuclear sites (Fig. 18) in addition to primary granules. Particularly, granule-poor areas and subplasma membrane areas are heavily labeled (Fig. 18) (4,5,10,15).

The morphology of activated eosinophils has been documented in a number of circumstances (Table 2). For example, tissue and body cavity fluid eosinophils that have migrated from the peripheral blood virtually always show some combination of changes reflecting activation (1) (or secretion—see later). In certain disorders—i.e., the hypereosinophilic syndrome (HES)—activation morphology has been documented in peripheral blood eosinophils (Figs. 7 and 16) (9,46,47) as well as in tissue eosinophils (Figs. 12 and 13) (10).

Activation morphology has been associated with cultured eosinophils in a variety of circumstances (Table 2). Initially, when human cord blood mononuclear cells were cultured in the presence of either rhIL-3 (2,3), rhIL-5 (2,3), or rhIL-5

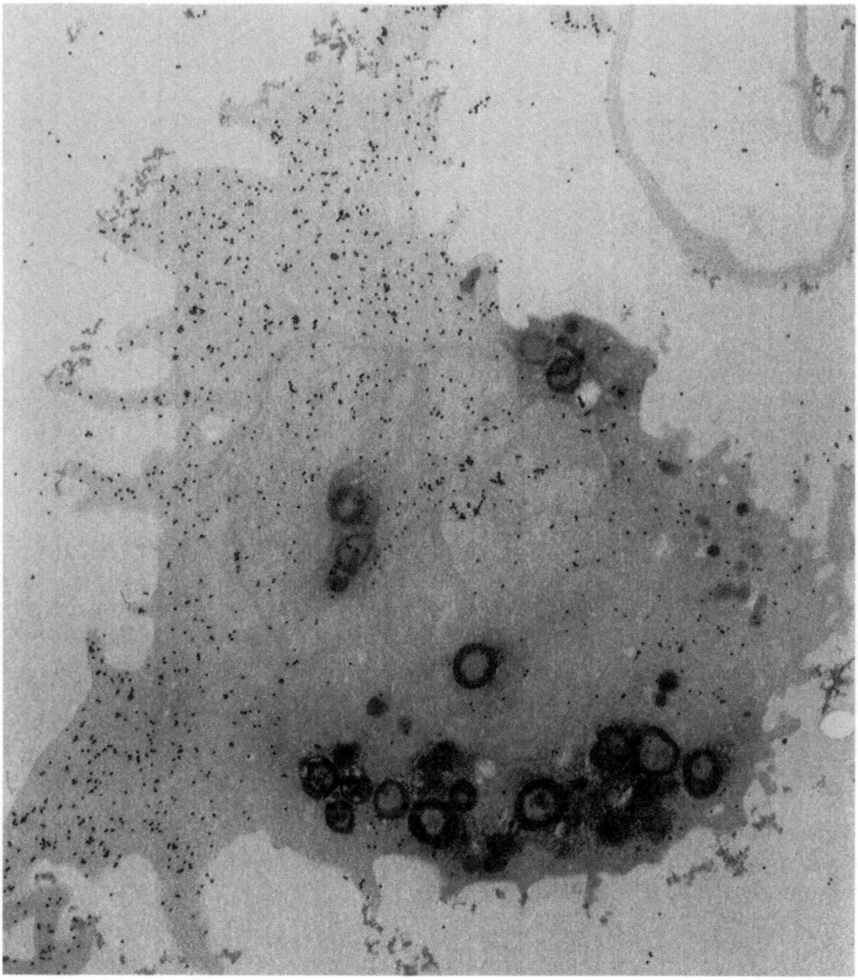

Figure 18 Activated eosinophil in rhIL-5-containing culture prepared with cytochemistry to demonstrate EPO in secondary granule matrix compartments (note lack of reaction in core compartments) and with immunogold to demonstrate CLC protein. CLC protein is diffusely present (20-nm gold particles) in the nucleus and cytoplasm. Note that CLC protein is present in the subplasma membrane area (but not on the plasma membrane) and that released dense aggregates of EPO are bound to the plasma membrane. ×10,920.

Table 2 Circumstances in Which Activation Morphology of Human Eosinophils Occurs

Circumstance	Ref.
In vivo	
Crohn's disease—gastrointestinal tract	11–14
Ulcerative colitis—gastrointestinal tract	Dvorak AM, unpublished data
Bacterial-infected continent pouches—gastrointestinal tract	Dvorak AM, unpublished data
Multiple disorders—esophageal, gastric, small and large intestinal diagnostic biopsies	Dvorak AM, unpublished data
Hodgkin's disease—lymph nodes	18
HES—blood, bone marrow, skin	10,46,47
Bullous pemphigoid—skin	49
Multiple disorders—diagnostic skin biopsies	Dvorak AM, unpublished data
Multiple disorders—lung biopsies	Dvorak AM, unpublished data
Pleural, peritoneal fluids—diagnostic collections	1; Dvorak AM, unpublished data
Primary, benign tumors; primary and metastatic malignant tumors—diagnostic biopsies from multiple tissue sites	Dvorak AM, unpublished data
Histiocytosis X—bone	Dvorak AM, unpublished data
In vitro	
Human cord blood cells cultured in rhIL-3	2,3
Human cord blood cells cultured in rhIL-5	2,3
Human cord blood cells cultured in rhIL-5 and IL-2-free fraction of PHA-stimulated human T-cell-conditioned medium	4,5
Human cord blood cells cultured in conditioned medium that contained IL-4, IL-5, and GM-CSF of Th 2 helper human T-cell clone	48
Peripheral blood eosinophils cultured in GM-CSF	46
Peripheral blood eosinophils cultured in GM-CSF and fibroblasts	46
Peripheral blood eosinophils cultured with endothelial cells	46

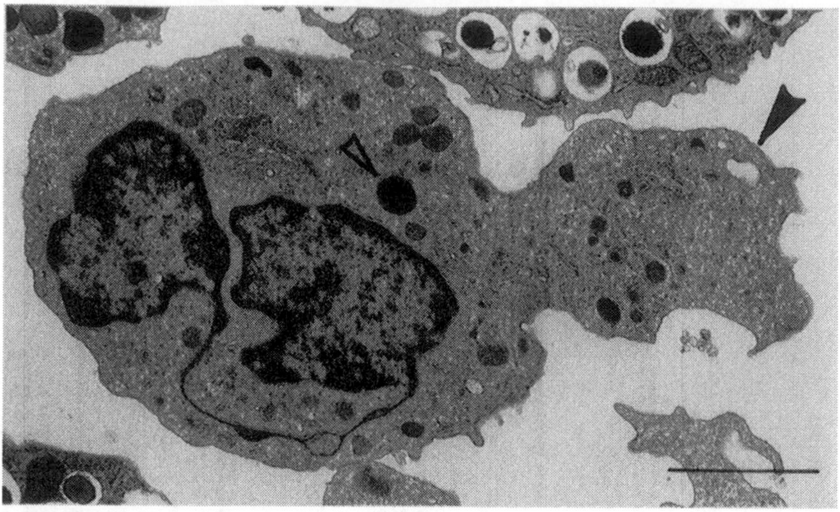

Figure 19 Mature eosinophil, which developed in a 5-week culture of cord blood mononuclear cells containing helper T-cell conditioned medium, shows activation morphology characterized by decreased specific granules (open arrowhead) and motile structure, or uropod (closed arrowhead). Bar = 2.4 μm. (With permission from Ref. 1.)

supplemented with an IL-2-free fraction of PHA-stimulated human-T-cell-conditioned medium (4,5), variable numbers of mature eosinophils expressed activation morphology. Similarly, when human cord blood cells were cultured in the IL-4, IL-5, and GM-CSF-containing conditioned media produced by a cloned Th 2 human helper T cell, mature eosinophils expressed activation morphology (48). Additionally, in this circumstance, many mature cells also expressed motility morphology, or uropods (Fig. 19) (48). Later, peripheral blood eosinophils were cultured with either GM-CSF, GM-CSF plus fibroblasts, or endothelial cells and found by morphometric analysis to express many features of activation morphology (46).

V. RELEASING EOSINOPHIL MORPHOLOGY

A. Cell Injury

Many reactive tissue eosinophils express classic morphological criteria of cell injury and cell death. We have found this to be the case in large numbers of tissue eosinophils present in skin biopsies of patients with bullous pemphigoid (BP) (49) and HES (Figs. 20 and 21) (10). More recently, we have noted the morphology of apoptosis and necrosis in eosinophils cocultured with 3T3 fibroblasts and the human immune deficiency virus (HIV) (50). Classic apoptosis is accompanied by extreme nuclear chromatin condensation and the formation of nuclear apoptotic bodies in the absence of plasma membrane damage. Plasma membrane damage results in cytoplasmic swelling and classical necrosis (51). Some authors describe these events as a continuum, such that early apoptosis may lead later to membrane damage and lysis of cells. This morphological continuum was present in eosinophils that were cocultured with 3T3 fibroblasts and HIV (50). In skin biopsies of patients with BP or HES, however, the primary process (at the time of biopsy) involving eosinophils was necrosis (Figs. 20 and 21) (10,49). That is, numerous eosinophils displayed centralization of granules, plasma and organelle membrane breakage, and chromatolysis of their nuclei (Fig. 21). The end result of these cell

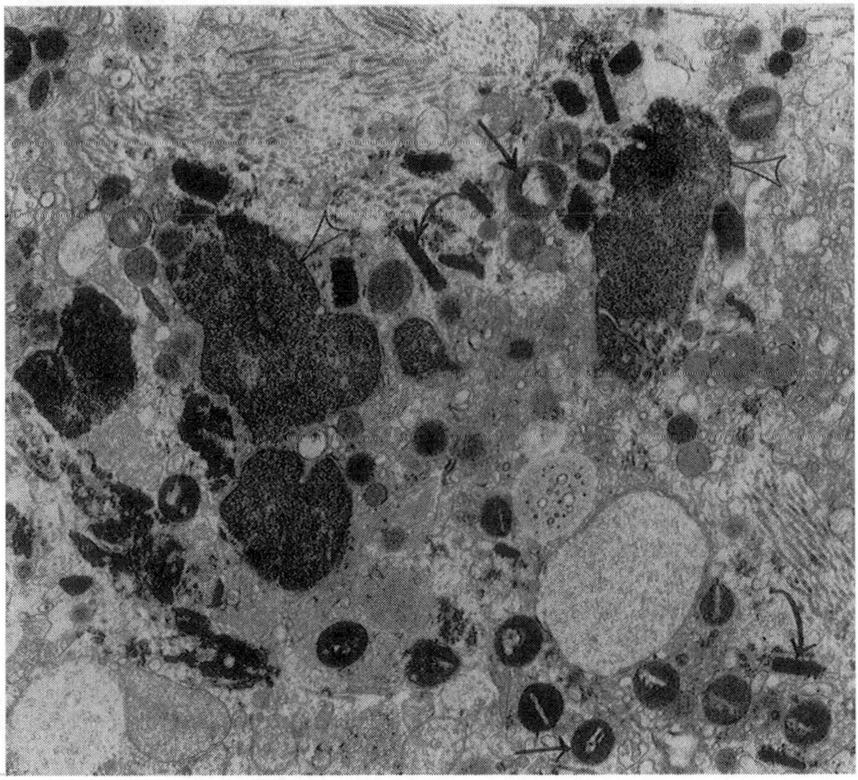

Figure 20 Skin biopsy of a patient with HES shows necrosis of eosinophils. Damaged eosinophil nuclei (open arrowheads), spilled membrane-bound secondary granules with core compartment losses (straight arrows), and free, dense cores (curved arrows) are present among dermal collagen arrays. ×7400.

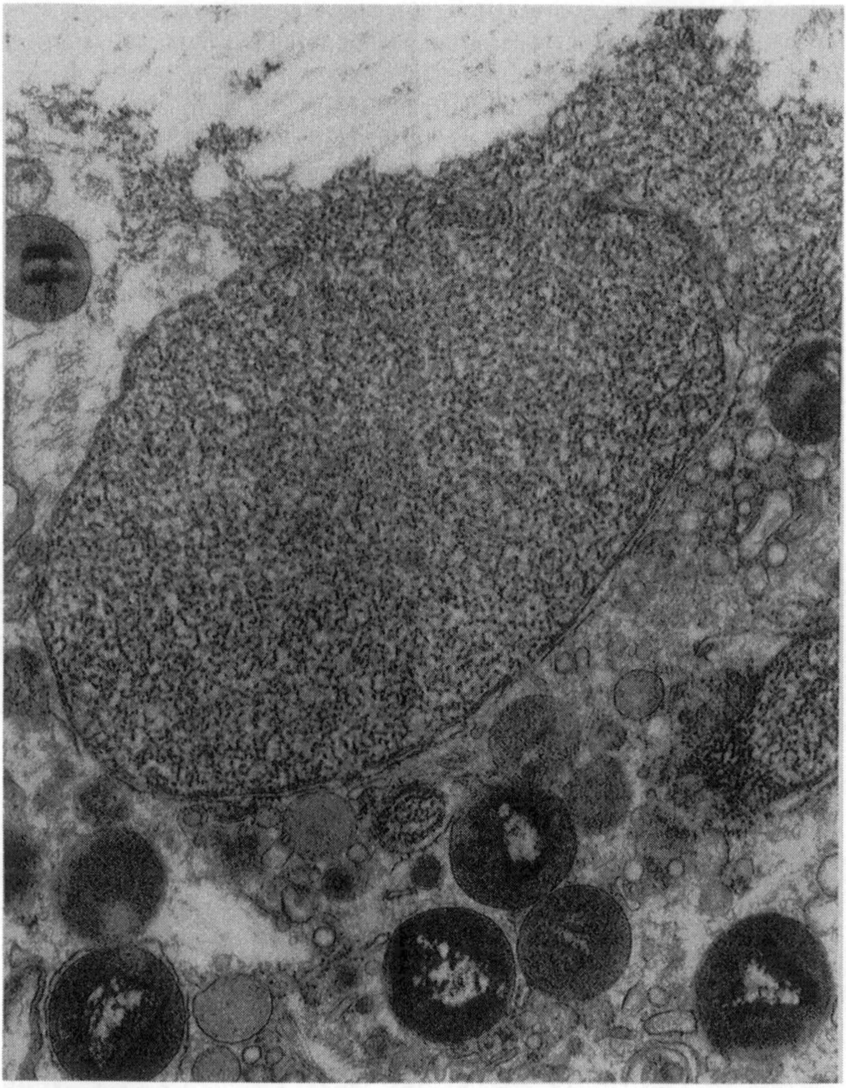

Figure 21 Higher magnification of similar biopsy shows eosinophil necrosis characterized by chromatolysis of the nucleus. Note streaming of damaged chromatin through the broken nuclear membrane into adjacent interstitial collagen beside a free, membrane-bound secondary granule. ×22,750.

injury patterns is to release membrane-bound or partially membrane-bound secretory granules into interstitial tissues (Fig. 20). Thus, although originating in cell injury patterns, these forms of eosinophil damage may contribute potent mediators of inflammation to inflammatory sites.

B. Cell Secretion

Eosinophils are classic secretory cells that store important cell synthetic products in membrane-bound cytoplasmic secretory granules. Potent cytotoxic granule proteins can be released to the microenvironment by extrusion of membrane-free granule contents (Fig. 22) (1; Dvorak AM, unpublished data). This anatomical process resembles that of regulated secretion of granulated secretory cells, in general, and specifically of anaphylactic degranulation of mast cells and basophils (6). It is effected in two ways. One, individual granule membranes and portions of the plasma membrane fuse around the periphery of cells, and specific granule matrix and core contents are extruded through these degranulation pores (Fig. 22); two, granule membranes may fuse within the cytoplasm to form degranulation chambers, within which membrane-free matrix and core components of specific granules are located (Fig. 23). Some of these chambers also open to the cell exterior through degranulation pores (Fig. 22). The end result of this form of eosinophil degranulation is the reduction in cytoplasmic-specific granules. While eosinophils are plentiful in many circumstances in biopsied tissues, the ultrastructural documentation of this event in vivo is rare. We recently have found this form of secretion from mature eosinophils to be prevalent in biopsies of ileal tissues from patients who were demonstrated to have inflammation and tissue-invasive bacterial infections (Figs. 22 and 23) (Dvorak AM, unpublished data). Similarly, we and others have found that immature eosinophils, or eosinophilic myelocytes, utilize a similar extrusion mechanism of membrane-free primary granules in vivo in biopsies of bone marrow (Fig. 3) (1,8,52–55). Peripheral blood eosinophils from normal and HES donors have been examined for ultrastructural mechanisms of regulated secretion stimulated by the calcium ionophore A23187 (56,57). In each case, extrusion of granules to the exterior, into enlarging cytoplasmic vacuoles, and diminished numbers of cytoplasmic granules were reported (56,57). Thus, it is clear that mature and immature human eosinophils in vivo and in vitro undergo classic regulated secretory events.

We have observed a continuum of morphologies associated with the release of specific granule contents from activated, undamaged tissue eosinophils. Initially, we observed ragged losses from the core compartment of secondary granules in mature tissue eosinophils (Fig. 24) in ileal samples from patients with Crohn's disease (11–14,58). Later, in in vivo samples, losses from the core compartment, from the matrix compartment, and from both compartments of secondary granules were noted (Fig. 25) (reviewed in 1). The end morphological result of such losses is

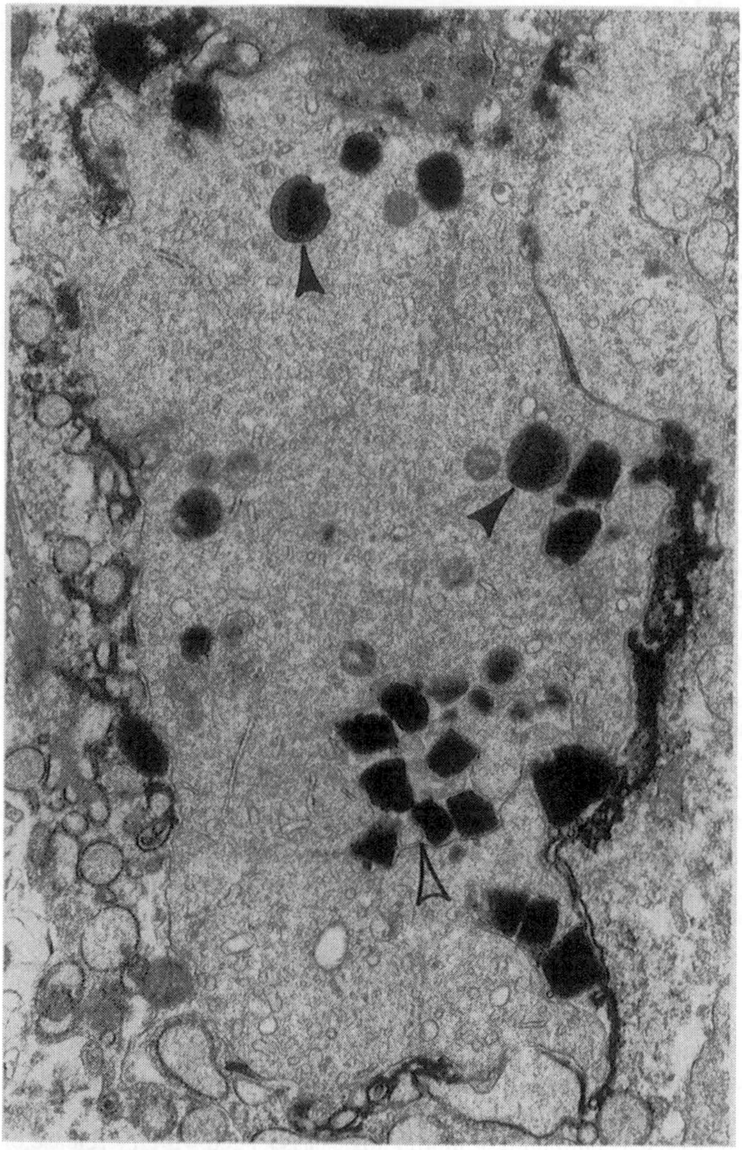

Figure 22 Bacteria-infected pouchitis patient with ulcerative colitis shows regulated secretion by extrusion of membrane-free secondary granules through multiple openings around the eosinophil. Both cores and diffuse linear matrix material surround the degranulating cell. Several unaltered secondary granules remain in the cytoplasm (closed arrowheads). A cytoplasmic degranulation chamber is filled with extruded granule cores (open arrowhead). ×15,000 (before reduction).

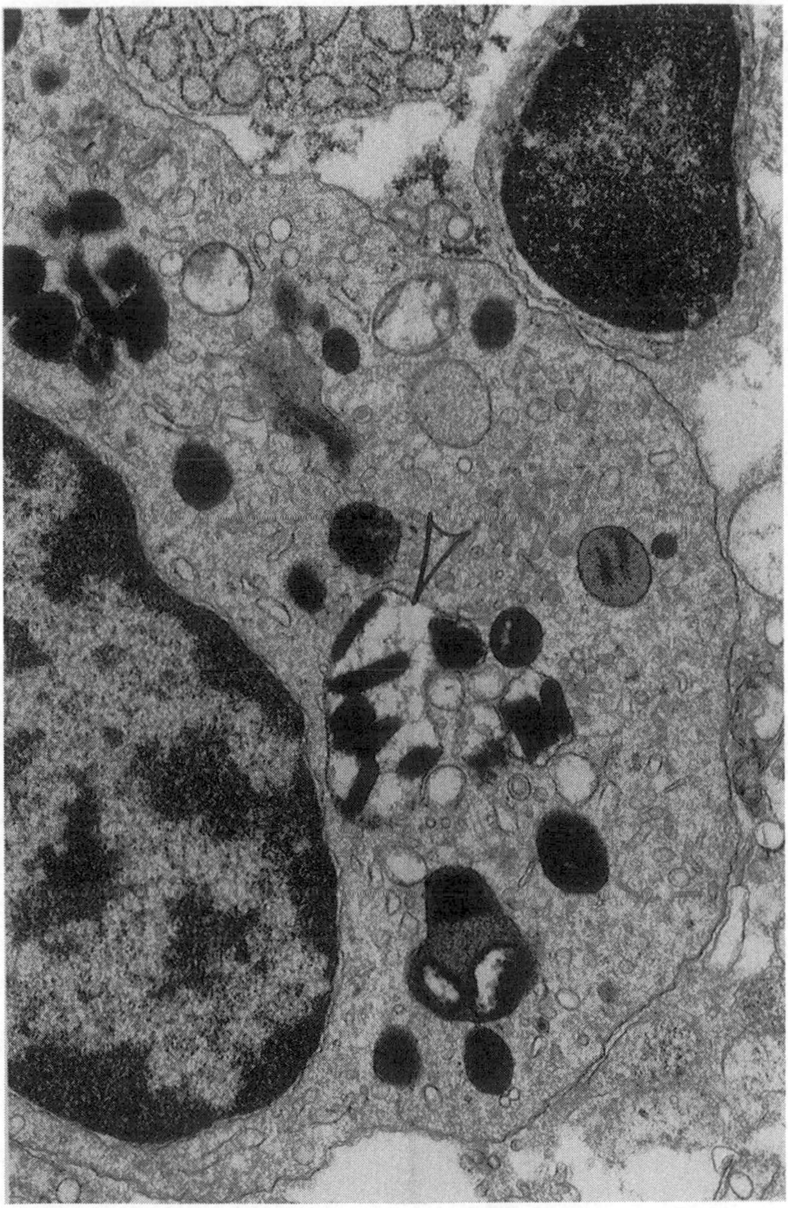

Figure 23 This eosinophil (similar source as Fig. 22) also shows a core-filled cytoplasmic degranulation chamber (arrowhead). ×17,000 (before reduction).

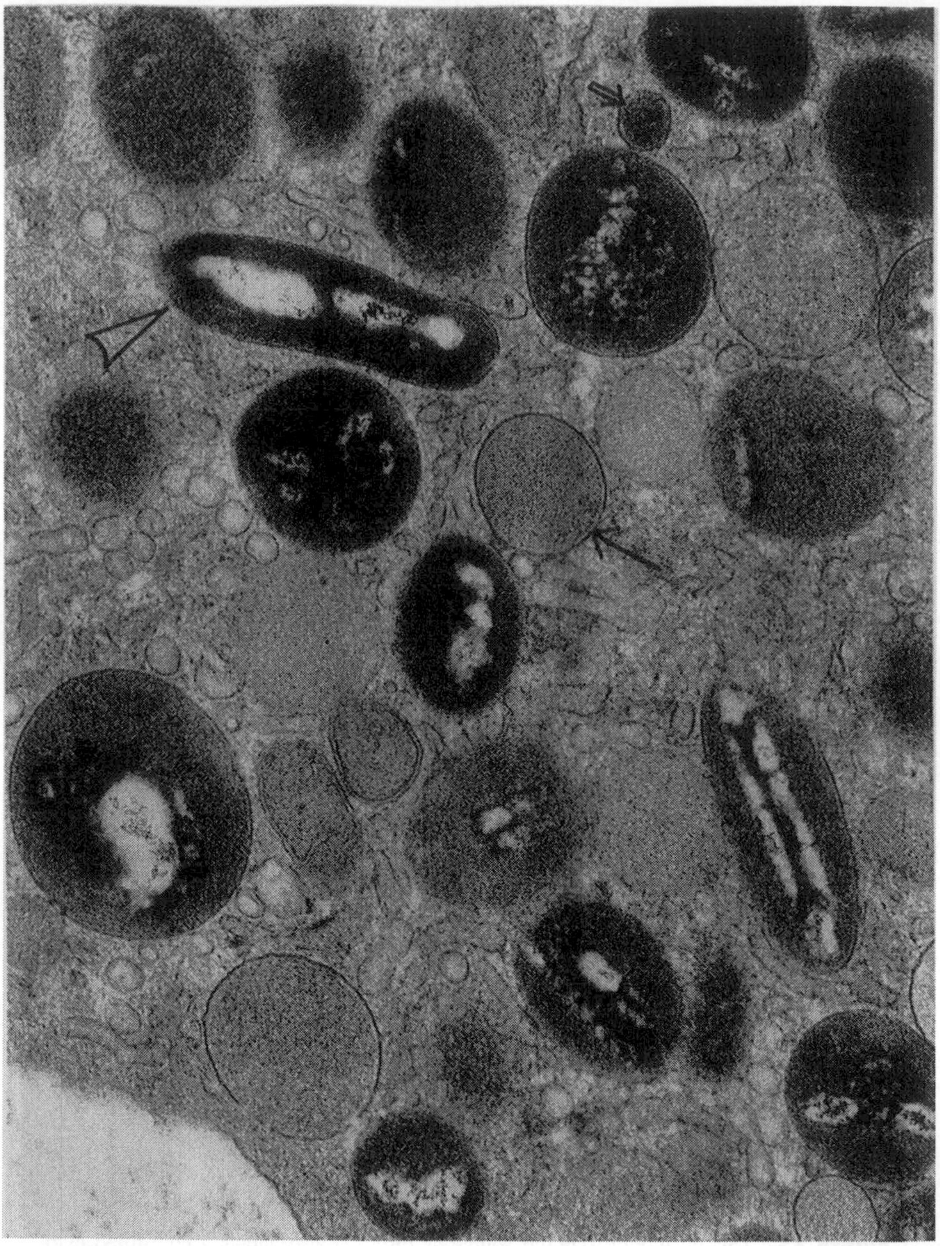

Figure 24 Tissue eosinophil (skin, HES) shows ragged losses from specific granule cores (arrowhead) and increased numbers of primary granules (closed arrow). A small granule (open arrow) and a number of cytoplasmic vesicles are present. ×52,000 (before reduction).

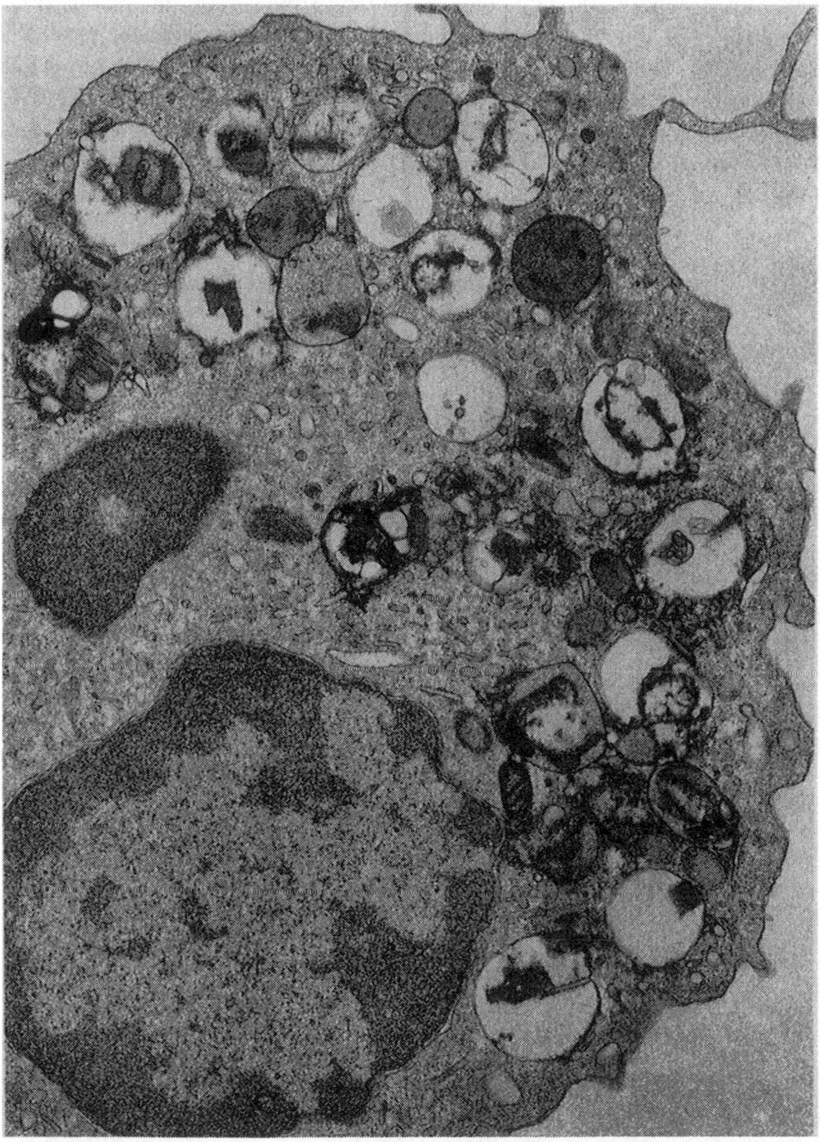

Figure 25 Well-preserved eosinophil, obtained from a malignant pleural effusion, showing in situ losses from secondary granule matrix and core compartments. Some granules are empty or nearly empty. ×18,500. (With permission from Ref. 1.)

the presence of empty granule containers in the cytoplasm of viable cells (Fig. 25)—a process analogous to piecemeal degranulation of basophils and mast cells (reviewed in 6). Other investigators have noted similar losses from the major granule population of human eosinophils in a number of circumstances (59). We noted a similar mechanism of secretion in mature human eosinophils arising in cultures of cord blood mononuclear cells that were supplemented with rhIL-5 and a

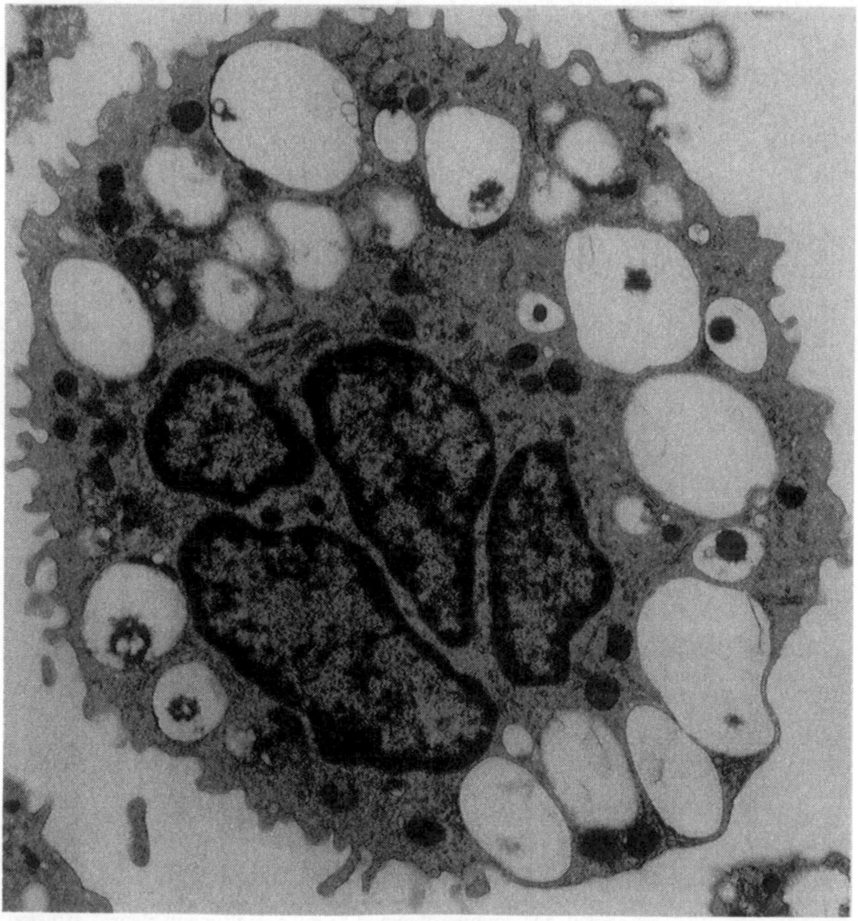

Figure 26 Mature eosinophil grown in rhIL-5 shows polylobed granulocyte nucleus and numerous enlarged, nonfused, empty and partially empty secondary granule containers typical of piecemeal degranulation. Partially empty secondary granule containers retain some dense granule material and vesicles. ×15,500 (before reduction). (With permission from Ref. 4.)

fraction of the culture supernatant of phytohemagglutinin-stimulated human T lymphocytes from which IL-2 was eliminated (4). Routine ultrastructural examination of these mature cells revealed extensive PMD (Fig. 26) characterized by partially empty and empty secondary granule chambers in the cytoplasm. Small, smooth membrane-bound vesicles were evident within empty granule chambers as well as adjacent to them (Fig. 27). Virtually all mature eosinophils in such cultures displayed PMD of secondary granule contents and assumed the morphology of numerous tissue eosinophils encountered in pathological biopsies or fluids from body cavities in vivo (Fig. 25) (1,4).

Little is known about the actual triggers that induce the secretory morphologies in vivo that we have discussed. Biochemical studies have shown that human eosinophils stimulated with specific immunoglobulin E (IgE) release peroxidase (60); others have demonstrated secretion of EDN in response to both IgA and IgG, but not to IgE, IgM, or IgD (61). More recently, the regulatory effect of cytokines on eosinophil secretion was studied (62). In these studies, the cytokines [rIL-1, rIL-2, rIL-3, rIL-4, rIL-5, rIL-6, interferon-γ (IFN-γ), granulocyte-macrophage CSF (GM-CSF), and tumor necrosis factor (TNF)] were tested individually or in conjunction with IgA- and IgG-induced secretion of EDN from purified, normodense peripheral blood eosinophils (62). These studies showed that rIL-5 was the most potent enhancer of Ig-induced release of EDN. GM-CSF and rIL-3 also enhanced Ig-induced EDN secretion, but less potently than rIL-5 (62). GM-CSF or rIL-5 alone induced EDN secretion from peripheral blood eosinophils (62). The remaining cytokines did not stimulate eosinophil secretion. This work provides strong evidence for rIL-5 as a significant trigger for the secretion of EDN from the matrix compartment of eosinophil-specific granules (62) and supports our observation of emptying and empty matrix compartments, or PMD, in mature eosinophils cultured in rhIL-5 (Fig. 26) (4).

VI. VESICULAR TRANSPORT OF SPECIFIC GRANULE MATRIX EOSINOPHIL PEROXIDASE—A MECHANISM FOR EFFECTING PIECEMEAL DEGRANULATION FROM MATURE AND IMMATURE HUMAN EOSINOPHILS

We investigated the mechanism of PMD by mature human eosinophils that developed in rhIL-5-containing conditioned medium from cultured human cord blood mononuclear cells by a cytochemical method to detect peroxidase (5). Vesicular transport of EPO from the specific granule matrix compartment to the cell surface was associated with PMD (Fig. 28). This process involved budding of eosinophil peroxidase-loaded vesicles and tubules from specific granules (Figs.

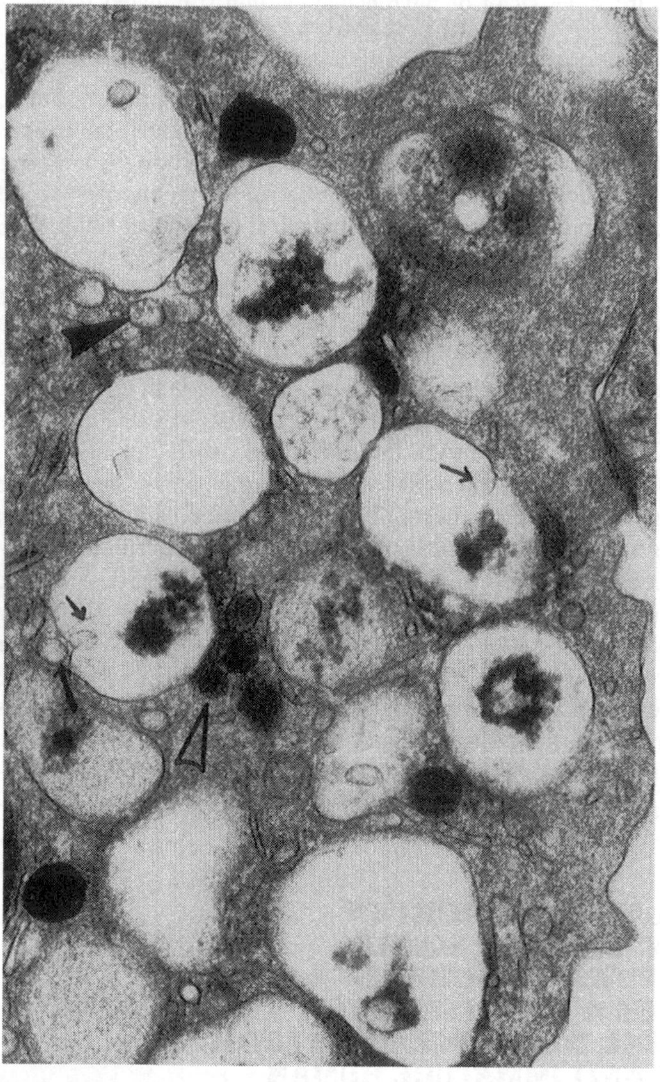

Figure 27 Piecemeal degranulation of mature eosinophil grown in rhIL-5. Note clusters of perigranular smooth vesicles, some of which are lucent (long arrow) or contain lightly dense (closed arrowhead) or very dense contents (open arrowhead). Some nonfused, empty secondary granule chambers contain residual dense granule material and lucent vesicles (short arrows). ×31,500. (With permission from Ref. 4.)

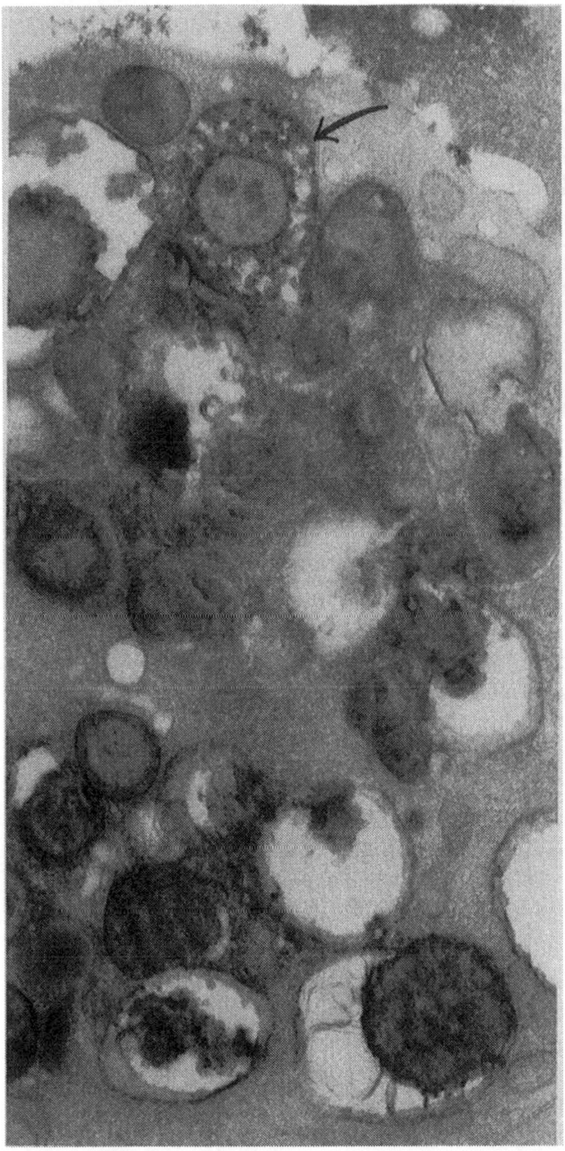

Figure 28 A mature eosinophil grown in rhIL-5 and reacted cytochemically to demonstrate peroxidase shows piecemeal degranulation. Note that some secondary granules do not contain peroxidase in their matrix compartment, whereas others do. Some granules (arrow) are enlarged with peroxidase in the expanded matrix surrounding a central core that does not contain peroxidase. The cell surface has dense peroxidase focally attached to it. ×27,000 (before reduction).

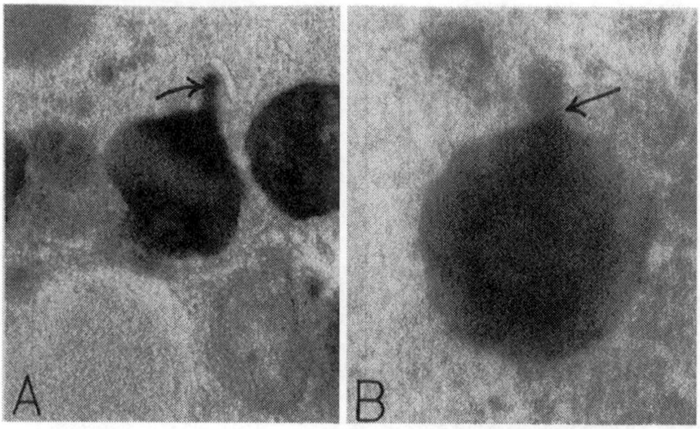

Figure 29 Similar cytochemical preparations show peroxidase-positive vesicles (arrows) attached to positive secondary granules. (A) ×36,000 (with permission from Ref. 5); (B) ×101,500.

29, 30). Some EPO that was released from eosinophils remained bound to the cell surface of releasing eosinophils (Figs. 28 and 31). These studies establish vesicular transport (Figs. 30 and 31) as a mechanism for emptying the specific eosinophil granule matrix compartment during IL-5-associated PMD (5).

Recently, immunogold studies of human eosinophils in duodenal biopsy tissues of a patient with eosinophilic gastroenteritis provided support for selective release of specific eosinophil granule contents, or PMD (63,64). For example, in our studies of granule core losses in Crohn's disease biopsy tissues, we also noted extensive intragranular and perigranular vesicular and membranous formations and associated, focally diffuse perigranular and subplasma membrane clouds of dense materials (Fig. 32) in the cytoplasm of eosinophils with core losses (13,14,58). We postulated that these tissue eosinophils were actively and selectively releasing the cytotoxic core protein, MBP, and that MBP might be responsible in part, or entirely, for the extensive axonal damage that is regularly seen in Crohn's disease (12,65–67). Using an immunogold labeling approach, these clouds and related membranous structures have been shown to contain MBP in eosinophils in which MBP was demonstrated to be absent from the core compartment (63,64). Conversely, these areas did not contain the secondary granule matrix proteins, EPO or ECP, when stained with specific antibodies and secondary gold-labeled antibodies (63,64). Thus, selective release of MBP from the core compartment, but not EPO or ECP from the matrix compartment, of reactive tissue eosinophils in eosinophilic gastroenteritis was shown (63,64). These studies

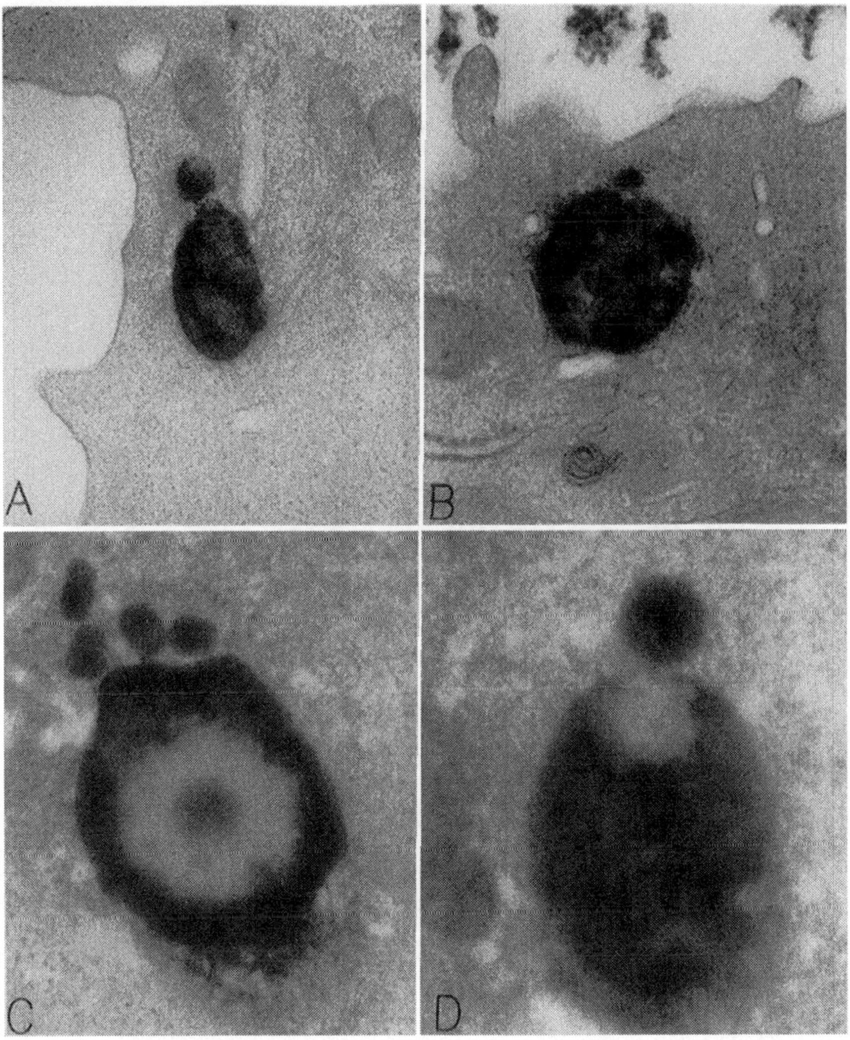

Figure 30 Similar cytochemical preparation of rhIL-5-containing eosinophil cultures show peroxidase-positive vesicles adjacent to peroxidase-positive secondary granules. Secreted, dense peroxidase aggregates are adjacent to the cell surface in (B). The peroxidase-negative central core is seen well in (C), and a focal "piece" of absent peroxidase in the matrix approximates the size of an adjacent peroxidase-loaded vesicle in (D). (A) ×43,000; (B) ×19,000; (C) ×60,000; (D) × 106,500 (magnifications before reduction).

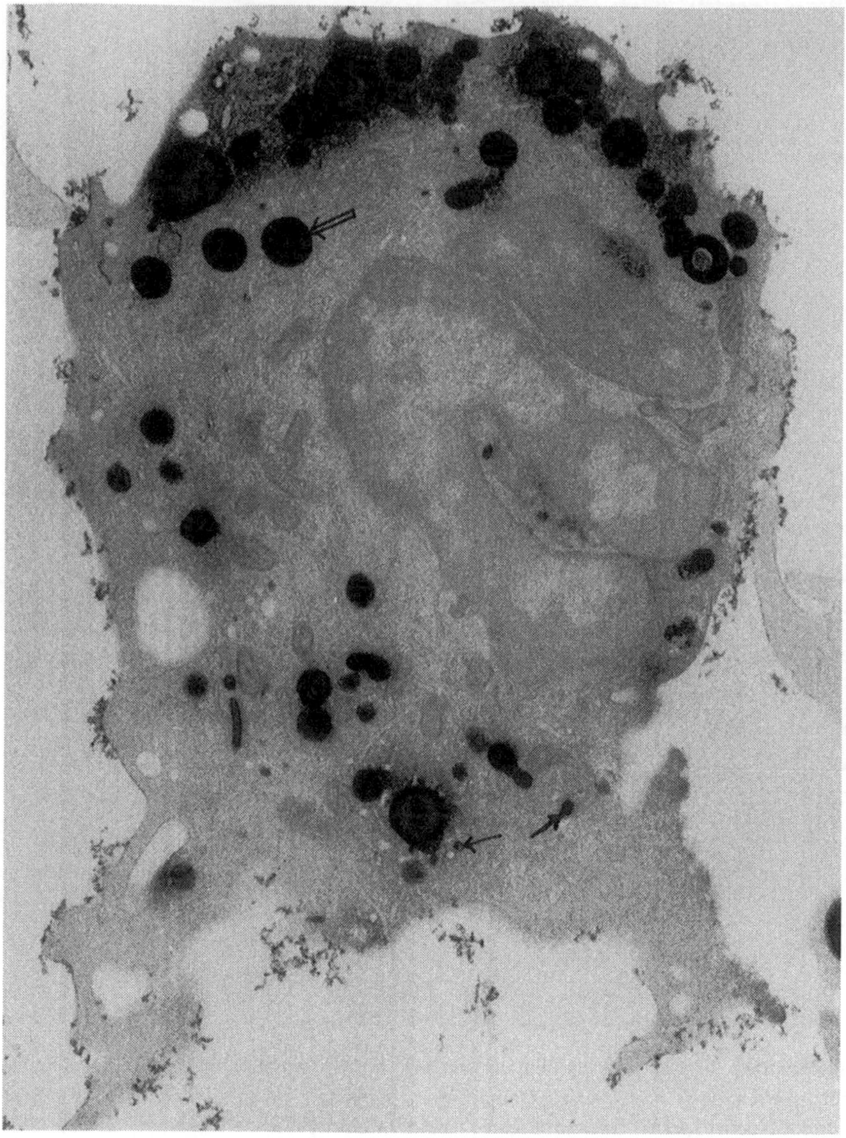

Figure 31 Cytochemical reaction for peroxidase (rhIL-5-grown mature human eosinophil) shows positive lipid bodies (open arrow), secondary granule matrix compartments, and cytoplasmic vesicles (closed arrows). Secreted peroxidase is bound to the plasma membrane. ×15,500. (With permission from Ref. 5.)

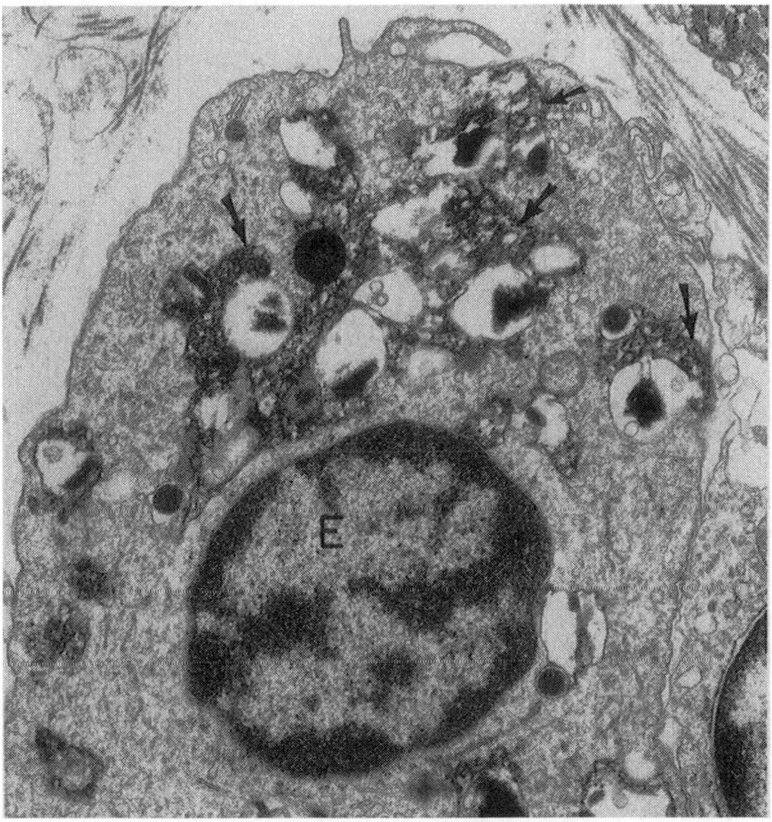

Figure 32 Tissue eosinophil (ileum, Crohn's disease) shows cytoplasmic clouds (arrows) of granule content surrounding partially empty granules. Some of these clouds have reached the subplasma membrane area. E = eosinophil. ×10,000. (With permission from Ref. 14.)

support the concept of PMD by reactive eosinophils as a secretory mechanism, in the absence of extrusion of granules in toto.

Eosinophilic myelocytes may also utilize vesicular transport of a matrix protein, EPO, to effect PMD under certain circumstances (Figs. 33 and 34) (7). For example, we recently examined developmentally arrested eosinophilic myelocytes that arose in 21-day cultures of human cord blood cells supplemented with a conditioned murine-growth-factor-containing medium known to contain IL-3, which did not support further maturation of eosinophils (7,68). We used a cytochemical method to detect peroxidase to examine these cells (7). Unlike

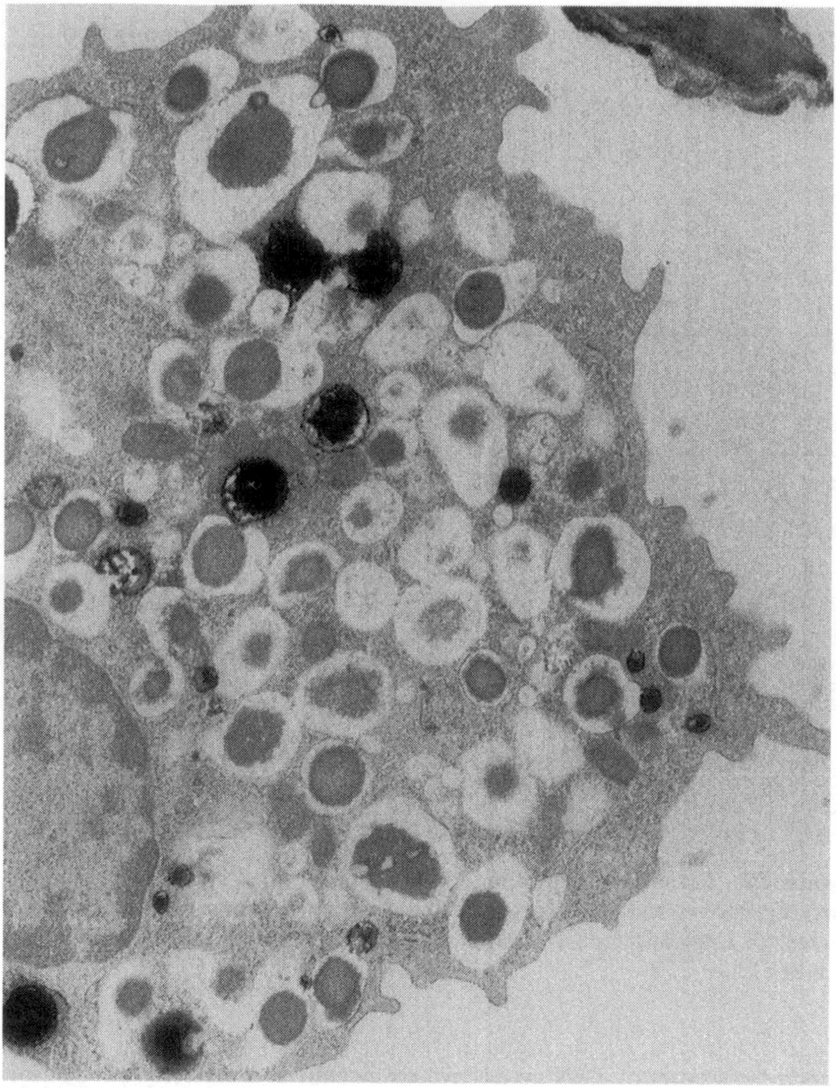

Figure 33 Developmentally arrested eosinophilic myelocyte (cultured from human cord blood cells in medium that does not permit full eosinophil maturation), reacted by cytochemistry to demonstrate peroxidase, has several positive granules and many positive perigranular vesicles. The matrix compartment of nearly all secondary granules is devoid of peroxidase or visible contents of any kind; central peroxidase- negative cores remain. Synthetic structures are also devoid of peroxidase. ×15,040. (With permission from Ref. 7.)

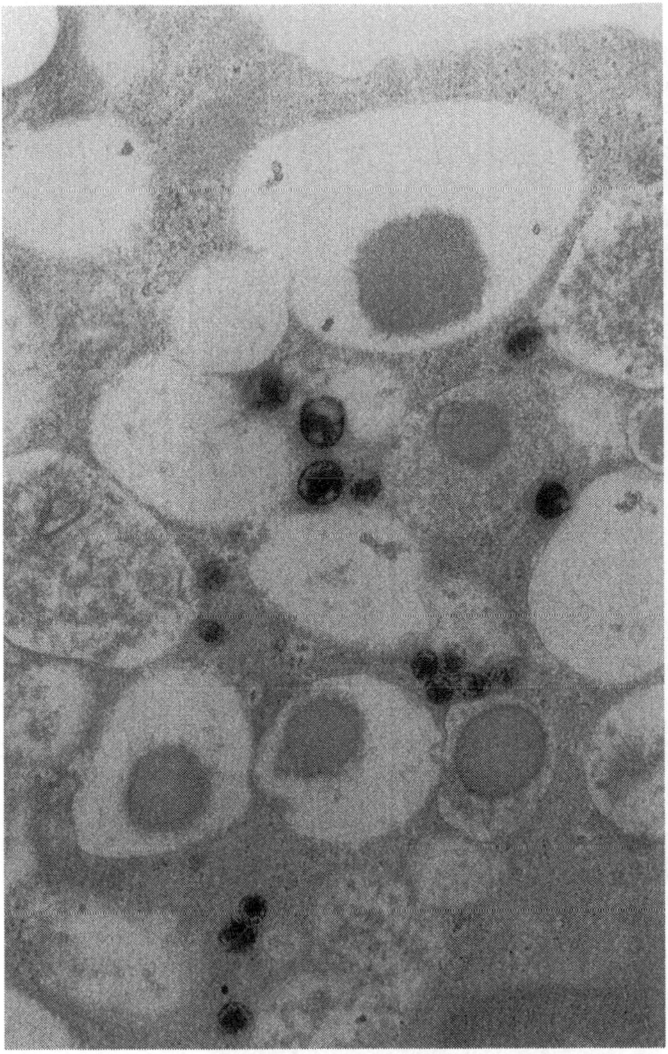

Figure 34 Preparation similar to Figure 33 shows a large number of peroxidase-positive cytoplasmic vesicles and empty matrix compartments of secondary, core-containing granules. ×27,000. (With permission from Ref. 7.)

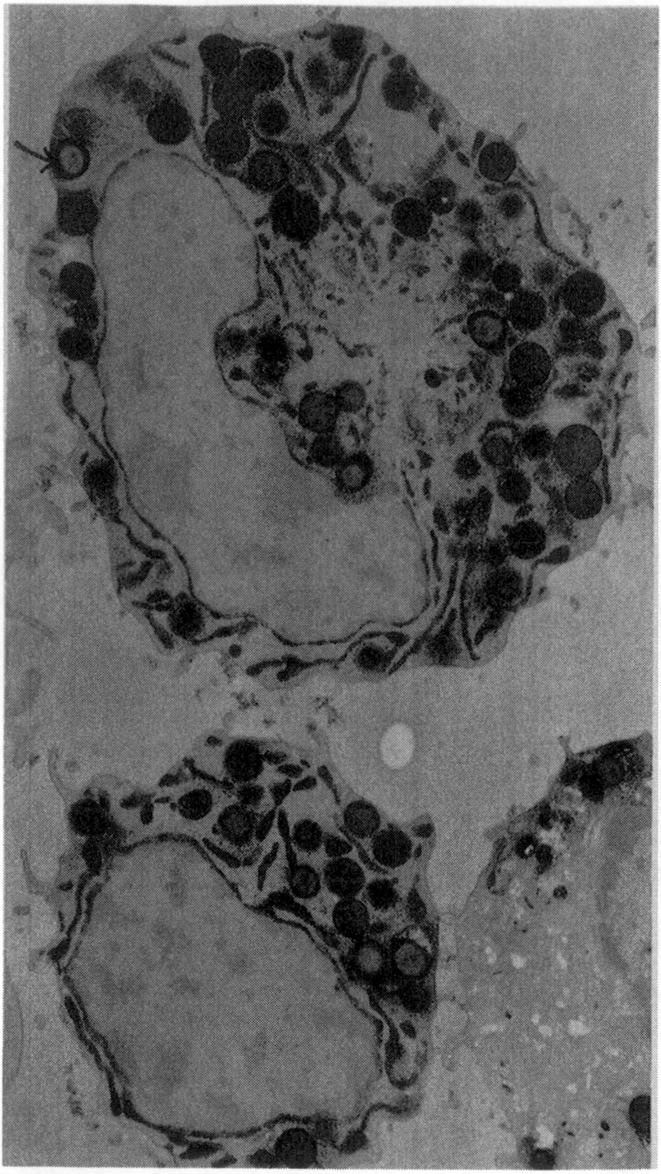

Figure 35 Normally developing eosinophilic myelocytes, cultured for 3 weeks in rhIL-5, show peroxidase cytochemical reaction product in immature granules, in the matrix compartment of mature secondary granules (arrow), in the perinuclear and rough endoplasmic cisterns, and in Golgi structures. ×8645. (With permission from Ref. 68.)

normally developing eosinophilic myelocytes, which contain peroxidase in synthetic organelles and in immature and mature granules (Fig. 35) (2), the developmentally arrested cells showed evidence of decreased synthesis and increased secretory transport of peroxidase (Fig. 33) (7). Thus, peroxidase was generally absent in the perinuclear and rough endoplasmic cisterns, in Golgi structures, in immature granules, and in the matrix compartment of most mature granules (Figs. 33 and 34) (7). Rather, bicompartmental specific granules displayed an empty, peroxidase-negative matrix compartment and central, peroxidase-negative core material. Peroxidase was present in numerous perigranular vesicles, some of which were attached to granules (Figs. 33 and 34) (7). These findings show that vesicle-mediated PMD is part of the secretory repertoire of immature human eosinophils and suggest that eosinophilic myelocytes may participate in important physiological and/or pathological events that require selective secretion from the specific granule matrix compartment in vivo.

VII. SUMMARY

Mature and immature human eosinophils are morphologically identifiable by their subcellular contents; mature eosinophilic granulocytes, in addition to polylobed nuclei, contain a mixture of crystalloid core-containing secondary granules, core-free primary granules, small granules, vesiculotubular structures, and lipid bodies, whereas immature precursors are largely distinguished by enhanced synthetic machinery and variable amounts of immature granules. Specialized ultrastructural tools have been used to localize key eosinophil products in certain subcellular organelles and their subcompartments (Table 1). For example, MBP is present in secondary granule crystalline cores; EDN and ECP are present in the matrix compartment of secondary granules; EPO is also in the matrix of secondary granules as well as in immature granules and synthetic machinery of immature eosinophils; acid phosphatase, arylsulfatase, and catalase are present in small granules; CLC protein is present in primary granules, cytoplasm, and nucleus; reactive oxygen products are present in vesiculotubular structures; arachidonic acid and prostaglandin endoperoxide synthase are present in lipid bodies. Mature eosinophils in body fluids, tissue biopsies, and developing in rhIL-3- or rhIL-5-supplemented cultures of human cord blood mononuclear cells show morphological changes of cellular activation. Generally, these include reduction in secondary granules, increases in primary granules, small granules, vesiculotubular structures, and lipid bodies, and enhanced nuclear and cytoplasmic expression of CLC protein and motile forms, or uropods.

Secretion morphology of mature and immature eosinophils includes extrusion of membrane-free granules into either enlarging cytoplasmic vacuoles or the extracellular milieu—a process identical to regulated secretion of a wide variety of epithelial and endocrine secretory cells and to anaphylactic degranulation of basophils and mast cells. This type of secretion has been documented in mature eosinophils in bacterial-infected ileal tissues in vivo and in isolated peripheral blood eosinophils stimulated with the calcium ionophore. Immature eosinophilic myelocytes have been shown to undergo similar granule extrusion secretory events of primary granules in vivo in human bone marrow samples.

Another secretory pattern, termed piecemeal degranulation, also occurs in human eosinophils. This process accounts for selective release of the contents of secondary granule matrix, or core compartments, as well as release of the contents of both secondary granule compartments, thereby giving rise to empty secondary granule containers in the cell cytoplasm. Vesicular transport of EPO has been documented as a mechanism for effecting PMD of mature eosinophils, which arise in rhIL-5-supplemented cultures of cord blood cells (cultures that permit complete maturation of the eosinophils lineage), as well as of immature eosinophilic myelocytes arising from similar cells in culture medium that does not permit further maturation of eosinophils. Thus, there is morphological evidence for at least two secretory mechanisms—granule exocytosis and PMD mediated by vesicular transport—in both mature eosinophils and their immature precursors.

ACKNOWLEDGMENTS

This work was supported by PHS Grant AI 33372 and the Lillian Gong Memorial Research Fund. We thank Peter K. Gardner for editorial assistance and Linda Letourneau for technical assistance.

REFERENCES

1. Dvorak AM, Ackerman SJ, Weller PF. Subcellular morphology and biochemistry of eosinophils. In: Harris JR, ed. Blood Cell Biochemistry. Vol. 2: Megakaryocytes, Platelets, Macrophages and Eosinophils. London: Plenum Press, 1991:237–344.
2. Saito H, Hatake K, Dvorak AM, Leiferman KM, Donnenberg AD, Arai N, Ishizaka K, Ishizaka T. Selective differentiation and proliferation of hematopoietic cells induced by recombinant human interleukins. Proc Natl Acad Sci USA 1988; 85:2288–2292.
3. Dvorak AM, Saito H, Estrella P, Kissell S, Arai N, Ishizaka T. Ultrastructure of eosinophils and basophils stimulated to develop in human cord blood mononuclear cell cultures containing recombinant human interleukin-5 or interleukin-3. Lab Invest 1989; 61:116–132.
4. Dvorak AM, Furitsu T, Letourneau L, Ishizaka T, Ackerman SJ. Mature eosinophils stimulated to develop in human cord blood mononuclear cell cultures supplemented with recombinant human interleukin-5. I. Piecemeal degranulation of specific granules and distribution of Charcot-Leyden crystal protein. Am J Pathol 1991; 138:69–82.
5. Dvorak AM, Ackerman SJ, Furitsu T, Estrella P, Letourneau L, Ishizaka T. Mature eosinophils stimulated to develop in human cord blood mononuclear cell cultures

supplemented with recombinant human interleukin-5. II. Vesicular transport of specific granule matrix peroxidase, a mechanism for effecting piecemeal degranulation. Am J Pathol 1992; 140:795–807.

6. Dvorak AM. Degranulation of basophils and mast cells. In: Dvorak AM (vol ed), Harris JR (series ed). Blood Cell Biochemistry. Vol. 4: Basophil and Mast Cell Degranulation and Recovery. London: Plenum Press, 1991: 101–275.
7. Dvorak AM, Estrella P, Ishizaka T. Cultured human eosinophilic myelocytes sustain extensive vesicle-mediated losses of peroxidase from the specific granule matrix compartment. Submitted.
8. Weller PF, Dvorak AM. Human eosinophils—development, maturation and functional morphology. In: Busse W, Holgate S, eds. Mechanisms in Asthma and Rhinitis. Implications for Diagnosis and Treatment. Cambridge, MA: Blackwell Scientific Publishers, 1993.
9. Dvorak AM, Letourneau L, Login GR, Weller PF, Ackerman SJ. Ultrastructural localization of the Charcot-Leyden crystal protein (lysophospholipase) to a distinct crystalloid-free granule population in mature human eosinophils. Blood 1988; 72: 150–158.
10. Dvorak AM, Weller PF, Monahan-Earley RA, Letourneau L, Ackerman SJ. Ultrastructural localization of Charcot-Leyden crystal protein (lysophospholipase) and peroxidase in macrophages, eosinophils and extracellular matrix of the skin in the hypereosinophilic syndrome. Lab Invest 1990; 62:590–607.
11. Dvorak AM. Ultrastructure of human gastrointestinal system. Interactions among mast cells, eosinophils, nerves and muscle in human disease. In: Snape WJ Jr, Collins SM, eds. Effects of Immune Cells and Inflammation on Smooth Muscle and Enteric Nerves. Boca Raton, FL: CRC Press, 1991:139–168.
12. Dvorak AM. Ultrastructural pathology of Crohn's disease. In: Goebell H, Peskar BM, Malchow M, eds. Inflammatory Bowel Diseases—Basic Research and Clinical Implications. Lancaster: MTP Press, 1988:3–41.
13. Dvorak AM, Monahan RA, Osage JE, Dickersin GR. Crohn's disease: transmission electron microscopic studies. II. Immunologic inflammatory response. Alterations of mast cells, basophils, eosinophils, and the microvasculature. Hum Pathol 1980; 11: 606–619.
14. Dvorak AM, Dickersin GR. Crohn's disease: electron microscopic studies. In: Sommers SC, Rosen PP, eds. Pathology Annual—Part 2. Vol. 14. New York: Appleton-Century-Crofts, 1979:259–306.
15. Dvorak AM, Letourneau L, Weller PF, Ackerman SJ. Ultrastructural localization of Charcot-Leyden crystal protein (lysophospholipase) to intracytoplasmic crystals in tumor cells of primary solid and papillary neoplasm of the pancreas. Lab Invest 1990; 62:608–615.
16. Komiyama A, Spicer SS. Microendocytosis in eosinophilic leukocytes. J Cell Biol 1975; 64:622–635.
17. Parmley RT, Spicer SS. Cytochemical and ultrastructural identification of a small type granule in human late eosinophils. Lab Invest 1974; 30:557–567.
18. Parmley RT, Spicer SS. Altered tissue eosinophils in Hodgkin's disease. Exp Mol Pathol 1975; 23:70–82.

19. Schaefer HE, Hubner G, Fischer R. Spezifische Mikrogranula in Eosinophilen. Eine vergleichende elektronenmikroskopische Untersuchung an verschiedenen Saugern zur Charakterisierung einer besonderen Granulationsform bei eosinophilen Granulozyten. Acta Haematol (Basel) 1973; 50:92–104.
20. Weller PF, Dvorak AM. Arachidonic acid incorporation by cytoplasmic lipid bodies of human eosinophils. Blood 1985; 65:1269–1274.
21. Weller PF, Monahan-Earley RA, Dvorak HF, Dvorak AM. Cytoplasmic lipid bodies of human eosinophils. Subcellular isolation and analysis of arachidonate incorporation. Am J Pathol 1991; 138:141–148.
22. Frigas E, Loegering DA, Gleich GJ. Cytotoxic effects of the guinea pig eosinophil major basic protein on tracheal epithelium. Lab Invest 1980; 42:35–43.
23. Gleich GJ, Frigas E, Loegering DA, Wassom DL, Steinmuller D. Cytotoxic properties of the eosinophil major basic protein. J Immunol 1979; 123:2925–2927.
24. Ackerman SJ, Gleich GJ, Loegering DA, Richardson BA, Butterworth AE. Comparative toxicity of purified human eosinophil granule cationic proteins for schistosomula of *Schistosoma mansoni*. Am J Trop Med Hyg 1985; 34:735–745.
25. Butterworth AE, Wassom DL, Gleich GJ, Loegering DA, David JR. Damage to schistosomula of *Schistosoma mansoni* induced directly by eosinophil major basic protein. J Immunol 1979; 122:221–229.
26. Kierszenbaum F, Ackerman SJ, Gleich GJ. Destruction of bloodstream forms of *Trypanosoma cruzi* by eosinophil granule major basic protein. Am J Trop Med Hyg 1981; 30:775–779.
27. Wassom DL, Gleich GJ. Damage to *Trichinella spiralis* newborn larvae by eosinophil major basic protein. Am J Trop Med Hyg 1979; 28:860–863.
28. Peters MS, Rodriguez M, Gleich GJ. Localization of human eosinophil granule major basic protein, eosinophil cationic protein, and eosinophil-derived neurotoxin by immunoelectron microscopy. Lab Invest 1986; 54:656–662.
29. Durack DT, Sumi SM, Klebanoff SJ. Neurotoxicity of human eosinophils. Proc Natl Acad Sci USA 1979; 76:1443–1447.
30. Fredens K, Dahl R, Venge P. The Gordon phenomenon induced by the eosinophil cationic protein and eosinophil protein X. J Allergy Clin Immunol 1982; 70:361–366.
31. Young JD-E, Peterson CGB, Venge P, Cohn ZA. Mechanism of membrane damage mediated by human eosinophil cationic protein. Nature 1986; 321:613–616.
32. Henderson WR, Jong EC, Klebanoff SJ. Binding of eosinophil peroxidase to mast cell granules with retention of peroxidatic activity. J Immunol 1980; 124:1383–1388.
33. Jong EC, Klebanoff SJ. Eosinophil-mediated mammalian tumor cell cytotoxicity: role of the peroxidase system. J Immunol 1980; 124:1949–1953.
34. Nathan CF, Klebanoff SJ. Augmentation of spontaneous macrophage-mediated cytolysis by eosinophil peroxidase. J Exp Med 1982; 155:1291–1308.
35. Nogueira NM, Klebanoff SJ, Cohn ZA. *T. cruzi*: sensitization to macrophage killing by eosinophil peroxidase. J Immunol 1982; 128:1705–1708.
36. Ramsey PG, Martin T, Chi E, Klebanoff SJ. Arming of mononuclear phagocytes by eosinophil peroxidase bound to *Staphylococcus aureus*. J Immunol 1982; 128:415–420.
37. Iozzo RV, MacDonald GH, Wight TN. Immunoelectron microscopic localization of catalase in human eosinophilic leukocytes. J Histochem Cytochem 1982; 30:697–701.

38. Ginsel LA, Onderwater JJM, Fransen JAM, Verhoeven AJ, Roos D. Localization of the low-M_r subunit of cytochrome b_{558} in human blood phagocytes by immunoelectron microscopy. Blood 1990; 76:2105–2116.
39. Curnutte JT, Babior BM. Chronic granulomatous disease. Adv Hum Genet 1987; 16: 229–297.
40. Weller PF, Ryeom SW, Dvorak AM. Lipid bodies: structurally distinct, nonmembranous intracellular sites of eicosanoid formation. In: Bailey JM, ed. Prostaglandins, Leukotrienes, Lipoxins and PAF. New York: Plenum Press, 1992:353–362.
41. Galli SJ, Dvorak AM, Peters SP, Schulman ES, MacGlashan DW Jr., Isomura T, Pyne K, Harvey VS, Hammel I, Lichtenstein LM, Dvorak HF. Lipid bodies—widely distributed cytoplasmic structures that represent preferential nonmembrane repositories of exogenous [^{3}H]arachidonic acid incorporated by mast cells, macrophages, and other cell types. In: Bailey JM, ed. Prostaglandins, Leukotrienes, and Lipoxins. Biochemistry, Mechanism of Action, and Clinical Applications. New York: Plenum Press, 1985:221–239.
42. Dvorak AM. Biochemical contents of granules and lipid bodies—two distinctive organelles found in basophils and mast cells. In: Dvorak AM (vol ed), Harris JR (series ed). Blood Cell Biochemistry. Vol. 4:Basophil and Mast Cell Degranulation and Recovery. London: Plenum Press, 1991:27–65.
43. Dvorak AM, Dvorak HF, Peters SP, Schulman ES, MacGlashan DW Jr, Pyne K, Harvey VS, Galli SJ, Lichtenstein LM. Lipid bodies: cytoplasmic organelles important to arachidonate metabolism in macrophages and mast cells. J Immunol 1983; 131:2965–2976 (republished: J Immunol 1984; 132:1586–1597.
44. Dvorak AM, Hammel I, Schulman ES, Peters SP, MacGlashan DW Jr, Schleimer RP, Newball HH, Pyne K, Dvorak HF, Lichtenstein LM, Galli SJ. Differences in the behavior of cytoplasmic granules and lipid bodies during human lung mast cell degranulation. J Cell Biol 1984; 99:1678–1687.
45. Dvorak AM, Morgan E, Schleimer RP, Ryeom SW, Lichtenstein LM, Weller PF. Ultrastructural immunogold localization of prostaglandin endoperoxide synthase (cyclooxygenase) to non-membrane-bound cytoplasmic lipid bodies in human lung mast cells, alveolar macrophages, type II pneumocytes and neutrophils. J Histochem Cytochem 1992; 40:759–769.
46. Caulfield JP, Hein A, Rothenberg ME, Owen WF, Soberman RJ, Stevens RL, Austen KF. A morphometric study of normodense and hypodense human eosinophils that are derived in vivo and in vitro. Am J Pathol 1990; 137:27–41.
47. Peters MS, Gleich GJ, Dunnette SL, Fukuda T. Ultrastructural study of eosinophils from patients with the hypereosinophilic syndrome: a morphological basis of hypodense eosinophils. Blood 1988; 71:780–785.
48. Jabara HH, Ackerman SJ, Vercelli D, Yokota T, Arai K-I, Abrams J, Dvorak AM, Lavigne MC, Banchereau J, De Vries J, Leung DYM, Geha RS. Induction of interleukin-4-dependent IgE synthesis and interleukin-5-dependent eosinophil differentiation by supernatants of a human helper T-cell clone. J Clin Immunol 1988; 8: 437–446.
49. Dvorak AM, Mihm MC Jr, Osage JE, Kwan TH, Austen KF, Wintroup BU. Bullous pemphigoid, an ultrastructural study of the inflammatory response: eosinophil,

basophil and mast cell granule changes in multiple biopsies from one patient. J Invest Dermatol 1982; 78:91–101.
50. Weller PF, Marshall WL, Lucey DR, Rand TH, Dvorak AM, Finberg RW. Infection, apoptosis and killing of mature human eosinophils by human immunodeficiency virus-1. Submitted.
51. Duvall E, Wyllie AH. Death and the cell. Immunol Today 1986; 7:115–119.
52. Scott RE, Horn RG. Fine structural features of eosinophil granulocyte development in human bone marrow. Evidence for granule secretion. J Ultrastruct Res 1970; 33: 16–28.
53. Hyman PM, Teichberg S, Starrett S, Vinciguerra V, Degnan TJ. Secretion of primary granules from developing human eosinophilic promyelocytes. Proc Soc Exp Biol Med 1978; 159:380–385.
54. Butterfield JH, Ackerman SJ, Scott RE, Pierre RV, Gleich GJ. Evidence for secretion of human eosinophil granule major basic protein and Charcot-Leyden crystal protein during eosinophil maturation. Exp Hematol 1984; 12:163–170.
55. Anteunis A, Astesano A, Robineaux R. Ultrastructural characteristics of developing eosinophil leukocytes in human bone marrow during acute leukemia. Evidence for extracellular granule release from human eosinophils. Inflammation 1977; 2:17–26.
56. Henderson WR, Harley JB, Fauci AS, Chi EY. Hypereosinophilic syndrome human eosinophil degranulation induced by soluble and particulate stimuli. Br J Haematol 1988; 69:13–21.
57. Henderson WR, Chi EY. Ultrastructural characterization and morphometric analysis of human eosinophil degranulation. J Cell Sci 1985; 73:33–48.
58. Dvorak AM. Ultrastructural evidence for release of major basic protein-containing crystalline cores of eosinophil granules in vivo: cytotoxic potential in Crohn's disease. J Immunol 1980; 125:460–462.
59. Tai P-C, Spry CJF. The mechanisms which produce vacuolated and degranulated eosinophils. Br J Haematol 1981; 49:219–226.
60. Khalife J, Capron M, Cesbron J-Y, Tai P-C, Taelman H, Prin L, Capron A. Role of specific IgE antibodies in peroxidase (EPO) release from human eosinophils. J Immunol 1986; 137:1659–1664.
61. Abu-Ghazaleh RI, Fujisawa T, Mestecky J, Kyle RA, Gleich GJ. IgA-induced eosinophil degranulation. J Immunol 1989; 142:2393–2400.
62. Fujisawa T, Abu-Ghazaleh R, Kita H, Sanderson CJ, Gleich GJ. Regulatory effect of cytokines on eosinophil degranulation. J Immunol 1990; 144:642–646.
63. Torpier G, Colombel JF, Mathieu-Chandelier C, Capron M, Dessaint JP, Cortot A, Paris JC, Capron A. Eosinophilic gastroenteritis: ultrastructural evidence for a selective release of eosinophil major basic protein. Clin Exp Immunol 1988; 74:404–408.
64. Capron M, Grangette C, Torpier G, Capron A. The second receptor for IgE in eosinophil effector function. Chem Immunol 1989; 47:128–178.
65. Dvorak AM, Silen W. Differentiation between Crohn's disease and other inflammatory conditions by electron microscopy. Ann Surg 1985; 201:53–63.
66. Dvorak AM, Osage JE, Monahan RA, Dickersin GR. Crohn's disease: transmission electron microscopic studies. III. Target tissues. Proliferation of and injury to smooth muscle and the autonomic nervous system. Hum Pathol 1980; 11:620–634.
67. Dvorak AM. What is the evidence for damaged enteric autonomic nerves in Crohn's

disease? In: MacDermott RP, ed. Inflammatory Bowel Disease: Current Status and Future Approach. Amsterdam: Elsevier, 1988:705–711.

68. Dvorak AM, Ishizaka T, Galli SJ. Ultrastructure of human basophils developing in vitro. Evidence for the acquisition of peroxidase by basophils and for different effects of human and murine growth factors on human basophil and eosinophil maturation. Lab Invest 1985; 53:57–71.

DISCUSSION (Speaker: A. Dvorak)

M. Capron: I am very impressed by your morphological studies suggesting that eosinophils can release their granule proteins by a vesicular transport mechanism or "piecemeal degranulation." This supports our results showing the differential release of EPO or ECP according to the slowness of activation (1,2). My question is—will you perform double labeling to investigate whether EPO and ECP could use different vesicles?

Dvorak: Yes, and thanks.

Spry: Is there a relationship between lipid bodies and primary granules? When the crystalloid is lost from cells leaving an "empty" ghost of itself behind, what do you think causes the "space" to remain? What might it be due to? Could the matrix be a sensitized structure that cannot occupy the space left by the crystalloid? Or is it that the EM procedure somehow "knocks the crystalloid out of the nest"?

Dvorak: We have no morphological relationship between the two very different organelles—lipid bodies and primary granules. Primary granules are clearly typical secretory granules and they contain the nonactivated eosinophil's store of Charcot-Leyden crystal protein; lipid bodies, on the other hand, contain arachidonic acid and prostaglandin endoperoxide synthase. Moreover, lipid bodies are ubiquitous in mammalian cells (which do not have CLC-protein-containing granules). The only other cell type that displays CLC protein in a granule location is the basophil in humans. I have also wondered how the core space stays in place after cores have left the specific granules. Your explanation is a possibility. Of course, the fact that this space is electron lucent is not proof that it is empty, it merely means that no electron-dense materials remain. I do not think the EM procedures either "reverse the staining of cores," "dissolve the cores," or "somehow knock them out of their nests," because we regularly image the core compartment in human eosinophils, often in the same section showing focal core losses.

Abrams: Could superantigen be involved in inducing eosinophil product release via the regulated pathway, given the association with staphylococcal and patient pouchitis biopsy samples you described?

Dvorak: Absolutely, I think experiments designed to test this possibility using the *Staphylococcus* superantigen and eosinophils are very important to explore.

Gleich: In experiments using light microscopy, we observed vacuoles in peripheral blood eosinophils of patients with the hypereosinophilic syndrome. By electron microscopy we did not see cytoplasmic vacuoles but rather normal numbers of small eosinophil-specific granules. Do you have any comments regarding these small granules?

Dvorak: In peripheral blood and tissues of patients with the HES we also have seen increased numbers of small granules, primary granules (which are generally smaller than specific granules), in addition to increased lipid bodies and some evidence of core losses from specific granules. Completely empty granules are exceedingly rare. So, small granules (hydrolytic enzymes) and primary granules (Charcot-Leyden crystal protein) are increased and specific granules with cores are decreased. These increases could be secondary to increased production of small granules, increased production and/or retention of primary granules, enhanced secretion of specific granules, or their contents, with subsequent reduction in their size.

Konig: In addition to staphylococcal superantigens, coagulase-positive staphylococci release various hemolysins—among these the α toxin induces histamine release from mast cells and basophils while the δ hemolysin modulates leukotriene and PAF formation from granulocytes.

Dvorak: I think it is entirely possible that the staphylococcal organisms (and/or other bacteria) that we have cultured in replicate biopsy samples of pouchitis (ileum) in which we have identified regulated eosinophil granule extension may be the stimulus for this secretory response of eosinophils.

Venge: When we separate human eosinophils on sucrose density gradients, we find a striking heterogeneity. One dense peak contains all four major eosinophil products, whereas other, less dense peaks seem to contain only ECP and EPX/EDN but no EPO. My problem is how to classify these granules in relation to your classification, and my question to you is whether you find a similar heterogeneity and whether your primary granules contain ECP and EPX/EDN.

Dvorak: So far, the only thing that is known to be in the primary granules is the Charcot-Leyden crystal protein. It will be difficult to demonstrate ECP and EDN in these granules since they are matrix proteins in the specific core-containing granules. Therefore, when an apparent core-free granule is labeled for ECP and EDN by immunogold, one will not be able to be certain that it represents a primary granule or a cut through a specific granule that does not pass through the core compartment. Ultimately, double and triple labeling of primary granules with different size gold-labeled secondary antibodies might help to resolve this issue.

Ackerman: A comment in response to Per Venge's point about finding granules (after subcellular fractionation) that contain only ECP/EDN but not other cationic proteins: I caution that isolated granules may be extremely leaky and against overinterpretation of such findings.

Dvorak: It sounds like it could be a problem to me although this does not represent my area of expertise.

Gleich: Have you observed Medusa cells in your specimens? On one occasion in conjunction with a clinical colleague we observed one such cell having a long tubular process (resembling an elephant's trunk).

Dvorak: I have never seen Medusa cells, perhaps because electron microscopic sections of 70 nm are so thin that they miss these processes that have been described by Hanker primarily in studies of whole mounts of cells.

Roos: The intracellular vesicles contain cytochrome b_{558}, a component of the NADPH-oxidase. Upon cell activation, these vesicles fuse with the plasma membrane and the cytochrome integrates with two cytosolic proteins into an active oxidase that generates superoxide at the plasma membrane or phagosome membrane. Thus, superoxide is not generated in these small "secretory vesicles" but at the cell surface. We have recently found that these vesicles also contain albumin (taken up from the extracellular medium), complement receptor type 3 (CD16/CD18), and a number of other proteins. These vesicles are only detected when blood is drawn in a fixative, because conventional cell purification already leads to fusion of these vesicles with the plasma membrane.

Dvorak: Your demonstration of the involvement of the vesiculotubular system with the generation of superoxide (by the localization with immunogold of cytochrome b_{558}) is interesting. I agree with your observation of the need for instantaneous fixation for their demonstration. For example, all peripheral blood, bone marrow, body cavity fluids, and tissue biopsies that we examine by electron microscopy are instantaneously placed in fixative within seconds of their removal from the patients. Under these circumstances large numbers of vesiculotubular structures are regularly imaged in eosinophils that express an activated morphological phenotype in a variety of human diseases. We regularly use their presence in granule-poor eosinophils for cell identification purposes since in our experience they are unique to the eosinophil lineage.

DISCUSSION REFERENCES

1. Capron M, Grangette C, Torpier G, Capron A. The second receptor for IgE in eosinophil effector function. Chem Immunol 1989; 47:128–178.
2. Tomassini M, Tsicopoulos A, Tai PC, Gruart V, Tonnel AB, Prin L, Capron A, Capron M. Release of granule proteins by eosinophils from allergic and non-allergic patients with eosinophils on immunoglobulin-dependent activation. J Allergy Clin Immunol 1991; 88:365.

10

Eosinophil Differentiation and Cytokine Networks in Allergic Inflammation

Judah A. Denburg, Jerry Dolovich, Jean Marshall, Isabelle Pin, Peter Gibson, Isao Ohno, Susetta Finotto, Fred Hargreave, and Manel Jordana
McMaster University, Hamilton, Ontario, Canada

I. INTRODUCTION

Asthma and related airways diseases are characterized by eosinophil, basophil, and mast cell prominence in tissues and biological fluids (1–3). This chapter will focus on different approaches our group has taken to understanding inflammatory cell accumulation in chronic airways disease. In particular, the issue of eosinophil, mast cell, and basophil differentiation as it relates to the process of cell recruitment will be discussed.

II. EOSINOPHILS AND MAST CELLS IN NASAL POLYPS

In light microscopic sections, activated eosinophils can be demonstrated by immunostaining for eosinophil cationic protein in nasal polyps (Fig. 1). Polyps represent a very interesting and convenient model to study the recruitment of eosinophils to airway tissues. Using primary cultures of nasal polyp structural cells, namely fibroblasts and epithelial cells, we have shown that both these cell types produce cytokines, including granulocyte-macrophage colony-stimulating factor (GM-CSF), interleukin (IL)-6, and IL-8 (4–6), which play roles in perpetuating the inflammatory process. In addition, other hemopoietic factors that aid and direct the differentiation of various myeloid cells, in particular basophils and

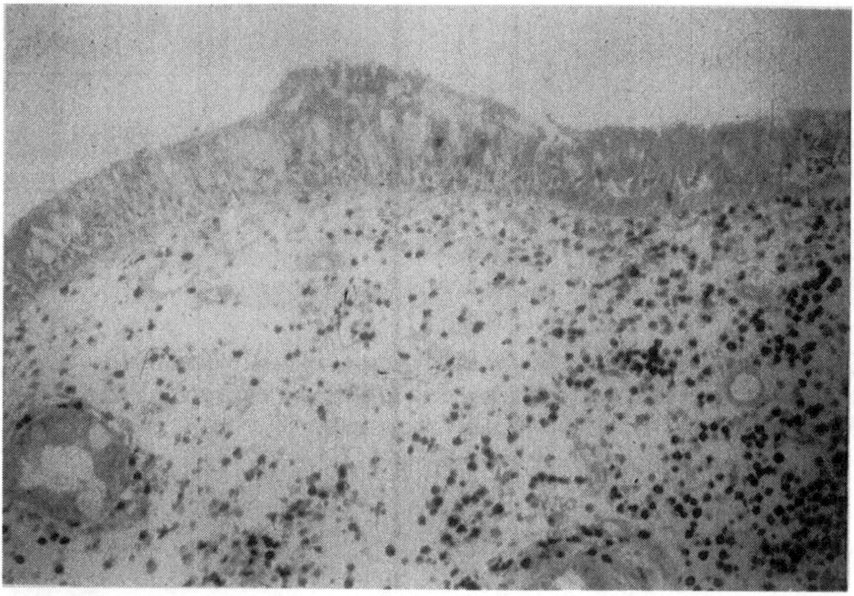

Figure 1 Immunostaining of nasal polyp section with EG2, an antibody to eosinophil cationic protein. Cells staining positive indicate activated eosinophils.

eosinophils, but also neutrophils and monocytes, are secreted by tissue structural cells from polyps (7–9). Among these are an as yet uncharacterized monocyte-macrophage factor derived from epithelial cells (10) and a fibroblast-derived basophilic differentiation factor that is distinct from GM-CSF (6,7).

III. REGULATION OF INFLAMMATORY CELL DIFFERENTIATION BY CORTICOSTEROIDS

We have studied pharmacological regulation by corticosteroids in vitro of a number of these processes. Budesonide, a topical corticosteroid, in addition to diminishing nasal polyp size and cellular infiltrates of eosinophils and mast cells in vivo, inhibits gene expression and production of GM-CSF, IL-6, and IL-8 by nasal epithelial cells and fibroblasts (6,7,11,12). In a recent set of experiments, dose-dependent, direct inhibitory effects of budesonide on progenitors for basophils and eosinophils, using HL-60 cells after alkaline passage and butyrate induction or methylcellulose cultures, were observed (Fig. 2). These data are in agreement with information in the literature on the effects of steroids on eosinophil differentiation in vitro (3,4). An opposite effect of steroids on neutrophil differentiation is

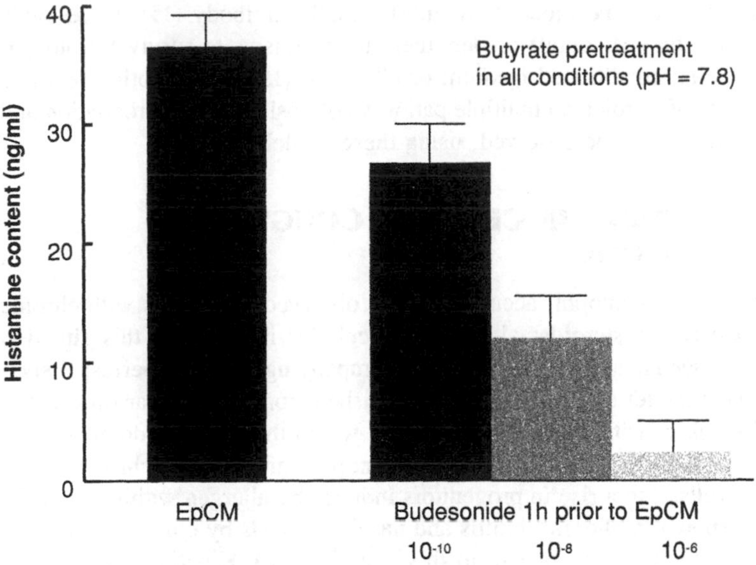

Figure 2 Direct inhibitory effect of budesonide on histamine production by alkaline-passaged, butyrate-induced HL-60 cells. The latter were incubated at varying doses of budesonide for 1 h prior to addition of epithelial conditioned medium (EpCM) for 5 days under alkaline conditions and in the presence of sodium butyrate (33,34,37,39).

consistently found due to direct and indirect stimulatory effects on granulopoiesis, leading to neutrophil progenitor differentiation (13,14). GM-CSF production and gene expression are both influenced by steroids. Both up-regulation by IL-1α and down-regulation by budesonide can be demonstrated in vitro (6,7,11,12). There may be several, interactive pharmacological effects of budesonide that eventuate in suppression or enhancement of inflammatory cell differentiation.

IV. EFFECTS OF CORTICOSTEROIDS ON EOSINOPHILS

There are also effects of budesonide and similar corticosteroids on the survival response of eosinophils in vitro. Nasal or bronchial epithelial cell conditioned media prolong eosinophil survival in vitro; this effect is mimicked by GM-CSF and can be shown to be blocked by anti-GM-CSF antibody (15). Budesonide preincubation with eosinophils renders them unresponsive to survival prolongation by epithelial conditioned medium or GM-CSF (15). Thus, both direct and indirect effects of steroids on multiple pathways of eosinophil differentiation and survival/activation can be observed, using these models.

V. EOSINOPHILS IN CHRONIC COUGH AND ASTHMA

Another model of eosinophil accumulation is observed in patients with chronic cough responsive to steroids; although the cellular infiltrate in this situation resembles that seen in asthma, there is no accompanying airway hyperresponsiveness (16). This model can be readily distinguished from the cellular differential profiles of smokers with chronic bronchitis in which there is a predominance of macrophages (17). It is hypothesized that eosinophil infiltration in chronic cough or asthma is related to a rise in progenitors induced by allergen within 24 h (18), as has been shown in allergic rhinitis and nasal polyposis by our group (19,20); neutrophil progenitors do not rise in this situation (18). Kinetic studies are in progress to examine progenitors for a longer time span after allergen provocation.

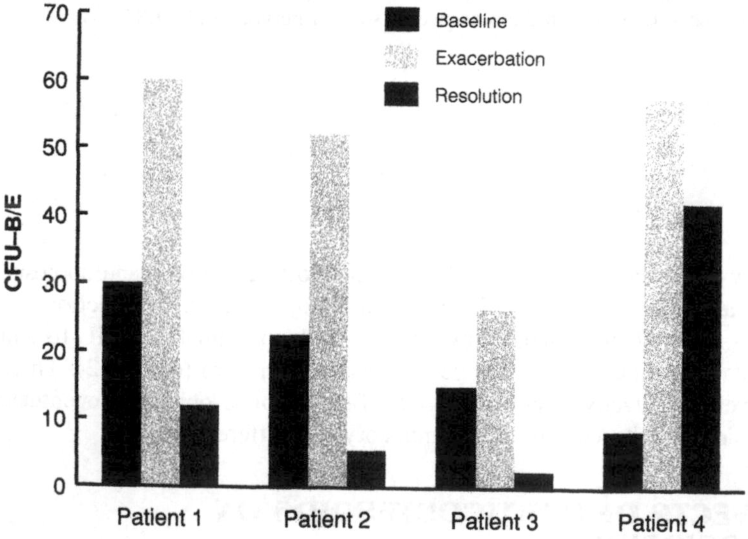

Figure 3 Basophil-eosinophil colony-forming unit (CFU) increases in individual patients (numbered 1–4) during a prospective, controlled, mild exacerbation of asthma. Colonies were enumerated as previously described (2,4,18–20,22,37); asthma exacerbations were as previously defined (2).

Sputum eosinophilia can also be examined using induced sputum samples from asthmatics (21); a rise both in eosinophils and, to a lesser extent, in metachromatic cells is seen up to 32 days post–allergen challenge; only slow restoration to normal levels of eosinophils occurs, correlating directly with the PC20 returning to baseline (21). In addition, basophil-eosinophil colony-forming units (CFU), a measure of a common, committed progenitor for these cells (22), rise during clinical exacerbations of asthma (2) and represent one of the earliest and most sensitive indices of an asthmatic attack (Fig. 3). This rise in progenitors can then be shown to return to baseline after inhaled corticosteroid.

VI. ENDOCRINE EFFECT OF CYTOKINES: AN HYPOTHESIS

These results suggest that topical corticosteroids may be inhibiting endocrine-like effects of cytokines derived from airways structural cells, leading to changes in bone marrow progenitor release and circulation. Relevant to this are primate studies of parenteral administration of IL-3 and GM-CSF leading to a rise in basophils, eosinophils, and their progenitors in circulation (23,24). One can thus speculate that allergen challenge leads to release of GM-CSF, IL-3, and/or IL-5 in vivo, causing basophil-eosinophil CFU to rise in circulation; this effect is mimicked when such cytokines are given parenterally. Whether this represents a primary mechanism for inflammatory cell progenitor fluctuations following allergen provocation or steroid treatment needs to be investigated.

VII. RELATIONSHIPS BETWEEN EOSINOPHIL AND BASOPHIL LINEAGES

Based on a compilation of data using immunophenotyping of basophilic/eosinophilic differentiating cell lines such as HL-60, KU812 (25), as well as the human mast cell line HMC-1 (26), we can propose a scheme of differentiation of the basophil, eosinophil, and, perhaps, mast cell lineages in humans (Fig. 4). Whether mast cells derive from a committed, intermediate, myeloid progenitor is undetermined. In the HL-60 system, YB5B8, originally described as a mast cell and myeloid antigen (27), which has now clearly been shown to be identical to the c-*kit* gene product (28), is found on HL-60 cells during induction toward both basophilic and eosinophilic differentiation (Fig. 4). Cells are present in these cultures which are both YB5B8-positive and BSP-1-positive (a basophil granule marker) (29,30). A BSP-1-positive, CD25-positive, CD11b-positive cell, bearing a high-affinity receptor for IgE, is probably the end-stage of this pathway (3). Factors that stimulate this differentiation pathway in the human include GM-CSF, IL-3, IL-5, and nerve growth factor (31–35).

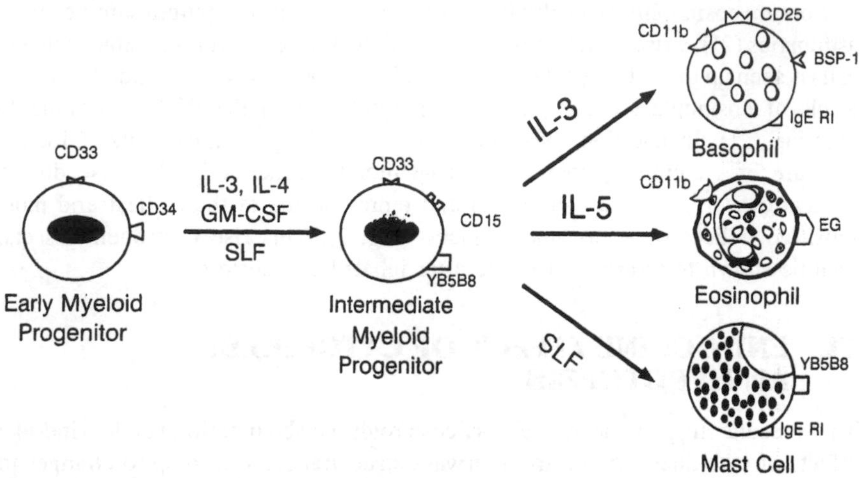

Figure 4 Schematic, hypothetical proposal of steps in human basophil, eosinophil, and mast cell differentiation. The immunophenotyping of HL-60 cells (30) undergoing basophilic differentiation (33,34,37,39) was utilized to construct this scheme, together with information reviewed from the literature (23–25). SLF, Steel Locus Factor.

With regard to effects of cytokines on progenitors, it has already been established that IL-5 affects eosinophil differentiation in animals and humans (36). We have recently shown that IL-5, using either a TF-1 assay or the HL-60 assay, also has effects in stimulating histamine cell-producing activity and, morphologically, basophil differentiation (37). IL-5 thus is also a basophilopoietin, a fact anticipated by the close relationship of these lineages (22), and by our original finding of basophilic differentiation of HL-60 cells, which can bear receptors for IL-5, under similar in vitro circumstances in which eosinophilic differentiation is induced (38,39).

VIII. BASOPHIL/EOSINOPHIL DIFFERENTIATION FACTORS

Another potential basophil/eosinophil differentiation factor may exist, as determined by a series of antibody inhibition experiments using conditioned medium from a human T-cell line (Mo-CM), a primary source of GM-CSF (34,35); the nature of this factor is not yet known. Several nasal polyp lavage fluids similarly have been shown to contain eosinophil survival-prolonging factors such as GM-CSF, IL-3, or IL-5 (Fig. 5), which also can induce basophil and eosinophil differentiation.

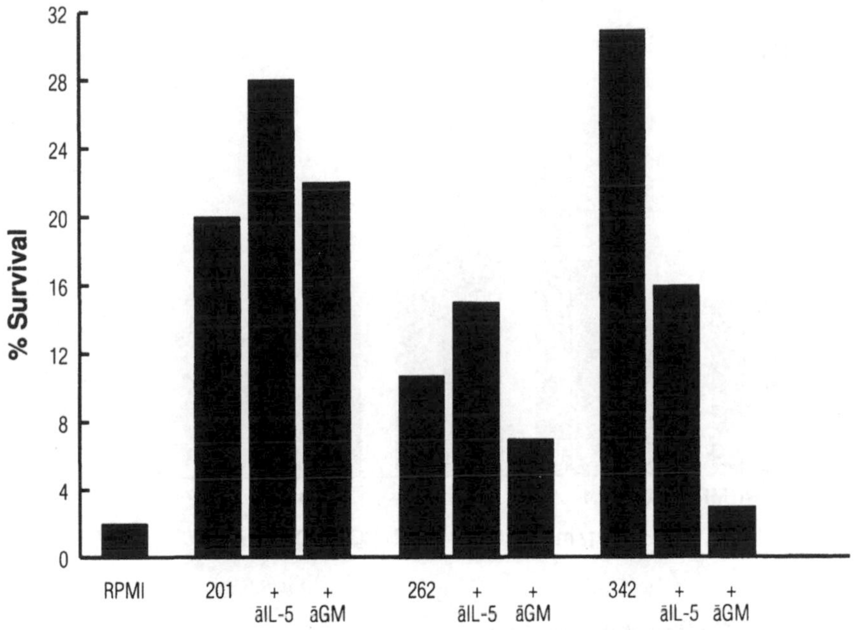

Figure 5 Detection of IL-5 and GM-CSF in nasal polyp lavage fluids by an antibody inhibition assay. An eosinophil survival assay was used to detect IL-5 and GM-CSF in three separate nasal lavage fluids: in #201, neither IL-5 nor GM-CSF was present; in #262, only GM-CSF was detected; and in #342, both were detected. Subsequent experiments have indicated that combinations of antibodies to IL-3, IL-5, and GM-CSF may be necessary to fully abrogate the eosinophil survival-enhancing activities in these airway lavage fluids.

IX. EOSINOPHIL CYTOKINE PRODUCTION AND CYTOKINE RESPONSIVENESS

Eosinophils themselves can express genes for and produce cytokines, as has been recently demonstrated for TGF-α (40). Nasal polyp sections stained with chromotope 2R and examined by in situ hybridization for TGFβ-1 gene expression reveal that the TGF-β-1 gene is expressed in cells that are also chromotope 2R-positive (i.e., eosinophils) (41); this can be confirmed by immunostaining. Similar data regarding GM-CSF and TNF-α in nasal polyp eosinophils have recently become available (42,43). To these data must be added recent findings on the presence of mRNA and/or product for IL-1, IL-3, IL-5, and IL-6 in human eosinophils (44–47). IL-8 must now also be added to this list (Hansel, Blaser, unpublished).

Thus, the potential involvement of eosinophils as both targets and producers of cytokines in allergic reactions may be modulated by corticosteroids. It is also

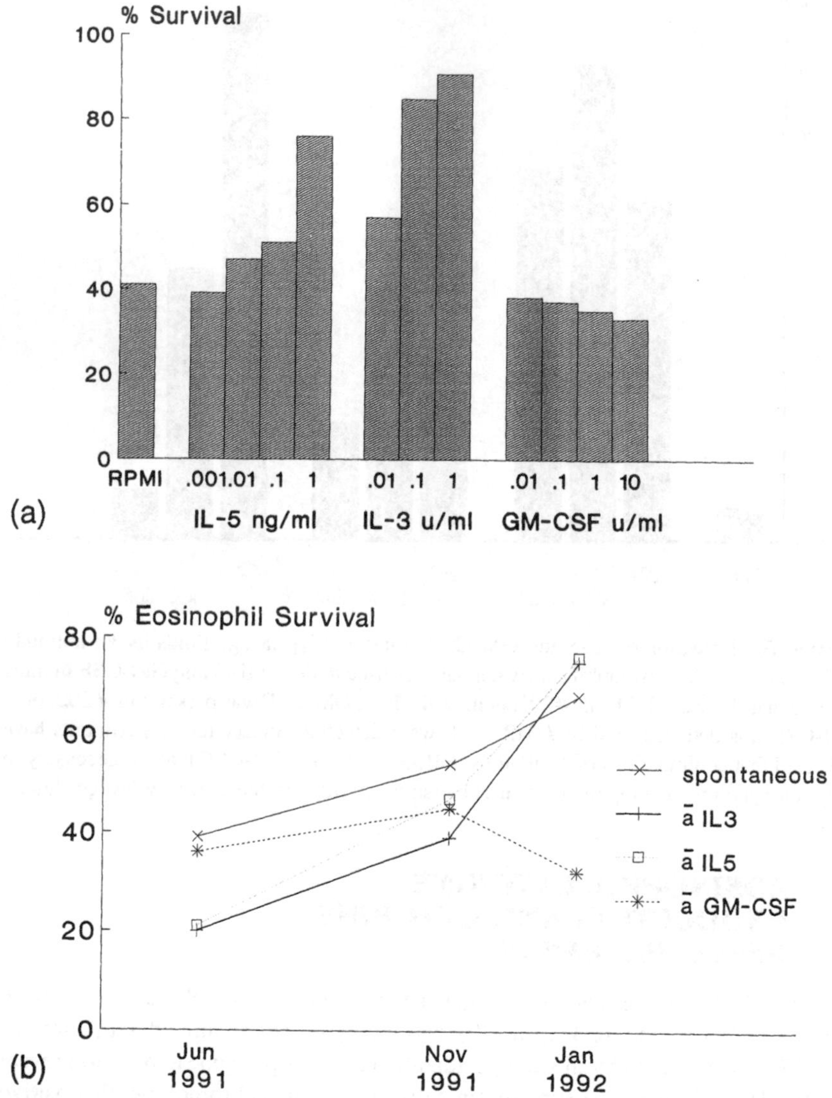

Figure 6 (a) Heterogeneity of cytokine-induced eosinophil survival prolongation responses in hypereosinophilic syndrome: lack of responsiveness to GM-CSF, in contrast to IL-3 and IL-5. (b) Change in spontaneous cytokine-dependent survival prolongation response profile of above hypereosinophilic syndrome patient eosinophils over time. The first time point (June 1991) is the profile shown above (a); this changes so that by January 1992, spontaneous survival was abrogated by antibodies to GM-CSF, but not by antibodies to IL-3 or IL-5.

interesting, in this regard, that eosinophils may be heterogeneous in their production of and/or responses to cytokines, as evidenced by variability in cytokine-induced eosinophil survival prolongation in hypereosinophilic syndrome (Fig. 6a and 6b). This accords with the biopsy studies of nasal polyp in which varying numbers of, but not all, eosinophils express mRNA for TGF-β, GM-CSF, and TNF-α (41– 43). How eosinophils control their own survival and/or differentiation in vivo in various allergic-type inflammatory states remains to be clarified.

X. SUMMARY

Microenvironmental molecular control of the inflammatory process is complex. Hemopoietic progenitors may be of key importance in this process, but a series of cell interactions also takes place, leading to accumulation of mature cells such as eosinophils. Eosinophils and other inflammatory cells, even at terminal stages of differentiation or activation, are capable of feeding back and enhancing the inflammatory process. Understanding the precise interactions among progenitors, inflammatory cells, and structural cells represents a major challenge toward deciphering the type and extent of chronic inflammatory processes in allergic and related conditions.

ACKNOWLEDGMENTS

This research was supported by grants from the Medical Research Council of Canada and AB Draco, Sweden.

REFERENCES

1. Kirby JG, Hargreave FE, Gleich GJ, O'Byrne PM. Bronchoalveolar cell profiles of asthmatic and nonasthmatic subjects. Am Rev Respir Dis 1987; 136:379–383.
2. Gibson PG, Dolovich J, Girgis-Gabardo A, Morris MM, Anderson M, Hargreave FE, Denburg JA. The inflammatory response in asthma exacerbation: changes in circulating eosinophils, basophils and their progenitors. Clin Exp Allergy 1990; 20:661–668.
3. Kawabori S, Denburg JA, Schwartz LB, Irani AA, Wong D, Jordana G, Evans S, Dolovich J. Histochemical and immunohistochemical characteristics of mast cells in nasal polyps. Am J Respir Cell Mol Biol 1992; 6:37–43.
4. Ohnishi M, Ruhno J, Bienenstock J, Milner R, Dolovich J, Denburg JA. Human nasal polyp epithelial basophil/mast cell and eosinophil colony-stimulating activity: the effect is T-cell-dependent. Am Rev Respir Dis 1988; 138:560–565.
5. Ohnishi M, Ruhno J, Bienenstock J, Dolovich J, Denburg JA. Hemopoietic growth factor production by cultured cells of human nasal polyp epithelial scrapings: kinetics, cell source and relationship to clinical status. J Allergy Clin Immunol 1989; 83:1091–1100.
6. Denburg JA, Dolovich J, Ohtoshi T, Cox G, Gauldie J, Jordana M. The microenviron-

mental differentiation hypothesis of airway inflammation. Am J Rhinol, 1990; 4: 29–32.
7. Denburg JA, Jordana M, Vancheri C, Gauldie J, Harnish D, Ohtoshi T, Tsuda T, Gibson P, Ruhno J, Bienenstock J, Hargreave FE, Dolovich J. Locally derived hemopoietic cytokines for human basophils, mast cells and eosinophils in respiratory disease. In: Galli SJ, Austen KF, eds. Mast Cell and Basophil Differentiation and Function in Health and Disease. New York: Raven Press, 1989:59–69.
8. Vancheri C, Gauldie J, Bienenstock J, Cox G, Scicchitano R, Stanisz A, Jordana M. Human lung fibroblast-derived (GM-CSF) mediates eosinophil survival in vitro. Am J Respir Cell Mol Biol 1989; 1:289–295.
9. Vancheri C, Ohtoshi T, Cox G, Xaubet A, Abrams JS, Gauldie J, Dolovich J, Denburg JA, Jordana M. Neutrophilic differentiation induced by human upper airway fibroblast-derived granulocyte/macrophage colony-stimulating factor (GM-CSF). Am J Respir Cell Mol Biol 1991; 4:11–17.
10. Ohtoshi T, Vancheri C, Cox G, Gauldie J, Dolovich J, Denburg JA, Jordana M. Monocyte-macrophage differentiation induced by human upper airway epithelial cells. Am J Respir Cell Mol Biol 1991; 4:255–263.
11. Ohtoshi T, Xaubet A, Andersson B, Vanzeigleheim M, Dolovich J, Jordana M, Denburg JA. Nasal inflammation mediated by human structural cell-derived GM-CSF; effect of budesonide (abstr). J Allergy Clin Immunol 1990; 85:297.
12. Ohtoshi T, Zhou X, Gauldie J, Andersson B, Vanzeigleheim M, Liehl E, Ceska M, Dolovich J, Jordana M, Denburg J. Human upper airway epithelial cell and fibroblast IL-8 production and its modulation by steroids (abstr). J Allergy Clin Immunol 1991; 87:174.
13. Butterfield JH, Ackerman SJ, Weiler D, Eisenbrey AB, Gleich GJ. Effects of glucocorticoids on eosinophil colony growth. J Allergy Clin Immunol 1986; 78: 450–457.
14. Bjornson BH, Harvey JM, Rose L. Differential effect of hydrocortisone on eosinophil and neutrophil proliferation. J Clin Invest 1985; 76:924–929.
15. Cox G, Ohtoshi T, Vancheri C, Denburg JA, Dolovich J, Gauldie J, Jordana M. Promotion of eosinophil survival by human bronchial epithelial cells and its modulation by steroids. Am J Respir Cell Mol Biol 1991; 4:525–531.
16. Gibson PG, Dolovich J, Denburg J, Ramsdale EH, Hargreave FE. Chronic cough: eosinophilic bronchitis without asthma. Lancet 1989; 1:1346–1348.
17. Gibson PG, Girgis-Gabardo A, Morris MM, Mattoli S, Kay JM, Dolovich J, Denburg JA, Hargreave FE. Cellular characteristics of sputum from patients with asthma and chronic bronchitis. Thorax 1989; 44:693–699.
18. Gibson PG, Manning PJ, O'Byrne PM, Girgis-Gabardo A, Dolovich J, Denburg JA, Hargreave FE. Allergen-induced asthmastic responses. Relationship between increases in airway responsiveness and increases in circulating eosinophils, basophils and their progenitors. Am Rev Respir Dis 1991; 143:331–335.
19. Otsuka H, Dolovich J, Befus AD, Telizyn S, Bienenstock J, Denburg JA. Basophilic cell progenitors, nasal metachromatic cells and peripheral blood basophils in ragweed-allergic patients. J Allergy Clin Immunol 1986; 78:365–371.
20. Otsuka H, Dolovich J, Befus AD, Bienenstock J, Denburg JA. Peripheral blood

basophils, basophil progenitors, and nasal metachromatic cells in allergic rhinitis. Am Rev Respir Dis 1986; 133:757–762.

21. Pin I, Gibson PG, Kolendowicz R, Girgis-Gabardo A, Denburg JA, Hargreave FE, Dolovich J. Induced sputum cell counts: a noninvasive method to investigate airway inflammation in asthma. Thorax 1992; 47(1):25–29.
22. Denburg JA, Telizyn S, Messner H, Lim B, Jamal N, Ackermann SJ, Gleich GJ, Bienenstock J. Heterogeneity of human peripheral blood eosinophil-type colonies: evidence for a common basophil-eosinophil progenitor. Blood 1985; 66:312–318.
23. Mayer P, Valent P, Schmidt G, Liehl E, Bettelheim P. The in vivo effects of recombinant human interleukin-3: demonstration of basophil differentiation factor, histamine-producing activity, and priming of GM-CSF responsive progenitors in nonhuman primates. Blood 1989; 74:613–621.
24. Donahue RE, Seehra J, Metzger M, Lefebvre D, Rock B, Carbone S, Nathan DG, Garnick M, Sehgal PK, Laston D, LaVallie E, McCoy J, Schendel PF, Norton C, Turner K, Yang Y-C, Clark SC. Human IL-3 and GM-CSF act synergistically in stimulating hematopoiesis in primates. Science 1988; 241:1820–1823.
25. Kishi K. A new leukemia cell line with Philadelphia chromosome characterized as basophil precursors. Leuk Res 1985; 9:381–390.
26. Butterfield JH, Weiler D, Dewald G, Gleich GJ. Establishment of an immature mast cell line from a patient with mast cell leukemia. Leuk Res 1988; 12:345–355.
27. Ashman LK, Gadd SJ, Mayrhofer G, Spargo LDJ, Cole SR. A murine monoclonal antibody to an acute myeloid leukemia-associated antigen identifies tissue mast cells. In: McMichel A, ed. Leukocyte Differentiation Antigens. London: Oxford University Press, 1987:726.
28. Lerner NB, Nocka KH, Cole SR, Qiu F, Strife A, Ashman LK, Besmer P. Monoclonal antibody YB5.B8 identifies the human c-kit protein product. Blood 1991; 77:1876–1883.
29. Bodger MP, Newton LA. The purification of human basophils: their immunophenotype and cytochemistry. Br J Haematol 1987; 67:281–284.
30. Wong DA, Kawabori S, Switzer J, Valent P, Bettelheim P, Dolovich J, Ishizaka T, Jordana M, Denburg J. Characterization of histamine-containing basophilic cells induced by in vitro induction of HL-60 cell line (abstr). J Allergy Clin Immunol 1990; 85:173.
31. Matsuda H, Coughlin MD, Bienenstock J, Denburg JA. Nerve growth factor promotes human hemopoietic colony growth and differentiation. Proc Natl Acad Sci USA 1988; 85:6508–6512.
32. Matsuda H, Switzer J, Coughlin MD, Bienenstock J, Denburg JA. Human basophilic cell differentiation promoted by 2.5S nerve growth factor. Int Arch Allergy Appl Immunol 1988; 86:453–457.
33. Tsuda T, Wong DA, Dolovich J, Bienenstock J, Marshall J, Denburg JA. Synergistic effects of nerve growth factor and granulocyte-macrophage colony-stimulating factor on human basophilic cell differentiation. Blood 1991; 77:971–979.
34. Tsuda T, Switzer J, Bienenstock J, Denburg JA. Interactions of hemopoietic cytokines on differentiation of HL-60 cells; nerve growth factor is a basophilic lineage-specific co-factor. Int Arch Allergy Appl Immunol 1990; 91:15–21.

35. Denburg JA. Cytokine-induced human basophil/mast cell growth and differentiation in vitro. Springer Semin Immunopathol 1990; 12:401–414.
36. Yokota T, Coffman RL, Hagiwara H, Rennick DM, Takebe Y, Yokota K, Gemmell L, Shrader B, Yang G, Meyerson P, Luh J, Hoy P, Pène J, Brière F, Spits H, Banchereau J, de Vries J, Lee FD, Arai N, Arai K-I. Isolation and characterization of lymphokine cDNA clones encoding mouse and human IgA-enhancing factor and eosinophil colony-stimulating factor activities: relationship to interleukin 5. Proc Natl Acad Sci USA 1987; 84:7388–7392.
37. Denburg JA, Silver JE, Abrams JS. Interleukin-5 is a human basophilopoietin: induction of histamine content and basophilic differentiation of HL-60 cells and of peripheral blood basophil-eosinophil progenitors. Blood 1991; 77:1462–1468.
38. Plaetinck G, Van der Heyden J, Tavernier J, Fachè I, Tuypens T, Fischkoff S, Fiers W, Devos R. Characterization of interleukin 5 receptors on eosinophilic sublines from human promyelocytic leukemia (HL-60) cells. J Exp Med 1990; 172:683–691.
39. Hutt-Taylor SR, Harnish D, Richardson M, Ishizaka T, Denburg JA. Sodium butyrate and a T lymphocyte cell line-derived differentiation factor induce basophilic differentiation of the human promyelocytic leukemia cell line HL-60. Blood 1988; 71:209–215.
40. Wong DTW, Weller PF, Galli SJ, Elovic A, Rand TH, Gallagher GT, Chiang T, Chou MY, Matossian K, McBride J, Todd R. Human eosinophils express transforming growth factor α. J Exp Med 1990; 172:673–681.
41. Ohno I, Lea RG, Flander KC, Clark DA, Banwatt D, Dolovich J, Denburg J, Harley CB, Gauldie J, Jordana M. Eosinophils in chronically inflamed human upper airway tissues express transforming growth factor β1 gene (TGFβ1). J Clin Invest 1992; 89: 1662–1668.
42. Ohno I, Lea R, Finotto S, Marshall J, Denburg J, Dolovich J, Gauldie J, Jordana M. Granulocyte/macrophage colony-stimulating factor (GM-CSF) gene expression by eosinophils in nasal polyposis. Am J Respir Cell Mol Biol 1991; 5:505–510.
43. Finotto S, Ohno I, Lea R, Marshall J, Denburg J, Dolovich J, Gauldie J, Jordana M. Tumor necrosis factorα (TNFα) gene expression by eosinophils in nasal polyp (NP) tissues (abstr). Am Rev Respir Dis 1992; 145:A440.
44. Kita H, Ohnishi T, Okubo Y, Weiler D, Abrams JS, Gleich GJ. Granulocyte/macrophage colony-stimulating factor and interleukin 3 release from human peripheral blood eosinophils and neutrophils. J Exp Med 1991; 174:745–748.
45. Moqbel R, Hamid Q, Ying S, Barkans J, Abrams JS, Kay AB. Human eosinophils as cytokine producing inflammatory cells (abstr). FASEB J 1992; 6:A1402.
46. Desreumaux P, Janin A, Colombel JF, Prin L, Plumas J, Emilie D, Torpier G, Capron A, Capron M. Interleukin 5 messenger RNA expression by eosinophils in the intestinal mucosa of patients with coeliac disease. J Exp Med 1992; 175:293–296.
47. Wong DTW, Elovic A, Matossian K, Nagura N, McBride J, Chou MY, Gordon JR, Rand TH, Galli SJ, Weller PF. Eosinophils from patients with blood eosinophilia express transforming growth factor β_1. Blood 1991; 78:2702–2707.

DISCUSSION (Speaker: J. Denburg)

Konig: Does TGFβ affect eosinophil differentiation and isotype switch?

Denburg: This is speculative at this point, but TGF-β may regulate subsequent differentiation of eosinophils and basophils, based on some as yet unpublished work by Sillaber et al. Whether TGF-β has effects also on IgG isotype switching in nasal polyps is not known.

Townley: Were the primates already antigen-sensitized, before antigen injection?

Denburg: No, the primates were not already sensitized. Two groups of investigators have shown that repeated injections of GM-CSF and IL-3 cause blood eosinophilia, basophilia, and increases in circulating basophil/eosinophil progenitors, similar to what we see in allergen-challenged asthmatics.

Ackerman: You have routinely demonstrated increased numbers of circulating eosinophil and basophil progenitors in association with allergic and related reactions. Could you speculate on the physiological relevance of this? What is the source of IL-5 in the bone marrow for eosinophil differentiation?

Denburg: The peripheral blood progenitors could be the most readily available repository for quick delivery to tissues where chronic or acute allergic inflammation is occurring. This needs to be proven directly, and we are developing an animal model to do this. I do not know what the source of IL-5 is in the bone marrow, but in allergic tissues clearly there are IL-5-gene-expressing and IL-5-producing cells.

Wardlaw: I have always considered attractive the concept that in situ differentiation of eosinophils from precursors in the tissues could explain, in part, eosinophil stimulation. Do you ever see eosinophil-like cells undergoing mitosis in nasal polyps or have you been able to take precursor-like cells from nasal polyps and induce eosinophil colony formation?

Denburg: We do see variations in nuclear morphology and degree of apparent maturation of eosinophilic and basophilic myelocytes in polyps and related tissues in humans. In a guinea pig model of basophilopoiesis, we did see mitotic basophilic myelocytes, which are not evident in human polyps. However, we may be dealing with terminal steps in differentiation and do not therefore expect mitotic figures. We have shown and published data on the presence of hemopoietic colony-forming cells, for basophils primarily, in nasal polyps.

Hansel: If we consider the spectrum of cytokines produced by eosinophils, this includes the possibility of TGF-α and -β, GM-CSF, TNF, IL-1, IL-3, IL-5, IL-6, and IL-8. With the exception of IL-5, this looks like a myeloid- or macrophage-like pattern. Is this an overgeneralization or could this concept be informative?

Denburg: I would not wish to characterize the pattern of cytokines as myeloid or other. It may be that we are simply seeing aspects of cell activation in the abundant expression of cytokine genes.

Takatsu: Mac-1 is usually expressed on macrophages. Do you have any explanation that eosinophil expresses Mac-1 antigen?

Denburg: As others have discussed, it may play a role in eosinophil tissue accumulation.

11

Eosinophil Membrane Receptors: Function of IgE- and IgA-Binding Molecules

Monique Capron, Marie-José Truong, Pierre Desreumaux, Bouchaïb Lamkhioued, Margherita Tomassini, and André Capron
Centre d'Immunologie et de Biologie Parasitaire, Institut Pasteur, Lille, France

I. INTRODUCTION

Over the past 20 years, several lines of evidence have raised growing interest in the effector function of eosinophils, regarding a variety of clinical states associated with hypereosinophilia, including parasitic and allergic diseases (1). The cytolytic potential of eosinophils has been extensively studied in cytotoxicity assays against parasite larvae. These mechanisms are strictly dependent on antibodies and/or complement, which interact with specific receptors on eosinophil membrane. Adhesion molecules allowing cellular interactions and tissue migration have also been described on eosinophils. Differentiation and activation of eosinophils are controlled by hemopoietic growth factors, namely IL-3, GM-CSF, and specifically IL-5, which bind to the corresponding membrane receptors. Indeed, eosinophils possess a large variety of membrane receptors, which belong to different gene superfamilies, including members of the *immunoglobulin family* (such as FcγR or FcαR), members of the *integrin family* (VLA4 as β1 integrin or LFA1/CD11a, CR3/Mac1/CD11b, CR4/p150,95/CD11c, and CD18 as β2 integrins), and members of the *cytokine receptor family* (IL-3, GM-CSF, or IL-5 receptor). Eosinophils also exhibit molecules with a *lectin domain* such as FcεRII/CD23 (C-type lectin) or, more recently, Mac2/ε binding protein (S-type lectin). Although the existence of lectins has been known for more than 100 years, the idea that they may act as recognition molecules is only recent, and the specificity of lectins proved to be more exquisite than originally assumed (2). Cell surface lectins may bind to soluble glycoproteins or may combine with carbohydrates of components of the extracellular matrix, promoting cell adhesion.

The aim of the present review is to highlight the structure and the functions of IgE- and IgA-binding molecules expressed on human eosinophil membrane, with particular emphasis on lectins.

II. IgE-BINDING MOLECULES

Immunoglobulin E is the class of immunoglobulin responsible for human anaphylaxis and allergic reactions, which exerts its biological function through interaction with cell surface proteins. Two main groups of IgE-binding molecules have been described. The first class of molecules, defined as IgE receptors with a membrane anchor domain, is found on most cells involved in the immune response. Mast cells and basophils present high-affinity Fc receptors for IgE (FcϵRI) (3), and lymphocytes and inflammatory cells present low-affinity Fc receptors for IgE (FcϵRII/CD23) (4). In addition to these IgE Fc receptors, several molecules present the ability to bind both IgE and selective carbohydrates. These molecules constitute a part of the family of lectins with specific carbohydrate-recognition domain (5).

In order to identify the molecules that bind IgE on human eosinophils, our first strategy was based on the described antigenic similarities between eosinophil IgE receptor and the B-cell FcϵRII/CD23 (6,7). This approach allowed us to demonstrate the variable membrane expression of several epitopes of CD23 on eosinophils from patients. Moreover, CD23 was shown to participate in IgE binding and IgE-mediated cytotoxicity of human eosinophils (7). However, the low amounts of mRNA hybridizing with the CD23 cDNA probe, found in a very small proportion of patients (7), suggested either a low level of expression of CD23 mRNA in eosinophils from patients (as shown for some granule protein mRNA) (8), the instability of this given mRNA species, or a low level of homology between the "CD23 like" expressed on eosinophils and the B-cell CD23. In addition, the screening of an eosinophil cDNA library with the CD23 probe did not permit the cloning of a cDNA related to CD23 (9), leading to the hypothesis that human eosinophils might express another type of IgE-binding molecule. A family of molecules has particularly drawn our attention: Mac-2, a murine macrophage cell surface protein (10) shown to be highly homologous to the carbohydrate binding protein 35 (CBP35) (11) and to the rat IgE-binding protein (ϵBP) (12). The human counterparts of ϵBP (hϵBP) and Mac-2 (hMac-2) cDNA have recently been cloned (13,14). The comparison between hMac-2 and hϵBP sequences reveals that hMac-2 is almost identical to hϵBP. These molecules, belonging to the S-type lectins (thiol-dependent) with a specific β-galactoside recognition domain (15), are detected on the cell surface, in the cytoplasm, and in the nucleus. They are expressed by various cell types, specifically after activation (16). The precise functions of these lectins are still unknown; however, ϵBP has recently been proposed as a cell adhesion protein (16).

We have investigated whether human eosinophils can express molecules of this lectin family, using binding of antibodies directed against Mac-2 or against hϵBP and Northern blot with the corresponding cDNAs. In addition, we have evaluated the involvement of these molecules in the IgE-dependent function of eosinophils. The expression of Mac-2 molecules on the eosinophil surface was first analyzed by flow cytometry and compared to the expression of IgE receptors. Independently of eosinophil origin and density, surface Mac-2 molecules were detected in about half of patients with eosinophilia ($n = 22$). The binding of IgE contrasting with the absence of surface Mac-2 on eosinophils from some patients suggested the existence of additional IgE-binding molecules such as CD23 (7). Northern blot analysis performed with eosinophil RNA and probed with hMac-2 or hϵBP cDNA revealed the same RNA band at 1.2 kb in all eosinophil preparations, expressing or not membrane Mac-2. Similar results have recently been reported for different mast cell lines, with variable ϵBP surface expression and comparable mRNA content (16). Our results suggest that the mechanism of surface expression of Mac-2/ϵBP can be differentially regulated according to the patients. Both anti-Mac-2 and anti-ϵBP detected a molecule at 29 kDa in eosinophil extracts by Western blot and immunoprecipitation. However, two smaller-MW molecules could be also detected at 20 kDa and 15 kDa. Further experiments are required to identify these smaller-MW molecules.

The involvement of Mac-2/ϵBP in the IgE-dependent functions of eosinophils was also investigated, since Mac-2/ϵBP has been shown to bind IgE through its carbohydrate moiety (13). Experiments of inhibition of radiolabeled IgE binding were performed to study whether Mac-2/ϵBP could be involved in IgE binding to eosinophils (Table 1). Anti-Mac-2 monoclonal antibody (MAb) significantly inhibited, in a dose-dependent manner, the binding of ^{125}I IgE, in contrast to control MAb. Moreover, in order to evaluate whether the molecules recognized by anti-Mac-2 MAb could participate in the IgE-dependent effector function of eosinophils, the effect of anti-Mac-2 MAb was investigated in a cytotoxicity assay involving eosinophils and polyclonal IgE antibodies from patients (Table 1). Anti-Mac-2 MAb strongly inhibited IgE-dependent cytotoxicity of hypodense eosinophils against schistosome larvae, by comparison with control antibodies. These results indicate the participation of Mac-2/ϵBP in IgE binding and IgE-dependent cytotoxicity mediated by eosinophils. FcϵRII/CD23, which presents significant sequence homology with a class of animal lectins (17), has been shown to be also involved in these functions (7). However, in the case of C-type (calcium-dependent-binding) lectins, IgE binding to FcϵRII is pH-dependent but independent of the carbohydrate structure of IgE (18,19). In contrast to FcϵRII/CD23, Mac-2 and hϵBP bind to IgE in a carbohydrate-dependent manner and show

Table 1 Inhibitory Effects of Anti-Mac 2 MAb on IgE Binding and IgE-Mediated Cytotoxicity by Eosinophils

	% inhibition[a]	
Final dilution of MAb supernatants	IgE binding[b]	IgE-dependent cytotoxicity[c]
Anti-Mac 2		
1/10	59.8 ± 9.6	86.2 ± 5.7
1/100	47.6 ± 9.7	85.4 ± 8.3
1/500	5.9 ± 5.9	8.1 ± 8.1
Control		
1/10	15.7 ± 15.7	11.3 ± 4.5
1/100	17.4 ± 0.9	12.0 ± 7.8
1/500	9.0 ± 9.0	2.3 ± 2.3

[a]Calculated by comparison with control values in the absence of supernatant (mean of four experiments ± SEM).
[b]Binding of ^{125}I-labeled IgE (PS myeloma protein).
[c]IgE-dependent eosinophil-mediated cytotoxicity toward *Schistosoma mansoni* schistosomula.

heterogeneity in the binding of various IgE glycoforms (13,20). The participation of the two groups of IgE-binding molecules (CD23 and Mac-2/εBP) in the cytotoxic function of eosinophils suggest an important role in the recognition of IgE by CRD domains of C- and S-type lectins. In addition to the role of Mac-2/εBP in IgE binding to eosinophils, the inhibition by anti-Mac-2 MAb of eosinophil adhesion to parasite targets also suggests the function of such molecules as cell adhesion proteins and should be explored in other adhesive properties of eosinophils.

Taken together, these results indicate that not only transmembrane receptors such as FcεRII/CD23, but also soluble molecules expressed on the surface of eosinophils, might participate in the IgE-dependent effector function.

III. IgA-BINDING MOLECULES

The first evidence of IgA binding to eosinophils was reported in 1988 by flow cytometry (21). Between 5 and 90% eosinophils from eosinophilic patients are able to bind serum IgA, as revealed by staining with fluorescein-labeled anti-human IgA. The large variations in the expression of IgA receptors (R) in individual patients suggest a modulation of this receptor, as already shown for IgE R. Moreover, cytophilic IgA could be detected by this assay in some patients,

Table 2 Specificity of the Binding of Radiolabeled Secretory Component (SC)

Inhibitor	% inhibition of binding of SC[a]
SC[b]	61.9 ± 8.1
IgG[b]	4.2 ± 4.2
Serum IgA[b]	17.0 ± 12.3
Secretory IgA[b]	50.9 ± 4.8
Galactose[c]	65.2 ± 11.2
Glucose[c]	70.9 ± 3.1
Mannose[c]	77.4 ± 0.7

[a]Calculated by comparison with control values without inhibitor (mean of three to seven experiments ± SEM).
[b]At a 100-fold molar excess dose.
[c]At 100 mM.

mainly with parasitic infections or allergic diseases. By various biochemical procedures, IgA R was characterized as a 55- to 60-kDa molecule, a MW closely similar to that of FcαR on neutrophils and monocytes. The recognition of eosinophil FcαR by the MAb that binds to neutrophil and monocyte FcαR indicated that eosinophil FcαR might be similar to the myeloid cell IgA R, a member of the Ig superfamily (22).

The increased binding of secretory IgA versus serum IgA, together with the detection of secretory component bound to eosinophil surface by flow cytometry, led us to suggest the existence of a binding site for secretory component on eosinophils. Flow cytometry and Scatchard analysis revealed the presence of a saturable binding site for purified secretory component on eosinophils from patients with eosinophilia (23). The competitive effect of carbohydrates suggested the lectin nature of the binding site (Table 2), which was further characterized as a 15-kDa molecule, eluted from a secretory component Sepharose column by carbohydrates (23).

The precise nature of this 15-kDa lectin is presently under investigation, especially its relationships with the 15-kDa smaller-MW component recognized by anti-Mac 2 MAb.

IV. ROLE OF IgE- AND IgA-BINDING MOLECULES IN SECRETION OF GRANULE PROTEINS

IgE and IgA antibodies have been shown to induce the cytotoxic properties of eosinophils against parasite larvae (21,24), suggesting that after binding of these antibody isotypes to their appropriate receptors, release of toxic granule proteins may occur. These membrane receptors can be triggered by antibodies bound to parasite larvae or by immune complexes (25). They can also be triggered by addition of insolubilized ligands such as Ig-coated Sepharose beads. Using this model, it has been shown, for instance, that IgG- or IgA-coated beads could stimulate the release of eosinophil-derived neurotoxin (EDN) (26). Interestingly, IgE-coated beads did not induce EDN release, a result that might be due to the eosinophil population used (normodense cells) (24). Secretory IgA provided the most potent stimulus for EDN release, but no explanation was really given (26). We have used an alternative experimental approach, based on the cross-linking of receptor bound ligands by addition of either antigens or of a second antibody.

Effector functions of eosinophils against parasites, as well as against host cells, appear to be mainly mediated by the release of cationic proteins, such as major basic protein (MBP), eosinophil cationic protein (ECP), EDN, and eosinophil peroxidase (EPO). All these preformed mediators display a strong cytotoxic potential (27), but the precise mechanisms leading to eosinophil degranulation are still poorly understood. Studies on parasite-infected patients have shown that not all the granule proteins were simultaneously released after the same stimulus of activation, suggesting a differential release of EPO and ECP (28,29). It was not clear, however, whether eosinophils isolated from allergic patients could degranulate on exposure to allergens. To answer these questions, we have evaluated the release of EPO and ECP by eosinophils from allergic subjects, after the addition of the specific allergen or antihuman IgE MAb, in comparison to eosinophils from patients with eosinophilia of nonallergic origin. In addition, to further investigate the recently described interactions between IgA and human eosinophils (21,26), we have monitored the release of granule proteins after exposure of purified eosinophils to antihuman IgE, antihuman IgA, and antihuman IgG monoclonal antibodies and provided evidence that these three isotypes can be involved, although in different ways, in the antibody-dependent mechanisms of eosinophil activation (30).

Eosinophils isolated from allergic patients were incubated with the specific allergens (*Dactylis glomerata* or *Dermatophagoides pteronyssinus*). Eosinophils from nonallergic patients with hypereosinophilia were tested in the same conditions and the release of EPO evaluated. Only eosinophils isolated from allergic patients released significant amounts of EPO after stimulation with the specific allergen. To investigate the specificity of allergen-induced EPO release, eosinophils from allergic patients were incubated with unrelated allergen; no EPO release was observed in such experiments. In the same groups of patients, EPO release was measured after the addition of antihuman IgE antibodies. Similarly to stimulation with allergen, eosinophils from allergic patients released higher levels of EPO than eosinophils from nonallergic patients. Moreover, there was a significant correlation between EPO release induced by allergen and by antihuman IgE

MAb ($r = 0.93$; $p < 0.001$), suggesting the participation of surface IgE in this mechanism of release. ECP was measured by a radioimmunoassay in aliquots of the same samples assayed for EPO. In none of the allergic patients did the addition of the specific allergen or antihuman IgE induce a significant release of ECP. A similar comparison between EPO and ECP release after stimulation with anti-IgE antibodies in a total of 16 patients revealed the absence of correlation between EPO and ECP release.

To investigate the IgA-dependent eosinophil activation, the release of EPO and ECP was evaluated after incubation of eosinophils with antihuman IgA MAb. The release of EPO and ECP was compared in 14 patients. Six of the 14 patients released significant levels of EPO after stimulation with antihuman IgA, whereas nine patients released significant amounts of ECP. Interestingly, the individual results of EPO and ECP release were significantly correlated ($r = 0.6$; $p < 0.02$). Similarly, the release of EPO and ECP was compared on different aliquots of the same supernatants of eosinophils purified from patients after incubation with antihuman IgG MAb. All the patient eosinophils released significantly high levels of ECP but did not release significant amounts of EPO, except for one patient, giving high levels of both EPO and ECP.

Taken together, these results suggest that EPO is not released from human eosinophils in the same conditions of activation as ECP and raises the interesting question of the mechanism of exocytosis of these granule proteins. We have no definitive arguments to answer this question, but electron microscopy studies performed on tissue eosinophils by immunogold labeling suggest that the granule proteins might leave the granule through the cytoplasm, by a "vesicular transport mechanism" similar to that described for delayed-type cutaneous basophil hypersensitivity (31). Eosinophil activation is probably a complex phenomenon depending not only on the stimulus used, but also on the state of activation or differentiation of cells.

Perhaps the most novel and interesting data are those concerning IgA-mediated eosinophil activation. Surface-bound IgA antibodies seemed to represent a potent stimulus, since both EPO and ECP could be released. These results confirm our previous findings on the interactions between IgA and eosinophils (21) and are in total agreement with the IgA-mediated release of EDN (26). To better understand the role of IgA in mediator release and the nature of the membrane receptors involved, we have compared the release of EPO and ECP induced by different molecular forms of IgA. Whereas incubation of eosinophils with secretory IgA alone induced release of EPO and ECP, the addition of monomeric or polymeric IgA1 or IgA2 did not lead to EPO or ECP release. These results suggest that the preferential mediator release induced by secretory IgA was not due to the increased proportion of IgA2 in secretory IgA nor to the presence of J chain in polymeric IgA. We thus tested the hypothesis that the main factor responsible could be the presence of secretory component (SC) in secretory IgA. Indeed, a successive incubation of eosinophils with SC and anti-SC antibodies led to highly significant amounts of EPO and ECP released. Further experiments are needed to investigate whether the 15-kDa molecule involved in SC binding might also be involved in mediator release induced by secretory IgA.

Taken together, these findings might be of relevance to the pathogenesis of a variety of human diseases characterized by eosinophilia involving mucosal surfaces. It should be noted that eosinophils are increased in the nasal mucosa of allergic patients during natural pollen exposure, and there is a relationship between eosinophil activation and allergen-induced local inflammation (32). However, there is a lack of correlation between the variable ECP levels in nasal lavages and the local hyperresponsiveness observed after nasal allergen challenge (33). In contrast, elevated blood levels of MBP, ECP, and EDN have been detected in patients with allergic asthma after bronchial challenge, whereas EPO levels were reduced (34). Interestingly, we recently demonstrated that not only granule proteins, but also cytokines, might be differentially released by eosinophils, according to the stimulus of activation. All these data point to the differential release of eosinophil mediators. However, it cannot be excluded that in vivo pathways of eosinophil secretion might differ from the in vitro situation, depending, for instance, on exposure to a particular local microenvironment. In this respect, the influence of cell–cell interactions, mediated by intercellular adhesion molecules (35), or the role of soluble factors such as cytokines has been recently suggested (see Chap. 6, this volume).

In conclusion, further studies are needed to investigate the relationships between the various receptors and molecules present on eosinophil membrane but also to identify the transduction signals induced by the interaction between the receptors and their ligands.

V. SUMMARY

Human eosinophils exhibit various membrane receptors belonging to different gene superfamilies, including members of the immunoglobulin, integrin, complement receptor, and cytokine receptor family. They also possess molecules with a lectin domain. In the present review, the structure and functions of IgE- and IgA-binding molecules are analyzed. In addition to CD23, eosinophils can express molecules of the S-type lectin family, namely Mac2/εBP, as shown by flow cytometry, Northern blot analysis, and immunoprecipitation. Mac2/εBP is involved in IgE binding and IgE-dependent cytoxicity mediated by eosinophils, suggesting that not only transmembrane receptors, but also cytosolic molecules bound to cell surface, might participate in IgE-dependent effector functions.

In addition to the classic FcαR, which binds to serum IgA, a saturable binding site for purified secretory component has been detected on eosinophils. Granule proteins can be released after exposure of eosinophils to antihuman IgE and antihuman IgA antibodies or to the specific antigen. However, no correlation was observed between EPO and ECP release, suggesting a differential release of these two mediators according to the stimulus of activation. Various experimental procedures indicate that the preferential mediator release induced by secretory IgA is due to the binding of secretory component and the subsequent activation of eosinophils. Taken together, these findings point to the effector function of eosinophils in local and mucosal immunity.

REFERENCES

1. Spry CJF. Eosinophils. A Comprehensive Review and Guide to the Scientific and Medical Literature. Oxford: Oxford Medical Publications, 1988.
2. Sharon N, Lis H. Lectins as cell recognition molecules. Science 1989; 246:227.
3. Metzger H, Alcaras G, Holman R, Kinet JP, Pribluda V, Quarto R. The receptor with high affinity for immunoglobulin E. Annu Rev Immunol 1986; 4:419.
4. Capron A, Dessaint JP, Capron M, Joseph M, Ameisen JC, Tonnel AB. From parasites to allergy: the second receptor for IgE (FcεR2). Immunol Today 1986; 7:15.
5. Drickamer K. Two distinct classes of carbohydrate-recognition domains in animal lectins. J Biol Chem 1988; 263:9557.
6. Grangette C, Gruart V, Ouaissi MA, Rizvi F, Delespesse G, Capron A, Capron M. IgE receptor on human eosinophils (FcεRII): comparison with B cell CD23 and association with an adhesion molecule. J Immunol 1989; 143:3580.
7. Capron M, Truong MJ, Aldebert D, Gruart V, Suemura H, Delespesse G, Tourvieille, Capron A. Heterogeneous expression of CD23 epitopes by eosinophils from patients. Relationships with IgE-mediated functions. Eur J Immunol 1991; 21:2423.
8. Gruart V, Truong MJ, Plumas J, Zandecki M, Kusnierz JP, Prin L, Capron A, Capron M. Decreased expression of eosinophil peroxidase and major basic protein mRNA during eosinophil maturation. Blood 1992; 79:2592.
9. Truong MJ, Gruart V, Capron A, Capron M, Tourvieille B. Cloning and expression of a cDNA encoding a non-classical MHC class I antigen (HLA-E) in eosinophils from hypereosinophilic patients. J Immunol 1992; 148:627.
10. Cherayil BJ, Weiner SJ, Pillai S. The Mac-2 antigen is a galactose-specific lectin that binds IgE. J Exp Med 1989; 170:1959.
11. Jia S, Wang JL. Carbohydrate binding protein 35. Complementary DNA sequence reveals homology with proteins of the heterogeneous nuclear RNP. J Biol Chem 1988; 263:6009.
12. Albrandt K, Orida NK, Liu FT. An IgE-binding protein with a distinctive repetitive sequence and homology with an IgG receptor. Proc Natl Acad Sci USA 1987; 84: 6859.
13. Robertson MW, Albrandt K, Keller D, Liu FT. Human IgE-binding protein: a soluble lectin exhibiting a highly conserved interspecies sequence and differential recognition of IgE glycoforms. Biochemistry 1990; 29:8093.
14. Cherayil BJ, Chaitivitz S, Wong C, Pillai S. Molecular cloning of a human macrophage lectin specific for galactose. Proc Natl Acad Sci USA 1990; 87:7324.

15. Paroutaud P, Levi G, Teichberg VI, Strosberg AD. Extensive amino-acid sequence homologies between animal lectins. Proc Natl Acad Sci USA 1987; 84:6345.
16. Frigeri LG, Liu FT. Surface expression of functional IgE-binding protein, an endogenous lectin, on mast cells and macrophages. J Immunol 1992; 148:861.
17. Ikuta K, Takami M, Kim CW, Honjo T, Miyoshi T, Tagaya Y, Kawabe T, Yodoi J. Human lymphocyte Fc receptor for IgE: sequence homology of its cloned cDNA with animal lectins. Proc Natl Acad Sci USA 1987; 84:649.
18. Vercelli D, Helm B, Marsh P, Padlan E, Geha RS, Gould H. The B cell binding site on human immunoglobulin E. Nature 1989; 338:649.
19. Richards ML, Katz DH. The binding of IgE to murine FcεRII is calcium-dependent but not inhibited by carbohydrate. J Immunol 1990; 144:2638.
20. Robertson MW, Liu FT. Heterogeneous IgE glycoforms characterized by differential recognition of an endogenous lectin (IgE-binding protein). J Immunol 1991; 147:3024.
21. Capron M, Tomassini M, Van der Vorst E, Kusnierz JP, Papin JP, Capron A. Existence et fonctions d'un récepteur pour l'immunoglobuline A sur les éosinophiles humains. CR Acad Sci/Immunol 1988; 307:397.
22. Maliszewski CR, March CJ, Schoenborn MA, Gimpel S, Shen L. Expression cloning of a human Fc receptor for IgA. J Exp Med 1990; 172:1665.
23. Capron M, Gruart V, Broussolle A, Capron A. Binding-site for secretory component on human eosinophils. FASEB J 1991; 5:A640.
24. Capron M, Spiegelberg HL, Prin L, Bennich HL, Butterworth AE, Pierce RJ, Ouaissi MA, Capron A. Role of IgE receptors in effector function of human eosinophils. J Immunol 1984; 232:462.
25. Capron M, Plumas J. Eosinophil membrane receptors. In: Makino S, Fukuda T, eds. Eosinophils: Biological and Clinical Aspects. (In press.)
26. Abu-Ghazaleh RI, Kujisawa T, Mestecky J, Kyle RA, Gleich GJ. IgA-induced eosinophil degranulation. J Immunol 1989; 142:2393.
27. Gleich GJ, Adolphson CR. The eosinophilic leukocyte: structure and function. Adv Immunol 1986; 39:177.
28. Khalife J, Capron M, Cesbron JY, Tai PC, Taelman H, Prin L, Capron A. Role of specific IgE antibodies in peroxidase (EPO) release from human eosinophils. J Immunol 1986; 137:1659.
29. Capron M, Tomassini M, Torpier G, Kusnierz JP, McDonald S, Capron A. Selectivity of mediators released by eosinophils. Int Arch Allergy Appl Immunol 1989; 88:54.
30. Tomassini M, Tsicopoulos A, Tai PC, Gruart V, Tonnel AB, Prin L, Capron A, Capron M. Release of granule proteins by eosinophils from allergic and non-allergic patients with eosinophils on immunoglobulin-dependent activation. J Allergy Clin Immunol 1991; 88:365.
31. Capron M, Grangette C, Torpier G, Capron A. The second receptor for IgE in eosinophil effector function. Chem Immunol 1989; 47:128.
32. Pipkorn U, Karlsson G, Enerback L. The cellular response of the human allergic mucosa to natural allergen exposure. J Allergy Clin Immunol 1988; 82:1046.
33. Andersson M, Andersson P, Venge P, Pipkorn U. Eosinophils and eosinophil cationic protein in nasal lavages in allergen-induced hyperresponsiveness: effect of topical glucocorticoid treatment. Allergy 1989; 44:342.

34. Durham SR, Loegering DA, Dunnette SL, Gleich GJ, Kay AB. Blood eosinophils and eosinophil-derived proteins in allergic asthma. J Allergy Clin Immunol 1989; 84: 931–936.
35. Wegner CD, Gundel RH, Reilly P, Haynes N, Letts LG, Rothlein R. Intracellular adhesion molecule-1 (ICAM1) in the pathogenesis of asthma. Science 1990; 247:456.

DISCUSSION (Speaker: M. Capron)

Ackerman: Your data on inhibition of sIgA binding to the 15-kDa putative receptor on eosinophils contradicts what is known regarding the specificity of lectins (S-type versus C-type) where S-type β-galactoside-binding lectins are inhibitable by lactose and galactose but not mannose, and C-type lectin activity is inhibitable by mannose, but not lactose and related sugars. Inhibition by glucose, galactose, and mannose would therefore appear to be nonspecific or unrelated to lectin-like binding. Please comment.

Capron: I have not claimed that the receptor for secretory component belonged to S- or C-type lectins. I only mentioned the elution of the 15-kDa component by various carbohydrates. We now have experiments of microsequencing in progress to try to identify this 15-kDa molecule. It is true that the inhibition of binding to lectins must be specific for given sugars. However, inhibition experiments on intact eosinophils and soluble ligands by various carbohydrates are very difficult to interpret.

Kay: How do you distinguish the binding affinity of IgE for Mac-2 and CD23?

Capron: We will use IgE synthetic peptides (lacking carbohydrates) to differentiate between IgE binding to CD23 or to Mac-2 lectins.

Holgate: Is anything known about the signal transduction mechanism via lectins?

Capron: There is no report in the literature of the biological function of Mac-2, at the level of signal transduction.

Wardlaw: It might be expected that neutrophils express Mac-2 but neutrophils do not bind IgE or have IgE-dependent functions. Do neutrophils express Mac-2, and if so, how do you reconcile this discrepancy?

Capron: We now have experiments in progress on the expression of Mac-2 by neutrophils. In this respect, it is interesting that Mac-2 is a cytosolic molecule that is not always expressed on the surface (as shown by our studies on eosinophils and also by published reports on mast cell lines).

Konig: Do patients with high SCD23 have high levels of SMac-2?

Capron: We have not yet studies soluble Mac-2 in the serum from eosinophilic patients (mainly because of the absence of a sensitive assay). However, we are now performing experiments to show that purified eosinophils can release SCD23 or SMac-2 in supernatants.

Butterworth: I would like to add a comment to your interesting results on IgA. Dave Dunne, Kareen Thorne, and I have been using Marianne Bruggeman's set of humanized chimeric anti-NIP monoclonals to test the capacity of antibodies of different isotypes but a single antigenic specificity to mediate eosinophil-dependent killing of schistosome larvae (using NIP-coated schistosomula as targets). We find that IgA is the most effective isotype in mediating killing by GM-CSF-activated eosinophils. In this case this was the equivalent of serum IgA produced by the hybridomas, without the secretory component.

Weller: The demonstration of IL-5 within eosinophils by immunocytochemistry is very exciting. With regard to the capacity of the eosinophil to secrete IL-5 and/or to utilize IL-5 for autocrine stimulation, it will be interesting to know how IL-5 is packaged within the eosinophil. Your immunocytochemical staining suggested focal staining of eosinophils. Are there any data available on where IL-5 is found within the cells, whether it be granules or other organelles?

Capron: To answer this question we have started experiments using electron microscopy and immunogold labeling.

Coffman: What are the kinetics of IL-5 release following IgA or IgG stimulation—was IL-5 message induced in patients or was it present already?

Capron: IL-5 was released after overnight incubation with IgA anti-IgA or with IgE anti-IgE. We have not yet performed kinetic studies. However, I can add that whereas peripheral blood eosinophils had IL-5 mRNA (without ex vivo stimulation), they need to be stimulated to express the protein (as detected by immunocytochemistry) or to secrete it in the supernatants.

Moqbel: It is known that IgA is a major immunoglobulin found in mucosal surfaces during enteral helminthic infections. Is there any evidence for in vivo killing of parasites via IgA or does IgA exert its protective properties suggested in your studies in combination with mucus by entrapping the worms?

Capron: We have previously published that human IgA antibodies could participate in ADCC reactions mediated by eosinophils against parasites (by inhibition experiments with aggregated IgA) (1). We have recently confirmed the same result with serum from SM 28 GST immunized rats and not eosinophils (2).

Spry: What is the relationship between intensity of IL-5 staining and type and status of patient disease?

Capron: We have evidence that IL-5 mRNA expression by eosinophils decreased with treatment of patients, either gluten restriction of patients with celiac disease or α-interferon therapy in patients with hypereosinophilic syndrome.

A. Capron: There is now increasing evidence, both in experimental models and in human populations, that IgA antibodies are a significant component of the protective immune response to schistosomes. We have in particular shown that IgA antibodies to the protective SM 28 GST neutralize the glutatione S transferase enzymatic activity of the molecule, leading to significant decrease in worm fecundity and egg viability. Monoclonal IgA antibodies recently produced in our laboratory exhibit strong protective effects in the mouse after passive transfer. Finally, in our collaborative studies with A. Butterworth in Kenya, we have observed that in addition to the already mentioned correlation between specific IgE response and acquisition of immunity in human populations, there was a significant association between IgA antibodies to schistosome antigens (and in particular for SM 28 GST) and the age-dependent decrease in egg output.

Vadas: Do parasites carry the signals that are recognized by lectins? If so, why is IgA necessary?

Capron: It is possible that eosinophils possess different lectin-like molecules, which might bind to the corresponding sugars on parasites, but such interactions involved in the adhesion step are probably not sufficient by themselves to induce killing. We have found differences using ELISA or a functional assay to titrate IL-5 especially when eosinophils were incubated with Ig and anti-Ig, but in the absence of further stimulus [i.e., ELISA (+) and proliferation assay (−)], one good explanation could be the use of a mouse B-cell line, IL-5 dependent, which is 100 times less sensitive than homologous cells.

Abrams: One more simple explanation, rather than involving soluble receptor (IL-5) secretion/shedding, might be that mouse B-cell target lines can be about 100 times less sensitive for human IL-5 than for mouse IL-5, i.e., read out in the homologous species.

Capron: This point is well taken. The best assay in a homologous situation would certainly be the CTLL-2 cell line transfected with the human IL-5R α chain.

Leiferman: Dermatitis herpetiformis (DH) is associated with gluten-sensitive enteropathy, celiac sprue with IgA deposition in skin and circulating IgA antibodies; where was the IL-5 localized in DH? Histopathologically DH is characterized by a neutrophilic infiltrate and eosinophils are not a prominent component of the infiltrate. Although we have not systematically studied DH lesions by immunofluorescence for eosinophil granule proteins, a few cases that we studied have not shown prominent eosinophil infiltration or granule protein deposition in lesions. Please comment.

Capron: We had only one patient with DH associated with celiac disease. Obviously, further studies are needed, especially comparison between IgA-associated DH and IgE-dependent atopic dermatitis, where the eosinophil infiltrate is more prominent.

DISCUSSION REFERENCES

1. Capron M, Tomassini M, Ven der Vorst E, Kusnierz JP, Pepin JP, Capron A. Existence et fonctions d'un recepteur pour l'immunoglobline A sur les eosinophiles humaines. CR Acad Sci/Immunol 1988; 307:397–402.
2. Grezel D, Capron M, Fontaine J, Lecocq JP, Capron A. Protective immunity induced in rat schistosomiasis by a single dose of the Sm 28 GST recombinant antigen: Effector mechanisms involving IgE and IgA antibodies. Eur J Immunol 1993; 23:455–460.

12

Cytokine Regulation of Eosinophil-Mediated Inflammatory Reactions by Modulation of Eosinophil Programmed Cell Death and Subsequent Priming for Augmented Function

William F. Owen, Jr. and K. Frank Austen
Harvard Medical School and Brigham and Women's Hospital, Boston, Massachusetts

I. INTRODUCTION

Of the granulocytes that participate in the effector limb of adaptive immunity, the eosinophil is most easily studied because the regulation of its functional and physical phenotype in the circulation and tissue microenvironment can be reproduced in vitro. These processes involve the exposure of isolated peripheral blood eosinophils to selected cytokines and comparison of the outcome phenotype with that of eosinophils associated with specific diseases. The pathobiology of disease processes associated with peripheral blood and tissue eosinophilia begins with those factors that regulate eosinophilopoiesis in the bone marrow and proceeds to the coordinated maintenance of cellular viability at extramedullary sites. Disorders that affect either or both of these processes result in abnormal quantities of eosinophils, which can then present a functional phenotype primed for increased expression of ligand-mediated eosinophil proinflammatory functions.

II. THE CYTOKINE-DEPENDENT REGULATION OF EOSINOPHIL VIABILITY IN VITRO BY ATTENUATION OF APOPTOSIS

More than 90% of the eosinophils purified from the peripheral blood of healthy donors or of patients with mild allergic rhinitis by the centrifugation of mixed leukocyte buffy coats through discontinuous gradients of metrizamide (18–24% w/v) are isolated from the 22/23 and 23/24% metrizamide interfaces and the cell pellet. These normal-density eosinophils (normodense) survive poorly in enriched medium alone (RPMI-1640 supplemented with 10% fetal bovine serum); after 48 h of culture, less than 20% of the original number of eosinophils plated are viable as determined by their exclusion of trypan blue (1–4) or their uptake and nuclear staining with acridine orange (Owen WF, Austen KF, unpublished results). As assessed by brightfield microscopy during the first 48 h of culture in enriched medium alone, the nuclei of the eosinophils lose their segmentation and become increasingly condensed and shrunken. After the eosinophils are stained with acridine orange, visualization with fluorescent microscopy reveals that the eosinophil nuclei progressively lose their intense green staining and become pyknotic. These morphological abnormalities, which are a consequence of internucleosomal cleavage of the chromatin, indicate programmed cell death (apoptosis) and have also been noted for neutrophils (5). These senescent cells are then eliminated by vitronectin receptor-mediated phagocytosis by macrophages (6).

We examined freshly isolated normodense eosinophils cultured in enriched medium alone for the development of internucleosomal DNA fragmentation. The cell DNA was extracted in sodium lauryl sarkosinate and proteinase K from freshly isolated normodense eosinophils and replicate eosinophils cultured for 10, 20, and 38 h. After electrophoresis in 2% agarose gels, individual extracts of DNA were assessed for the extent of fragmentation and determination of the size of the fragments by direct visualization with ethidium bromide. The DNA extracts from normodense eosinophils exhibited prominent fragmentation by 20 h of culture in enriched medium alone, and the DNA fragments were multiples of approximately 200 bp (7). The DNA extracted from eosinophils cultured for an equivalent length of time in RPMI-1640 that had been boiled or was rendered acidic did not exhibit this orderly pattern of internucleosomal fragmentation. Instead, the DNA was digested into innumerable sizes, which rendered a smear on agarose gel electrophoresis instead of the characteristic step-ladder appearance (Fig. 1). Thus, under in vitro conditions programmed cell death could be distinguished from cell necrosis. These findings have been extended by the recent observation that senescent, but not freshly isolated, eosinophils undergo phagocytosis by macrophages (8). Because mature eosinophils contain potentially cytotoxic granule proteins that appear to function without further processing (9), the constitutive programmed cell death of eosinophils and their subsequent ingestion by macrophages fulfill the biologically necessary task of preventing unregulated cell survival and ineffective removal of toxic granule constituents.

Eosinophil survival in vitro is extended optimally if picomolar quantities of granulocyte-macrophage colony-stimulating factor (GM-CSF), interleukin (IL)-3, or IL-5 are included in the culture with enriched medium (1–3). Whereas no eosinophils survive in culture for 7 days in enriched medium alone, 40–80% of the

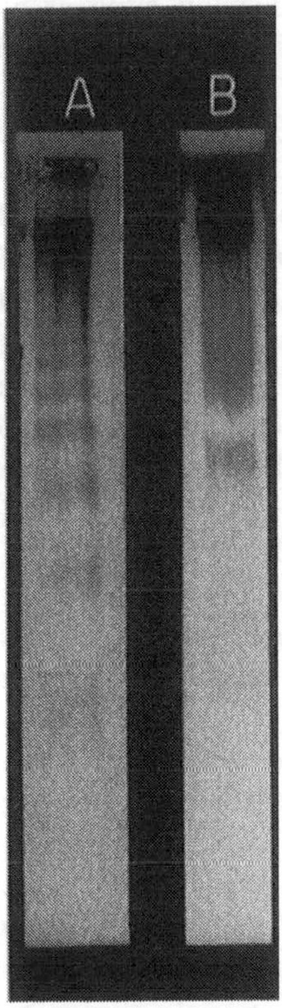

Figure 1 DNA fragmentation of eosinophils cultured for 24 h in enriched medium alone (lane A) or in enriched medium acidified with HCl to pH 7.0 (lane B). DNA was extracted from 1.5×10^6 replicate normodense eosinophils. The ethidium bromide-stained DNA was visualized under ultraviolet light after agarose gel electrophoresis.

cells survive in enriched medium supplemented with 10 pM GM-CSF or IL-3 or 1 pM IL-5. If the eosinophils are exposed to GM-CSF, IL-3, or IL-5 for a defined interval, such as 48 h, and then are transferred to fresh medium without an added cytokine, their survival is extended by the interval of exposure. For optimal eosinophil survival in vitro, the cells must be continuously exposed to the cytokine. The effect of these three hematopoietins on the survival of mature granulocytes is relatively selective for eosinophils, and other cell types such as neutrophils do not survive.

Because the constitutive loss of eosinophil viability was mediated by a time-dependent internucleosomal DNA cleavage, which was prominent by 20 h, we analyzed the DNA from normodense eosinophils cultured in GM-CSF, IL-3, or

IL-5 to demonstrate that these cytokines maintained eosinophil survival by attenuating the cleavage process. DNA extracted from replicate freshly isolated normodense eosinophils, cultured for various intervals in enriched medium supplemented with 10 pM GM-CSF, 10 pM IL-3, or 1 pM IL-5, did not undergo internucleosomal fragmentation after 38 h of culture (7).

Tumor necrosis factor (TNF) primes normodense eosinophils for enhanced antibody-dependent cytotoxicity against *Schistosoma mansoni* larvae (10,11) and for the generation of augmented quantities of superoxide and hypobromous acid (12); however, it does not increase eosinophil viability or attenuate constitutive internucleosomal DNA fragmentation (2,7). In contrast, interferon-γ, which somewhat improves the in vitro survival of normodense eosinophils (13), also partially attenuates internucleosomal DNA fragmentation, but these effects are less sustained than those of GM-CSF with replicate cells (Owen WF, Austen KF, unpublished results). Thus, the greater capacity of cytokines IL-5, IL-3, and GM-CSF to maintain the in vitro survival of normodense eosinophils parallels their capacity to prevent programmed cell death. Moreover, these three cytokines recognize a different binding unit expressed by receptors that use the same partner protein for signal transduction (14,15).

Two steps of activation are suggested by the observed prerequisite for a cytokine-dependent cytoprotective effect to precede cytokine-dependent priming for augmented ligand-mediated functions. As the second is not maximal until about 7 days of cytokine action in vitro, enhancement of signal transduction for the generation of superoxide or leukotriene C_4 (LTC_4) or for target cell cytotoxicity is a consequence of a change in phenotype and not merely of persistence of function because of sustained viability.

III. EFFECT OF A CHANGE IN EOSINOPHIL PHENOTYPE IN VIVO ON PROGRAMMED CELL DEATH

Peripheral blood eosinophils, isolated by density gradient centrifugation from three patients with glucocorticoid-unresponsive idiopathic hypereosinophilic syndrome (IHES), were segregated into less dense (hypodense) and normal density fractions of relatively high purity (greater than 80%). In comparison to eosinophils derived from the normodense fractions, the hypodense eosinophils exhibited a three- to fourfold extended in vitro survival in the absence of an added cytokine (16). Based on antibody neutralization of viability-sustaining activity in serum or plasma of patients with IHES, the eosinophils isolated from such patients had been exposed to abnormally elevated levels of IL-5. These observations suggested that as a consequence of their cytokine exposure in IHES, the hypodense eosinophils were relatively resistant to programmed cell death and primed for ligand-mediated responses.

Therefore, we examined the influence of prolonged exposure to an eosinophil-directed hematopoietin on the programmed cell death of eosinophils rendered hypodense in vitro as defined by gradient sedimentation. Replicate normodense eosinophils were cultured in enriched medium alone or in medium supplemented with 10 pM GM-CSF in the presence of a monolayer of mouse 3T3 fibroblasts. Previously, we demonstrated that after 7 days, this coculture system produces hypodense eosinophils that are 100% pure and more than 95% viable (1–3). After 7 days of coculture, these hypodense eosinophils were washed and transferred to enriched medium alone. At defined intervals, the original normodense eosinophils and the in vitro–generated hypodense replicates were assessed for internucleosomal DNA fragmentation by their uptake and staining with acridine orange and for viability by their exclusion of trypan blue. As compared to the normodense eosinophils, the replicate hypodense eosinophils did not exhibit an equivalent degree of programmed cell death until more than 24 h later (Fig. 2). As analyzed by agarose gel electrophoresis to validate that internucleosomal fragmentation (apoptosis, not necrosis) was the mechanism of programmed cell death, much less fragmentation was observed in the replicate hypodense eosinophils.

These studies demonstrate that prolonged cytokine exposure alone is sufficient to alter the kinetics of eosinophil programmed cell death. Thus, in clinical disorders associated with the presence of increased numbers of hypodense eosino-

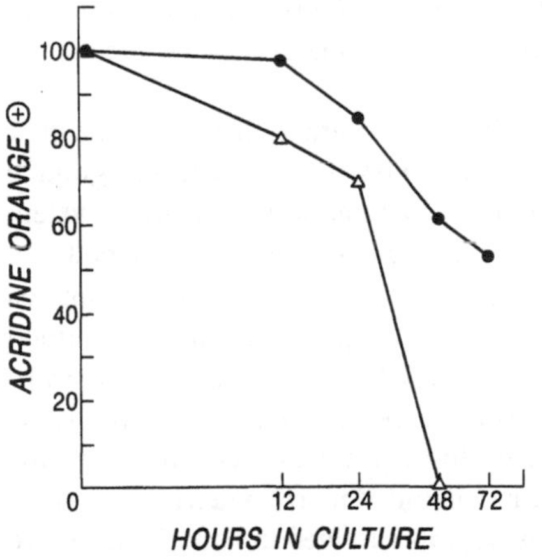

Figure 2 Kinetics of programmed cell death for normodense eosinophils (△) and for replicate hypodense eosinophils (●). The extent of internucleosomal DNA fragmentation was quantitated by the uptake and nuclear staining with acridine orange (% positive) as visualized by fluorescent microscopy. Only cells that excluded trypan blue were analyzed. Hypodense eosinophils were generated by the coculture of normodense eosinophils for 7 days in 10 pM GM-CSF with 3T3 fibroblasts.

phils, the eosinophilia is a consequence of both cytokine-dependent, accelerated intramedullary eosinophilopoiesis and the conversion of the peripheral blood eosinophils to a long-lived hypodense phenotype with an acquired relative resistance to programmed cell death. Furthermore, these studies demonstrate that the acute cytoprotective role provided by the eosinophil-directed hematopoietins can be extended even in the absence of more cytokine. Such a mechanism would allow the eosinophil to maintain an optimal state of viability to manifest its proinflammatory function in the tissue microenvironment where the levels of the cytokines may be diminished. However, because constitutive programmed cell death is delayed but not absent in hypodense eosinophils, the duration of an eosinophil-mediated inflammatory reaction would be self-limited and the consequences of eosinophil cytolysis detoxified by monocyte-macrophage uptake upon the signal of programmed cell death.

IV. CYTOKINE-DEPENDENT MECHANISMS OF PHENOTYPIC ALTERATION

After the eosinophil-directed hematopoietins have exerted an acute cytoprotective effect by the prevention of constitutive programmed cell death, the eosinophils can participate optimally in the subsequent development of an eosinophilic inflammatory lesion by transendothelial migration before or after conversion to the primed hypodense sedimenting phenotype. The selective adherence of eosinophils to vascular endothelium is mediated by the interaction of the IL-4-induced vascular cell adhesion molecule-1 (VCAM-1) and the eosinophil integrin, very late activation antigen-4 (VLA-4) (17–19).

The phenotypic conversion of eosinophils has been characterized based on their relative density gradient centrifugation, surface expression of selected granule proteins, and primed proinflammatory functions. With an in vitro culture system for eosinophils developed in this laboratory, it has been established that hypodense sedimenting eosinophils are derived from normodense eosinophils by continuous exposure to any of three eosinophil-directed hematopoietins with a common signal-transduction partner. During culture with GM-CSF, IL-3, and IL-5, normodense eosinophils become progressively less dense and are virtually all hypodense after 7 days of culture (1–3). Although a uniform physical mechanism for the hypodense sedimenting phenotype is unclear, a detailed morphometric analysis of normodense eosinophils and hypodense eosinophils (derived in vitro from culture with cytokines or in vivo from patients with IHES) suggested that a composite of factors, including cell swelling, a decrease in the volume of cytoplasm occupied by granules, and an increase in granule lucency, contribute physically to the hypodense phenotype (20). Despite the ultrastructural observation that cell swelling was present in the hypodense eosinophils, the water volume of the hypodense eosinophils derived in vitro by analysis with radiolabeled intracellular solutes was unchanged from that of replicate normodense eosinophils (unpublished results). Based on the selected ability of monoclonal antibodies directed against eosinophil cationic protein (ECP) to recognize peripheral blood eosinophils from patients with IHES or from healthy donors after priming in vitro with an incompletely characterized monokine, it has been suggested that the surface expression immunogenicity of this granule protein may be an indicator of cytokine priming (21).

Cationic granule proteins such as ECP may be noncovalently associated with negatively charged, cell-associated macromolecules such as proteoglycans. Because alterations in the structure of cell-associated proteoglycans may influence the surface expression of the associated granule protein, we examined the capacity of selected cytokines to modify eosinophil proteoglycans. A 12-h exposure of normodense eosinophils to GM-CSF, IL-3, or IL-5 resulted in a fourfold increase in the uptake of sulfate, which was associated with the synthesis of Pronase-resistant, cell-associated proteoglycans of an unprecedented size (22). The cell-associated proteoglycans purified from normodense eosinophils were approximately 80,000 MW, whereas the proteoglycans from cytokine-primed eosinophils were 300,000 MW. This increased size was a consequence of a ~ fourfold increase in the size of the glycosaminoglycan side chains. Furthermore, the cytokine-stimulated eosinophils synthesized the unique proteoglycan containing chondroitin sulfate E glycosaminoglycan. These data suggest that a contributor to the apparent altered cell surface expression of selected eosinophil granule proteins may be the difference in the structure of the intragranular proteoglycan/protein complex.

Eosinophil heterogeneity has also been characterized in studies of cell functions, such as antibody-dependent cytotoxicity, ligand-stimulated superoxide generation, and leukotriene synthesis. As is true for the alteration in physical phenotype and for proteoglycan biosynthesis, these functions are regulated by interactions with the eosinophil-directed hematopoietins. Exposure of normodense eosinophils to GM-CSF, IL-3, or IL-5 during the assay augments antibody-dependent cytotoxicity against the larvae of *S. mansoni* (1–3,10). After eosinophils are exposed for a prolonged period to cytokine together with a soluble constitutive activity derived from mouse 3T3 fibroblasts (3) or from human HF-15 fibroblasts (Owen WF, Austen KF, unpublished results), the expression of antibody-dependent cytotoxicity is independent of their being maintained with GM-CSF, IL-3, or IL-5 during assay. Therefore, constitutive cell components of the microenvironment, such as fibroblasts, function in concert with the hematopoietic cytokines to prime the eosinophil functional phenotype for signal transduction.

Normodense eosinophils do not elaborate LTC_4 or superoxide when stimulated with formyl-methionyl-leucyl-phenalanine (FMLP). However, during a 7-day interval of coculture with GM-CSF and 3T3 fibroblasts, the eosinophils become hypodense and respond to FMLP as a soluble ligand with the generation of superoxide and the formation of LTC_4 (23). The quantity of LTC_4 generation on a per cell basis by FMLP stimulation of these phenotypically altered eosinophils is as great as that elaborated by freshly isolated normodense eosinophils stimulated with an optimal concentration of calcium ionophore A23187 (21). As was observed for antibody-dependent cytotoxicity, FMLP-stimulated LTC_4 generated was negligible if the hypodense sedimenting phenotype was achieved by the cytokine-dependent culture of normodense eosinophils in the absence of fibroblasts. The

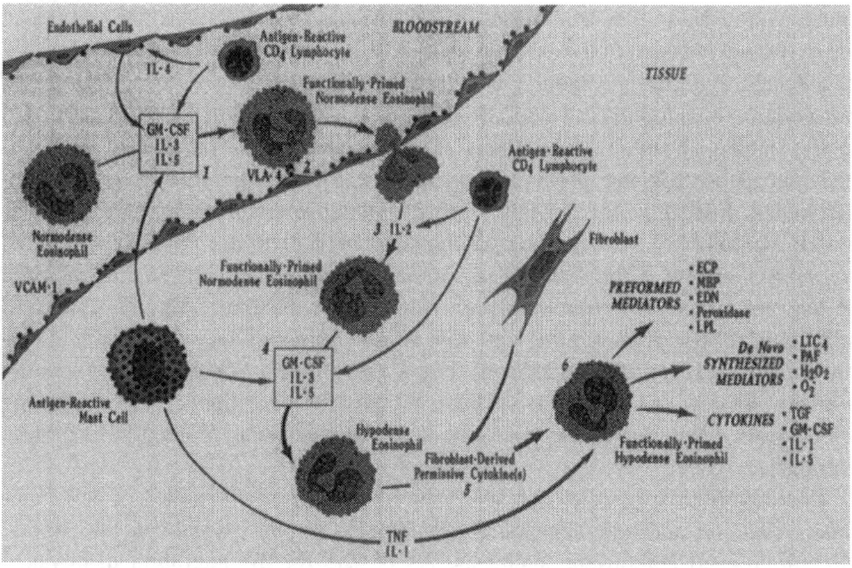

Figure 3 TH-2-class cytokines regulate eosinophil programmed cell death and eosinophil phenotypes. Step 1: Under the influence of GM-CSF, IL-3, or IL-5, constitutive programmed cell death is prevented, and normodense eosinophils become functionally primed without a change in centrifugation density. Step 2: Cytokine-primed normodense eosinophils express VLA-4 on their membrane; endothelial cell expression of its ligand VCAM-1 is enhanced by IL-4. Step 3: IL-2, which is a potent eosinophil chemoattractant, contributes to the directed migration of the cells from the vascular space into the tissue microenvironment. Step 4: Continued exposure to the eosinophil-directed hematopoietins induces a decrease in the relative centrifugation density of the cells without continued functional priming. Step 5: Persistence of enhanced function for the hypodense phenotype requires continued eosinophil exposure to a soluble, permissive activity derived from fibroblasts. Step 6: The resulting functionally primed hypodense eosinophils exhibit their proinflammatory function by the release of granule and cytosolic proteins, the synthesis of lipid mediators and reactive oxygen species, and the elaboration of selected cytokines. ECP, eosinophil cationic protein; MBP, major basic protein; EDN, eosinophil-derived neurotoxin; LPL, lysophospholipase; PAF, platelet-activating factor; TGF, transforming growth factor.

capacity of hypodense eosinophils to elaborate LTC_4 in response to stimulation with FMLP was not related either to an increase in the activity of the microsomal glutathionyl-*S*-transferase, LTC_4 synthase (23), or to an alteration in the carrier-mediated LTC_4 export system (24,25; Owen WF, Austen KF, unpublished results). Therefore, the duration of cytokine priming in a connective tissue microenvironment can profoundly influence the pattern of eosinophil responsiveness to soluble ligands.

The TH-2 class of cytokines exerts a fundamental regulatory role for all phases of eosinophil-mediated inflammatory processes (Fig. 3). Eosinophilopoiesis depends on the interaction of eosinophil precursors with IL-5 and GM-CSF (or IL-3). Constitutive programmed cell death of the mature normodense eosinophil is prevented by continued exposure to these eosinophil-directed hematopoietins. Selective endothelial adhesion, a mandatory anticipatory step to eosinophil migration into the tissue microenvironment, is augmented by another TH-2 class cytokine, IL-4, which induces the expression of VCAM-1 on endothelial cells. Finally, the conversion of normodense eosinophils to the long-lived, proinflammatory hypodense phenotype and the ligand responsiveness of these cells are determined by their continued exposure to selected TH-2-class cytokines. Although TH-2 CD4+ T cells are the accepted source of these cytokines, in allergic reactions it may be important to consider an alternative or supplementary source, namely IgE-activated mast cells, based on studies with mouse and human mast cells (26).

V. SUMMARY

When maintained ex vivo in enriched medium alone, freshly isolated, normodense human peripheral blood eosinophils survive poorly as defined by trypan blue exclusion, acridine orange uptake, and LDH release. The constitutive loss of eosinophil viability is preceded by the development of internucleosomal DNA degradation, as defined by the degradation of cellular DNA into fragments of approximately 200-bp multiples in size (apoptosis). In contrast, the DNA isolated from replicate normodense eosinophils that were cultured in acidified medium was degraded into fragments of innumerable sizes (necrosis). The maintenance of eosinophil survival in vitro by GM-CSF, IL-3, and IL-5 is associated with the prevention of internucleosomal DNA degradation. Tumor necrosis factor-α, which primes selected eosinophil functions but does not maintain their survival in culture, did not prevent DNA fragmentation. The capacity of GM-CSF, IL-3, and IL-5 to maintain the survival of normodense eosinophils was antagonized by glucocorticoids and cycloheximide. In comparison to their normodense counterparts, the survival of hypodense eosinophils, generated in vitro by the 7-day culture of normodense eosinophils with GM-CSF in the presence of 3T3 fibroblasts, was increased in the absence of an added cytokine. This enhanced survival was accompanied by a diminished constitutive rate of DNA fragmentation. Therefore, the eosinophil-directed hematopoietins exert an acute cytoprotective effect, which selectively maintains the survival of eosinophils by preventing constitutive internucleosomal DNA degradation. This process, which is distinguishable from cellular necrosis, is promoted by glucocorticoids and by the inhibition of protein synthesis. The attenuated kinetics of programmed cell death exhibited by hypodense eosinophils provides a mechanism for the optimal expression of eosinophil proinflammatory functions in the tissue microenvironment where the levels of cytokine may be attenuated.

REFERENCES

1. Owen WF, Rothenberg ME, Silberstein DS, Gasson JC, Stevens RL, Austen KF, Soberman RJ. Regulation of eosinophil viability, density, and function by granulocyte/macrophage colony-stimulating factor in the presence of 3T3 fibroblasts. J Exp Med 1987; 166:129–141.
2. Rothenberg ME, Owen WF, Silberstein DS, Woods J, Soberman RJ, Austen KF, Stevens RL. Human eosinophils have prolonged survival, enhanced functional properties, and become hypodense when exposed to human interleukin 3. J Clin Invest 1988; 81:1986–1992.
3. Rothenberg ME, Petersen J, Stevens RL, Silberstein DS, McKenzie DT, Austen KF, Owen WF. IL-5-dependent conversion of normodense eosinophils to the hypodense phenotype uses 3T3 fibroblasts for enhanced viability, accelerated hypodensity, and sustained antibody-dependent cytotoxicity. J Immunol 1989; 143:2311–2316.
4. Rothenberg ME, Owen WF, Silberstein DS, Soberman RJ, Austen KF, Stevens RL. Eosinophils cocultured with endothelial cells have increased survival and functional properties. Science 1987; 237:645–647.
5. Savill JS, Wyllie AH, Henson JE, Walport MJ, Henson PM, Haslett C. Macrophage phagocytosis of aging neutrophils in inflammation. Programmed cell death in the neutrophil leads to its recognition by macrophages. J Clin Invest 1989; 83:865–875.
6. Savill J, Dransfield I, Hogg N, Haslett C. Vitronectin receptor-mediated phagocytosis of cells undergoing apoptosis. Nature 1990; 343:170–173.
7. Her E, Frazer J, Austen KF, Owen WF. Eosinophil hematopoietins antagonize the programmed cell death of eosinophils. Cytokine and glucocorticoid effects on eosinophils maintained by endothelial cell-conditioned medium. J Clin Invest 1991; 88:1982–1987.
8. Stern M, Meagher L, Savill J, Haslett C. Apoptosis in human eosinophils: programmed cell death in the eosinophil leads to phagocytosis by macrophages and is modulated by interleukin-5. J Immunol. In press.
9. Weller PF. The immunobiology of eosinophils. N Engl J Med 1991; 324:1110–1118.
10. Silberstein DS, David J. Tumor necrosis factor enhances eosinophil cytotoxicity to Schistosoma mansoni larvae. Proc Natl Acad Sci USA 1986; 83:1055–1059.
11. Silberstein DS, Owen WF, Gasson JC, Dipersio J, Golde DW, Bina JC, Soberman RJ,

Austen KF, David JR. Regulation of human eosinophil function by granulocyte-macrophage colony-stimulating factor. J Immunol 1986; 137:3290–3294.

12. Slungaard A, Vercellotti GM, Walker G, Nelson RD, Jacob HS. Tumor necrosis factor alpha/cachectin stimulates eosinophil oxidant production and toxicity towards human endothelium. J Exp Med 1990; 171:2025–2041.
13. Valerius T, Repp R, Kalden JR, Platzer E. Effects of IFN on human eosinophils in comparison with other cytokines. A novel class of eosinophil activators with delayed onset of action. J Immunol 1990; 145:2950–2958.
14. Lopez AF, Vadas MA, Woodcock JM, Milton SE, Lewis A, Elliott MJ, Gillis D, Ireland R, Olwell E, Park LS. Interleukin-5, interleukin-3, and granulocyte-macrophage colony-stimulating factor cross-compete for binding to cell surface receptors on human eosinophils. J Biol Chem 1991; 266:24741–24747.
15. Chihara J, Plumas J, Gruart V, Tavernier J, Prin L, Capron A, Capron M. Characterization of a receptor for interleukin 5 on human eosinophils: variable expression and induction by granulocyte/macrophage colony-stimulating factor. J Exp Med 1990; 172:1347.
16. Owen WF, Rothenberg ME, Petersen J, Weller P, Silberstein DS, Sheffer AL, Stevens RL, Soberman RJ, Austen KF. Interleukin 5 and phenotypically altered eosinophils in the blood of patients with the idiopathic hypereosinophilic syndrome. J Exp Med 1989; 170:343–348.
17. Dobrina A, Menegazzi R, Carlos TM, Nardon E, Cramer R, Zacchi T, Harlan JM, Patriarca P. Mechanisms of eosinophil adherence to cultured vascular endothelial cells. Eosinophils bind to the cytokine-induced endothelial ligand vascular cell adhesion molecule-1 via the very late activation antigen-4 integrin receptor. J Clin Invest 1991; 88:202–206.
18. Weller PF, Rand TH, Goelz SE, Chi-Rosso G, Lobb RR. Human eosinophil adherence to vascular cell adhesion molecule 1 and endothelial leukocyte adhesion molecule 1. Proc Natl Acad Sci USA 1991; 88:7430–7433.
19. Schleimer RP, Sterbinsky SA, Kaiser J, Bickel CA, Klunk DA, Tomioka K, Newman W, Luscinskas FW, Gimbone MA, McIntyre BW, Bochner BS. IL-4 induces adherence of human eosinophils and basophils but not neutrophils to endothelium. Association with expression of VCAM-1. J Immunol 1992; 148:1086–1092.
20. Caulfield JP, Hein A, Rothenberg ME, Owen WF, Soberman RJ, Stevens RL, Austen KF. A morphometric study of normodense and hypodense human eosinophils that are derived in vivo and in vitro. Am J Pathol 1990; 137:27–41.
21. Spry CJF. Synthesis and secretion of eosinophil granule substances. Immunol Today 1985; 6:332–335.
22. Rothenberg ME, Pomerantz JL, Owen WF, Avraham S, Soberman RJ, Austen KF, Stevens RL. Characterization of a human eosinophil proteoglycan and augmentation of its size by interleukin 3, interleukin 5, and granulocyte-macrophage colony-stimulating factor. J Biol Chem 1988; 263:13901–13908.
23. Owen WF, Petersen J, Austen KF. Eosinophils primed and phenotypically altered by culture with granulocyte-macrophage colony-stimulating faction and 3T3 fibroblasts generate LTC_4, in response to FMLP. J Clin Invest 1991; 87:1958–1963.

24. Owen WF, Soberman RJ, Yoshimoto T, Sheffer A, Lewis RA, Austen KF. Synthesis and release of LTC_4 by human eosinophils. J Immunol 1987; 138:532–538.
25. Lam B, Owen WF, Austen KF, Soberman RJ. The identification of a distinct export step following the biosynthesis of leukotriene C_4 by human eosinophils. J Biol Chem 1989; 264:12885–12889.
26. Plaut M, Pierce JH, Watson CJ, Hanley-Hyde J, Nordan RP, Paul WE. Mast cell lines produce lymphokines in response to cross-linkage of $Fc_{\epsilon}RI$ or to calcium ionophores. Nature (Lond) 1989; 339:64.

DISCUSSION (Speaker: W. Owen)

Holgate: Could you tell us more about the receptors expressed on eosinophils and macrophages associated with the engulfment of senescent cells?

Owen: It has been reported in abstract form by Stern et al. (1) that senescent human eosinophils undergo phagocytosis by monocytes/macrophages. Antibodies to the vitronectin receptor block this process and EGD competes against eosinophil phagocytosis. If eosinophils that are senescent undergo cytolysis and release their granule proteins, this interaction with macrophages provides a potentially protective mechanism for the host.

Abrams: What can you tell us about the biology and requirements for 3T3 fibroblasts cocultured in your eosinophil survival bioassay system? Is the requirement satisfied by conditioned medium from the fibroblast indicating a soluble factor?

Owen: Normodense human peripheral blood eosinophils survive and become hypodense when cultured in GM-CSF, IL-3, or IL-5 alone. However, such eosinophils are not functionally primed. We reported that normodense eosinophils cocultured with murine 3T3 fibroblasts in the presence of cytokine have greater survival, become hypodense more rapidly, and remain functionally upregulated. These effects are observed even if the eosinophils and fibroblasts are physically separated by a semipermeable filter. This soluble factor(s) is extremely short-lived and has therefore been exceedingly difficult to purify. Thus far, neutralizing antibodies against a survey of cytokines that are fibroblast-derived have been nondiagnostic.

Busse: You have stated that some cytokines, such as IL-3, IL-5, and GM-CSF, affect cell density and survival, whereas others, such as TNFα, enhance eosinophil function but do not change cell density or survival. Have you done experiments in which both IL-5 and IL-3 or GM-CSF were cocultured with eosinophils and have you studied the effect this had on both density and function?

Owen: These experiments have not been performed.

Hansel: You elegantly discuss the prevention of eosinophil apoptosis by cytokines. However, have you studied cytokines or other stimuli that may accelerate apoptosis?

Owen: Currently we are screening a repertoire of cytokines looking for factors that will induce apoptosis, such as has been reported for IL-4 and human monocytes.

Austen: Please comment further on the antagonistic effects of steroids on cytokine (GM-CSF, IL-5, and IL-3) prevention of apoptosis.

Owen: Glucocorticoids such as dexamethasone antagonize the viability-sustaining effect of GM-CSF, IL-3, and IL-5. This is accomplished by the steroid-inducing internucleosomal DNA degradation, despite the presence of the cytokine. The capacity of the glucocorticoids to antagonize the maintenance of eosinophil viability is correlated with their anti-inflammatory potential (dexamethasone > prednisone > hydrocortisone; testosterone without effect). However, if the concentration of GM-CSF, IL-3, or IL-5 is increased to greater than or equal to 10 pM, the eosinophils are resistant to this glucocorticoid effect.

Kay: It is important to appreciate that "eosinophil survival factors" delay (not prevent) apoptosis.

Denburg: Have you tried to abrogate spontaneous survival of hypodense eosinophils using monoclonal antibodies to IL-3, IL-5, or GM-CSF? I ask this because we have recently reported on a changing profile over time of autocrine, cytokine-induced eosinophil survival in a patient with HES. Although initially antibodies to IL-3 and IL-5, but not to GM-CSF, abrogated spontaneous survival, later the profile reversed and antibody to GM-CSF, but not to IL-3 or IL-5, abrogated spontaneous survival.

Owen: We have not performed such longitudinal studies.

Ackerman: Could you comment on whether you believe that apoptosis represents a physiologically significant mechanism for eosinophil secretion of lipid and granule protein mediators of inflammation in tissues or has more relevance for the removal of senescent cells?

Owen: Clearly, programmed cell death provides a mechanism for terminating eosinophil-mediated inflammatory reactions, and for the clearance of senescent eosinophils. I am uncertain of its precise role in granule secretion. However, I do not feel that apoptic cell death is a major mechanism for granule secretion. This is because programmed cell death in eosinophils is a relatively slow process—internucleosomal DNA degradation takes hours and precedes membrane disruption, whereas eosinophil degranulation occurs minutes after the addition of an agonist. Alternatively, I suspect that degranulation stimulates programmed cell death. Has Dr. Dvorak observed apoptosis in degranulating eosinophils?

Dvorak: Eosinophil apoptotic nuclei do occur in in vivo biopsy samples in disease characterized by a high degree of eosinophil injury. As originally described by Wyllie, apoptosis and necrosis are a morphological continuum, with apoptosis often preceding the morphology of necrosis. More recently we have noted the morphology of apoptosis in eosinophils cultured with the HIV virus (Weller P et al, unpublished data). Regardless of which morphology prevails in biopsy samples (i.e., apoptosis or necrosis), some disorders have sufficient amounts of damaged eosinophils to effectively exceed the capacity of tissue histiocytes to internalize the debris. Thus, in bullous pemphigoid and HES, large amounts of eosinophil-derived granules, nuclear and cytoplasmic debris are present in the tissue microenvironment. In HES, tissue macrophages (in addition to internalizing visible membrane-bound eosinophil granules that are released by necrosis, by phagocytosis) also take up soluble Charcot–Leyden crystal protein of eosinophil origin by endocytosis as identified by immunogold ultrastructural label.

Gleich: Two comments—First, macrophages may play a critical role in the control of eosinophil-associated inflammation and we have observed MBP localization in alveolar macrophages from patients with asthma and others have found intact eosinophil granules in macrophages (by EM). Second, eosinophil granule proteins seem to correlate with eosinophil density; hypodense cells have less MBP than normodense cells, and one can show that approximately 50% of the four granule proteins are lost from eosinophils cultured with IL-5.

Owen: With respect to the initial comment, it validates the in vitro observation of macrophage phagocytosis of apoptic eosinophils. Such a mechanism would provide a means of terminating eosinophil-mediated inflammatory reactions. As for the second point, I would like to remind you of our morphometric study of eosinophils with John Caulfield (2) in which we examined hypodense and normodense eosinophils derived in vitro or isolated in vivo. We observed that the granule volume and granule integrity varied depending on the cell culture condition. Thus observations made concerning MBP content alterations with cell culture should not be extended to other situations. Further, no uniform morphological correlate of hypodensity was observed. We reported a composite of factors including cell swelling, a decrease in the volume occupied by granules, and an increase in granule lucency.

Spry: Dr. Tai was the first to show in 1976 that hypogranular eosinophils could survive for over 2 weeks in T-cell-conditioned medium. She was also the first to show very similar observations to those you reported today. She made one additional observation—that there was considerable intrapatient variation in their capacity to survive in the presence of cytokines. I summarized her work in a posthumous paper (3). Is it also your experience that there is considerable variation in the survival in cultured eosinophils from patients with HES to growth factors?

Owen: We have observed that the hypodense eosinophils from one IHES donor were not sustained in vitro by GM-CSF but were maintained by IL-5. In marked contrast, the normodense eosinophils from this donor were maintained in culture by both GM-CSF and IL-5. We have no explanation for this dichotomy of effects.

Gleich: How marked are the differences between normodense eosinophils of normal individuals in their survival with cytokines? Our laboratory has studied the survival of eosinophils, mediated by IL-5, in over 100 persons (mainly normals) and we have always seen increased responsiveness to IL-5. Therefore, I am interested in the degree of survival differences you have observed.

Owen: Among normal donors, the normodense eosinophils exhibit differing sensitivities to GM-CSF, IL-3, and IL-5 but they always respond with some degree of augmented survival ex vivo.

Barnes: Have you examined the molecular mechanisms involved in eosinophil survival—and, in particular, the role of protein kinase C and tyrosine kinase?

Owen: These studies are ongoing and therefore are incomplete.

Lee: Neutrophils have programmed cell death and have receptors for GM-CSF. Do you have any insights into the mechanism of the selectivity of eosinophils over neutrophils in enhanced survival to GM-CSF?

Owen: The kinetics of internucleosomal DNA degradation is much more rapid in neutrophils than in eosinophils. Whereas DNA degradation is prominent in 20–24 h for eosinophils in medium alone, it is notable for neutrophils in 4–6 h. Therefore, it appears that programmed cell death is a more rapid process in neutrophils than eosinophils. The mechanism that accounts for the failure of GM-CSF to abrogate the endonuclease in neutrophils is unclear. In contrast, we reported that the conditioned medium from HF15 human fibroblasts maintains the in vitro survival of human neutrophils for at least 72 h, and this augmented ex vivo survival is accompanied by a complete abolition of internucleosomal DNA degradation. We are actively attempting to isolate this factor.

DISCUSSION REFERENCES

1. Stern M, Meagher L, Savill J, Ren Y, Haslett C. Eosinophils undergo apoptosis and are recognized and ingested as whole cells by macrophages. Am Rev Respir Dis 1991; 143:A331.
2. Caulfield JP, Hein A, Rothenberg ME, Owen WF, Soberman RJ, Stevens RL, Austen KF. A morphometric study of normodense and hypodense human eosinophils that are derived in vivo and in vitro. Am J Pathol 1990; 137:27–41.
3. Tai P-C, Sun L, Spry CJ. Effects of IL-5, granulocyte/macrophage colony-stimulating factor (GM-CSF) and IL-3 on the survival of human blood eosinophils in vitro. Clin Exp Immunol 1991; 85:312–316.

13

Mast Cell and Eosinophil Cytokines in Allergy and Inflammation

Stephen J. Galli, John R. Gordon, Barry K. Wershil, and John J. Costa
Beth Israel Hospital and Harvard Medical School, Boston, Massachusetts

Aram Elovic and David T. W. Wong
Harvard School of Dental Medicine, Boston, Massachusetts

Peter F. Weller
Beth Israel Hospital and Harvard Medical School, Boston, Massachusetts

I. INTRODUCTION

Mast cells and eosinophils not only represent critical effector cells in allergic reactions, but also participate in a wide variety of biological responses in which immunoglobulin E (IgE) has no demonstrable role (reviewed in 1–8). However, certain clinically important aspects of allergic reactions, such as the striking leukocyte recruitment and fibrosis that occur in some of these disorders, are incompletely understood in mechanistic terms. The functions of mast cells and eosinophils in responses not involving IgE are even more obscure.

This review focuses primarily on recent evidence indicating that mast cells or eosinophils can produce multifunctional cytokines and that such molecules may represent important mediators in biological responses associated with the activation of these cells. Space limitations do not permit adequate discussion of several important related issues, which have been reviewed elsewhere. These include detailed consideration of the natural history of mast cells (4,5) or eosinophils (6–8), or the role of multifunctional cytokines in regulating mast cell, basophil, or eosinophil proliferation, maturation, phenotype, or function (8–10).

II. DISTRIBUTION AND FUNCTION OF MAST CELLS, BASOPHILS, AND EOSINOPHILS

Mammalian mast cells and basophils are similar in that both: (1) express the high-affinity receptor for IgE ($Fc_{\epsilon}RI$) (11), (2) release upon IgE-dependent or other forms of stimulation a similar, though not identical, panel of biologically active mediators (1–3), and (3) participate in IgE-dependent and certain IgE-independent biological responses (1–5). However, mast cells and basophils differ in many aspects of their natural history. For example, the numbers of tissue mast cells can increase (or decrease) in association with certain biological responses, but even normal vascularized tissues have a resident population of mature mast cells, whose numbers and phenotypic characteristics vary according to anatomical site (5). And many of these mast cells normally reside in close proximity to such potential targets of mediator/cytokine action as vascular endothelial cells, fibroblasts, epithelial cells, or nerves (5). Mast cells are derived from precursors that reside in the bone marrow but do not circulate in mature form (4,5). By contrast, mature basophils are granulocytes that normally circulate in the blood, albeit in very low numbers, but do not normally reside in the peripheral tissues (2,5). On the other hand, a wide variety of biological responses are associated with elevated levels of circulating basophils and/or the recruitment of basophils into the affected tissues (2).

Eosinophils are a distinct lineage of granulocytes, which, in humans, usually account for ~3% of circulating leukocytes (6,8). However, unlike basophils, large numbers of eosinophils normally are present within certain nonhematopoeitic tissues. In fact, the peripheral tissues rather than the blood and bone marrow contain the majority of mature eosinophils, which are especially abundant near the mucosal surfaces of the gastrointestinal, respiratory, and genitourinary tracts (6). Greatly increased numbers of eosinophils can appear in the blood and tissues in association with a variety of immune responses, such as immune reactions to helminthic parasites, in which eosinophils function as effectors of host defense (6–8). In other settings associated with tissue eosinophil infiltration, such as asthma, eosinophil-derived mediators probably contribute to the pathogenesis of disease (6–8). In yet other reactions, such as host responses to neoplasms (reviewed in 12,13), the specific role of the eosinophil remains obscure.

III. MAST CELLS AS A SOURCE OF MULTIFUNCTIONAL CYTOKINES

The first direct evidence that mast cells could produce a well-characterized cytokine was derived from studies of mast cell populations maintained in vitro. Chung et al. (14) demonstrated that certain Abelson murine leukemia virus (A-

MuLV)-transformed tumorigenic mouse mast cell lines constitutively contained granulocyte-macrophage colony-stimulating factor (GM-CSF) mRNA and released GM-CSF bioactivity (9). Brown et al. (15) reported that the majority of several spontaneously or A-MuLV-transformed mouse mast cell lines constitutively expressed mRNA for interleukin (IL)-4 and that some of them constitutively released IL-4 bioactivity, whereas only a few of the lines contained mRNA for IL-3 or GM-CSF. Brown et al. (15) also detected low levels of IL-4 mRNA in each of five IL-3-dependent mouse mast cell lines, but these cells did not release detectable IL-4 bioactivity. Humphries et al. (16) showed that some A-MuLV-transformed mast cells constitutively produced IL-3 and GM-CSF mRNA and product and also produced a bioactivity similar to that of IL-6.

The first cytokine bioactivity to be clearly associated with normal mast cells was tumor necrosis factor-alpha (TNF-α)/cachectin. Several groups demonstrated that in vitro–derived, IL-3-dependent mouse mast cells, rat basophilic leukemia (RBL) cells, or freshly isolated mouse or rat peritoneal mast cells expressed cytolytic activity against certain cellular targets, and some examples of this mast-cell-dependent cytotoxicity exhibited partial inhibition by antibodies to TNF-α (reviewed in 17). Young et al. (18) showed that mouse peritoneal mast cells, as well as IL-3-dependent and -independent mouse mast cells generated in vitro, contained a cytoplasmic granule-associated cytokine with immunological and cytotoxic activities very similar to those of TNF-α. Young et al. (18) also showed that mouse mast cells released TNF-α-like bioactivity upon appropriate stimulation in vitro. Gordon and Galli (19) showed that this TNF-α-like cytotoxic mediator was a product of the TNF-α/cachectin gene, and that unstimulated mouse peritoneal mast cells constitutively contained approximately twice as much TNF-α bioactivity as was detected in LPS-stimulated mouse peritoneal macrophages. This finding illustrates the important point that the biology of some mast-cell-associated cytokines may differ significantly from that of the same cytokines in other cell types. Thus, macrophages, T cells, and B cells contain little or no preformed TNF-α bioactivity, whereas mouse peritoneal mast cells contain substantial preformed stores of TNF-α available for immediate release upon appropriate stimulation of the cell (reviewed in 17,19).

IV. IgE-DEPENDENT REGULATION OF MAST CELL CYTOKINE PRODUCTION

IgE-dependent stimulation of murine mast cells results in increased levels of mRNA for many cytokines and also results in the release of some cytokine products. Richards et al. (20) showed that RBL tumor cells released a TNF-α-like cytotoxic activity upon challenge with anti-IgE. Subsequently, three groups reported at about the same time that IL-3-dependent in vitro–derived mast cells activated via the $Fc_{\epsilon}RI$ represented a source of several multifunctional cytokines.

Plaut et al. (21) demonstrated that IgE-dependent stimulation of long-term IL-3-dependent mouse mast cell lines resulted in increased levels of mRNA for IL-3, IL-4, IL-5, and IL-6 and the release of bioactivity for IL-3, IL-4, and IL-6. Wodnar-Filipowicz et al. (22) reported that primary cultures of IL-3-dependent, bone-marrow-derived cultured mast cells (BMCMC) stimulated with IgE and specific antigen exhibited increased levels of IL-3 and GM-CSF mRNA and released IL-3 and GM-CSF bioactivity. Burd et al. (23) showed that long-term IL-3-dependent or -independent mouse mast cell lines activated by IgE and antigen exhibited increased levels of mRNA for IL-1α, IL-3, IL-5, IL-6, GM-CSF, and four members of the macrophage inflammatory protein (MIP)-1 gene family: TCA3, JE, MIP-1α, and MIP-1β, and two of the four mast cell lines tested exhibited IgE-dependent augmentation of levels of mRNA for interferon-γ (IFN-γ). Burd et al. (23) also showed that IgE-dependent activation of mouse mast cell lines or primary cultures of BMCMC resulted in release of bioactivity for IL-1, IL-4, and IL-6. In contrast to mRNA for other cytokines, transforming growth factor (TGF)-β mRNA was readily detectable in unstimulated mast cells and only slightly elevated by IgE-dependent activation of mast cell lines or BMCMC (24). Gurish et al. (25) showed that IgE-dependent activation of BMCMC resulted in a preferential increase in mRNA for cytokines relative to those for secretory granule proteins, indicating that the production of these distinct mediators may be differentially regulated.

V. TNF-α AS AN EXAMPLE OF A NEW CLASS OF MAST-CELL-ASSOCIATED MEDIATORS

Mast-cell-associated mediators conventionally are divided into the preformed mediators, such as histamine and the proteases, which are stored in the cells' cytoplasmic granules and are released upon cell activation, and the newly synthesized mediators, such as PGD_2 and the sulfidopeptide leukotrienes, which are not stored but are produced and secreted only upon appropriate stimulation of the cells (2,3). Gordon and Galli (26) proposed that TNF-α may be representative of a new class of mast-cell-associated mediators, in that IgE-dependent activation of mouse mast cells results in the rapid release of preformed stores of TNF-α bioactivity, as well as the induction of increased levels of TNF-α mRNA associated with the sustained release of large quantities of newly synthesized TNF-α bioactivity (Fig. 1). Thus, for both in vitro-derived cultured mouse mast cells and purified mouse peritoneal mast cells, the total amount of TNF-α bioactivity present in the cells and their supernatant 1–2 h after IgE-dependent activation exceeded that present in the unstimulated cells (19,26). The sustained release of newly synthesized TNF-α by appropriately stimulated mast cells represents a mechanism by which mast-cell-derived TNF-α can exert its actions on leukocyte emigration and

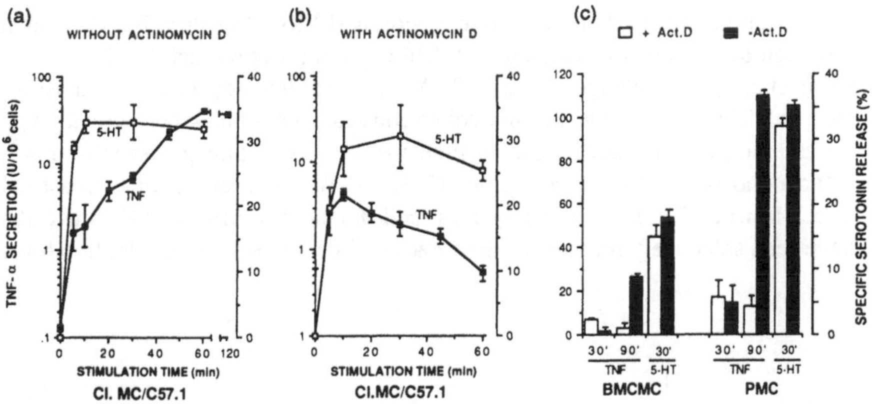

Figure 1 Kinetics of release of preformed and newly synthesized TNF-α from mouse mast cells stimulated via the $Fc_{\epsilon}RI$. Mast cells were sensitized with a monoclonal mouse anti-DNP IgE and then challenged with $DNP_{30\text{-}40}$-HSA in the presence (+) or absence (−) of the transcription inhibitor actinomycin D (Act. D) at 10 μg/ml. In some experiments, the cells were incubated with ^{3}H-5HT (1 μCi/ml) during the IgE-sensitization period, to permit the cells to incorporate ^{3}H-5HT into their cytoplasmic granules prior to washing and challenge with specific antigen (see Ref. 21). (a) Kinetics of secretion of ^{3}H-5HT and TNF-α from Cl.MC/C57.1 cells in the absence of Act. D. (b) Kinetics of secretion of ^{3}H-5HT and TNF-α from Cl.MC/C57.1 cells in the presence of Act. D. (c) Effect of transcription inhibition with Act. D on the secretion of ^{3}H-5HT and TNF-α from primary cultures of mouse bone-marrow-derived mast cells (BMCMC) and from purified freshly isolated mouse peritoneal mast cells (PMC). Data for TNF-α are mean ± SEM (n = 4/point); data for ^{3}H-5HT are mean ± SEM (n = 2–4/point). Values for TNF-α release in the absence of Act. D are significantly different ($p < 0.01$) than values for the same cell types in the presence of Act. D at all intervals ≥30 min after initiation of stimulation for Cl.MC/C57.1 cells and at 90 min after initiation of stimulation for BMCMC and PMC. [From Gordon and Galli (26).]

activation, and other effects, for prolonged periods during IgE-dependent and perhaps other biological responses.

VI. IgE-DEPENDENT LEUKOCYTE RECRUITMENT IN MOUSE SKIN REQUIRES MAST CELLS AND IS DEPENDENT ON TNF-α

Although the ability of mast cells to generate cytokines may represent an important facet of this cell's contribution to a wide variety of biological responses (17,27), much of the interest in the clinical significance of mast cell cytokine production has focused on late-phase reactions (LPRs). In LPRs, IgE-dependent mast cell activation is associated with leukocyte infiltration that develops within hours of antigen challenge (2,3,27,28). Notably, the leukocyte infiltration associated with LPRs, rather than the immediate consequences of mast cell activation per se, is thought to be largely responsible for the tissue damage associated with asthma and other allergic conditions (2,3,28). We have shown that "mast cell knock-in mice," i.e., genetically mast-cell-deficient WBB6F$_1$-*W*/*W*^{v} mice that have been selectively repaired of their mast cell deficiency by the adoptive trans-

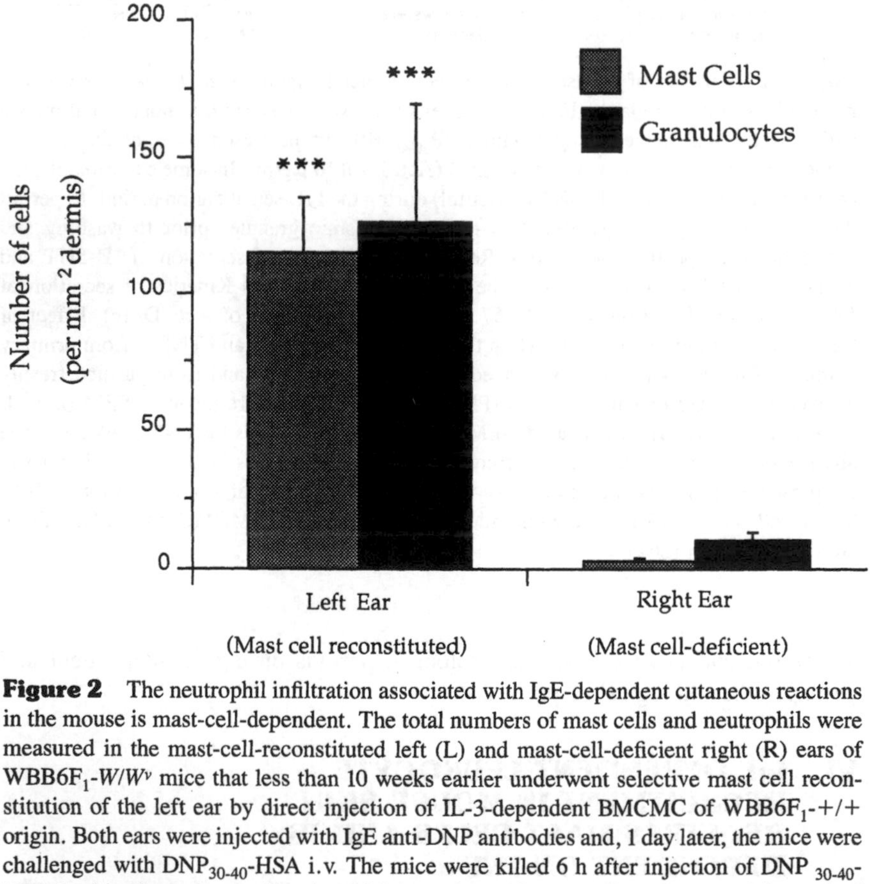

Figure 2 The neutrophil infiltration associated with IgE-dependent cutaneous reactions in the mouse is mast-cell-dependent. The total numbers of mast cells and neutrophils were measured in the mast-cell-reconstituted left (L) and mast-cell-deficient right (R) ears of WBB6F$_1$-*W*/*W*^{v} mice that less than 10 weeks earlier underwent selective mast cell reconstitution of the left ear by direct injection of IL-3-dependent BMCMC of WBB6F$_1$-+/+ origin. Both ears were injected with IgE anti-DNP antibodies and, 1 day later, the mice were challenged with DNP$_{30\text{-}40}$-HSA i.v. The mice were killed 6 h after injection of DNP$_{30\text{-}40}$-HSA for histological quantification of the numbers of mast cells and neutrophils/mm^2 of dermis. Data represent mean ± SEM (n = 4/point). Significant differences between the left and right ears are shown as ***$p < 0.001$ by the paired Student's t-test, two-tailed. [From Wershil et al. (29).]

fer of immature mast cells derived in vitro from the bone marrow cells of the congenic normal ($WBB6F_1$-+/+) mice, can be used to identify and quantify the specific contributions of mast cells and mast-cell-derived cytokines to individual biological responses (4,5,29). Our studies have provided several lines of evidence indicating that mast cells and TNF-α significantly contribute to the expression of IgE-dependent inflammation elicited in mouse skin: (1) mast cells are essential for the leukocyte infiltration associated with IgE-dependent cutaneous reactions, which do not occur in the absence of this cell type (29) (Fig. 2); (2) the amount of TNF-α mRNA detectable in skin challenged with IgE and antigen is much higher at sites containing mast cells than at mast-cell-deficient sites (26); (3) local administration of an antiserum to TNF-α reduces by ~50% the leukocyte infiltration associated with IgE-dependent cutaneous inflammation (29) (Fig. 3).

These findings strongly indicate that the production of TNF-α by mast cells importantly contributes to the leukocyte infiltration observed at IgE-dependent reactions in mice. However, it is unlikely that TNF-α is the only mast cell-associated mediator that influences leukocyte recruitment in this setting; moreover, mast cell activation may also result in the recruitment of additional sources of TNF-α at these sites (reviewed below and in 29).

VII. CYTOKINE PRODUCTION BY MOUSE BASOPHILS

Ben-Sasson et al. (30) reported that mouse splenic non-B, non-T cells can produce IL-4 and IL-3 in response to cross-linkage of high-affinity $Fc_\epsilon R$. Conrad et al. (31) demonstrated that the IL-4-producing potential of splenic or bone marrow non-B, non-T cells markedly increased in mice that had been infected with *Nippostrongylus brasiliensis* or had been injected with antibodies to mouse IgD. Recently, Seder et al. (32) reported that the splenic or bone marrow non-B, non-T cell populations from anti-IgD-injected mice that express $Fc_\epsilon R$ and IL-4-producing capacity are highly enriched in basophils. This finding, and others discussed in Ref. 32, suggest that the mouse basophil may be able to produce IL-4 and perhaps IL-3 upon IgE-dependent and other forms of activation. However, the basophil-enriched populations also contained rare immature cells in the mast cell lineage and numerous mononuclear cells lacking cytoplasmic granules. Thus, the precise identity of the IL-4-producing cell (or cells) in these preparations remains to be determined.

VIII. CYTOKINE PRODUCTION BY HUMAN MAST CELLS AND BASOPHILS

Steffen et al. (33) reported that cells with features of basophils or mast cells that had been generated in vitro from human hematopoietic cells expressed TNF-α mRNA by in situ hybridization with a murine cDNA probe and identified a

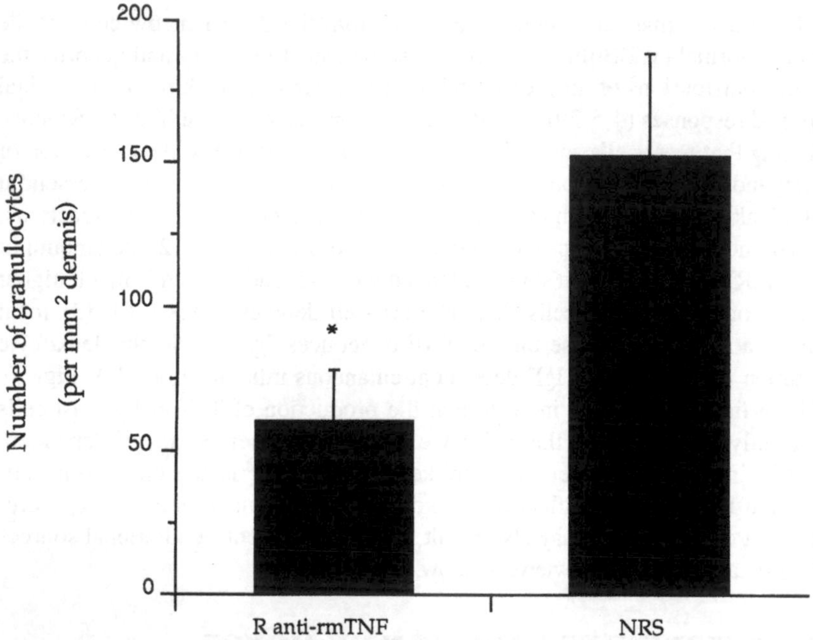

Figure 3 Inhibition of the neutrophil infiltration associated with IgE-dependent reactions by a rabbit antirecombinant mouse TNF-α antiserum (R anti-rmTNF). Both ears of C57BL/6 mice were injected with IgE anti-DNP antibodies. The next day the left ears were injected with 20 μl of a 1:100 dilution of R anti-rm TNF antiserum and the right ears were injected with 20 μl of a 1:100 dilution of normal rabbit serum (NRS) in the same vehicle. (Note: This amount of R anti-rmTNF antiserum inhibited the neutrophil accumulation observed 6 h after the i.d. injection of 40 U of rmTNF-α by 84% compared to values in contralateral ears injected with 40 U of rmTNF-α and a 1:100 dilution of NRS.) The mice were immediately challenged with $DNP_{30\text{-}40}$-HSA i.v. and sacrificed 6 h later for measurement of neutrophil infiltration. Data represent mean ± SEM ($n = 8$). Significant differences between the left and right ears are shown as $*p < 0.05$ by the paired Student's t-test, two-tailed. [From Wershil et al. (29).]

cytotoxic factor in the supernatants of these cells whose bioactivity could be partially blocked by an anti-TNF-α antibody. However, it was not determined whether the cells identified by in situ hybridization with the TNF-α probe were basophils or mast cells. Using an organ culture system, Klein et al. (34) demonstrated that IgE-dependent activation of human foreskin mast cells resulted in the expression of a TNF-α-inducible leukocyte adherence molecule (ELAM-1) on adjacent vascular endothelial cells and showed that antibodies to TNF-α inhibited the development of augmented expression of ELAM-1 in tissue fragments challenged with agents to induce mast cell activation. Subsequently, this group demonstrated the presence of both immunoreactive TNF-α and TNF-α mRNA in human skin mast cells (35). In contrast to the findings of Klein et al. (34), Leung et al. (36) reported that inhibition of the allergen-induced augmented ELAM-1 expression that was observed in a skin organ culture system required simultaneous treatment with antisera to both TNF-α and IL-1. The results of Leung et al. (36) are consistent with evidence indicating that IL-1 can contribute to leukocyte recruitment at sites of allergic diseases through induction of leukocyte adherence structures on vascular endothelial cells (37) and demonstrating that IL-1 is released at early stages in the development of cutaneous allergic responses (38). Taken together, these findings support the possibility that mast-cell-derived TNF-α may contribute to the leukocyte infiltration observed in IgE-dependent cutaneous LPRs in humans, but indicate that IL-1 (from mast cell and/or other sources) may also have an important role.

Mast-cell-derived TNF-α may also contribute to LPRs expressed in human lung. We reported preliminary evidence that purified human lung mast cells can release TNF-α upon stimulation via the $Fc_{\epsilon}RI$ (39). And Wegner et al. (40) showed in a primate model of asthma that a monoclonal antibody to intercellular adhesion molecule (ICAM-1), a leukocyte adherence structure whose expression on the surface of vascular endothelial cells can be augmented by IL-1, TNF-α, or IFN-γ, attenuated both airway eosinophilia and airway hyperresponsiveness.

IX. CYTOKINE PRODUCTION BY EOSINOPHILS

Del Pozo et al. (41) reported that peritoneal eosinophils purified (to 94% purity) from mice infected with *Mesocestoides corti* and then stimulated with LPS were positive for IL-1 mRNA by in situ hybridization and released IL-1 bioactivity when stimulated with LPS. By contrast, no IL-1 was detected in lysates of unstimulated eosinophils. Although many other cell types also can produce IL-1, this was the first report that eosinophils can produce mRNA and product of a well-characterized cytokine.

Elovic et al. (12) showed that infiltrating eosinophils, but not resident mast cells, represented the major source of TGF-α mRNA in experimentally induced carcinomas of the hamster cheek pouch and that ~40% of these eosinophils contained TGF-α product by immunohistochemistry. Approximately 70% of the bone marrow eosinophils in normal hamsters were also positive for TGF-α mRNA, and ~20% of the bone marrow eosinophils were positive for product by immunohistochemistry. These findings demonstrated that hamster eosinophils represent a potentially significant source of TGF-α, a cytokine that binds to the receptor for the epidermal growth factor and is both a potent mitogen for epithelial cells and an angiogenic factor (42). The findings also suggested that the expression of TGF-α by hamster eosinophils may be regulated during eosinophil maturation and/or by factors at sites of eosinophil infiltration.

In the first demonstration that human eosinophils represent a source of a multifunctional cytokine, Wong et al. (13) showed that the great majority of eosinophils infiltrating the interstitial tissues adjacent to two colonic adenocarcinomas and two oral squamous cell carcinomas labeled specifically by in situ hybridization with a ^{35}S-riboprobe for human TGF-α (hTGF-α). No other identifiable leukocytes in these lesions contained detectable hTGF-α mRNA. Wong et al. (13) also examined leukocytes purified from a patient with the idiopathic hypereosinophilic syndrome. Eighty percent of these eosinophils, but none of the patient's neutrophils or mononuclear cells, were positive for hTGF-α mRNA by in situ hybridization, and 55% of these eosinophils were positive by immunohistochemistry with a monoclonal antibody directed against the COOH terminus of the mature hTGF-α peptide. The identification of the purified eosinophil-associated transcript as hTGF-α was confirmed by polymerase chain reaction product restriction enzyme analysis followed by Southern blot hybridization (Fig. 4). In addition, TGF-α was detected by radioimmunoassay (RIA) or by an epidermal growth factor (EGF) receptor radioassay in the cell-free supernatants of purified eosinophils isolated from two patients with the idiopathic hypereosinophilic syndrome. In contrast to eosinophils from the patient with the hypereosinophilic syndrome, the peripheral blood eosinophils from only two of seven normal donors had detectable TGF-α mRNA, and none of these eosinophils contained immunohistochemically detectable TGF-α product. Taken together, these findings suggested that the expression of TGF-α by eosinophils in humans may be under microenvironmental regulation at sites of eosinophil infiltration and also may be influenced by factors associated with hypereosinophilia.

Demonstration of TGF-α production by eosinophils at sites of tissue pathology, or by eosinophils in patients with the idiopathic hypereosinophilic syndrome, identifies a novel mechanism by which eosinophils might contribute to the pathology associated with these disorders. It will be of interest to determine whether eosinophils infiltrating sites of such allergic disorders as asthma or atopic dermatitis also produce TGF-α, since this cytokine may significantly contribute to the bronchial epithelial or keratinocyte proliferation that can occur in association with these disorders (43–45). In collaboration with Mark Liu, we have demonstrated by in situ hybridization that the eosinophils present in the bronchoalveolar lavage (BAL) fluids obtained from allergic subjects 18 h after segmental bronchial challenge with ragweed antigen are strongly positive for TGF-α mRNA (46). That study provided the first evidence that eosinophils recruited to sites of experimentally induced allergen-dependent mast cell activation represent a potential source of cytokines, in this case, TGF-α.

However, in other disorders characterized by extensive tissue infiltrates of eosinophils, such as onchocerciasis and the pulmonary and cardiac lesions of

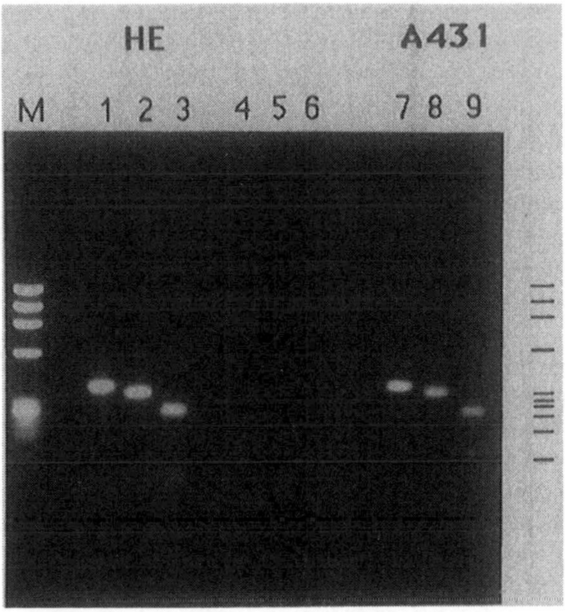

Figure 4 Human TGF-α mRNA phenotyping and characterization of total RNA isolated from human eosinophils (74% eosinophils, 26% neutrophils) purified from the blood of a patient with the idiopathic hypereosinophilic syndrome. (Lanes 1–3) HE, human eosinophils; (lanes 4–6) negative control (no RNA templates); (lanes 7–9) A431, human epidermoid carcinoma cell line (positive control for TGF-α). (Lanes 1, 4, and 7) PCR products of respective samples. (Lanes 2, 5, and 8) PvuII digestion of respective PCR products. (Lanes 3, 6, and 9) PstI digestion of respective PCR products. M, ϕχ-174 HaeIII-digested molecular weight DNA markers electrophoresed at the same time on this 2% agarose gel (from top to bottom: 1353,1078,872,603,310,281,271,234,194,118,and 72 bp). [From Wong et al. (13)]

patients with the idiopathic hypereosinophilic syndrome, eosinophil infiltration is associated with significant interstitial or endomyocardial fibrosis (47). We therefore investigated whether eosinophils could express TGF-β, a multifunctional cytokine that can potently promote extracellular matrix formation (48,49). Northern blot analysis detected the 2.5-kb TGF-β_1 (50) transcript in RNA isolated from eosinophils purified from a patient with the idiopathic hypereosinophilic syndrome (51). We then examined leukocytes from two patients with the idiopathic hypereosinophilic syndrome and two patients with blood eosinophilia due to other causes. All the blood eosinophils of these patients exhibited TGF-β_1 mRNA by in situ hybridization and TGF-β_1 protein by immunohistochemistry. No other cell type in these preparations contained TGF-β_1 mRNA by in situ hybridization whereas, as expected, other leukocytes did contain TGF-β_1 protein by immunohistochemistry. Evaluation of leukocytes from four normal donors revealed a lack of detectable TGF-β_1 protein in all the eosinophils examined, whereas eosinophils from two of these four normal donors labeled weakly for TGF-β_1 mRNA by in situ hybridization.

These results demonstrated that eosinophils in the peripheral blood of patients with blood eosinophilia can express TGF-β_1, whereas lower or undetectable levels of expression of TGF-β_1 were observed in the blood eosinophils of normal subjects. Subsequently, we (unpublished data) and Ohno et al. (52) demonstrated that many eosinophils infiltrating nasal polyps are also positive for TGF-β_1 mRNA and product.

Recent studies indicate that, in addition to TGF-α and TGF-β_1, human eosinophils represent potential sources of GM-CSF (53,54), IL-1α (55), IL-3 (53), IL-5 (56), IL-6 (57), IL-8 (58), MIP-1α (59,60), and TNF-α (59,60). These findings are potentially of great interest, in that they permit much speculation about new mechanisms by which eosinophils might express their function in health and disease. On the other hand, the factors that regulate eosinophil cytokine expression in vivo are not understood. Moreover, many cells other than the eosinophil represent potential sources of each eosinophil-associated cytokine. It seems reasonable to suspect that eosinophils will represent particularly important sources of cytokines in settings characterized by leukocytic infiltrates containing large numbers of this cell type. Nevertheless, much additional work will be required to identify the particular patterns of eosinophil cytokine expression in individual immune responses or diseases associated with hypereosinophilia and/or prominent tissue infiltrates of eosinophils.

X. MAST CELL–LEUKOCYTE CYTOKINE CASCADES IN INFLAMMATION AND IMMUNITY

When one considers the diversity of multifunctional cytokines detected in different mouse mast cell populations, the broad spectrum of biological responses that can be influenced by these cytokines, and the possibility that the expression of cytokine bioactivity by mast cell populations can be differentially regulated by various stimuli of mast cell activation (reviewed in 17,27), one might reasonably propose an enormous number of *potential* roles for mast cell cytokines in adaptive or pathological responses. Consideration of all the biological effects of each mouse mast-cell-associated cytokine would require a separate review for each molecule. However, the known effects of these agents include regulation of IgE production (IL-4, IFN-γ), regulation of mast cell proliferation and phenotype (IL-3, IL-4, GM-CSF, IFN-γ, TGF-β), modulation of leukocyte (basophil, eosinophil, neutrophil, and monocyte) effector function (IL-3, IL-4, GM-CSF, IFN-γ, TNF-α), and numerous actions in inflammation, clotting, angiogenesis, wound repair, tissue remodeling, and the development of pathological fibrosis (reviewed in 4,5,9,10, 17,27,61–67). As reviewed elsewhere (4,5,10,17,27,67,68), mast cells have been implicated in many of the same responses.

However, most analyses of mast cell cytokine production have been based on in vitro–derived mouse mast cells. Indeed, except in the case of TNF-α, there is virtually no information about cytokine production by freshly isolated mast cells, which can express multiple phenotypic differences from in vitro–derived mast cells (4,5). Moreover, in many biological responses, including those dependent on IgE, mast cells or basophils will represent just one of many potential sources of cytokines, and many of these mast-cell-derived cytokines have complex activities which overlap, synergize, and/or antagonize those of other cytokines or inflammatory mediators (reviewed in 4,17,27). In addition, it is not yet certain to what extent findings based on analyses of mouse mast cell cytokine production apply to human mast cells. As a result, a clear picture of the actual roles of mast cell (or basophil) cytokine production in the diverse biological responses thought to be influenced by these cells will require much additional study.

Nevertheless, data derived from both murine and human studies strongly suggest that mast-cell-derived TNF-α importantly contributes to the expression of IgE-dependent inflammation and leukocyte recruitment. In the mouse, IgE-dependent cutaneous reactions contain predominantly neutrophils (29), yet another cell type that can itself produce TNF-α (69,70). In humans, IgE-dependent cutaneous reactions, as well as the infiltrates associated with asthma, also contain large numbers of eosinophils. Eosinophils represent a potential source of some cytokines that may also be produced by mast cells, such as IL-1 (41,55), GM-CSF (53,54), IL-3 (53), IL-5 (56), IL-6 (57), MIP-1α (59,60), TNF-α (59,60), and TGF-β_1 (51,52), and at least one cytokine that apparently is not produced by mast cells, TGF-α (12,13).

Based on findings such as these, we have proposed that the expression of many IgE-dependent reactions, as well as other responses in which mast cells play an important role, reflect the activities of a "mast cell–leukocyte cytokine cascade" (4,17,27). In this hypothesis, mast cell activation has an essential or important role in the initiation of the response in part through the release of TNF-α and other cytokines that can influence the recruitment and/or function of additional effector cells (neutrophils, eosinophils, basophils, lymphocytes, monocytes, and platelets). These recruited cells then importantly influence the progression of the response by providing additional sources of certain cytokines also produced by mast cells and new sources of some cytokines (e.g., TGF-α) that are not produced by the mast cell. As the reactions progress further, cytokines from mast cells and other resident cells, and from recruited (leukocyte) sources, exert complex effects on resident cells (vascular endothelial cells, fibroblasts, epithelial cells, nerves, mast cells), which may importantly contribute to the vascular and epithelial changes, as well as to the tissue remodeling, angiogenesis, and fibrosis, which are so prominent in many disorders associated with mast cell activation and leukocyte infiltration. At certain points in the natural history of these processes, mast-cell- or eosinophil-derived cytokines may contribute to the down-regulation of the response. To cite just some among many possible examples, TGF-β_1 can inhibit the proliferation and/or functional properties of lymphocytes (48,49), macrophages (71), neutrophils (72), and mast cells (73).

XI. CONCLUSIONS

Despite the current gaps in our understanding of mast cell and eosinophil cytokine production, we feel that the available data support a few broad conclusions or generalizations. First, it is likely that the predominant actions of mast-cell- or eosinophil-derived cytokines ordinarily will be expressed in the immediate vicinity of the cells (reviewed in 7,27). In addition to paracrine or autocrine effects mediated by soluble cytokines, certain cytokine-dependent signals may be transmitted through a "juxtacrine" mechanism, involving direct contact between cells expressing cell-membrane-associated forms of a cytokine and adjacent cells bearing the appropriate cytokine receptors (74,75). Such "juxtacrine" effects have been reported for cells expressing biologically active forms of TGF-α (74,75). If eosinophils express TGF-α associated with the cell surface, which has not yet been determined, this would permit the eosinophil to deliver TGF-α-dependent signals directly to EGF-receptor-bearing cells present at sites of eosinophil infiltration. On the other hand, effects of mast-cell- or eosinophil-derived cytokines might become widespread under certain conditions, such as after parenteral exposure to specific antigen or during systemic mastocytosis in the case of mast-cell-derived cytokines (17,27), or during the idiopathic hypereosinophilic syndrome or in other disorders associated with increased circulating or tissue levels of eosinophils in the case of eosinophil-derived cytokines.

Second, the extent to which mast cells or eosinophils, as opposed to other cell types, represent an important source of a particular cytokine may vary greatly, depending on the characteristics of the biological response under investigation. In responses that generate stimuli capable of activating multiple sources of cytokines other than mast cells or eosinophils, the specific contribution of mast-cell- or eosinophil-derived cytokines may be relatively unimportant or difficult to discern. Thus, TNF-α appears to be critical for the expression of contact sensitivity (CS) reactions in mouse skin (76), yet most analyses have detected no impairment of the ability of genetically mast-cell-deficient mice to express CS (reviewed in 4,5). On the other hand, mast-cell- or eosinophil-derived cytokines may have significant roles in responses that are associated with the relatively selective activation of mast cells and/or the relatively selective recruitment or activation of eosinophils. For example, studies in both mice and humans indicate that mast cells may represent by far the most critical initial source of TNF-α and other proinflammatory cytokines in IgE-dependent LPRs (29,35,36).

Third, many allergic reactions and other biological responses may depend on a mast cell–leukocyte cytokine cascade in which mast-cell-derived cytokines contribute to an influx of neutrophils, eosinophils, and mononuclear cells, which in turn provide additional cytokine activities important to the progression of the response. In many other settings, eosinophil infiltration may occur by mast-cell-independent mechanisms. However, assessment of the relative importance of mast cells, eosinophils, and other sources of cytokines in specific immunological or pathological responses (or in normal tissues) will require direct analysis of the cells present in these settings. In situ hybridization and immunohistochemical approaches may be of great value in determining which among many potential cellular sources of a particular cytokine actually generates that product during individual biological responses. For example, TGF-α can be produced by many cell types in addition to the eosinophil (reviewed in 12,13,42). Yet our in situ hybridization and immunohistochemical studies indicate that eosinophils can represent a major source of TGF-α in several settings associated with eosinophil infiltration, including colon or oral carcinoma (12), sites of allergen challenge in allergic subjects (46), inflammatory bowel disease (77), and nasal polyps (77). We have demonstrated that genetically mast-cell-deficient and mast-cell-reconstituted mast-cell-deficient mice (mast cell–knock in–mice) can be used to identify and characterize the roles of mast cells and mast-cell-derived cytokines in specific biological responses in vivo (4,5,29). This work shows that mast cells are essential for IgE-dependent tissue swelling and leukocyte infiltration in mouse skin and that TNF-α significantly contributes to this inflammatory response. Unfortunately, we are not aware of mutant mice that exhibit a selective deficiency of eosinophils.

Fourth, the appreciation that certain of the biologically and clinically significant consequences of mast cell activation or eosinophil infiltration may reflect the ability of these cells to produce cytokines suggests new perspectives into the therapy of disorders influenced by these cell types. The IgE-dependent induction of increased levels of cytokine mRNAs in in vitro–derived mouse mast cells can be inhibited by cyclosporin A (23) or dexamethasone (78). Dexamethasone also can diminish significantly the IgE-dependent release of TNF-α by mouse mast cells and the tissue swelling and leukocyte infiltration associated with IgE-dependent reactions in mouse skin (78). In vitro studies demonstrate that cyclosporin A can inhibit IL-3 and GM-CSF release from human eosinophils (53). Corticosteroids can markedly diminish many of the clinical and histological manifestations of cutaneous LPRs and other allergic responses in the skin and other organs and can inhibit mediator release from human basophils, yet corticosteroids do not inhibit IgE-dependent release of histamine from human mast cells (reviewed in 2,9) or mouse mast cells (78). Finally, a recent report indicates that treatment with cyclosporin A may be beneficial in patients with corticosteroid-dependent asthma (79). In light of these findings, it will be of interest to evaluate the possibility that one of the mechanisms by which corticosteroids interfere with the expression of allergic responses (and perhaps other pathological reactions) is by suppressing cytokine production by mast cells, as well as by interfering with cytokine production and the release of other mediators by eosinophils and other cells present at or recruited to sites of mast cell activation.

XII. SUMMARY

It has recently become apparent that two of the cell types that participate in many allergic and inflammatory conditions, mast cells and eosinophils, can each produce a variety of cytokines. While the spectrum of signals that regulate cytokine production by these cells remains to be fully defined, several lines of evidence indicate that mast cells activated by IgE-dependent mechanisms can initiate leukocyte infiltration in allergic reactions in part through the production of TNF-α and perhaps other cytokines. Moreover, eosinophils and other cells recruited to these reactions represent additional sources of cytokines capable of influencing several aspects of both the amplification and resolution of these responses. While such mast cell–leukocyte cytokine cascades may be especially important in the development of IgE-dependent allergic reactions and immune responses, cytokine production by mast cells and/or eosinophils probably contributes to the pathogenesis of a broad range of disorders associated with the activation or infiltration of these cells. Emerging evidence indicates that cytokine production by mast cells or eosinophils can be suppressed by corticosteroids or cyclosporin A. There is no doubt that many cell types in addition to mast cells and eosinophils contribute to the pathogenesis of allergic and inflammatory reactions involving mast cell activation or eosinophil infiltration, and corticosteroids or cyclosporin A may suppress such responses through multiple effects on diverse cellular targets. Nevertheless, it will be of some interest to determine whether one mechanism by which these agents diminish allergic or inflammatory responses in vivo is through suppression of cytokine production by mast cells and eosinophils.

ACKNOWLEDGMENTS

This work was supported in part by United States Public Health Service Grants AI-20241, AI-22674, AI-23990, AI-31982, AI-33372, HL-46563, PO1 DK-33506 (subproject 6), DE-08680, DE-00318, Physician Scientist Award DE-00312 (AE), a Scholar in Allergy Award (JJC) from the American Academy of Allergy and Immunology and Marion-Merrell Dow, a Cancer Research Scholar Award (DTWW) from the American Cancer Society, Massachusetts Division, and the Beth Israel Hospital Pathology Foundation.

REFERENCES

1. Ishizaka T. Mechanisms of IgE-mediated hypersensitivity. In: Middleton E Jr., Reed CE, Ellis EF, Adkinson NF Jr., Yuninger JW, eds. Allergy: Principles and Practice. 3d ed. St. Louis: Mosby, 1988:71–93.
2. Galli SJ , Lichtenstein LM. Biology of mast cells and basophils. In: Middleton E Jr., Reed CE, Ellis EF, Adkinson NF Jr., Yuninger JW, eds. Allergy: Principles and Practice. 3d ed. St. Louis: Mosby, 1988:106–134.
3. Holgate ST, Robinson C, Church MK. Mediators of immediate hypersensitivity. In: Middleton E Jr., Reed CE, Ellis EF, Adkinson NF Jr., Yuninger JW, eds. Allergy: Principles and Practice. 3d ed. St. Louis: Mosby, 1988:135–163.
4. Galli SJ, Geissler EN, Wershil BK, Gordon JR, Tsai M, Hammel I. Insights into mast cell development and function derived from analyses of mice carrying mutations at *beige, W/c-kit* or *Sl/SCF* (c-*kit* ligand) loci. In: Kaliner MA, Metcalfe DD eds. The Role of the Mast Cell in Health and Disease. New York: Marcel Dekker, 1992: 129–202.
5. Galli SJ. New insights into "the riddle of the mast cells": microenvironmental regulation of mast cell development and phenotypic heterogeneity. Lab Invest 1990; 62:5–33.
6. Spry C. Eosinophils. New York: Oxford University Press, 1989:484 pp.
7. Gleich GJ, Loegering DA. Immunobiology of eosinophils. Annu Rev Immunol 1984; 2:429–459.
8. Weller PF. The immunobiology of eosinophils. N Engl J Med 1991; 324:1110–1118.
9. Schleimer RP, Derse CP, Friedman B, Gillis S, Plaut M, Lichtenstein LM, MacGlashan DW Jr. Regulation of human basophil mediator release by cytokines. I. Interaction with antiinflammatory steroids. J Immunol 1989; 143:1310–1317.
10. Costa JJ, Burd PR, Metcalfe DD. Mast cell cytokines. In: Kaliner MA, Metcalfe DD eds. The Role of the Mast Cell in Health and Disease. New York: Marcel Dekker, 1992:443–466.
11. Kinet J-P. The high affinity receptor for IgE. Curr Opinion Immunol 1990; 2: 499–505.
12. Elovic A, Galli SJ, Weller PF, Chang ALC, Chiang T, Chou MY, Donoff RB, Gallagher GT, Matossian K, McBride J, Tsai M, Todd R, Wong DTW. Production of transforming growth factor-alpha by hamster eosinophils. Am J Pathol 1990; 137: 1425–1434.
13. Wong DTW, Weller PF, Galli SJ, Elovic A, Rand TH, Gallagher GT, Chiang T, Chou MY, Matossian K, McBride J, Todd R. Human eosinophils express transforming growth factor-alpha. J Exp Med 1990; 172:673–681.
14. Chung SW, Wong PMC, Shen-Ong G, Ruscetti S, Ishizaka T, Eaves CJ. Production of granulocyte-macrophage colony-stimulating factor by Abelson virus-induced tumorigenic mast cell lines. Blood 1986; 68:1074–1081.
15. Brown MA, Pierce JH, Watson CJ, Falco J, Ihle JN, Paul WE. B cell stimulatory factor-1/interleukin-4 mRNA is expressed by normal and transformed mast cells. Cell 1987; 50:809–818.
16. Humphries K, Abraham S, Krystal G, Lansdorp P, Lemoine F, Eaves CJ. Activation of

multiple hemopoietic growth factor genes in Abelson virus–transformed myeloid cells. Exp Hematol 1988; 16:774–781.
17. Gordon JR, Burd PR, Galli SJ. Mast cells as a source of multifunctional cytokines. Immunol Today 1990; 11:458–464.
18. Young JD-E, Liu C, Butler G, Cohn ZA, Galli SJ. Identification, purification, and characterization of a mast cell-associated cytolytic factor related to tumor necrosis factor. Proc Natl Acad Sci USA 1987; 84:9175–9179.
19. Gordon JR, Galli SJ. Mast cells as a source of both preformed and immunologically inducible TNF-α/cachectin. Nature 1990; 346:274–276.
20. Richards AL, Okuno T, Takagaki Y, Djeu JY. Natural cytotoxic cell-specific cytotoxic factor produced by IL-3-dependent basophilic/mast cells. Relationship to TNF. J Immunol 1988; 141:3061–3066.
21. Plaut M, Pierce JH, Watson CJ, Hanley-Hyde J, Nordan RP, Paul WE. Mast cell lines produce lymphokines in response to cross-linkage of $Fc_{\epsilon}RI$ or to calcium ionophores. Nature 1989; 339:64–67.
22. Wodnar-Filipowicz A, Heusser CH, Moroni C. Production of the haemopoietic growth factors GM-CSF and interleukin-3 by mast cells in response to IgE receptor-mediated activation. Nature 1989; 339:150–152.
23. Burd PR, Rogers HW, Gordon JR, Martin CA, Jayaraman S, Wilson SD, Dvorak AM, Galli SJ, Dorf ME. Interleukin 3-dependent and -independent mast cells stimulated with IgE and antigen express multiple cytokines. J Exp Med 1989; 170:245–257.
24. Tsai M, Gordon JR, Galli SJ. Mast cells constitutively express transforming growth factor β mRNA (abstr). FASEB J 1990; 4:A1944.
25. Gurish MF, Ghildyal N, Arm J, Austen KF, Avraham S, Reynolds D, Stevens RL. Cytokine mRNA are preferentially increased relative to secretory granule protein mRNA in mouse bone marrow–derived mast cells that have undergone IgE-mediated activation and degranulation. J Immunol 1991; 146:1527–1533.
26. Gordon JR, Galli SJ. Release of both preformed and newly synthesized tumor necrosis factor α (TNF-α)/cachectin by mouse mast cells stimulated by the $Fc_{\epsilon}RI$. A mechanism for the sustained action of mast cell-derived TNF-α during IgE-dependent biological responses. J Exp Med 1991; 174:103–107.
27. Galli SJ, Gordon JR, Wershil BK. Cytokine production by mast cells and basophils. Curr Opinion Immunol 1991; 3:865–873.
28. Lemanske RF Jr., Kaliner MA. Late phase allergic reactions. In: Middleton E Jr., Reed CE, Ellis EF, Adkinson NF Jr., Yuninger JW, eds. Allergy: Principles and Practice. 3d ed. St Louis: Mosby, 1988:224–246.
29. Wershil BK, Wang Z, Gordon JR, Galli SJ. Recruitment of neutrophils during IgE-dependent cutaneous late phase responses in the mouse is mast cell dependent: partial inhibition of the reaction with antiserum against tumor necrosis factor-alpha. J Clin Invest 1991; 87:446–453.
30. Ben-Sasson SZ, LeGros G, Conrad DH, Finkelman FD, Paul WE. Cross-linking Fc receptors stimulate splenic non-B, non-T cells to secrete interleukin 4 and other lymphokines. Proc Natl Acad Sci USA 1990; 87:1421–1425.
31. Conrad DH, Ben-Sasson SZ, LeGros G, Finkelman FD, Paul WE. Infection with *Nippostrongylus brasiliensis* or injection of anti-IgD antibodies markedly enhances

Fc-receptor-mediated interleukin 4 production by non-B, non-T cells. J Exp Med 1990; 171:1497–1508.

32. Seder RA, Paul WE, Dvorak AM, Sharkis SJ, Kagey-Sobotka A, Niv Y, Finkelman FD, Barbieri SA, Galli SJ, Plaut M. Mouse splenic and bone marrow cell populations that express high-affinity Fc_{ϵ} receptors and produce interleukin 4 are highly enriched in basophils. Proc Natl Acad Sci USA 1991; 88:2835–2839.
33. Steffen M, Abboud M, Potter GK, Yung YP, Moore MAS. Presence of tumor necrosis factor or a related factor in human basophil/mast cells. Immunology 1989; 66:445–450.
34. Klein LM, Lavker RM, Matis WL, Murphy GF. Degranulation of human mast cells induces an endothelial antigen central to leukocyte adhesion. Proc Natl Acad Sci USA 1989; 86:8972–8976.
35. Walsh LJ, Trinchieri G, Waldorf HA, Whitaker D, Murphy GF. Human dermal mast cells contain and release tumor necrosis factor α which induces endothelial leukocyte adhesion molecule-1. Proc Natl Acad Sci USA 1991; 88:4220–4224.
36. Leung DYM, Pober JS, Cotran RS. Expression of endothelial-leukocyte adhesion molecule-1 in elicited late phase allergic reactions. J Clin Invest 1991; 87:1805–1809.
37. Bochner BS, Luscinskas FW, Gimbrone MA Jr., Newman W, Sterbinsky SA, Derse-Anthony CP, Klunk D, Schleimer RP. Adhesion of human basophils, eosinophils, and neutrophils to interleukin 1-activated human vascular endothelial cells: contributions of endothelial cell adhesion molecules. J Exp Med 1991; 173:1553–1556.
38. Bochner BS, Charlesworth EN, Lichtenstein LM, Derse CP, Gillis S, Dinarello CA, Schleimer RP. Interleukin-1 is released at sites of human cutaneous allergic reactions. J Allergy Clin Immunol 1990; 86:830–839.
39. Gordon JR, Post T, Schulman ES, Galli SJ. Characterization of mouse mast cell TNF-α induction in vitro and in vivo, and demonstration that purified human lung mast cells contain TNF-α. FASEB J 1991; 5:A1009.
40. Wegner CD, Gundel RH, Reilly P, Haynes N, Letts LG, Rothlein R. Intracellular adhesion molecule-1 (ICAM-1) in the pathogenesis of asthma. Science 1990; 247: 456–459.
41. Del Pozo V, De Andres B, Martin E, Maruri N, Zubeldia JM, Palomino P, Lahoz C. Murine eosinophils and IL-1: αIL-1 mRNA detection by in situ hybridization. J Immunol 1990; 144:3117–3122.
42. Derynck R. Transforming growth factor-α. Cell 1988; 54:593–595.
43. Slifman NR, Adolphson CR, Gleich GJ. Eosinophils: biochemical and cellular aspects. In: Middleton E Jr., Reed CE, Ellis EF, Adkinson NF Jr., Yuninger JW, eds. Allergy: Principles and Practice. 3d ed. St. Louis: Mosby, 1988:179–205.
44. Thurlbeck WM, Hogg JC. The pathology of asthma. In: Middleton E Jr., Reed CD, Ellis EF, Adkinson NF Jr., Yuninger JW, eds. Allergy: Principles and Practice. 3d ed. St. Louis: Mosby, 1988:1008–1017.
45. Hanifin J. Atopic dermatitis. In: Middleton E Jr., Reed CE, Ellis EF, Adkinson NF Jr., Yuninger JW, eds. Allergy: Principles and Practice. 3d ed. St. Louis: Mosby, 1988: 1403–1428.
46. Liu M, Matossian K, Wong DTW, Weller PF, Galli SJ. Expression of mRNA for transforming growth factor-α (TGF-α) by eosinophils at sites of segmental airway

challenge with antigen in allergic asthmatic subjects (abstr). Am Rev Respir Dis 1992; 145:A452.
47. Nutman TB, Ottesen EA, Cohen SG. Eosinophilia and eosinophil-related disorders. In: Middleton E Jr., Reed CE, Ellis EF, Adkinson NF Jr., Yuninger JW, eds. Allergy: Principles and Practice. 3d ed. St. Louis: Mosby, 1988:861–890.
48. Sporn MB, Roberts AB, Wakefield LM, de Crombrugghe B. Some recent advances in the chemistry and biology of transforming growth factor-beta. J Cell Biol 1987; 105:1039–1045.
49. Bernard JA, Lyons RM, Moses HL. The cell biology of transforming growth factor beta. Biochim Biophys Acta 1990; 1032:79–87.
50. Derynck R, Jarrett JA, Chen EY, Eaton DH, Bell JR, Assoian RK, Roberts AB, Sporn MB, Goeddel DV. Human transforming growth factor-beta complementary DNA sequence and expression in normal and transformed cells. Nature 1985; 316:701–705.
51. Wong DTW, Elovic A, Matossian K, Nagura N, McBride J, Chou MY, Gordon JR, Rand TH, Galli SJ, Weller PF. Eosinophils from patients with blood eosinophilia express transforming growth factor β1. Blood 1991; 78:2702–2707.
52. Ohno I, Lea RG, Flanders KC, Clark DA, Banwatt D, Dolovich J, Denburg J, Harley CB, Gauldie J, Jordana M. Eosinophils in chronically inflamed human upper airway tissue express transforming growth factor β1 gene (TGFβ1). J Clin Invest 1992; 89: 1662–1668.
53. Kita H, Ohnishi T, Okubo Y, Weiler D, Abrams JS, Gleich GJ. Granulocyte/macrophage colony-stimulating factor and interleukin 3 release from human peripheral blood eosinophils and neutrophils. J Exp Med 1991; 174:745–748.
54. Moqbel R, Hamid Q, Ying S, Barkans J, Hartnell A, Tsicopoulos A, Wardlaw AJ, Kay AB. Expression of mRNA and immunoreactivity for the granulocyte/macrophage colony-stimulating factor in activated human eosinophils. J Exp Med 1991; 174: 749–752.
55. Weller PF, Rand TH, Barrett T, Elovic A, Wong DTW, Finberg RW. Accessory cell function of human eosinophils: HLA-DR-dependent, MHC-restricted antigen presentation and interleukin-1α expression. J Immunol 1993; 150:2554–2562.
56. Desreumaux P, Janin A, Colombel JF, Prin L, Plumas J, Emilie D, Torpier G, Capron M. Interleukin 5 messenger RNA expression by eosinophils in the intestinal mucosa of patients with coeliac disease. J Exp Med 1992; 75:293–296.
57. Hamid Q, Barkans J, Ying S, Abrams JS, Kay AB, Moqbel R. Human eosinophils synthesize and secrete interleukin-6, in vitro. Blood 1992; 80:1496–1501.
58. Hansel TT, Braun RK, Erard F, Rihs S, DeVries IJM, Blaser K, Walker C. Human peripheral blood eosinophils produce and release interleukin-8 on stimulation with calcium ionophore. Eur J Immunol 1993; 23:956–960.
59. Costa JJ, Matossian K, Beil WJ, Wong DTW, Gordon JR, Dvorak AM, Weller PF, Galli SJ. Human eosinophils can express the cytokines TNF-α and MIP-1α (abstr). J Allergy Clin Immunol 1993: 91:A174.
60. Costa JJ, Matossian K, Resnick MB, Beil WJ, Wong DTW, Gordon JR, Dvorak AM, Weller PF, Galli SJ. Human eosinophils can express the cytokines TNF-α and MIP-1α. J Clin Invest. In press.
61. Beutler B, Cerami A. The biology of cachectin/TNF—a primary mediator of the host response. Annu Rev Immunol 1989; 7:625–655.

62. Arai K-i, Lee F, Miyajima A, Miyatake S, Arai N, Yokota T. Cytokines: coordinators of immune and inflammatory responses. Annu Rev Biochem 1990; 59:783–836.
63. Johnston SL, Holgate ST. Cellular and chemical mediators—their roles in allergic diseases. Curr Opinion Immunol 1990; 2:513–524.
64. Wolpe SD, Cerami A. Macrophage inflammatory proteins 1 and 2: members of a novel superfamily of cytokines. FASEB J 1989; 3:2565–2573.
65. Finkelman FD, Katona IM, Urban JF Jr., Paul WE. Control of in vivo IgE production in the mouse by interleukin 4. In: Chadwick D, Evered D, Whelan J, eds. IgE, Mast Cells and the Allergic Response. Ciba Foundation Symposium No. 147. Chichester, UK: Wiley, 1989:3–22.
66. Balkwill FR, Burke F. The cytokine network. Immunol Today 1989; 10:299–304.
67. Claman HN. Mast cells, T cells and abnormal fibrosis. Immunol Today 1985; 6: 192–194.
68. Selye H. The Mast Cells. Washington, DC: Butterworths, 1965.
69. Dubravec DB, Spriggs DR, Mannick JA, Rodrick ML. Circulating human peripheral blood granulocytes synthesize and secrete tumor necrosis factor α. Proc Natl Acad Sci USA 1990; 87:6758–6761.
70. Djeu JY, Serbousek D, Blanchard DK. Release of tumor necrosis factor by human polymorphonuclear leukocytes. Blood 1990; 76:1405–1409.
71. Nelson BJ, Ralph P, Green SJ, Nacy CA. Differential susceptibility of activated macrophage cytokine effector reactions to the suppressive effects of transforming growth factor-β1. J Immunol 1991; 146:1849–1857.
72. Gamble JR, Vadas MA. Endothelial adhesiveness for blood neutrophils is inhibited by transforming growth factor-β. Science 1988; 242:97–99.
73. Brodie DH, Wasserman SI, Alvaro-Gracia J, Zvaifler NJ, Firestein GS. Transforming growth factor-β_1 selectively inhibits IL-3-dependent mast cell proliferation without affecting mast cell function or differentiation. J Immunol 1989; 143:1591–1597.
74. Wong ST, Winchell LF, McCune BK, Earp HS, Herman B, Lee DC. The TGF-α precursor expressed on the cell surface binds to the EGF receptor on adjacent cells, leading to signal transduction. Cell 1989; 56:495–506.
75. Anklesaria P, Teixido J, Laiho M, Pierce JH, Greenberger JS, Massague J. Cell-cell adhesion mediated by binding of membrane-anchored transforming growth factor-α to epidermal growth factor receptors promotes cell proliferation. Proc Natl Acad Sci USA 1990; 87:3289–3293.
76. Piguet PF, Grau GE, Hauser C, Vassalli P. Tumor necrosis factor is a critical mediator in hapten-induced irritant and contact hypersensitivity reactions. J Exp Med 1991; 173:673–679.
77. Emery S, Elovic A, Wong DTW, Weller PW, Galli SJ. Expression of transforming growth factor-alpha (TGF-α) mRNA by human eosinophils in diverse pathological processes (abstr). FASEB J 1991; 5:A1695.
78. Wershil BK, Gordon JR, Wang Z-S, Lavigne JA, Galli SJ. Dexamethasone (DEX) inhibits IgE-dependent TNF-α production by mouse mast cells (MCs) in vitro-and suppresses IgE-, mast cell- and TNF-α-dependent inflammation in vivo (abstr). FASEB J 1992; 6:A1716.
79. Alexander AG, Barnes NC, Kay AB. Trial of cyclosporin in corticosteroid-dependent chronic severe asthma. Lancet 1992; 339:324–328.

DISCUSSION (Speaker: S. Galli)

Kita: Some comments on production of IL-3 and GM-CSF from eosinophils. We cultured human peripheral blood eosinophils with a calcium ionophore, ionomycin. Ionomycin induced a prolonged survival of eosinophils, which peaks at 1 μM. We took these supernatants and put them into culture of fresh isolated unstimulated eosinophils. These supernatants enhanced survival of fresh eosinophils, and the effects were completely abolished by treatment of the supernatants with combination of anti-IL-3 and anti-GM-CSF antibody, suggesting that IL-3 and GM-CSF are contained in the supernatants. Cytokines produced were also measurable by collaboration with Dr. John Abrams. Eosinophils produced one-half as much IL-3 and one-fifth as much GM-CSF as mononuclear cells. Production of cytokines from eosinophils was blocked by cyclosporin A. These data suggest eosinophils have a capacity to produce and release IL-3 and GM-CSF.

Moqbel: We have already reported that human eosinophils express mRNA for GM-CSF after stimulation with either calcium ionophore or IFN-γ. This message was shown to be translated in the cell, and indeed the elegant parallel study of Drs. Kita and Gleich demonstrated the release of this cytokine as well as IL-5 after activation. In a recent study, conducted with Dr. Qutayba Hamid using the technique of in situ hybridization, we were able to demonstrate that peripheral blood human eosinophils expressed mRNA for the Th2-cytokine IL-6. Approximately 20% of eosinophils incubated with medium alone over a period of 24 h were mRNA+ for IL-6 antisense. After a similar period of incubation with IFN-γ, the percentage of cells expressing mRNA increased to almost 50%. No hybridization signals were observed with the sense probe. We were careful to colocalize the hybridization signals to human eosinophils, and for this we used simultaneous in situ hybridization and immunocytochemistry using EG2 as a phenotypic marker for eosinophils as well as carbol chromotrope 2R. The message was shown to be translated in these cells using a polyclonal anti-IL-6 antibody to detect IL-6 immunoreactivity by immunocytochemistry. In collaboration with Dr. John Abrams in DNAX, concentrations of IL-6 were measured in the supernatants of both control and activated eosinophils using an immunoenzymatic assay for IL-6.

Hansel: We have used an immunomagnetic selection technique to prepare eosinophils of >99.5% purity from the blood of noneosinophilic normal subjects. Culture of pure eosinophils with calcium ionophore resulted in release of IL-8 measurable by ELISA, and IL-8 mRNA was detectable after PCR amplification. Finally, blood eosinophils were shown to contain preformed cytoplasmic IL-8 by immunohistochemistry (APAAP). Eosinophil production of IL-8 may influence the localization and activation of tissue eosinophils.

Gleich: A comment and a question. First, eosinophil infiltration and degranulation are seen in a variety of syndromes associated with fibrosis, including sclerosing cholangitis, retroperitoneal fibrosis, and mediastinal fibrosis. Second, the recent observation that the $Fc_{\epsilon}RI$ (high-affinity IgE receptor) is present on Langerhans cells suggests that one need not attribute allergen-stimulated reactions to the mast cell. Could you comment?

Galli: The types and amount of cytokines produced by eosinophils in specific examples of the many disorders characterized by eosinophil infiltration and fibrosis should be determined on a case-by-case basis, since the pattern of cytokine production by eosinophils may vary in these different settings. Nevertheless, it seems reasonable to suggest that TGF-β, TGF-α and TNF-α, as well as other cytokines, may represent mediators by which eosinophils can promote fibrosis and other tissue changes in a wide variety of diseases, immunological reactions, and other biological responses.

With regard to the role of the mast cell in IgE-dependent reactions to allergens, we have demonstrated using our "mast cell-knock in mice" that essentially all of the early or late tissue swelling, or leukocyte infiltration, associated with IgE-dependent cutaneous reactions in mouse skin is mast-cell-dependent (1,2). Genetically mast-cell-deficient $WBB6F_1-W/W^v$ mice have normal numbers of Langerhans cells (3), although I am not sure that it has yet been determined whether mouse Langerhans cells express the $Fc_\epsilon RI$. Nor is it yet known, even in human Langerhans cells, whether these cells can release cytokines in response to signaling through the $Fc_\epsilon RI$. We and others have shown that W/W^v or Sl/Sl^d mast-cell-deficient mice can express fatal active anaphylaxis, which in at least some instances may be mediated by IgE antibodies (reviewed in 4). I certainly believe that in some settings allergens can induce IgE-dependent reactions that reflect important contributions by effectors other than the mast cell, including Langerhans cells, or cells that express the $Fc_\epsilon RII$ (CD23). Indeed, one of the attractive characteristics of mast cell-knock in mice is that these mice permit direct assessment of the importance of the mast cell, as opposed to other cell types, for the various specific consequences of allergen challenge in different experimental model systems.

Townley: Do you think, or have any evidence, that mast cell cytokine release affects Th2 cells to stimulate cytokine release and thus link immediate and late allergic reactions?

Coffman: IL-4 is, perhaps, the best candidate for a mast cell cytokine that can have a profound influence on T-cell-subset differentiation and function.

Galli: In vitro–derived mouse mast cells certainly can produce IL-4 upon IgE-dependent activation (5,6). On the other hand, genetically mast-cell-deficient mice have no obvious impairment in their ability to synthesize IgE in response to antigen challenge and can have baseline levels of IgE significantly higher than those in the congenic normal mice (reviewed in 4). Finally, recent evidence indicates that basophils, which are present in mast-cell-deficient mice, may represent yet another source of IL-4 in IgE-dependent reactions (7). As a result, the biological significance of mast cells as a source of IL-4 is not yet clear, even in mice.

Butterworth: We have heard about a series of autocrine loops, potentially leading to recruitment of eosinophils, prolongation of their survival, and enhanced functional activity; but we have heard of no mechanisms for switching off the reaction once the initiating lesion has been controlled. Is there any evidence for inhibitory cytokines or other mediators being produced during the later stages of eosinophil activation (other than simply cell death)?

Galli: I am not aware of any direct evidence that eosinophils can down-regulate their own functional activity. However, TGF-β can inhibit adhesiveness of blood neutrophils for vascular endothelial cells (8) and can inhibit IL-3-dependent mouse mast cell proliferation (9). Although I am not aware of whether such studies already have been performed, it would be of interest to know whether TGF-β might inhibit interactions between vascular endothelial cells and eosinophils or influence cytokine-dependent eosinophil survival in vitro.

Such effects could represent examples of mechanisms by which eosinophil-derived TGF-β might help to dampen an inflammatory response.

M. Capron: In relation to the differences between human and mouse systems, it is interesting to notice that mouse eosinophils do not express IgE receptors of the $Fc_{\epsilon}RIII$ CD23 type. Indeed, the recent cloning of murine CD23 has shown that only $Fc_{\epsilon}RIIa$ species was present and not the $Fc_{\epsilon}RIIb$ (which is expressed on cytokine-induced monocytes or eosinophils).

In relation to cytokine production by eosinophils, different studies have used only the ionomycin or calcium ionophore as stimuli for the production of IL-3, GM-CSF, IL-6, or IL-8. However, in contrast, IL-5 is not secreted after stimulation with Ca^{2+} ionophore but only after IgE- or IgA-dependent activation.

Galli: It will be of considerable interest to characterize further the patterns of cytokine production by eosinophils in situ in different biological responses or challenge in vitro with various stimuli of activation.

Spry: In relation to your comment that there are no good models of inflammation in eosinophil-depleted animals, Cheever et al. (10) have shown that anti-IL-5-treated mice did not produce eosinophils in response to *Schistosoma japonicus* infection. Although the granulomas were one-third smaller than in untreated mice, the fibrotic response to eggs in the liver was fully retained. This suggested that eosinophils are not essential for the development of egg granulomas.

Galli: Thank you, Dr. Spry, for reminding us of one of the several different approaches that have been taken to deplete eosinophils in mice or other experimental animals. I will revise the text of my chapter to state what I should have said, which is that I am not aware of the description of mutant mice with a selective deficiency in eosinophils. Eliminating IL-5 or antagonizing its effect may influence biological responses either because eosinophils are reduced in number or for reasons unrelated to the eosinophil. If a biological response, such as *S. japonicum*–associated hepatic fibrosis, is unaffected by the elimination of a particular cell type (whether eosinophils, mast cells, or other cells), one can conclude that the eliminated cell was not essential for the response. However, the cell in question might still represent one of a limited number of cell types that provide similar or identical function in that response. Such redundancy of functional activity probably is not uncommon in immunological or inflammatory responses.

Durham: We have heard convincing evidence that both mast cells and eosinophils produce a range of cytokines. However, the lesions discussed (i.e., celiac disease, late nasal responses) are also characterized by far greater numbers of activated T lymphocytes, which must be at least as likely a source for these cytokines. Colocalization of phenotypic markers and cytokines is difficult. Have you used combined immunostaining/in situ hybridization to identify the cell source of cytokines?

Galli: We have not used combined immunohistochemistry and in situ hybridization. However, mast cells can be identified in our in situ hybridization specimens by the purple staining of their cytoplasmic granules with Geimsa stain (11). Moreover, we reported that eosinophils can be identified specifically in our in situ hybridization preparations by their

Geimsa-enhanced autofluorescence when they are viewed under ultraviolet light with a rhodamine filter (11,12). In immunohistochemical preparations, eosinophils can be identified after staining with 0.2% aniline blue by their blue fluorescence when viewed under ultraviolet light with a DAPI filter (11,12).

Bochner: What cytokines are consistently *not* expressed by eosinophils?

Galli: Although I'd rather not present any more of our unpublished data at this time, I don't think that eosinophils will be found to represent a significant source of every cytokine.

DISCUSSION REFERENCES

1. Wershil BK, Mekori YA, Murakami T, Galli SJ. ^{125}I-fibrin deposition in IgE-dependent immediate hypersensitivity reactions in mouse skin. Demonstration of the role of mast cells using genetically mast cell-deficient mice locally reconstituted with cultured cells. J Immunol 1987; 139:2605–2614.
2. Wershil BK, Wang Z-S, Gordon JR, Galli SJ. Recruitment of neutrophils during IgE-dependent cutaneous late phase responses in the mouse is mast cell dependent: partial inhibition of the reaction with antiserum against tumor necrosis factor-alpha. J Clin Invest 1991; 87:446–453.
3. Askenase PW, Van Loveren H, Kraueter-Kops S, Yacov Y, Meade R, Theoharides TC, Nordlung JJ, Scovern H, Gershon MD, Ptak W. Defective elicitation of delayed-type hypersensitivity in W/W^{v} and Sl/Sl^{d} mast cell-deficient mice. J Immunol 1983; 131:2687–2694.
4. Takeishi T, Martin TR, Katona IM, Finkelman FD, Galli SJ. Differences in the expression of the cardiopulmonary alterations associated with anti-immunoglobulin E-induced or active anaphylaxis in mast cell-deficient and normal mice. Mast cells are not required for the cardiopulmonary changes associated with certain fatal anaphylactic responses. J Clin Invest 1991; 88:598–608.
5. Plaut M, Pierce JH, Watson CJ, Hanley-Hyde J, Nordon RP, Paul WE. Mast cell lines produce lymphokines in response to cross-linkage of Fc-epsilon RI or to calcium ionophores. Nature 1989; 339:64–67.
6. Burd PR, Rogers HW, Gordon JR, Martin CA, Jayaraman S, Wilson SD, Dvorak AM, Galli SJ, Dorf ME. Interleukin-3-dependent and independent mast cells stimulated with IgE and antigen express multiple cytokines J Exp Med 1989; 170:245–257.
7. Seder RA, Paul WE, Dvorak AM, Sharkis SJ, Kagey-Sobotka A, Niv Y, Finkelman FD, Barbieri SA, Galli SJ, Plaut M. Mouse splenic and bone marrow cell populations that express high-affinity Fc-epsilon receptors and produce interleukin 4 are highly enriched in basophils. Proc Natl Acad Sci USA 1991; 88:2835–2839.
8. Gamble JR, Vadas MA. Endothelial adhesiveness for blood neutrophils is inhibited by transforming growth factor-β. Science 1988; 242:97–99.
9. Brodie DH, Wasserman SI, Alvaro-Gracia J, Zvaifler NJ, Firestein GS. Transforming growth factor-β1 selectively inhibits IL-3-dependent mast cell proliferation without affecting mast cell function or differentiation. J Immunol 1989; 143:1591–1597.
10. Cheever AW, Xu YH, Sher A, Macedonia JG. Analysis of egg granuloma formation in

Schistosoma japonicum–infected mice treated with antibodies to interleukin-5 and gamma interferon. Infect Immun 1991; 59:4071–4074.
11. Elovic A, Galli SJ, Weller PF, Chang ALC, Chiang T, Chou MY, Donoff RB, Gallagher GT, Matossian K, McBride J, Tsai M, Todd R, Wong DTW. Production of transforming growth factor-alpha by hamster eosinophils. Am J Pathol 1990; 137: 1425–1434.
12. Wong DTW, Weller PF, Galli SJ, Elovic A, Rand TH, Gallagher GT, Chiang T, Chou MY, Matossian K, McBride J, Todd R. Human eosinophils express transforming growth factor-alpha. J Exp Med 1990; 172:673–681.

14

Eosinophils in Immunological Reactions

Peter F. Weller
Beth Israel Hospital and Harvard Medical School, Boston, Massachusetts

I. INTRODUCTION

Eosinophils are normal cellular components within mucosal tissues, as well as being participants in varied allergic, immunological, and other disease processes (1). Increases in blood eosinophilia may accompany these diseases, but it is in tissues, normal mucosal tissues of the gastrointestinal and respiratory tracts, as well as mucosal and other tissue sites in association with specific disease processes, that eosinophils characteristically are found. It is in these normal and diseased tissues that eosinophils function and contribute to the immunopathogenesis of various diseases (2). The functions of eosinophils may be complex. On the one hand, eosinophils are capable of acute cellular responses. These acute responses include degranulation, oxidative burst activity, and the elaboration of lipid mediators, notably leukotriene C_4(LTC_4) and platelet-activating factor (PAF). On the other hand, eosinophils can interact collaboratively with other cellular elements. Eosinophils may respond to cytokines elaborated by other cell types. Some of these cytokines, such as GM-CSF, IL-3, and IL-5, prolong eosinophil viability and modulate the acute responses of eosinophils. Other cytokines, including some lymphokines, stimulate eosinophils without heightening acute cellular responses. Moreover, eosinophils themselves may both elaborate cytokines and provide other mechanisms for stimulating adjacent cells. In this chapter, we will review studies related to both aspects of eosinophil functioning. First, we will consider some of the aspects pertinent to the intracellular sites and regulation of eosinophil eicosanoid formation. Second, we will consider findings pertinent to intercellular interactions between lymphocytes and eosinophils.

II. LIPID BODIES AS SITES OF EOSINOPHIL EICOSANOID FORMATION

A. Eosinophil Lipid Bodies as Storage Sites of Esterified Arachidonate

In all cells, substrate arachidonate is not present to any large degree as free fatty acid (3), but rather is esterified within glycerolipids and is mobilized principally from classes of phospholipids by the actions of phospholipases in order to initiate the cascades that eventuate in eicosanoid formation. Since eicosanoid precursors containing arachidonyl-phospholipids are widely assumed to reside within cell membranes, it is cellular membranes in eosinophils as in other cells that are assumed to constitute the sole sites of eicosanoid formation. Prostaglandin H_2 (PGH) synthase (cyclooxygenase) is believed to reside within membranes (4,5), perhaps, as interpreted by some (6) but not all (7), to have a conventional hydrophobic transmembrane spanning region. In other cell types, PGH synthase has been localized to the endoplasmic reticulum, to the nuclear membrane, and infrequently to the plasma membrane (5). The key proximal enzyme in the synthesis of leukotrienes, 5-lipoxygenase, is cytosolic but can undergo translocation to membranes where it acts to form 5-HPETE and LTA_4 (8). In neutrophils, a membrane-bound protein, 5-lipoxygenase activating protein (FLAP), is involved in the translocation of cytosolic 5-lipoxygenase to membranes (9,10). Another enzyme in the eosinophil involved in the synthesis of LTC_4, LTC_4 synthase, is believed to reside at microsomal membranes, based on studies in other cells.

In addition to cell membranes, an alternative intracellular domain potentially involved in eicosanoid formation in various cells, especially eosinophils, is the lipid body. Cytoplasmic lipid bodies, morphologically distinct structures, are non-membrane-bound, roughly spherical, 0.5–2-μm-diameter accumulations of lipid and protein. Lipid bodies, which form in a wide diversity of cells (11), including neutrophils (12) and eosinophils (13), are characteristically sparse in normal cells but increase in numbers and size in cells associated with inflammation. Normal blood neutrophils contain an average of <1 lipid body/cell, whereas eosinophils from normal donors routinely contain more lipid bodies, an average of ~ 5 lipid bodies/cell (12–14).

The solubility of lipid bodies in conventional alcohol-based hematological stains, e.g., Wright's and Giemsa stains, causes lipid bodies to be dissolved and missed. In contrast, if lipid is first preserved by exposure to osmium-tetroxide prior to staining or if staining is effected with the lipophilic fluorescent dye Nile red, then lipid body accumulations within leukocytes can be recognized and enumerated. With such specific staining, lipid body numbers can be recognized by light or electron microscopy to be increased in leukocytes associated with various inflammatory and immunological reactions (11). Eosinophil lipid bodies have an ultrastructural morphology similar to those found within neutrophils but are characteristically more electron dense following osmium staining, like lipid bodies in mast cells (12,13,15). Whether the heightened osmiophilic nature of eosinophil lipid bodies reflects preferential associations of specific types or quantities of osmiophilic phospholipids or proteins with these structures is not known. While lipid bodies lack a delimiting membrane, they often possess a more electron-dense peripheral shell and are found enmeshed in cytoskeletal elements (14). Eosinophil lipid bodies, with their dark osmiophilic staining, their comparable size to granules, and their heightened numbers in activated eosinophils, have at times been misidentified as cytoplasmic granules.

In human eosinophils, lipid bodies can serve as sites of deposition of arachidonyl-lipids. Ultrastructural autoradiographic studies by Weller and Dvorak localizing cell-incorporated ^{3}H-arachidonate within eosinophils demonstrated that lipid bodies are a predominant intracellular site of deposition of ^{3}H-arachidonate (13), as also found in the studies of Dvorak and her colleagues in other cells (11,12,16–18). The ^{3}H-arachidonate present within lipid bodies was not free fatty acid, as indicated by parallel analyses of total cellular ^{3}H-arachidonyl-lipids (13). Incorporated ^{3}H-arachidonate was not present as free fatty acid, but rather was esterified within glycerolipids, with most ^{3}H-arachidonate present in classes of phospholipids. By inference, lipid bodies would have contained esterified arachidonate. More direct evidence that eosinophil lipid bodies contain pools of arachidonyl-phospholipids was obtained after methods to successfully isolate eosinophil lipid bodies free of cellular membranes were developed (15). From eosinophils previously incubated overnight with ^{3}H-arachidonic acid to achieve isotopic equilibrium, the ^{3}H-arachidonate-containing lipids from isolated lipid bodies were principally phospholipids, with ^{3}H-arachidonate present in phosphatidylcholine, phosphatidylinositol, and phosphatidylethanolamine/serine in the same proportions as found in eosinophil membranes (Table 1) (15). In recovered lipid bodies, little ^{3}H-arachidonate was present as free fatty acid; some ^{3}H-arachidonate was found in triglycerides and a larger amount was present in diglycerides, possibly due to the actions of lipolytic enzymes occurring during lipid body isolation. Although it was possible to recover eosinophil lipid bodies free of contaminating cellular membranes (15), it has not been possible to quantitate the recovery of lipid bodies, because no protein marker for this structure has been identified, as utilized to monitor the distribution and recovery of other

Table 1 ^{3}H-Arachidonate Incorporation into Lipids of Isolated Eosinophil Lipid Bodies and Microsomal Membranes

		Total lipid classes (%)					% of total phospholipid			
	^{3}H-CPM	DG	TG	AA	Chol ester	PL	PI	PC	PE/PS	Late
Lipid bodies	387,650	2 ± 1	10 ± 4	5 ± 0	0 ± 0	58 ± 4	28 ± 3	32 ± 2	32 ± 2	8 ± 2
Membranes	2,252,950	6 ± 0	5 ± 2	1 ± 0	0 ± 0	88 ± 2	29 ± 2	36 ± 0	33 ± 2	2 ± 0

DG, diglycerides; TG, triglycerides; AA, arachidonic acid; Chol ester, cholesterol ester; PL, phospholipids; PT, phosphatidyl inositol; PC, phosphatidyl choline; PE/PS, phosphatidyl ethanolamine/phosphatidyl serine; Late, Late-clotting minor lipids.
Source: Adapted from Ref. 15.

cellular organelles during subcellular isolations. Nevertheless, eosinophil lipid bodies clearly constitute a structurally distinct pool of potential eicosanoid substrate containing arachidonyl-phospholipids.

B. Eosinophil Lipid Bodies as Sites of Enzyme Localization

Specific enzymes can be associated with lipid bodies, as indicated by histochemical studies localizing enzymes with nonspecific esterase and peroxidase activities to lipid bodies within eosinophils (18). Eosinophil lipid bodies have peroxidase activity, possibly, in part at least, due to peroxidase activity of an enzyme involved in eicosanoid synthesis (e.g., PGH synthase). Esterase and alkaline phosphatase activities have also been demonstrated at eosinophil lipid bodies (15,18). More directly pertinent to the processes involved in eicosanoid formation, by studying the localization of PGH synthase in various cells, including fibroblasts, monocytes, and eosinophils, we have now established that PGH synthase is localized at lipid bodies. Such localization can be demonstrated by immunohistochemical, immunogold ultrastructural, and indirect immunofluorescent localizations within intact, permeabilized eosinophils and by demonstrating immunochemically and by direct enzyme assay that PGH synthase is associated with isolated lipid bodies (19,20). Thus, lipid bodies appear to serve as distinct intracellular domains at which both pools of substrate containing arachidonyl-phospholipids and key eicosanoid-forming enzymes are localized.

C. Lipid Body Formation

Lipid bodies are prevalent in vivo in cells associated with various immune and inflammatory responses and are prominent in eosinophils from patients with eosinophilia, including blood and tissue eosinophils in those with the idiopathic hypereosinophilic syndrome and lesional or tissue eosinophils in varied conditions and responses (18). In vitro, lipid body formation can be elicited within 15–30 min by exposures of cells to *cis*-polyunsaturated fatty acids, including arachidonic acid (12,14). That this lipid body formation is not a manifestation of cellular injury or simply attributable to excess substrate arachidonyl-fatty acid has been indicated by showing that lipid body formation is stereochemically dependent on the structures of exogenous fatty acids (14). Unsaturated fatty acids with *trans* geometries of their double bonds are not effective stimuli of lipid body formation, although they enter cells and may be localized within lipid bodies (12). Rather unsaturated fatty acids with *cis* geometries to their double bonds are effective with potencies as lipid body inducers correlated with increasing numbers of *cis* double bonds. These *cis*-unsaturated fatty acids can stimulate protein kinase C, and other protein kinase C activators, including specific phorbol esters and cell-permeant diglycerides, also promote lipid body formation (14). Inhibitors of protein kinase

C, including 1-hexadecyl-2-O-methyl-glycerol, H-7 and staurosporine, inhibit lipid body formation elicited by *cis*-fatty acids, diglycerides, and phorbol esters (14). These findings indicate that protein kinase C activation is involved in lipid body formation and that the role of protein kinase C is not simply to contribute to the intracellular release of fatty acids that themselves cause lipid body formation. Thus, lipid body formation represents a coordinated cellular response, mediated by protein kinase C activation, that eventuates in the deposition of lipid and proteins in discrete, ultrastructurally defined intracellular domains (14).

With their prominence in eosinophils in association with inflammation, lipid bodies constitute specialized sites at which eicosanoid formation could occur for the heightened generation of eosinophil eicosanoid mediators of inflammation. This compartmentalization of eicosanoid formation at lipid bodies would provide a nonmembrane pool of arachidonate whose metabolic utilization could occur without perturbation of membranes if membranes were the sole stores of substrate fatty acid utilized for quantities of eicosanoids synthesized as paracrine mediators of inflammation. Moreover, lipid bodies would serve as sites at which the coordinated and regulated enzymatic events involved in arachidonate mobilization and oxidative metabolism could occur.

III. EOSINOPHIL-LYMPHOCYTE INTERACTIONS

A. Eosinophil CD4 Function and Lymphocyte Chemoattractant Factor

CD4, in addition to being present on helper T lymphocytes, is expressed on some cells of myeloid lineage, including monocytes and macrophages. We previously demonstrated that human eosinophils express CD4 (21). Human eosinophils express CD4 demonstrable by flow cytometry, at levels lower than on T helper cells (21). Since CD4 is expressed on eosinophils from normal donors as well as from hypereosinophilic donors, CD4 expression is not restricted to those eosinophils with an "activated" phenotype, as found in the blood of eosinophilic patients (2). Purified eosinophils, when cultured with 3T3 fibroblasts and granulocyte-macrophage colony stimulating factor (GM-CSF) and pulse-labeled ^{35}S-methionine/cysteine, synthesized immunoprecipitable ^{35}S-labeled CD4 (21). Thus, mature, blood-derived human eosinophils express CD4 and retain the capacity to synthesize new CD4 protein.

Three different ligands that interact with CD4 were used to investigate the functions of CD4 on eosinophils (22). The first CD4-binding ligand was the lymphokine lymphocyte chemoattractant factor (LCF), a 56-kDA homotetrameric glycoprotein elaborated by CD8+ human T lymphocytes (23,24). One agonist that stimulates the synthesis and release of LCF from T cells is histamine, acting via H2 receptors (25). LCF acts on CD4+ T lymphocytes and CD4+ monocytes (23) to stimulate cellular migration, enhance expression of class II major histocompatibility complex (MHC) antigens, increase interleukin-2 (IL-2) receptor expression and elicit fluxes in the intracellular messengers calcium and inositol phosphates (23–27). Several lines of evidence indicate that CD4 on T lymphocytes and monocytes is the cell receptor for LCF. First, all LCF-induced functional responses are blocked by Fab fragments of the anti-CD4 monoclonal antibody OKT4. Second, LCF binds to CD4 affinity columns. Third, anti-CD4 antibody (OKT4) inhibits LCF binding to CD4+ cells. Fourth, all LCF functional responses are mediated by cloned recombinant LCF, and finally, responsiveness to LCF is conferred by transfection of human CD4 into recipient cells (23,24).

LCF is a chemoattractant for CD4+ human eosinophils, as it is for CD4+ T cells and monocytes (22). Eosinophil chemotaxis was assayed by enumerating populations of eosinophils migrating at several depths within micropore filters (28). LCF uniformly stimulated eosinophil migration with eosinophils derived from both normal and eosinophilic donors. With standard checkerboard analyses, LCF was both chemokinetic and chemotactic for eosinophils (22). The concentrations of LCF eliciting half-maximal chemoattractant responses, the ED_{50}, ranged from about 10^{-11} to 10^{-12} M (22). To evaluate the potency of LCF as an eosinophil chemoattractant, the migration of eosinophils in response to LCF was compared to that elicited by PAF and the complement fragment C5a. PAF and C5a are potent eosinophil chemoattractants (29), although each also stimulates neutrophil migration. LCF, in studies with eosinophils from several donors, was 100- to 1000-fold more potent as an eosinophil chemoattractant than either C5a or PAF (22). LCF did not stimulate migration of neutrophils, which lack CD4 expression. Hence, LCF differed from C5a and PAF in being much more potent as an eosinophil chemoattractant and in being specific for eosinophils in comparison with neutrophils (22).

Two other CD4-binding ligands that stimulate migration of CD4+ mononuclear leukocytes (23,30) also stimulated migration of eosinophils (22). The anti-CD4 monoclonal antibody, OKT4 (and not control irrelevant or cell-surface class I MHC-directed antibodies), specifically stimulated eosinophil migration (22). Likewise recombinant HIV gp120, which binds to CD4 on eosinophils (21), stimulated eosinophil migration (22). Second, CD4 blocking antibodies (Fab fragments of OKT4, which themselves did not stimulate eosinophil migration) competitively inhibited eosinophil migration elicited by LCF (22). Thus, LCF-induced eosinophil migration is mediated by CD4 expressed on eosinophils.

Although CD4 expression on eosinophils is low in comparison to that on helper T lymphocytes and overlaps background fluorescence of eosinophils analyzed alone or after staining with IgG subclass control myeloma proteins (21), two experimental findings indicate that all eosinophils express functional CD4 (22). First, the pattern of CD4 expression found on eosinophils from all normal and eosinophilic donors, examined after staining with various anti-CD4 monoclonal antibodies, consistently yielded a single unimodal peak of CD4 staining with no bimodal pattern suggestive of separate CD4+ and CD4− subpopulations of eosinophils (21,22). Second, analysis of eosinophils migrating at different levels into chemotactic filters in response to LCF provided no evidence that only subpopulations of eosinophils were responding to LCF (22). Moreover, in assays with eosinophils from six different donors, the net numbers of eosinophils migrating in response to LCF were comparable to the net eosinophil migration elicited by PAF and C5a. Since receptors for C5a (31) and PAF are expressed in large numbers on all eosinophils, the comparable migration of eosinophils elicited by LCF and by C5a or PAF further indicated that functional CD4 is expressed on all, or virtually all, human eosinophils.

While LCF was a potent eosinophil chemoattractant, LCF did not stimulate some other effector functions of eosinophils (22). LCF itself neither stimulated eosinophil respiratory burst activity nor did it prime for heightened respiratory burst activity in response to other agonists. LCF did not cause eosinophil degranulation. Further, LCF neither directly stimulated eosinophil leukotriene C_4 release nor did it enhance the capacity of eosinophils to generate leukotriene C_4 in response to stimulation by submaximal concentrations of calcium ionophore A23187 (22).

Thus, CD4 expressed on human eosinophils functions as a signal-transducing transmembrane protein, and a natural ligand binding to CD4, namely LCF, is capable of stimulating eosinophil migration (22).

B. Interleukin-2 and Eosinophil Function

IL-2 receptors on lymphocytes and monocytes are composed of at least two noncovalently associated subunits, p55 (CD25) and p75. Association of both subunits is required to constitute a high-affinity IL-2 receptor, whereas individually, p55 and p75 bind IL-2 with only low and intermediate affinities, respectively. Eosinophils express CD25 detectable by flow cytometry (32–34), and CD25 is present on eosinophils without requiring in vitro activation (32). Northern blotting has established that eosinophils contain mRNA transcripts for IL-2 receptor component p55 (32). In contrast, binding of anti-p75 antibody to eosinophils was not detectable by flow cytometry, and in IL-2 ligand-binding assays, high-affinity IL-2 receptors were below the limits of detection on eosinophils from individuals with eosinophilia (32).

Eosinophils migrate in response to IL-2, and recombinant IL-2 was a potent chemoattractant for eosinophils with an ED_{50} for migration of about 10^{-12} M (32). Although the p55, but not the p75, subunit of the IL-2 receptor was detectable on eosinophils, two findings indicate that migratory responses of eosinophils are apparently mediated by high-affinity IL-2 receptors. First, the molar potency of

IL-2 as an eosinophil chemoattractant, with responses elicited by 10^{-12} M IL-2, is compatible with responses being mediated by a high-affinity receptor (32). Second, IL-2-induced eosinophil migration is competitively inhibited by monoclonal antibodies to either the p55 or p75 IL-2 receptor subunits (32). IL-2, which is principally chemokinetic and not chemotactic for eosinophils, is about 100- to 1000-fold more potent as an eosinophil chemoattractant than either PAF or C5a. Thus, IL-2 constitutes another lymphokine with potent activity in eliciting eosinophil migration.

C. Eosinophil Class II MHC-Dependent Antigen Presentation and Lymphocyte Stimulation

Although immature eosinophils developing within the marrow, like other immature myeloid cells, express class II MHC proteins, peripheral blood eosinophils express little of the class II MHC proteins (35). Even phenotypically activated eosinophils, obtained from the blood of patients with eosinophilia, do not display elevated amounts of these MHC proteins (35). Mature eosinophils, however, can be induced to express the class II MHC protein HLA-DR when HLA-DR$^-$, peripheral blood-derived eosinophils are maintained in culture with 3T3 fibroblasts and GM-CSF (35). This increased cell-surface expression of HLA-DR was not attributable solely to mobilization of latent, intracellular pools of HLA-DR, since eosinophils, cultured at ~100% purity with GM-CSF and 3T3 fibroblasts, synthesize new ^{35}S-methionine/cysteine-labeled heterodimeric HLA-DR. Thus, mature eosinophils have the capacity to synthesize and express HLA-DR (35).

To determine if eosinophil HLA-DR expression is functional, we investigated whether eosinophils could serve as antigen-presenting cells in stimulating lymphocyte-proliferative responses. Our findings indicate that eosinophils, induced in vitro to express HLA-DR by incubation with GM-CSF, can function as HLA-DR-dependent, MHC-restricted antigen-presenting cells (36). Eosinophils can process antigen and present it to the CD4+ lymphocytes in the context of the class II MHC protein HLA-DR and specifically stimulate lymphocyte proliferation. Expression of HLA-DR on eosinophils in vivo has been noted with eosinophils recovered from the sputum of asthmatics (37) and from the peritoneal cavity (38).

For optimal functioning as antigen-presenting cells, other antigen-presenting cell types also provide costimulatory signals, based on cell-surface proteins or elaborated cytokines. The capacity of human eosinophils to elaborate cytokines was first demonstrated for transforming growth factor-α (TGF-α) (39). The elaboration of IL-1-α by human eosinophils, as previously shown for murine eosinophils (40), could provide a costimulator of lymphocyte proliferation. Indeed, human eosinophils were shown to contain transcripts for IL-1-α and to contain immunochemically detectable IL-1-α protein. Thus, eosinophils have the potential capability to process and present locally encountered antigens to CD4+ lymphocytes and to initiate antigen-dependent lymphocyte responses.

IV. CONCLUSION

The functions of eosinophils in immunological reactions may be multiple, reflecting not only the diverse capabilities of eosinophils but also the heterogeneity of immunopathological alterations encountered in various eosinophil-related diseases. The effector functions of eosinophils utilize acute, potentially agonal, cellular responses, such as degranulation, oxidative burst activity, and eicosanoid release. Mechanisms underlying the regulated release of eicosanoids include the formation of intracytoplasmic lipid bodies, the association of eicosanoid-forming enzymes at these intracellular structures, and the metabolic transformation of arachidonate present within phospholipid and glycerolipid stores of lipid bodies.

The capacity of eosinophils to engage in other types of cellular responses suggests roles for eosinophils based on their capabilities to collaboratively interact with lymphocytes and other immunological and mesenchymal cells. Notably, two lymphocyte-derived cytokines are extraordinarily potent as eosinophil chemoattractants. Utilizing CD4 as a receptor, eosinophils respond to CD8+ lymphocyte-derived LCF lymphokine. Analogously, eosinophils migrate in response to IL-2. Just as mature eosinophils can respond to specific cytokines, e.g., IL-3, IL-5, and GM-CSF, with augmented effector functions, lymphokines and other cytokines may promote other activities of eosinophils. While migration responses have been used to determine the activities of these two lymphokines on eosinophils, it is possible that each of these lymphokines may modulate other, as yet unevaluated, noneffector functions of eosinophils. In addition to being able to respond to lymphocyte-derived cytokines, eosinophils may also be able to stimulate lymphocyte responses. Eosinophils can express cell surface proteins that enable them to interact with lymphocytes. HLA-DR binds to CD4 on lymphocytes, and the ability of eosinophils, following in vitro induction of HLA-DR expression, to function as antigen-presenting cells indicates that eosinophils are able to interact with CD4+ lymphocytes to elicit antigen-specific lymphocyte responses. If eosinophils exhibit the same function within tissues, eosinophils may have roles in initiating antigen-specific lymphocyte responses for those antigens which eosinophils preferentially encounter at mucosal sites or other locales. Other direct cell-cell interactions between eosinophils and other cell types are feasible based on eosinophil expression of CD4. Finally, cytokines elaborated by eosinophils could affect other cells proximate to eosinophils. The involvement of eosinophils in varied immunological reactions, therefore, may be governed by diverse mechanisms and reflect the many functional capabilities of eosinophils.

V. SUMMARY

Eosinophils have multiple functional capabilities. Eosinophils can oxidatively metabolize arachidonic acid to form eicosanoids derived from the cyclooxygenase and lipoxygenase pathways. Intracellular inclusions, termed lipid bodies, are prominent in eosinophils associated with inflammation. Lipid bodies have roles as sites of eicosanoid formation. Eosinophils also respond to and migrate to the lymphokines, lymphocyte chemoattractant factor, and IL-2, utilizing CD4 and apparent high-affinity IL-2 receptors for each, respectively. Eosinophils also can function as antigen-presenting cells to stimulate lymphocyte responses. Thus, eosinophils have the capacity to interact collaboratively with lymphocytes and to elaborate cytokines and other mediators pertinent to the multiple activities of eosinophils in immunological reactions.

REFERENCES

1. Beeson PB, Bass DA. The Eosinophil. Philadelphia: Saunders, 1977.
2. Weller PF. The immunobiology of eosinophils. N Engl J Med 1991; 324:1110–1118.
3. Irvine RF. How is the level of free arachidonic acid controlled in mammalian cells? Biochem J 1982; 204:3–16.
4. Rollins TE, Smith WL. Subcellular localization of prostaglandin-forming cyclooxygenase in Swiss mouse 3T3 fibroblasts by electron microscopic immunocytochemistry. J Biol Chem 1980; 255:4872–4875.
5. Smith WL. Prostaglandin biosynthesis and its compartmentalization in vascular smooth muscle and endothelial cells. Annu Rev Physiol 1986; 48:251–262.
6. Merlie JP, Fagan D, Mudd J, Needleman P. Isolation and characterization of the complementary DNA for sheep seminal vesicle prostaglandin endoperoxide synthase (cyclooxygenase). J Biol Chem 1988; 263:3550–3553.
7. DeWitt DL, Smith WL. Primary structure of prostaglandin G/H synthase from sheep vesicular gland determined from the complementary DNA sequence. Proc Natl Acad Sci USA 1988; 85:1412–1416.
8. Rouzer CA, Kargman S. Translocation of 5-lipoxygenase to the membrane in human leukocytes challenged with ionophore A23187. J Biol Chem 1988; 263:10980–10988.
9. Dahlen S-E, Franzen L, Raud J, Serhan CN, Westlund P, Wikstrom E, Bjorck T, Matsuda H, Webber SE, Veale CA, Puustinen T, Haeggstrom J, Nicolaou KC, Samuelsson B. Actions of lipoxin A_4 and related compounds in smooth muscle preparations and on the microcirculation in vivo. In: Wong P-YK, Serhan CN, eds. Lipoxins: Chemistry, Biosynthesis and Biological Activities. New York: Plenum Press, 1988:107–130.
10. Dixon RA, Diehl RE, Opas E, Rands E, Vickers PJ, Evans JF, Gillard JW, Miller DK. Requirement of a 5-lipoxygenase-activating protein for leukotriene synthesis. Nature 1990; 343:282–284.
11. Galli SJ, Dvorak AM, Peters SP, Schulman ES, MacGlashan DW Jr., Isomura T, Pyne K, Harvey VS, Hammel I, Lichtenstein LM, Dvorak HF. Lipid bodies: widely

distributed cytoplasmic structures that represent preferential non-membrane repositories of exogenous [^{3}H]-arachidonic acid incorporated by mast cells, macrophages and other cell types. In: Bailey JM, ed. Prostaglandins, Leukotrienes, and Lipoxins. New York: Plenum Press, 1985:221–239.

12. Weller PF, Ackerman SJ, Nicholson-Weller A, Dvorak AM. Cytoplasmic lipid bodies of human neutrophilic leukocytes. Am J Pathol 1989; 135:947–959.
13. Weller PF, Dvorak AM. Arachidonic acid incorporation by cytoplasmic lipid bodies of human eosinophils. Blood 1985; 65:1269–1274.
14. Weller PF, Ryeom SW, Picard ST, Ackerman SJ, Dvorak AM. Cytoplasmic lipid bodies of neutrophils: formation induced by *cis*-unsaturated fatty acids and mediated by protein kinase C. J Cell Biol 1991; 113:137–146.
15. Weller PF, Monahan-Earley RA, Dvorak HF, Dvorak AM. Cytoplasmic lipid bodies of human eosinophils: subcellular isolation and analysis of arachidonate incorporation. Am J Pathol 1991; 138:141–148.
16. Dvorak AM, Dvorak HF, Peters SP, Schulman ES, MacGlashan DW Jr., Pyne K, Harvey VS, Galli SJ, Lichtenstein LM. Lipid bodies: cytoplasmic organelles important to arachidonate metabolism in macrophages and mast cells. J Immunol 1983; 131:2965–2976.
17. Dvorak AM, Hammel I, Schulman ES, Peters SP, MacGlashan DW Jr., Schleimer RP, Newball HH, Pyne K, Dvorak HF, Lichtenstein LM, Galli SJ. Differences in the behavior of cytoplasmic granules and lipid bodies during human lung mast cell degranulation. J Cell Biol 1984; 99:1678–1687.
18. Dvorak AM, Ackerman SJ, Weller PF. Subcellular morphology and biochemistry of eosinophils. In: Harris JR, ed. Blood Cell Biochemistry: Megakaryocytes, Platelets, Macrophages and Eosinophils. London: Plenum Press, 1990:237–344.
19. Weller PF, Ryeom SW. Lipid bodies: focal, non-membranous intracellular domains for eicosanoid formation. Submitted.
20. Dvorak AM, Morgan E, Schleimer RP, Ryeom SW, Lichtenstein LM, Weller PF. Ultrastructural immunogold localization of prostaglandin endoperoxide synthase (cyclooxygenase) to nonmembrane-bound cytoplasmic lipid bodies in human lung mast cells, alveolar macrophages, type II pneumocytes and neutrophils. J Histochem Cytochem 1992; 40:759–769.
21. Lucey DR, Dorsky DI, Nicholson-Weller A, Weller PF. Human eosinophils express CD4 protein and bind HIV-1 GP120. J Exp Med 1989; 169:327–332.
22. Rand TH, Cruikshank WW, Center DM, Weller PF. CD4-mediated stimulation of human eosinophils: lymphocyte chemoattractant factor and other CD4-binding ligands elicit eosinophil migration. J Exp Med 1991; 173:1521–1528.
23. Cruikshank WW, Berman JS, Theodore AC, Bernardo J, Center DM. Lymphokine activation of T4+ lymphocytes and monocytes. J Immunol 1987; 138:3817–3823.
24. Cruikshank WW, Berman JS, Theodore AC, Bernardo J, Center DM. Lymphocyte chemoattractant factor (LCF) induces CD4-dependent intracytoplasmic signalling in lymphocytes. J Immunol 1991; 146:2928–2934.
25. Center DM, Cruikshank WW, Berman JS, Beer DJ. Functional characteristics of histamine receptor-bearing mononuclear cells. I. Selective production of lymphocyte chemoattractant lymphokine utilizing histamine as a ligand. J Immunol 1983; 131: 1854–1859.

26. Berman JS, Beer DJ, Cruikshank WW, Center DM. Chemoattractant lymphokines specific for the helper/inducer T-lymphocyte subset. Cell Immunol 1985; 95:105–112.
27. Center DM, Cruikshank W. Modulation of lymphocyte migration by human lymphokines I. Identification and characterization of chemoattractant activity for lymphocytes from mitogen-stimulated mononuclear cells. J Immunol 1982; 128:2569–2571.
28. McCrone EL, Lucey DR, Weller PF. Fluorescent staining for leukocyte chemotaxis: eosinophil-specific staining with aniline blue. J Immunol Methods 1988; 114:79–88.
29. Wardlaw AJ, Moqbel R, Cromwell O, Kay AB. Platelet activating factor. A potent chemotactic and chemokinetic factor for human eosinophils. J Clin Invest 1986; 78: 1701–1706.
30. Kornfeld H, Cruikshank WW, Pyle SW, Berman JS, Center DM. Lymphocyte activation by HIV-I envelope glycoprotein. Nature 1988; 335:445–448.
31. Gerard NP, Hodges MK, Drazen JM, Weller PF, Gerard C. Characterization of a receptor for C5a anaphylatoxin on human eosinophils. J Biol Chem 1989; 264:1760–1766.
32. Rand TH, Silberstein DS, Kornfeld H, Weller PF. Human eosinophils express functional interleukin 2 receptors. J Clin Invest 1991; 88:825–832.
33. Plumas J, Gruart V, Aldebert D, Truong MJ, Capron M, Capron A, Prin L. Human eosinophils from hypereosinophilic patients spontaneously express the p55 but not the p75 interleukin 2 receptor subunit. Eur J Immunol 1991; 21:1265–1270.
34. Riedel D, Lindemann A, Brach M, Mertelsmann R, Hermann F. Granulocyte-macrophage colony-stimulating factor and interleukin-3 induce surface expression of interleukin-2 receptor p55-chain and CD4 by human eosinophils. Immunology 1990; 70:258–261.
35. Lucey DR, Nicholson-Weller A, Weller PF. Mature human eosinophils have the capacity to express HLA-DR. Proc Natl Acad Sci USA 1989; 86:1348–1351.
36. Weller PF, Rand TH, Finberg RW. Human eosinophils function as HLA-DR dependent, MHC-restricted antigen-resenting cells FASEB J 1991; 5:A640.
37. Hansel TT, Braunstein JB, Walker C, Balser K, Bruijnzeel PLB, Virchow JC Jr., Virchow C. Sputum eosinophils from asthmatics express ICAM-1 and HLA-DR. Clin Exp Immunol 1991; 86:271–277.
38. Roberts RL, Ank BJ, Salusky IB, Stiehm ER. Purification and properties of peritoneal eosinophils from pediatric dialysis patients. J Immunol Methods 1990; 126: 205–211.
39. Wong DTW, Weller PF, Galli SJ, Rand TH, Elovic A, Chiang T, Chou MY, Gallagher GT, Matossian K, McBride J, Todd R. Human eosinophils express transforming growth factor α. J Exp Med 1990; 172:673–681.
40. del Pozo V, de Andres B, Martin E, Maruri N, Zubeldia JM, Palomino P, Lahoz C. Murine eosinophils and IL-1: αIL-1 mRNA detection by in situ hybridization. Production and release of IL-1 from peritoneal eosinophils. J Immunol 1990; 144: 3117–3122.

DISCUSSION (Speaker: P. Weller)

Konig: Is 5-lipoxygenase biologically active in lipid bodies? Does one see an interaction with FLAP? How is the 5-LO translocated from the lipid body to the membrane-bound FLAP?

Weller: The 5-LO, like the 15-LO, associated with lipid bodies is biologically active as evidenced by the finding that lipid bodies, isolated from human eosinophils by subcellular fractionation, express catalytically active 5- and 15-LO activities in appropriate assays with exogenous arachidonic acid. A possible association of FLAP with lipid bodies or with lipid-body-associated 5-LO remains to be evaluated, as does any potential translocation.

Gleich: Do peripheral blood eosinophils from normal persons express HLA-DR and CD4?

Weller: Eosinophils from the blood of normal and eosinophilic donors generally express little HLA-DR as quantitated by flow cytometry (1). Recent findings, however, that only several hundred HLA-DR molecules may need to be expressed for functional antigen presentation might make it feasible for eosinophils with even low-level HLA-DR expression to function as antigen-presenting cells (2,3). CD4 is definitely present on normal eosinophils. While the levels of CD4 on eosinophils are low compared to CD4+ T cells, the uniform capacity of eosinophils to migrate in response to LCF indicates that all eosinophils express functional CD4 (4).

Kay: How does LCF compare with PAF or C5a in relation to priming or other eosinophil functions? Also, does it recruit eosinophils when administered in vivo?

Weller: LCF does not directly stimulate, or prime for enhanced, degranulation, oxidative burst activity, or LTC_4 generation (4). We have not tested LCF in vivo.

Denburg: What is the role of CD4 on eosinophils in AIDS? Is there evidence from the pathology in AIDS that eosinophils may play an active role?

Weller: We have shown that mature CD4+, peripheral blood-derived eosinophils can be infected with HIV (5). Such infection in vitro can be associated with cytotoxicity for eosinophils. Thus, one might speculate that HIV-induced eosinophil cytotoxicity might cause release of toxic eosinophil proteins. Perhaps such cytotoxicity might contribute to organ dysfunction, perhaps in the gastrointestinal (GI) tract, contributing to the noninfectious component of HIV-associated gastroenteropathy.

Spry: In answer to Dr. Denburg's question I should like to draw your attention to our work on CD4 expression on marrow eosinophils and HIV infection of marrow eosinophils (6). The study showed that CD4 was expressed most strongly on bone marrow eosinophils and decreased in culture. Productive HIV infection was achieved 14 days after incubation of cultured normal marrow.

I have a naïve view that lipid bodies are sites of phospholipid retention in cells that require a mechanism to alter their surface area/volume ratio, in order to move and change shape. Is there any work to support this view? In relation to the capacity of eosinophils to present antigens, which you have shown so well, how do eosinophils compare with other "professional" antigen-presenting cells.

Weller: I suspect that lipid bodies may serve several functions, in addition to participating as domains for eicosanoid formation. Lipid bodies, for instance, participate as storage sites for cholesterol esters in some cell types. We do not have any evidence relating lipid bodies to membrane remodeling, although such might be possible.

Eosinophils in our in vitro assays based on presentation of a specific antigen, tetanus toxoid, were as effective as peripheral blood monocytes (7). Otherwise, we have no data to compare eosinophils with other antigen-presenting cells. I speculate that the potential role of eosinophils as antigen-presenting cells might be related to the distribution or localization of the eosinophils, such as in submucosal sites of the respiratory or GI tracts where eosinophils might encounter foreign antigens, or in sites of granulomatous inflammation surrounding helminthic parasites.

Hanel: Could you comment on the greater potency of GM-CSF compared with IFN-γ for the induction of HLA-DR on eosinophils?

It is exciting that eosinophil antigen presentation was comparable with that of monocytes. However, do eosinophils and monocytes express equivalent amounts of HLA-DR?

Weller: In our hands so far, GM-CSF has stimulated a greater proportion of eosinophils to express HLA-DR than has IFN-γ. The differences, as assessed by flow cytometry, are not great. Levels of HLA-DR on eosinophils are lower than on monocytes (1), but low-level expression of class II MHC proteins is not necessarily required for effective antigen presentation.

Barnes: Several priming stimuli increase lipid mediator secretion in eosinophils. Do you see evidence of increased lipid bodies and increased PGH synthase/5-LO/15-LO immunoreactivity after such stimuli? Are similar increases seen in hypodense eosinophils?

Weller: Certain of these priming stimuli elicit lipid body formation. Arachidonic acid, other *cis* polyunsaturated fatty acids, cell permeant diglycerides, and specific phorbol esters all promote lipid body formation, by means of PKC activation. Immunoreactivity for the several eicosanoid-forming enzymes is associated with the lipid bodies, as judged by immunocytochemistry. Quantitation of these enzymes has not yet been done. Hypodense eosinophils do have more lipid bodies, as judged by evaluation of eosinophils isolated from eosinophilic donors and resolved on Percoll gradients.

Coffman: Do you have evidence that eosinophil granule proteins interfere with the APC ability of eosinophils? Is this why you fix the cells?

Weller: Paraformaldehyde fixation of eosinophils was done for two reasons. First, as shown by Moreno and Lipsky (8), such fixation of mononuclear antigen-presenting cells can block mixed-leukocyte-type reactions, allowing studies of the MHC restriction of eosinophil antigen presentation. Second, fixation was done because eosinophils during coculture with lymphocytes, as done with eosinophil-sustaining cytokines, died and did not act as APCs. Such death of eosinophils during coculture with lymphocytes did not suppress basal incorporation of ^{3}H-thymidine into lymphocytes and hence was probably not significantly cytotoxic to lymphocytes under the culture conditions employed.

Lopez: In our experience, IL-3, IL-5, and GM-CSF tend to have the same spectrum of activity, yet you have shown that in your system GM-CSF, but not IL-3 or IL-5, enhances HLA-DR expression on eosinophils. Is this surprising and are there other systems where only one of these factors can be stimulatory on eosinophils?

Weller: We do not have an explanation why GM-CSF, and not the same concentrations of IL-3 or IL-5, stimulated eosinophil HLA-DR expression. GM-CSF will stimulate class II MHC expression on mononuclear leukocytes. Higher concentrations of IL-3 will stimulate eosinophil HLA-DR expression, so the differences may be due to differences in effective concentrations.

Abrams: Any evidence that LCF may be produced differentially by CD8 subsets? Any evidence from cloned CD8 T cells?

Weller: The molecular probes for LCF will allow a more precise evaluation of the cell types capable of elaborating LCF. These studies have become feasible only recently, so data on whether LCF production is restricted to CD8+ cells, or subsets, are not yet available.

M. Capron: How do you explain the inhibitory effects of anti-p75 IL-2R antibodies, since no p75 is detected in eosinophils? Did you use the more sensitive in situ hybridization with the p75 cDNA probe?

Weller: I suspect that the p75 component of the IL-2R is present on eosinophils but at levels that are not discernible by flow cytometry above the baseline fluorescence of eosinophils. We have not done in situ hybridization with a p75 probe.

Schleimer: If you modulate lipid body number with PKC activators, do you see a corresponding change in cyclooxygenase activity? Also, do steroids, which have been reported to modify cyclooxygenase activity in other cell types, alter lipid body number of cyclooxygenase levels in eosinophils?

Weller: PKC activation will prime cells, including eosinophils, for heightened lipoxygenase pathway activity, but we have not yet tested these for increased cyclooxygenase activity, nor have we as yet evaluated the effects of corticosteroids.

DISCUSSION REFERENCES

1. Lucey DR, Nicholson-Weller A, Weller PF. Mature human eosinophils have the capacity to express HLA-DR. Proc Natl Acad Sci USA 1989; 86:1348–1351.
2. Harding CV, Unanue ER. Quantitation of antigen-presenting cell MHC class II/peptide complexes necessary for T-cell stimulation. Nature 1990; 346:574–576.
3. Demotz S, Grey HM, Sette A. The minimal number of class II MHC-antigen complexes needed for T cell activation. Science 1990; 249:1028–1030.
4. Rand TH, Cruickshank WW, Center DM, Weller PF. CD4-mediated stimulation of human eosinophils: lymphocyte chemoattractant factor and other CD4-binding ligands elicit eosinophil migration. J Exp Med 1991; 173:1521–1528.
5. Weller PF, Lucey DR, Finberg RW. HIV-1 infection of human eosinophils. Clin Res 1989; 27:568A.
6. Freedman AR, Gibson FM, Fleming SC, Spry CJ, Griffin GE. Human immunodeficiency virus infection of eosinophils in human bone marrow culture. J Exp Med 1991; 174:1661–1664.
7. Weller PF, Rand TH, Barrett T, Elovic A, Wong DTW, Finberg RW. Accessory cell function of human eosinophils: HLA-DR dependent, MHC-restricted antigen-presentation and interleukin-1α expression. J Immunol 1993; 150:456–468.
8. Moreno J, Lipsky PE. Differential ability of fixed antigen-presenting cells to stimulate nominal antigen reactive and alloreactive T4 lymphocytes. J Immunol 1986; 136: 3579–3587.

15

Eosinophils and Platelet-Activating Factor

Michela Blom, Anton T. J. Tool, Arthur J. Verhoeven, and Dirk Roos
Central Laboratory of the Netherlands Red Cross Blood Transfusion Service and University of Amsterdam, Amsterdam, The Netherlands

Leo Koenderman
Academic Hospital, University of Utrecht, Utrecht, The Netherlands

I. INTRODUCTION

Eosinophilia is associated with a number of disease processes, including helminthic parasitic infections, allergic diseases, and a variety of other diseases with less defined causes (1–3). In contrast to the abundance of data describing the role of the eosinophil in inflammation, little is known about the regulation of this cell's functional activity. In this chapter, the regulatory effect of platelet-activating factor (PAF) on human eosinophils will be discussed.

PAF is a potent phospholipid mediator, secreted by many cells involved in inflammatory processes (4,5). Human eosinophils synthesize PAF upon stimulation with various stimuli, such as eosinophilic chemotactic factor of anaphylaxis (ECF-A), the chemotactic tripeptide fMLP, complement fragment C5a, the calcium ionophore A23187 (6), unopsonized zymosan (7), and IgG-Sepharose (8). PAF can directly activate (9,10) or prime the eosinophil. A cell function is said to be primed when this function is not induced by the priming agent itself, but shows enhanced activity after subsequent stimulation with a heterologous agonist. In our laboratory we have focused on the mechanism of PAF priming of responses induced by opsonized yeast particles (serum-treated zymosan, STZ) (11).

II. RESPIRATORY BURST ACTIVITY

STZ can activate the NADPH oxidase, resulting in enhanced oxygen consumption and production of toxic oxygen metabolites. The STZ-induced respiratory burst in eosinophils is almost completely blocked by monoclonal antibody (MAb) B2.12 directed against the iC3b-binding site on the complement receptor type 3 (CR3) (12). This illustrates the dominant role of the CR3 in the STZ-induced activation of the respiratory burst, despite the presence of IgG on the STZ particles.

Addition of STZ to human eosinophils isolated from the peripheral blood of normal donors (13) leads to a moderate activation of the respiratory burst, and prolonged activation is necessary before the maximal rate of oxygen consumption is achieved (Fig. 1, trace c). The enhanced oxygen consumption is almost completely inhibited by preincubation with the PAF-receptor antagonist WEB 2086 (Fig. 1, trace d). Furthermore, supernatant collected 10 min after STZ stimulation (maximal rate of oxygen consumption is reached at this time) contains PAF. STZ-stimulated eosinophils generate about 1800 pg of PAF/10^6 cells. Most

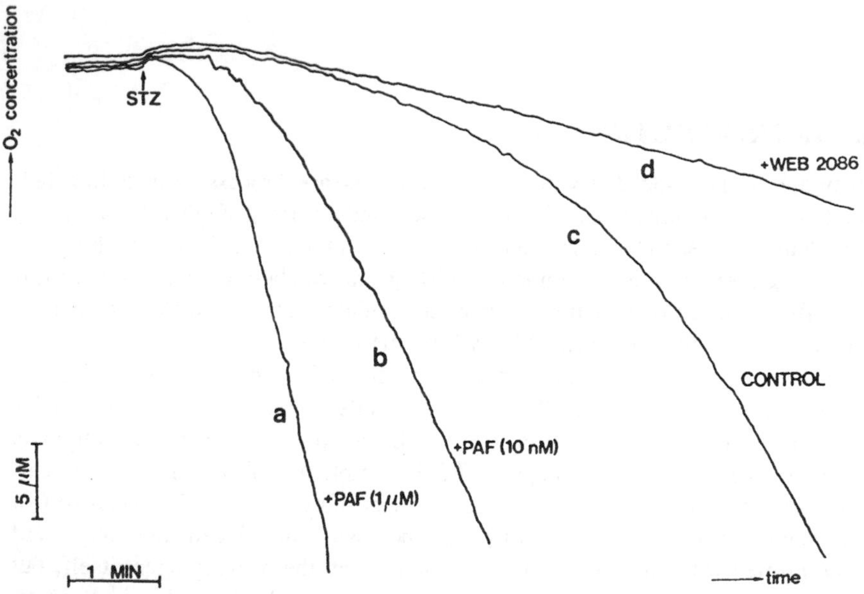

Figure 1 Influence of PAF and the PAF-receptor antagonist WEB 2086 on the respiratory burst of human eosinophils induced by STZ. After preincubation of the cells for 5 min at 37°C, either PAF (10 nM), PAF (1 μM), WEB 2086 (2.5 μM), or buffer (control) was added to human eosinophils (2 × 10^6/ml), followed by STZ (1 mg/ml) 2 min later (arrow). The results shown are representative for eight experiments. In the lower part of the figure the oxygen concentration is still >75% of the initial value. (From Ref. 14.)

of the PAF synthesized (about 65%) is not cell-associated, but is released into the extracellular medium (14). This is the first observation of a functional role of PAF during eosinophil activation with STZ. We calculated that the concentration of PAF in the extracellular medium of 2×10^6 eosinophils per ml can become as high as 10 nM. This concentration of PAF is sufficient to prime eosinophils, albeit suboptimally (Fig. 1, trace b). The supernatant of STZ-stimulated eosinophils is indeed able to prime fresh eosinophils, and this effect is again inhibited by WEB 2086 (14).

In conclusion, this study shows that upon stimulation with STZ, human eosinophils synthesize and release PAF. The release of PAF is an important factor in the ongoing activation of the respiratory burst by STZ.

III. SIGNAL TRANSDUCTION

The PAF-primed response induced by STZ is accompanied by changes in signal transduction. In untreated eosinophils, the accumulation of diglycerides (DAG) occurs rather late after stimulation with STZ, and the cytosolic free Ca^{2+} concentration remains unchanged (15). The propagation of the response is accompanied by accumulation of diglycerides. In PAF-primed eosinophils, already during the initiation phase a significant accumulation of diglycerides is observed after STZ stimulation (15). Also, a rapid increase in cytosolic free Ca^{2+} is found after STZ addition to PAF-primed eosinophils (Fig. 2). Furthermore, the changed signal transduction after priming is accompanied by an altered sensitivity for the PKC inhibitor staurosporine. In contrast to untreated cells, the initiation phase of the respiratory burst by PAF-primed cells is partially sensitive for inhibition by staurosporine. This result, together with the finding that DAG is accumulating in the first phase of the primed response to opsonized particles, illustrates that PKC participates in the signal initiation sequence only after priming with PAF.

So far, we have focused on the effect of PAF on intracellular messengers in the activation of human eosinophils. However, studies of homotypic aggregation responses and STZ binding of human eosinophils have shown some indications of important changes on the cell surface.

IV. HOMOTYPIC AGGREGATION

CR3 (CD11b/CD18) has a dual function: it binds the opsonic complement fragment iC3b, and after activation, CR3 also binds to "counterstructures" on other cells. This last phenomenon results in homotypic aggregation of eosinophils or in adherence of eosinophils to, for example, endothelial cell monolayers. Eosinophils isolated from the peripheral blood of normal donors have a low capacity to aggregate after activation with STZ. When eosinophils are primed with PAF, which by itself does not induce aggregation, the aggregation response, like the

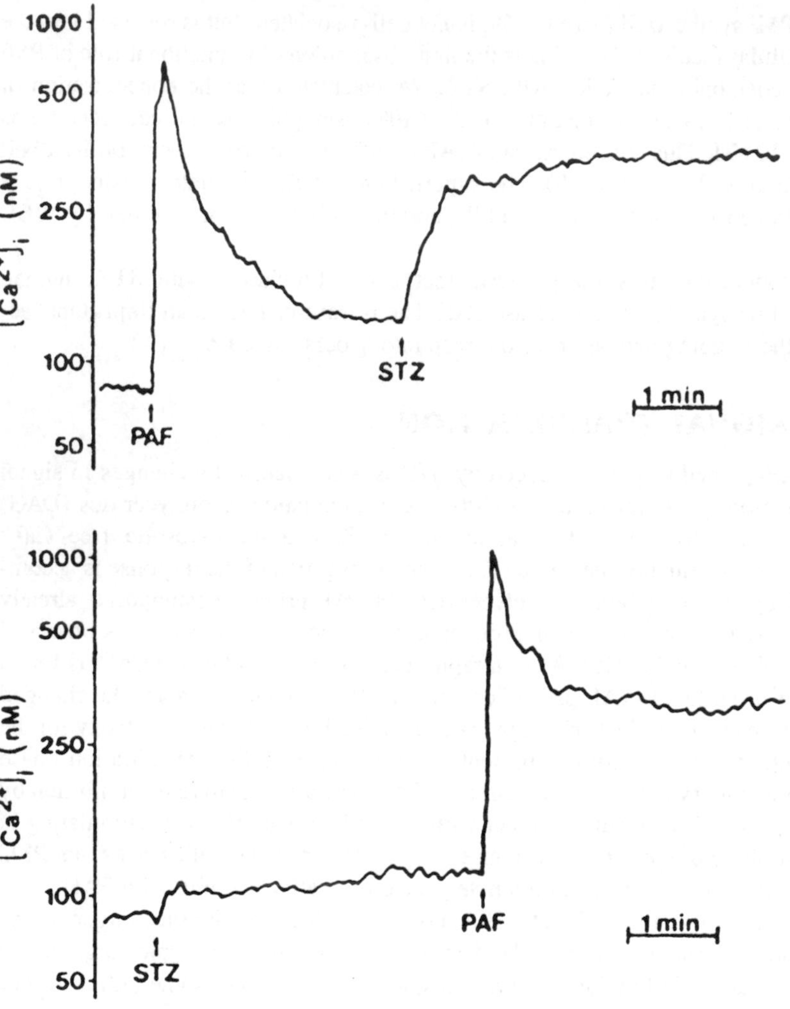

Figure 2 Effect of PAF treatment on changes in intracellular free Ca^{2+} concentration induced by STZ. After preincubation of indo-1-loaded eosinophils at 37°C, PAF (1 μM) was added and the measurement was continued. At the time indicated, STZ (1 mg/ml) was added to PAF-primed eosinophils (top) or unprimed eosinophils (bottom), and the incubation was continued. After 3 min, PAF was added to STZ-stimulated cells (bottom). The experiment shown is representative for four additional experiments. (From Ref. 15.)

respiratory burst, is markedly enhanced. The dose of 1 μM PAF, which is needed for optimal priming of the respiratory burst (Fig. 1, trace a), is also needed for a maximal priming of the homotypic aggregation induced by STZ. In contrast, the PMA-induced (100 ng/ml) aggregation is insensitive for priming (16). Furthermore, the PMA-induced aggregation is inhibited by a CD18 MAb, indicating that under these conditions the aggregation response is mediated via thc β2-integrin complex of adhesion molecules (LFA-1, CR3, and p150,95). Of the MAb against the different α chains of this family of proteins, only CD11b MAb proved to be inhibitory. These findings illustrate that the aggregation response induced by PMA is presumably mediated by a binding site on the α chain (CD11b) and the β chain (CD18) of CR3. The STZ-induced homotypic aggregation is almost *insensitive* for saturating amounts of CD18 but is inhibited in the presence of MAb B2.12 (directed against the iC3b-binding site on CR3). This finding, together with the size of the STZ particles, indicates that the STZ particles form bridges between opposing cells by binding to the ic3b-binding sites of CR3 on the cells. However, aggregation can also be activated by monovalent iC3b binding to CR3, as demonstrated by the aggregation response induced by iC3b-latex. This aggregation is enhanced by PAF priming and is, like the PMA-induced aggregation response, inhibited in the presence of CD18 (16). These results are summarized in Table 1.

The sensitivity for priming by PAF of the aggregation response induced by STZ as well as by iC3b-latex (Fig. 3), while the PMA aggregation response is not affected, indicates that the action of PAF priming is restricted to the iC3b-binding site on the CR3.

Possibly, PAF causes an intramolecular change in CR3 resulting in an affinity increase for iC3b. This hypothesis is strengthened by the finding that upregulation of CR3 during the 2-min treatment of eosinophils with PAF hardly occurs (16). To test this hypothesis, we investigated the possibility that PAF might act at an early step in the STZ-induced activation, e.g., the binding of STZ particles to CR3 on the eosinophils. For this purpose an STZ binding assay was developed.

Table 1 Effect of CD11b (B2.12), CD18, and PAF on the Homotypic Aggregation Induced by STZ Particles, iC3b-Latex, and PMA

	CD18	B2.12	PAF
PMA	Inhibits	Inhibits	No effect
STZ	No effect	Inhibits	Primes
iC3b-latex	Inhibits	Inhibits	Primes

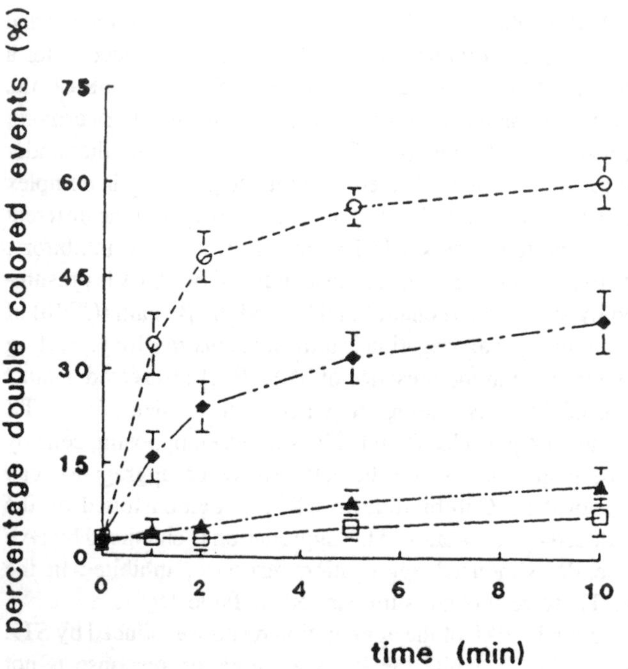

Figure 3 Effect of priming by PAF on the homotypic aggregation response of human eosinophils. Eosinophils were stained red or green fluorescent and were mixed. After 5 min preincubation of the eosinophils at 37°C, PAF (1 μM) was added and the incubation was continued for 2 min. Thereafter, STZ (1 mg/ml) or iC3b-latex (2×10^9 particles/ml) was added, and the incubation was continued for 10 min. Samples were taken and analyzed by flow cytometry. Double-colored events were scored and depicted as percentage of total colored events (16). iC3b latex-stimulated cells (□), PAF-primed, iC3b-latex-stimulated cells (♦), STZ-stimulated cells (▲), and PAF-primed, STZ-stimulated cells (○). Results are expressed as mean ± SEM of five different experiments. (From Ref. 16.)

V. INTERACTION OF STZ PARTICLES WITH EOSINOPHILS

The study of the direct interaction of STZ particles with human eosinophils has yielded further insight into the mechanism of priming by PAF. When the binding is determined by flow cytometric analysis, only a fraction of the eosinophils appears able to bind STZ particles. Treatment of the eosinophils with PAF strongly enhances the rate of particle binding and also doubles the percentage of eosinophils binding STZ (Fig. 4). This effect of PAF is strongly diminished in the presence of MAb B2.12 but is not affected by MAb IV.3 against $Fc_{\gamma}RII$ (17), indicating that the binding of iC3b to CR3 is an important step in the response of

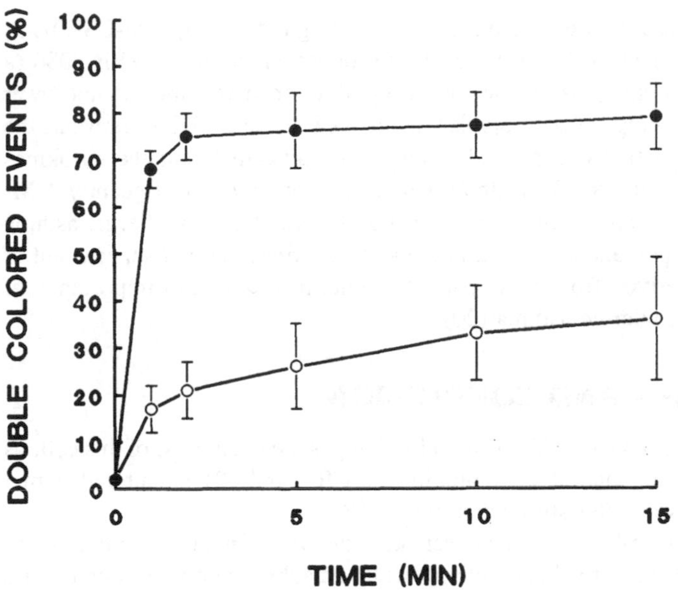

Figure 4 Effect of PAF priming on the binding of fluorescent STZ particles to human eosinophils. After 5 min preincubation of (red fluorescent) eosinophils at 37°C, PAF (1 μM) (●) or buffer (○) was added, and the incubation was continued for 2 min. Thereafter, green fluorescent STZ particles (1 mg/ml) were added and samples were taken at the times indicated. Double-colored events were measured by flow cytometry and depicted as percentage of total red events. Results are expressed as means ± SEM of five different experiments. (From Ref. 19.)

primed eosinophils to STZ. The IgG present on the STZ is not contributing, probably because the iC3b fixed on the Fc portion of the IgG is blocking the binding to $Fc_{\gamma}RII$. These experiments support the hypothesis that priming by PAF can induce the proper configuration of CR3 on human eosinophils to bind iC3b.

To couple the observed enhancement of the percentage of eosinophils binding STZ particles to a functional response, the respiratory burst activity on single-cell level was evaluated. The results obtained with the microscopic nitroblue tetrazolium test (19), in which individual cells with respiratory burst activity can be distinguished from cells lacking this activity, are in agreement with the results obtained in the STZ binding assay. Thus, priming by PAF enhances the percentage of eosinophils with respiratory burst activity (18).

STZ-induced responses in human eosinophils are primed not only by PAF, but also the cytokines granulocyte-macrophage colony-stimulating factor (GM-CSF), interleukin-3 (IL-3), and interleukin- 5 (IL-5) enhance the percentage of eosinophils binding STZ. These cytokines greatly potentiate the STZ-induced PAF synthesis and release. However, the enhanced binding induced by GM-CSF, IL-3, or IL-5 is independent of the increased PAF production, because WEB 2086 (a PAF-receptor antagonist), as well as the PLA_2 inhibitor mepacrine, do not lower the enhanced percentage of eosinophils binding STZ particles. B2.12 prevents the binding of STZ particles to primed eosinophils, indicating that the cytokines change the affinity of CR3 for iC3b (manuscript in preparation). Recently, GM-CSF, IL-3, and IL-5 have been demonstrated in the circulation of allergic asthma patients, and the spontaneous secretion of these cytokines into the supernatant of T lymphocytes derived from these patients points to the in vivo importance of these cytokines in allergic asthma (20).

VI. SUMMARY AND CONCLUSION

Priming of eosinophils with PAF profoundly changes the phenotype of this cell, as expressed by (1) an altered configuration of CR3 and (2) an altered signal transduction pathway after stimulation with STZ.

The first point is reflected by the observation that PAF enhances the number of human eosinophils that are able to interact with and subsequently respond to STZ particles. This is of great importance for the interpretation of results obtained in experiments with cell suspensions. Our experiments show that priming of eosinophil responses by PAF can, at least in part, be explained by a recruitment of cells able to respond to STZ. Furthermore, during STZ stimulation, the release of PAF by the eosinophils themselves is responsible for the participation of additional eosinophils able to respond to STZ. The PAF-receptor antagonist WEB 2086 inhibits not only the STZ-induced respiratory burst, but also the homotypic aggregation (not shown). Likewise, the percentage of unprimed eosinophils binding STZ particles is diminished in the presence of WEB 2086.

The second point is demonstrated by a PAF-induced increase in sensitivity for the PKC inhibitor staurosporine of the STZ-induced respiratory burst. Also the observation that only after PAF priming of the eosinophils is STZ able to induce a rise in cytosolic free Ca^{2+} indicates an altered signal transduction. The mechanism responsible for the change in signal transduction induced by PAF priming is not well understood. Possibly, coupling of a phospholipase(s) to CR3 is established or intensified without activation of these phospholipase(s).

ACKNOWLEDGMENTS

These studies were financially supported by Grant 900-512-079 from the Division of Health Research TNO and Grant 89.14.25 from the Netherlands Asthma Foundation. A. J. Verhoeven is a research fellow of the Royal Netherlands Academy of Sciences.

REFERENCES

1. Butterworth AE, Sturrock RF, Houba V, Mahmoud AAF, Sher A, Rees PK. Eosinophils as mediators of antibody-dependent damage to schistosomula. Nature 1975; 256:727–729.
2. Venge P, Hakansson L. Current understanding of the role of the eosinophilic granulocyte in asthma. J Clin Exp Allergy 1991; 21:31–37.
3. Weller PF. The immunobiology of eosinophils. N Engl J Med 1991; 324:1110–1118.
4. Barnes PJ. Inflammatory activities. Nature 1991; 349:284–285.
5. Braquet P, Rola-Pleszczynski M. Platelet-activating factor and cellular immune responses. Immunol Today 1987; 8:345–350.
6. Lee TC, Lenihan DJ, Malone B, Roddy LL, Wasserman SI. Increased biosynthesis of platelet-activating factor in activated human eosinophils. J Biol Chem 1984; 259: 5526–5530.
7. Burke LA, Crea AEG, Wilkinson JRW, Arm JP, Spur BE, Lee TH. Comparison of the generation of platelet-activating factor and leukotiene C_4 in human eosinophils stimulated by unopsonized zymosan and by the calcium ionophore A23187. J Allergy Clin Immunol 1990; 85:26–35.
8. Cromwell O, Wardlaw AJ, Champion A, Moqbel R, Osei D, Kay AB. IgG-dependent generation of platelet-activating factor by normal and low density human eosinophils. J Immunol 1990; 145:3862–3868.
9. Kroegel C, Yukawa T, Dent G, Venge P, Chung KF, Barnes PJ. Stimulation of degranulation from human eosinophils by platelet-activating factor. J Immunol 1989; 142:3518–3526.
10. Kroegel C, Pleass R, Yukawa T, Chung KF, Westwick J, Barnes PJ. Characterization of platelet-activating factor-induced elevation of cytosolic free calcium concentration in eosinophils. FEBS Lett 1989; 243:41–46.
11. Hoogerwerf M, Weening RS, Hack CE, Roos D. Complement fragments C3b and iC3b coupled to latex induce a respiratory burst in human neutrophils. Mol Immunol 1990; 27:159–167.
12. Van der Reyden HJ, van Rhenen DJ, Lansdorp PM, van 't Veer MB, Langenhuysen MMAC, Engelfriet CP, Kr. von dem Borne AEG. A comparison of surface marker analysis and FAB classification in acute myeloid leukemia. Blood 1983; 61:443–448.
13. Koenderman L, Kok PTM, Hamelink ML, Verhoeven AJ, Bruijnzeel PLB. An improved method for the isolation of eosinophilic granulocytes from peripheral blood of normal individuals. J Leukocyte Biol 1988; 44:79–86.
14. Tool ATJ, Koenderman L, Kok PTM, Blom M, Roos D, Verhoeven AJ. Release of platelet-activating factor is important for the respiratory burst induced in human eosinophils by opsonized particles. Blood 1992; 79:2729–2733.
15. Koenderman L, Tool ATJ, Roos D, Verhoeven AJ. Priming of the respiratory burst in human eosinophils is accompanied by changes in signal transduction. J Immunol 1990; 145:3883–3888.
16. Koenderman L, Kuijpers TW, Blom M, Tool ATJ, Roos D, Verhoeven AJ. Characteristics of CR3-mediated aggregation in human eosinophils: Effect of priming by platelet-activation factor. J Allergy Clin Immunol 1991; 87:947–954.

17. Looney RJ, Ryan DH, Takahashi K, Fleit HB, Cohen HJ, Abraham GN, Anderson CL. Identification of a second class of IgG F_c receptors on human neutrophils. A 40-kD molecule also found on eosinophils. J Exp Med 1986; 163:826–836.
18. Meerhof LJ, Roos D. Heterogeneity in chronic granulomatous disease detected with an improved nitroblue tetrazolium slide test. J Leukocyte Biol 1986; 39:699–711.
19. Blom M, Tool ATJ, Roos D, Verhoeven AJ. Priming of human eosinophils by platelet-activating factor enhances the number of cells able to bind and respond to opsonized particles. J Immunol 1992; 149:3672–3677.
20. Walker C, Virchow JC, Bruijnzeel PLB, Blaser K. T cell subsets and their soluble products regulate eosinophilia in allergic and non-allergic asthma. J Immunol 1991; 149:1829–1835.

DISCUSSION (Speaker: D. Roos)

Abrams: Do you have any dose-response data for the effect of cytokine induction of eosinophil CR3 measured by fluorescent serum-treated zymosan by FACs (fluorescent activated cell sorter) double staining? How sensitive is the effect?

Roos: My studies had 2.5 ng of GM-CSF/ml, 2.5 ng of IL-3, and 2.5 ng of IL-5/ml as lowest concentrations. Judging from these effects, I expect 1 ng of GM-CSF/ml, 0.5 ng of IL-3/ml, and 0.1 ng of IL-5/ml to be the lowest concentrations of these cytokines that can be measured by CR3 activation in our assay.

Busse: We have also found PAF primes blood eosinophils, and this process is associated with a transient intracellular calcium elevation. Similarly, in airway eosinophils we see an enhanced, or primed, response to the chemotactic peptide FMLP. In contrast, airway eosinophils show a desensitized response to PAF. This suggests the possibility of PAF priming. Where is the migration process of eosinophils to airway? Where in this process do you think that priming occurs?

Roos: From our work with neutrophil transmigration through endothelial cells pretreated with IL-1, we know that these endothelial cells express PAF on their surface (1,2). This could be the source of the eosinophil priming during the migration process.

Egan: PAF priming caused intracellular [Ca^{2+}] changes. Cytokine priming of eosinophils seems to be a similar phenomenon. Did you see [Ca^{2+}] changes during cytokine priming? [Ca^{2+}] was also increased when PAF was added after STZ. Was that accompanied by increased O_2 uptake?

Roos: Treatment of eosinophils with priming concentrations of GM-CSF, IL-3, or IL-5 does not induce a change in the concentration of free intracellular calcium (Koenderman et al., personal communication). When PAF is given after STZ, an enhanced oxygen uptake is seen, because rapid priming takes place during this respiratory burst.

Lopez: We showed that GM-CSF caused synthesis of PAF on neutrophils but this was released only after subsequent stimulation with other stimuli. Do eosinophils behave like neutrophils in this respect? Mepacrine was used to inhibit PLA_2. How specific is this compound and which type of PLA_2 is inhibited?

Roos: GM-CSF caused increased PAF synthesis only after subsequent stimulation of the eosinophils with other stimuli, such as STZ. Thus, in our hands, no PAF synthesis was detected after treatment of the cells with GM-CSF alone. The total amount of PAF produced in cells treated with GM-CSF (30 min) + STZ (10 min) was secreted for about 75%. Mepacrine totally inhibited PAF synthesis under these conditions. The PLA_2 inhibition to be used varies per cell type, and mepacrine proved to be a good inhibitor for the relevant PLA_2 in human eosinophils.

Thomas: Does PAF priming have a similar effect on neutrophil CR3?

Roos: In neutrophils, the iC3b-binding site of CR3 seems to be in an active state always. Thus, the response of neutrophils to STZ is not primed by any agent known. In contrast, the counterstructure-binding site of the neutrophil CR3 is not normally in an active state. Binding of SLex structures on neutrophils to ELAM-1 on endothelial cells activates this site (1).

Ackerman: In neutrophils there are intracellular stores of Mac-1 (CD11b) that can be rapidly expressed on the cell surface, even as a result of purifying the cells from the blood. Are similar intracellular pools present in eosinophils? Are they involved in the CD11b expression you have described? Do conformational changes in CD11b increase its affinity for its counterligand in eosinophils, as has been suggested by T. Springer for neutrophils?

Roos: There are, indeed, intracellular vesicles containing CD11b in eosinophils as indicated by immunoelectron microscopy. Over the time course of priming by cytokines or PAF that we have studied, these vesicles do not degranulate because cell-surface expression of CD11b hardly changes. We therefore propose that a conformational change in CR3 is responsible for the increase in affinity for iC3b.

Kay: PAF production by normal-density eosinophils following IgG-dependent or ionophore stimulation is considerably higher than in hypodense cells (3). Have you considered the responsiveness of eosinophils of different densities in your systems?

Roos: No, we have always worked with the total eosinophil population from human blood. These were isolated either by the 10 nM FMLP method of Koenderman et al. (4) or by CD16 immunodepletion method (5) with essentially the same results.

Schleimer: Dr. Kenichi Tomioka has shown that while fresh eosinophils do not exhibit increased CD11b in response to FMLP or PAF, eosinophils primed with GM-CSF for 2–4 days show a clear increase in CD11b. The fresh cells show no elevations in cytoplasmic calcium in response to FMLP while the cultured cells show a clear elevation of calcium. Have you tested longer incubations with cytokines for effects in your aggregation models?

Roos: We agree with the observations you mentioned. With freshly isolated eosinophils (carefully handled to avoid priming), we find no up-regulation of CD11b by FMLP or PAF, although longer incubations with PAF (30 min) lead to a small increase in CD11b expression. We reported earlier (4) that FMLP has very little effect on cytoplasmic calcium, although 1 μM FMLP leads to an increase of about 100 nM Ca^{2+}. We have not tested longer incubations with cytokines in our aggregation assay.

Wardlaw: We have found that 24-h culture of eosinophils with IL-3 results in a marked increase in CR3 expression, which is protein-synthesis-dependent and much more impressive than the short-term up-regulation from intracellular stores. However, we have great difficulty in finding any functional correlate for this increase, emphasizing that it is the conformational change in the CR3 which leads to its activation rather than increased expression. In this regard, the recent paper in *Cell* by Sam Wright's group demonstrating a neutrophil-derived lipid that can activate purified CR3 is very interesting (6). Do you know if eosinophils can generate this lipid?

Roos: We have written to Dr. Sam Wright for a collaborative study on the involvement of this same lipid in activating the CR3 on eosinophils.

Townley: Does PAF cause PAF release from eosinophils and how did you measure PAF?

Roos: Yes, PAF does cause PAF release, probably by priming mechanism by increased activity of PAF receptor response rather than by increased numbers of PAF receptors, as response is very rapid. PAF was measured by several methods (1) platelet aggregation, (2) GC-mass spec, (3) radioimmunoassay.

Morley: Please comment on the importance of PAF in vivo compared with antigen.

Roos: The ability of PAF and GM-CSF, IL-3, and IL-5 to prime eosinophils was demonstrated (1) with human eosinophils and (2) on several functions other than cell migration. However, Koenderman and colleagues in Utrecht have shown that IL-3, IL-5, and GM-CSF prime human eosinophils for chemotaxis in gradients of LTB_4, PAF, and FMLP, but not C5a (7). Eosinophils from the blood of patients with allergic asthma are primed for chemotaxis toward PAF, FMLP, or IL-8, but not toward C5a. Whether this also applies to transmigration over endothelial cell monolayers is not known, but Moser et al. (8) have found that eosinophils from such patients show enhanced transmigration over IL-1-pretreated endothelial cells.

Hansel: It is interesting that activated eosinophils can express increased amounts of surface CD11b without a concomitant increase in CD18. This has been described for low-density eosinophils by Hartnell et al. (9) and also for sputum eosinophils (10). Could you please comment on the possible functional significance of increased CD11b without CD18?

Roos: Increase in CD11b without concomitant increase in CD18 might be caused by (1) nonlinear binding of CD18 antibodies with increasing amounts of CD18 antigen or (2) a relative abundance of β_2 integrin subunits as compared to CD11b antigens.

DISCUSSION REFERENCES

1. Kuijpers TW, Hakkert BC, Hoogerwerf M, Leeuwenberg JFM, Roos D. Role of endothelial leukocyte adhesion molecule-1 and platelet-activating factor in neutrophil adherence to IL-1-prestimulated endothelial cells. J Immunol 1991; 147:1369–1376.
2. Kuijpers TW, Hakkert BC, Roos D. Neutrophil migration across monolayers of cytokine prestimulated endothelial cells: a role for endothelial cell-associated platelet-activating factor and for IL-8. J Cell Biol 1992; 117:565–572.

3. Cromwell O, Wardlaw AJ, Champion A, Moqbel R, Osei D, Kay AB. IgG-dependent generation of platelet-activating factor by normal and low density human eosinophils. J Immunol 1990; 145:3862–3868.
4. Koenderman L, Kok PTM, Hamelink ML, Verhoeven AJ, Bruijnzeel PLB. An improved method for the isolation of eosinophilic granulocytes from peripheral blood of normal individuals. J Leuk Biol 1988; 44:79–86.
5. Hansel TT, Pound JD, Pilling D, Kitas GD, Salmon M, Gentle TA, Lee SS, Thompson RA. Purification of human blood eosinophils by negative selection using immunomagnetic beads. J Immunol Methods 1989; 122:97–102.
6. Hermanowski-Vosatka A, Van Strijp JAG, Swiggard WJ, Wright SD. Integrin modulating factor-1: a lipid that alters the function of leukocyte integrins. Cell 1992; 68: 341–352.
7. Warringa RAJ, Koenderman L, Kok PTM, Kreukniet J, Bruijnzeel PLB. Modulation and induction of eosinophil chemotaxis by granulocyte-macrophage colony-stimulating factor and interleukin-3. Blood 1991; 77:2694–2700.
8. Moser R, Fehr J, Olgiati L, Bruijnzeel PLB. Migration of primed human eosinophils across cytokine-activated endothelial cell monolayers. Blood 1992; 79:2937–2945.
9. Hartnell A, Moqbel R, Walsh GM, Bradley B, Kay AB. Fc-gamma and CD11/CD18 receptor expression on normal density and low density human eosinophils. Immunology 1990; 69:264–270.
10. Hansel T, Braunstein JB, Walker C, Blaser K, Bruijnzeel PLB, Virchow JC Jr., Virchow C. Sputum eosinophils from asthmatics express ICAM-1 and HLA-DR. Clin Exp Immunol 1991; 86:271.

16

Pharmacological Control of Eosinophil Activation and Secretion

Peter J. Barnes and Mark A. Giembycz
National Heart and Lung Institute, London, England

I. INTRODUCTION

Since eosinophils may play a critical role in the pathophysiology of airway hyperresponsiveness and asthma, it is surprising that so little is known of their pharmacology and biochemistry (1,2). Although this may be due in part to diversion of attention to other inflammatory cells, such as mast cells, it is largely related to the fact that it has proved difficult to obtain these cells in sufficient number and purity for detailed studies of their function. Furthermore, the process of purification and the effects of prior therapy may lead to alterations in cell function, making interpretation difficult. It is important to understand the mechanisms of stimulus-response coupling in eosinophils and the pharmacological means by which these processes may be modulated as a possible therapeutic approach to asthma. Even the effects of drugs currently used to treat asthma are largely unknown.

A. Eosinophil Preparation

Peritoneal eosinophils from guinea pigs are a useful source of cells, since eosinophils may be prepared in large numbers with good viability and with a high degree of purity. Eosinophils may be obtained from male Dunkin-Hartley guinea pigs treated with either polymyxin B or human serum by weekly peritoneal lavage (3). Eosinophils are then purified using a discontinuous density gradient of isomolar Percoll solutions (4). It is routine to obtain eosinophils with a purity of >97% and viability of >99% from fractions 3–5. These fractions are pooled and washed in Hanks' balanced salt solution. Eosinophil numbers of $>10^8$ may be obtained by pooling cells from several animals, making it feasible to carry out pharmacological and biochemical studies that have previously been severely restricted by low cell numbers.

Human eosinophils may be prepared from blood taken from normal, atopic, asthmatic, or hypereosinophilic donors. Eosinophils are purified from buffy coat cells on metrizamide or Percoll density gradients. Eosinophils with a purity of >85% and viability of >95% are collected from the 22, 23, and 24% metrizamide density interfaces. The contaminating cells are predominantly neutrophils, and this may interfere with the interpretation of results. The number of cells available for study is usually $<10^7$, unless blood from patients with the hypereosinophilic syndrome is used, in which case there are doubts about the normality of cell function.

II. EOSINOPHIL ACTIVATION

Eosinophils may be activated by a number of stimuli, including IgG (5), IgA (6), IgE (7), opsonized zymosan (8), the complement fragment C5a (9), and the lipid mediators platelet-activating factor (PAF) (10,11) and leukotriene (LT)B_4 (12,13). In addition, eosinophils may be "primed" or activated by several cytokines, including interleukin (IL)-5 (14), granulocyte-macrophage colony-stimulating factor (GM-CSF) (15), and tumor necrosis factor-α (TNF-α) (16), all of which stimulate specific surface receptors. In addition, eosinophils may be activated by calcium ionophore A23187, which translocates calcium into the cell (17), by charged peptides, such as melittin and substance P (18), and by phorbol esters, such as phorbol myristate acetate (PMA), which activates protein kinase C (19).

In attempting to study the mechanisms involved in activation of eosinophils, we have concentrated initially on investigating the interaction between PAF and eosinophils, because PAF activates a relatively well-characterized receptor (20), which has recently been cloned (21,22). Furthermore, several potent antagonists of PAF receptors are now available (23,24).

A. Activation by PAF

The interaction between PAF and eosinophils may be of relevance in allergic diseases, since PAF may be one of the mediators that play a role in selective accumulation of eosinophils in the airways and may be involved in their activation. PAF is potently chemotactic for human eosinophils in vitro and has a greater effect on eosinophils than neutrophils (25,26). It also acts as a selective chemoattractant in vivo, causing accumulation of eosinophils into the skin of atopic individuals (27) and into the airways of primates and guinea pigs following both local and systemic administration (28–30). PAF also increases the adherence of eosinophils to human umbilical vein endothelial cells (31,32). PAF induces a shift in density in guinea pig peritoneal eosinophils, with an increased proportion of hypodense cells (33). Similar changes have been reported with human eosinophils (34). PAF also induces marked shape changes in human eosinophils, with elongation, degranulation, and induction of pseudopodia (Kroegel C et al., unpublished observations).

PAF has a potent secretory effect on eosinophils. Thus PAF potently induces the secretion of LTC_4 by human eosinophils (35) and of thromboxane, LTB_4, and, to a lesser extent, PGE_2 from guinea pig cells (36,37). This presumably involves activation of phospholipase A_2, since these mediators are not stored in eosinophils. PAF less potently stimulates the release of superoxide anions (O_2^-), and concentrations over 100-fold higher than necessary for chemotaxis or degranulation are required (10,38). PAF increases the expression of IgE receptors on eosinophils (39).

PAF induces degranulation from human and guinea pig eosinophils, although the extent to which this occurs may depend on whether the eosinophils have been primed. This may be demonstrated by a concentration-dependent release of granule contents, such as eosinophil peroxidase (EPO), eosinophil cationic protein (ECP), β-glucuronidase, alkaline and acid phosphatases, and arylsufatase B (10,11). PAF presumably leads to discharge of both large (specific) granules, which contain the eosinophil basic proteins, and small granules, which contain acid phosphatase and arylsufatase B (40). The degranulation response of eosinophils is presumably important in the helminthicidal effects of eosinophils (40), and in the toxic effects of eosinophils on airway epithelium (41–43).

PAF receptors have been identified on eosinophils by direct receptor-binding studies. [^{3}H]PAF has proved to be an unsatisfactory ligand, because it is rapidly metabolized and taken up by eosinophils and rapidly leads to down-regulation of receptors (20). Labeled antagonists, such as [^{3}H]WEB 2086, have proved to be much more satisfactory in identifying PAF receptors on a variety of cell types (44–46). Using this ligand, a single population of PAF receptors has been identified on both guinea pig and human eosinophils, with approximately 35,000 sites per cell, respectively (47). Studies in an eosinophil line indicate that exposure to IL-5 causes an increase in PAF receptor mRNA (22), suggesting that certain cytokines may increase the expression of PAF receptors.

Whether different subtypes of PAF receptors exist has not yet been resolved (48), but there are indications that there may be different subtypes of PAF receptors in eosinophils or, alternatively, that the PAF receptor may exist in different affinity states (38).

B. Activation by LTB_4

LTB_4 is a potent activator of guinea pig peritoneal and alveolar eosinophils (13,49,50) and also has effects on human blood eosinophils (51). LTB_4 stimulates chemotaxis and the oxidative burst, but not degranulation. Binding studies with

[^{3}H]LTB_4 indicate that both high- and low-affinity receptor exist on alveolar eosinophils (13). It is likely that the high-affinity receptor mediates chemotaxis, whereas the low-affinity receptor mediates the respiratory burst response (13).

C. Role of Ca^{2+}

A rise in the intracellular calcium ion concentration ($[Ca^{2+}]_i$) precedes secretion in all secretory cells, but the mechanism of this rise in $[Ca^{2+}]$ may vary from cell to cell, and even in the same cell type may vary with different activating stimuli. Using the fluorescent dye fura-2 to indicate $[Ca^{2+}]$, we have demonstrated that activation of guinea pig eosinophils with PAF results in a rapid and transient rise in $[Ca^{2+}]_i$ (12). The inactive precursor of PAF, lyso-PAF, is without effect, and the response to PAF is inhibited in a concentration-dependent manner by the PAF antagonist WEB 2086, indicating that it is mediated by surface receptors. It is inhibited by pertussis toxin, suggesting that an inhibitory G-protein is involved in coupling the receptor to the secretory response. A similar rise in $[Ca^{2+}]$ is also seen after activation with LTB_4. The fura-2 response to PAF is rapidly tachyphylactic, although the response to LTB_4 is unchanged, indicating down-regulation of PAF receptors (12). The bacterial chemotactic peptide formyl-Met-Leu-Phe (fMLP) is extremely weak in stimulating a fura-2 response, in marked contrast to neutrophils (18).

The fura-2 response to PAF is dependent on extracellular Ca^{2+}, since it is reduced by bathing the cells in buffer without Ca^{2+} or in EDTA, which chelates extracellular Ca^{2+}, but is not inhibited by the dihydropyridine calcium antagonist nimodipine, indicating that Ca^{2+} entry via voltage-dependent channels is not involved. The PAF-induced Ca^{2+} transient is, however, inhibited by nickel ions, which block Ca^{2+} entry via all channels, suggesting that occupation of PAF receptors opens receptor-operated channels, which allow the entry of Ca^{2+}, resulting in cell activation. The concentration-response curve for the fura-2 response is similar to that for EPO release, suggesting that the rise in $[Ca^{2+}]_i$ may be linked to degranulation, possibly via contractile myofilaments that are necessary for granule extrusion.

A part of the rise in $[Ca^{2+}]$ is not dependent on extracellular Ca^{2+} and depends on release of Ca^{2+} from intracellular stores. The mechanism of intracellular calcium release linked to cell activation depends on phosphoinositide hydrolysis, due to receptor-coupled activation of phosphoinositidase C (PIC), which generates inositol-(1,4,5)-trisphosphate (IP_3), which in turn releases Ca^{2+} from intracellular stores (52). In guinea pig eosinophils, PAF stimulates the incorporation of [^{3}H]inositol into membrane phosphoinositides, but using the conventional method for measuring phosphoinositide (PI) turnover, it was not possible to measure the generation of labeled inositol phosphates, probably because eosinophil basic proteins interfere with the separation of labeled inositol phosphates from the negatively charged stationary phase of the chromatography column (53). These problems have been overcome by using a competitive protein-binding assay that measures the absolute mass of IP_3 in cells (54). PAF stimulates a rapid rise in intracellular IP_3 with a maximal rise at 5s, which is transient (53). The rise in fura-2 response occurs *after* the rise in IP_3 concentration, and there is a positive correlation between the peak increase in IP_3 and the subsequent peak in the fura-2 response. The IP_3 response to PAF is less sensitive than the fura-2 response, indicating that higher concentrations of PAF are necessary to activate PIC and PI hydrolysis than are required for opening receptor-operated channels in the cell membrane. The PI response to PAF is inhibited by WEB 2086, indicating that PAF receptors are involved in this response. Other receptor-mediated stimuli, such as LTB_4 and C5a, also stimulate IP_3 formation in guinea pig eosinophils, but the phorbol ester PMA, which directly activates PKC, has no effect, as expected (53).

Several isoforms of PIC have now been recognized (55), and a PIC has been identified in guinea pig eosinophil-washed membranes that is exquisitely sensitive to Ca^{2+} and may be activated with LTB_4 and $GTP_{\gamma}S$ in the presence of GTP (Perkins R, Giembycz M, Barnes PJ, unpublished observations).

D. Protein Kinase C

PI hydrolysis also leads to the formation of 1,2-diacylglycerol, which activates PKC (56). PKC activation with phorbol esters results in release of O_2^- (11,38), suggesting that PKC is involved in the activation of NADPH-dependent respiratory burst oxidase. The PKC inhibitors Ro 31-8220, which acts at the ATP-binding site, and AMG-C_{16}, which blocks the diglyceride-binding site, both inhibit LTB_4 and PAF-induced O_2^- generation (57) and are potentiated by the diacylglycerol kinase inhibitor R 59022 (58).

PKC may also play a modulatory role in eosinophil activation, since PMA has an inhibitory effect on PAF-stimulated rise in $[Ca^{2+}]_i$ and on EPO release (59). This effect of PMA is inhibited by the antagonist staurosporine, indicating that PKC activation is likely to be involved. A similar feedback inhibitory mechanism has also been described in other cells (60) and may represent a mechanism for limiting the secretory response. Several isoenzymes of PKC have now been distinguished (61), and it is possible that these different effects of PKC activation may be mediated by different isoenzymes.

E. Phospholipase D

There is increasing evidence that phospholipase D (PLD) may play a role in signal transduction of inflammatory cells (62,63). PLD hydrolyzes predominantly phosphatidylcholine producing free choline and phosphatidic acid, which may then form diacylglycerol. Diacylglycerol can activate PKC, which in turn may activate PLD, providing a mechanism for perpetuation of responses. PLD may also be activated directly by certain receptors, including those that have intrinsic tyrosine kinase activity (64). In neutrophils PLD appears to be important in "priming" of the cells for increased responsiveness (65). PLD activity has also been demonstrated in human eosinophils (66) but does not appear to be important in oxidative burst responses to PAF or LTB_4. PLD is activated by phorbol esters in guinea pig eosinophils, but not by PAF or LTB_4 (Perkins R, Giembycz M, Barnes PJ, unpublished observations). In view of the potential importance of PLD in cell priming, it is important to investigate the effect of cytokines on PLD activation. Currently available PLD inhibitors such as wortmannin are nonspecific, and more selective inhibitors are needed; such drugs may have important therapeutic potential since they may reduce the priming of inflammatory cells and downregulate inflammation.

F. G-Proteins

G-Proteins are involved in coupling surface receptors to second messenger systems. There has been little research on the nature of G-proteins in eosinophils. PAF stimulates GTPase activity in eosinophils, which is inhibited by pertussis toxin (67). G-proteins are also involved in eosinophil degranulation, and a novel G-protein associated with exocytosis, G_E, appears to be involved (68,69).

G. Stimulus-Response Coupling

A picture of stimulus-response coupling in eosinophils is now evolving (Fig. 1). Binding of an agonist to its surface receptor on eosinophils may lead to the opening of a receptor-operated calcium channel via a pertussis-sensitive G-protein

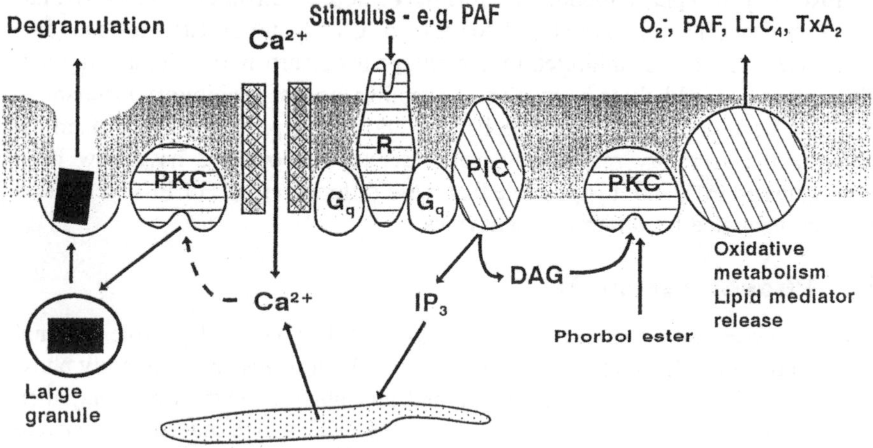

Figure 1 Stimulus-response coupling in eosinophils.

(G_i, G_q, or G_o). The entry of Ca^{2+} leads to degranulation with release of basic proteins and enzymes from granules. At the same time, receptor activation may stimulate phospholipase C via G_q, generating IP_3, which releases Ca^{2+} from intracellular stores, thus contributing to the rise in intracellular Ca^{2+}. PLC activation also generates diacylglycerol, which activates PKC. PKC activates the respiratory bust oxidase in eosinophils, generating O_2^-, which is immediately converted to H_2O_2. H_2O_2 interacts with EPO released from granules to form the highly toxic hypohalous anions (70,71), which may lead to the shedding of airway epithelium (72). Activation of PKC also leads to inhibition of cell activation, thereby limiting the activation mechanism. The rise in intracellular Ca^{2+} and possibly also PKC activation may result in the activation PLA_2 and the generation of arachidonic acid, which is converted to either cyclooxygenase (guinea pig) or lipoxygenase (human) products.

Whether stimulus-response coupling may be altered in disease states is not yet certain. The concept of priming in inflammatory cells is now well established; a subthreshold concentration of one agent markedly enhances the response induced by a second agent. This phenomenon has frequently been observed with various cytokines. For example, GM-CSF and other cytokines increase the responsiveness of eosinophils to PAF and other stimuli (15,73), and in vivo GM-CSF increases the recruitment of eosinophils into the lungs induced by PAF (74). The biochemical basis of priming in eosinophils is still not certain, but may involve PLD activation via cytokine receptors. Further research in this area is warranted, since eosinophils from patients with asthma appear to show exaggerated oxidative burst responses to activating agents such as PAF (75) and increased degranulation responses to serum opsonized Sephadex (76). This appears to occur at a late stage in stimulus-response coupling, since responses to PMA are also increased, and suggests that it may occur at the level of PKC.

H. Activation by Peptides

Eosinophils may be degranulated by a variety of basic peptides, including the bee venom peptide melittin, the neuropeptide substance P (SP), and the bacterial chemotactic peptide fMLP (18,77). There appears to be selective release of granule products, without the concomitant release of inflammatory mediators, and this appears to occur in a noncytotoxic manner. The activation by SP occurs only with high concentrations of the peptide ($>10^{-5}$ M) and is mediated by the N-terminal end of the molecule since $SP_{1\text{-}4}$ is also active, but $SP_{4\text{-}11}$ is not. The closely related tachykinin neurokinin A is also inactive, suggesting that the response to SP is not mediated via a tachykinin receptor, and may be related to the physical characteristics of the peptide. A similar response to SP is seen in cutaneous mast cells, which degranulate (histamine release), but do not release other mediators (such as PGD_2 or LTC_4) in response to SP (78). Mast cells are also degranulated by melittin, and presumably there is a similar mechanism of action to that which occurs in eosinophils. Recent evidence suggests that the response to these amphiphilic peptides is mediated via a direct activation of G-proteins, and this interaction may be blocked by pertussis toxin (79). It is unlikely that the activation of eosinophils by SP has any physiological relevance, since eosinophils would not be exposed to such high concentrations of SP released from sensory nerves. However, it is possible that melittin and the related peptide mastoparan in wasp venom may degranulate both eosinophils and mast cells in the skin during a sting.

The bacterial chemotactic peptide fMLP is a weak stimulant of guinea pig eosinophils (18,36,77), and there is a low density of binding sites (80). However coculture of eosinophils with monocyte-conditioned medium increases the efficacy of fMLP and suggests that priming may occur, although the nature of the priming agent is not yet determined (80). The actions of fMLP, in contrast to those of substance P and melittin, are mediated via surface receptors since the effect is blocked by the fMLP-receptor antagonist BOC-fLMP (18).

I. Cytokines

Several cytokines, including IL-2, IL-3, IL-5, GM-CSF, TNF, and interferon, may exert profound effects on eosinophil function, including their survival, proliferation, differentiation, density, and ability to undergo degranulation and to release mediators (81). The molecular pathways involved are uncertain. Most cytokine receptors are single polypeptide chains that cross the cell membrane once; cytokines bind to the large extracellular domain (82), and the intracellular domain leads to activation of transcription factors (such as AP-1), which influence the transcription of target genes (83). More research on the molecular mechanisms of cytokine effects on eosinophils is required.

III. MODULATION OF RESPONSES

If the eosinophil plays an important role in the pathophysiology of airway hyperresponsiveness and asthma, then it follows that it may be beneficial to develop drugs that either inhibit the infiltration of eosinophils into the airways or block their activation once attracted into the airways. In either case, this may involve the development of drugs that inhibit the stimulus-response mechanisms involved in eosinophil activation. Although there are no drugs that have been specifically designed to inhibit eosinophil responses, some of the drugs already used in asthma therapy may interact with these cells.

A. β-Adrenoceptor Agonists

β-Agonists are the most effective bronchodilators available and act predominantly by stimulating β_2-receptors on airway smooth muscle cells. Whether β-agonists have additional anti-inflammatory effects is now greatly debated (84). Using [^{125}I]pindolol, we have identified β-receptors on guinea pig peritoneal human blood eosinophils (approximately 7000 and 4000 receptors per cell), which have the pharmacological characteristics of β_2-receptors (85). After incubation with the β_2-agonist salbutamol for 30 min, however, there is no evidence for an inhibitory effect on O_2^- release, whether the eosinophils were stimulated with either opsonized zymosan or PMA in guinea pig or human eosinophils (85). With shorter incubation periods, an inhibitory effect on oxidative burst and thromboxane release is seen with the β_2-agonist salbutamol, formoterol, and procaterol (50,86). β-Agonists also have a weak inhibitory effect on immunoglobulin-induced degranulation of human eosinophils in the presence of a phosphodiesterase inhibitor (87). The lack of inhibitory effect observed with the longer incubation with salbutamol is almost certainly explained by tachyphylaxis and down-regulation of β-receptors, as has been observed in neutrophils (88). This has recently been confirmed in eosinophils by direct binding studies with [^{125}I]pindolol (Dent G et al, unpublished observations). The inhibitory effect of β-agonists is inversely related to how partial the agonist is. Thus the greatest degree of inhibition (>90%) is seen with the almost full agonist procaterol, less inhibition with formoterol (50%) and least inhibition with salbutamol (<40%) (89). The long-acting β_2-agonist salmeterol, which has only <30% agonist activity, has no inhibitory effect and even behaves like a β-blocker (90). Although the binding studies have clearly identified β_2-receptors on eosinophils, the inhibition of oxidative burst may be mediated through "atypical" β-receptors at which propranolol has a relatively low affinity. Whether these receptors correspond to β_3-receptors is currently under investigation using selective agonists and molecular probes.

The clinical significance of these findings may relate to the effect of β-agonists on airway hyperresponsiveness. With the rapid development of tachyphylaxis, it is unlikely that regular use of β-agonists would exert a significant inhibitory effect on eosinophil function. This is compatible with observations that demonstrate that regular use of β_2-agonists is not associated with any reduction in airway responsiveness (91,92) and may even be associated with an increased responsiveness and asthma symptoms (93,94). It is possible that the regular use of steroids may prevent the rapid tachyphylaxis of eosinophil β-receptors, thus allowing β-agonists to exert some inhibitory effect on eosinophil function, since steroids increase the transcription of β-receptors (95–97). These possibilities are currently under investigation.

B. Theophylline

The effects of theophylline on leukocyte function are complex because of the multiple pharmacological effects of this drug. In guinea pig and human eosinophils, a high concentration of theophylline (10^{-3} M) causes inhibition of O_2^-

release and thromboxane synthesis, which may be due to inhibition of phosphodiesterase (PDE) (86,98). High concentrations of theophylline inhibit PDE in eosinophil membranes and elevate cyclic AMP (99,100). At lower concentrations (within the therapeutic range) there is a paradoxical *increase* in O_2^- release, which appears to be due to adenosine receptor antagonism, since it has also been with 8-phenyltheophylline, which is a more potent adenosine antagonist that lacks PDE inhibitory activity. Interestingly, no inhibition of O_2^- release occurs even with high concentrations of 8-phenyltheophylline, supporting the view that the inhibition seen with high concentrations of theophylline is due to PDE inhibition. Exogenous adenosine has an inhibitory effect on O_2^- release, and 5′-*N*-ethylcarboxamide adenosine (NECA) has a greater inhibitory effect than R-*N*-phenylisopropyl adenosine (R-PIA), indicating that A_2-adenosine receptors are involved. These studies suggest that endogenous adenosine normally exerts an inhibitory effect on eosinophils and that theophylline in therapeutic concentrations results in potentiation of mediator release. A similar potentiating effect of theophylline on O_2^- release has been described in human neutrophils (101).

C. Selective Phosphodiesterase Inhibitors

As discussed above, high concentrations of theophylline have an inhibitory effect on eosinophil function through inhibition of the cytoplasmic enzyme PDE, which breaks down cyclic nucleotides. Theophylline is a weak, nonselective PDE inhibitor, but several isoenzymes have now been recognized (102,103). Using selective PDE inhibitors, we have recently investigated which isoenzymes are responsible for the inhibitory effects in eosinophils. PDE IV enzyme inhibitors, such as rolipram and denbufylline, are very effective in inhibiting O_2^- (measured fluorimetrically as H_2O_2) and thromboxane release from guinea pig eosinophils, whether induced by soluble (LTB_4, PAF) or particulate (opsonized zymosan) stimuli (99,100). By contrast, PDE III (SK&F 94120) and PDE V inhibitors (zaprinast) are ineffective (99,100). PDE IV activity has been directly measured in guinea pig eosinophils, and there are indications that high- and low-affinity components exist (99,100). This suggests either that more than one subtype of PDE IV is expressed or that a single isoform may undergo modification, such as phosphorylation. Similar measurements have been made in human eosinophils (Giembycz et al, unpublished). The nonselective PDE inhibitor IBMX inhibits immunoglobulin-mediated degranulation of human eosinophils (87). The combined PDE III/IV inhibitor zardaverine is also an effective inhibitor in human eosinophils (104). Combined PDE III/IV inhibitors may be particularly useful in asthma therapy, since PDE IV inhibitors may have an anti-inflammatory effect, whereas PDE III inhibition relaxes airway smooth muscle (105,106). In experimental animals, combined type III/IV inhibition blocks inflammatory cell infiltration after allergen and reduces airway hyperresponsiveness (107).

The inhibitory effect of PDE IV inhibitors is due to an increase in intracellular cyclic AMP concentration. A similar inhibitory effect is seen with the lipophilic stable cyclic AMP analogs dibutyryl cyclic AMP and 8-bromo cyclic AMP (87,99). Similarly, cholera toxin, which stimulates the stimulatory G-protein (G_s), irreversibly leads to increased cyclic AMP concentration and inhibition of degranulation responses in human eosinophils (87). By contrast, dibutyryl cyclic GMP is ineffective in human and guinea pig eosinophils.

D. Prostaglandins

Prostaglandins have an inhibitory effect on eosinophil function, and PGD_2 appears to be more potent than PGE_2 or $PGF_{2\alpha}$, suggesting that a DP-receptor is involved (108). There is no evidence for the presence of IP, FP, or TP receptors (36). Furthermore, pretreatment of eosinophils with a cyclooxygenase inhibitor does not affect eosinophil activation induced by LTB_4 or PAF, indicating that endogenously produced eicosanoids such as TxA_2 and PGE_2 do not have a modulatory role (36).

E. Corticosteroids

Corticosteroids are the most effective anti-inflammatory drugs currently used in the treatment of asthma (109), and they reduce airway hyperresponsiveness (110). Corticosteroids reduce the numbers of circulating eosinophils (111), and inhaled steroids reduce the number of hypodense eosinophils in the circulation of asthmatics (112), suggesting that steroids inhibit the local production of cytokines that are necessary for the formation of eosinophils in the bone marrow (IL-3, IL-5, GM-CSF, and others) (113). Corticosteroids also have direct effects on eosinophil function. Steroids block GM-CSF-induced eosinophil survival and the characteristic shift to a hypodense phenotype (114) and block the priming effect of cytokines on eosinophils (115). This suggests that eosinophils express glucocorticosteroid receptors. In low concentrations they inhibit the release of ECP (116). However, immunoglobulin-induced degranulation is only weakly inhibited by high concentrations of steroids, even when enhanced degranulation is produced by exposure to IL-5 (117). In high concentrations, dexamethasone inhibits the release of O_2^- from human eosinophils, an effect that is mediated by steroid receptors since it is inhibited by the steroid receptor antagonist RU 38486 (118).

F. Cromones

Sodium cromoglycate and nedocromil sodium have an inhibitory effect on chemotaxis of eosinophils induced by PAF and LTB_4 (119), but do not inhibit secretory responses (120). The effects of these two agents appear to be variable and may depend on the state of cell activation.

G. Chloride Channel Blockers

Recent studies have demonstrated that the loop diuretic frusemide (furosemide) is an effective inhibitor of various indirect bronchoconstrictor challenges in asthma, including exercise, allergen, metabisulfite, and adenosine challenges, yet is ineffective against histamine and methacholine challenges, which act directly on airway smooth muscle (121). This cannot be explained by inhibition of the $Na^+/K^+/2Cl^-$ co-transporter by frusemide, since the more potent cotransport inhibitor bumetanide is ineffective in the same challenges (122). A more likely explanation is that frusemide is inhibiting a certain type of Cl^- channel or exchange mechanism (121). Frusemide, but not bumetanide, is effective in inhibiting LTB_4-induced release of O_2^- from guinea pig and human eosinophils, and a similar inhibitory effect is seen by reducing the extracellular concentration of chloride ions and by the Cl^- channel blockers diisothiocyanato-stilbene-2,2′-disulfonic acid (DIDS) and 5-nitro-2-(3-phenylpropylamino)-benzoic acid (NPPB) (123). The effect of Cl^- channel blockers on opsonized zymosan-induced secretion is less marked, suggesting that Cl^- are involved in the activation of eosinophils by certain stimuli only. Similar effects of Cl^- channel blockers have also been reported in neutrophils.

The role of other ions in eosinophil secretion is not yet certain, although Mg^{2+} appears to be necessary for PAF-induced O_2^- secretion (38).

IV. SUMMARY AND CONCLUSIONS

An understanding of the molecular pharmacology of eosinophils has only just begun. There are close parallels between the pharmacology of human and guinea pig eosinophils where such comparisons have been made, and also close similarities in many of the mechanisms involved in neutrophil activation (51,124). The stimuli that activate eosinophils differ from those that activate neutrophils, however, due to differences in the receptors expressed by these cells. Little is known about mechanisms that change the responsiveness of eosinophils, but it is likely that exposure to cytokines may enhance responsiveness to certain other stimuli and that this heightened responsiveness must be mediated by an amplification of components of stimulus-response coupling. Since increased responsiveness of eosinophils may be important in the inflammation of asthma and other allergic diseases, it is possible that drugs will be developed that specifically block this effect. The refinement of pharmacological and biochemical assays, which may be suitable for smaller numbers of cells, and in some cases for assays of single cells, and the application of cDNA probes to investigate factors that regulate transcription of receptors, G-proteins, and ion channels will certainly lead to a greater understanding of eosinophil pharmacology in the future and may lead to the development of new therapeutic strategies for asthma.

ACKNOWLEDGMENTS

We thank the Medical Research Council (UK) and the Wellcome Trust for supporting the studies discussed in this chapter.

REFERENCES

1. Barnes PJ, Kroegel C, Yukawa T, Dent G, Chung KF. Pharmacology of eosinophils. In: Kay AB, ed. Eosinophils, Allergy and Asthma. Oxford: Blackwell, 1990: 144–157.
2. Giembycz MA, Barnes PJ. Stimulus-response coupling in eosinophils: receptors, signal transduction and pharmacological manipulation. In: Smith H, Cooke D, eds. Immunopharmacology of Eosinophils. London: Academic Press. In Press.
3. Pincus SH. Production of eosinophil-rich guinea pig peritoneal exudates. Blood 1978; 52:127–135.
4. Gartner I. Separation of human eosinophils in high density gradients of polyvinyl pyrolidone-coated silica gel (Percoll). Immunology 1990; 40:133–136.
5. Thorne KJ, Free J, Franks D, Oliver RC. The mechanisms of fc-mediated interaction with eosinophils with immobilized immune complexes. II. Identification of two membrane proteins modified by the interaction. J Cell Sci 1982; 56:357–369.
6. Abu-Ghazaleh RI, Fujisawa T, Mestecky R, Kyle RA, Gleich GJ. IgA-induced eosinophil degranulation. J Immunol 1989; 142:2393–2399.
7. Capron M, Spiegelberg HL, Prin L, et al. Role of IgE receptors in effector function of human eosinophils. J Immunol 1984; 132:462–468.
8. Winquist I, Olofsson T, Olsson I. Mechanisms for eosinophil degranulation: release of eosinophil cationic protein. Immunology 1984; 51:1–8.
9. De Simone C, Ferrari M, Pugnaloni L, Ferrarelli Rumi C, Sorfice F. Eosinophil-mediated cellular cytotoxicity induced by zymosan-activated serum. Immunol Lett 1986; 12:37–41.
10. Kroegel C, Yukawa T, Dent G, Chanez P, Chung KF, Barnes PJ. Platelet activating factor induces eosinophil peroxidase release from human eosinophils. Immunology 1988; 64:559–562.
11. Kroegel C, Yukawa T, Dent G, Venge P, Chung KF, Barnes PJ. Stimulation of degranulation from human eosinophils by platelet activating factor. J Immunol 1989; 142:3518–3526.
12. Kroegel C, Pleass R, Yukawa T, Chung KF, Westwick J, Barnes PJ. Characterization of platelet-activating factor-induced elevation of cystolic free calcium concentration in eosinophils. FEBS Lett 1989; 243:41–46.
13. Maghni K, de Brum-Fernanders AJ, Foldes-Filep E, Gaudry M, Burgeat P, Sirois P. Leukotriene B_4-receptors on guinea pig alveolar eosinophils. J Pharmacol Exp Ther 1991; 258:784–789.
14. Yamaguchi Y, Hayashi Y, Sugama Y, et al. High purified murine interleukin 5 (IL-5) stimulates eosinophil function and prolongs in vitro survival. J Exp Med 1988; 167: 1737–1742.
15. Owen WF, Rothenberg ME, Silberstein DS, et al. Regulation of human eosinophil

viability, density and function by granulocyte/macrophage colony stimulating factor in the presence of 3T3 fibroblasts. J Exp Med 1987; 166:129–141.
16. Whitcombe EA, Wesolek JH, Pincus SH. Modulation of human peripheral blood eosinophil function by tumor necrosis factor-alpha. Int Arch Allergy Appl Immunol 1989; 89:250–255.
17. Fukuda T, Ackerman SJ, Reed CE, Peters MS, Dunnette SL, Gleich GJ. Calcium ionophore A 23187 calcium dependent cytolytic degranulation of human eosinophils. J Immunol 1985; 135:1349–1356.
18. Kroegel C, Giembycz MA, Barnes PJ. Characterization of eosinophil activation by peptides. Differential effects of substance P, melittin and f-Met-Leu-Phe. J Immunol 1990; 145:2581–2587.
19. Yazdanbakhsh M, Eckman CM, De Boer M, Roos D. Purification of eosinophils from normal human blood, preparation of eosinoplasts and characterization of their response to virus stimuli. Immunology 1987; 60:123–129.
20. Dent G, Ukena D, Barnes PJ. PAF Receptors. In: Barnes PJ, Page CP, Henson PM, eds. Platelet Activating Factor and Human Disease. Oxford: Blackwell, 1989: 58–81.
21. Honda Z, Nakamura M, Miki I, et al. Cloning by functional expression of guinea pig lung platelet activating factor (PAF) receptor. Nature 1991; 349:342–346.
22. Nakamura M, Honda Z, Isumi T, et al. Molecular cloning and expression of platelet-activating factor receptor from human leukocytes. J Biol Chem 1991; 226:20400–204005.
23. Hosford D, Page CP, Barnes PJ, Braquet P. PAF-receptor antagonists. In: Barnes PJ, Page CP, Henson PM, eds. Platelet Activating Factor and Human Disease. Oxford: Blackwell. 1989:82–116.
24. Alabaster V. PAF antagonists and asthma. In: Barnes PJ, ed. New Drugs for Asthma 2. London: IBC Publications. 1992.
25. Wardlaw AJ, Moqbel R, Cromwell O, Kay AB. Platelet activating factor. A potent chemotactic and chemokinetic factor from human eosinophils. J Clin Invest 1986; 78:1701–1706.
26. Sigal CE, Valone FH, Holtzman J, Goetzl EJ. Preferential human eosinophil chemotactic activity of the platelet activating factor (PAF): 1-*o*-hexadecyl-2-acetyl-*sn*-glyceryl-3-phosphocholine (AGEPC). J Clin Immunol 1987; 7:179–188.
27. Henocq E, Vargaftig BB. Skin eosinophils in atopic patients. J Allergy Clin Immunol 1988: 81:691–695.
28. Arnoux B, Page CP, Denjean A, Nolibe D, Morley J, Benveniste J. Accumulation of platelets and eosinophils in baboon lung after PAF-acether challenge. Am Rev Respir Dis 1988; 137:855–860.
29. Denjean A, Arnoux B, Benveniste J. Long-lasting effect of intratracheal administration of PAF-acether in baboons. Am Rev Respir Dis 1988; 137:283.
30. Lellouch-Tubiana A, Lefort J, Simon M-T, Pfister A, Vargaftig BB. Eosinophil recruitment into guinea pig lungs after PAF-acether and allergen administration. Modulation by prostacyclin, platelet depletion and selective antagonists. Am Rev Respir Dis 1988; 137:948–954.
31. Kimani G, Tonnesen MG, Henson PG. Stimulation of eosinophil adherence to

human vascular endothelial cells in vitro by platelet activating factor. J Immunol 1988; 140:3161–3166.
32. Lamas AM, Mulroney CM, Schleimer RP. Studies of the adhesive interaction between purified human eosinophils and cultured vascular endothelial cells. J Immunol 1988; 140:1500–1510.
33. Yukawa T, Kroegel C, Evans P, Fukuda T, Chung KF, Barnes PJ. Density heterogeneity of eosinophil leukocytes: induction of hypodense eosinophils by platelet activating factor. Immunology 1989; 68:140–143.
34. Klopprogge E, de Leeuw AJ, De Monchy JGR, Kauffan AF. Hypodense eosinophilic granulocytes in normal individuals and patients with asthma: generation of hypodense cell populations in vitro. J Allergy Clin Immunol 1989; 83:393–400.
35. Bruijnzeel PLB, Koenderman L, Kok PTM, Hanelink ML, Verhagen JL. Platelet activating factor (Paf-acether) induced leukotriene C_4 formation and luminal dependent chemiluminescence of human eosinophils. Pharmacol Res Commun 1986; 18:61–69.
36. Giembycz MA, Kroegel C, Barnes PJ. Platelet activating factor stimulates cyclooxygenase activity in guinea-pig eosinophils. J Immunol 1990; 144:3489–3497.
37. Sun FF, Czuk CI, Taylor BM. Arachidonic acid metabolism in guinea-pig eosinophils: synthesis of thromboxane B_2 and leukotriene B_4 in response to soluble or particulate activators. J Leukocyte Biol 1989; 46:152–160.
38. Kroegel C, Yukawa T, Westwick J, Barnes PJ. Evidence for two platelet activating receptors on eosinophils: dissociation between PAF induced intracellular calcium mobilization, degranulation and superoxide anion generation. Biochem Biophys Res Commun 1989; 162:511–521.
39. Moqbel R, Walsh GM, Nagakura T, et al. The effect of platelet-activating factor on IgE binding to, and Ige-dependent biological properties of, human eosinophils. Immunology 1990; 70:251–257.
40. Gleich GJ, Adolphson CR. The eosinophilic leukocyte: structure and function. Adv Immunol 1986; 39:177–253.
41. Gleich GJ, Flavahan NA, Fujisawa T, Vanhoutte PM. The eosinophil as a mediator of damage to respiratory epithelium: a model for bronchial hyperreactivity. J Allergy Clin Immunol 1988; 81:776–781.
42. Yukawa T, Read RC, Kroegel C, et al. The effects of activated eosinophils and neutrophils on guinea pig airway epithelium in vitro. Am J Resp Cell Mol Biol 1990; 2:341–354.
43. Motojima S, Frigas E, Loegering DA, Gleich GJ. Toxicity of eosinophil cationic proteins for guinea pig tracheal epithelium in vitro. Am Rev Respir Dis 1989; 139: 801–805.
44. Ukena D, Dent G, Birke FW, Robaut C, Sybrecht GW, Barnes PJ. Radioligand binding of antagonists of platelet-activating factor to intact human platelets. FEBS Lett 1988; 228:285–289.
45. Dent G, Ukena D, Chanez P, Sybrecht GW, Barnes PJ. Characterization of PAF receptors on human neutrophils using the specific antagonist, WEB 2086: correlation between receptor binding and function. FEBS Lett 1989; 244:365–368.
46. Dent G, Ukena D, Sybrecht GW, Barnes PJ. [^{3}H]WEB 2086 labels platelet activating factor receptors in guinea pig and human lung. Eur J Pharmacol 1989; 169:313–316.

47. Ukena D, Kroegel C, Yukawa T, Sybrecht G, Barnes PJ. PAF-receptors on eosinophils: identification with a novel ligand [^{3}H]WEB 2086. Biochem Pharmacol 1989; 38:1702–1705.
48. Barnes PJ. Molecular biology: inflammatory activities. Nature 1991; 349:284–285.
49. Ng CP, Sun FF, Taylor BM, Wolin MS, Wong PY-K. Functional properties of guinea-pig eosinophil leukotriene B_4 receptor. J Immunol 1991; 147:3096–3103.
50. Rabe K, Dent G, Giembycz M, Ravenall S, Barnes PJ, Chung KF. Effects of long-lasting β-adenoceptor agonist formeterol vs albutamol on LTB_4 induced Ca^{2+} mobilisation, H_2O_2 and TxB_2 release in guinea pig eosinophils. Am Rev Respir Dis 1990; 141:A29.
51. Palmblad J, Gyllenhammer H, Lindgren JA, Malmsten CL. Effects of leukotrienes and f-Met-Leu-Phe on oxidative metabolism of neutrophils and eosinophils. J Immunol 1984; 132:3041–3045.
52. Berridge MJ. Inositol trisphosphate and diacylglycerol: two interacting second messengers. Annu Rev Biochem 1987; 56:159–193.
53. Kroegel C, Chilvers ER, Giembycz MA, Challiss RA, Barnes PJ. Platelet-activating factor stimulates a rapid accumulation of inositol (1,4,5)trisphosphate in guinea-pig eosinophils: relationship to calcium mobilization and degranulation. J Allergy Clin Immunol 1991; 88:114–124.
54. Chilvers ER, Challiss RAJ, Barnes PJ, Nahorski SR. Mass changes of inositol (1,4,5)trisphosphate in trachealis muscle following agonist stimulation. Eur J Pharmacol 1989; 164:587–590.
55. Majerus PW, Ross TS, Cunningham TW, Caldwell KK, Bennett-Jefferson A, Bansal VS. Recent insights into phosphatidylinositol signalling. Cell 1990; 63:459–465.
56. Nishizuka Y. Studies and perspectives of protein kinase C. Science 1986; 233: 305–312.
57. Rabe KF, Giembycz MA, Dent G, Barnes PJ. Activation of guinea-pig eosinophil respiratory burst by leukotriene B_4: role of protein kinase C. Fund Clin Pharmacol 1992; 6:353–358.
58. Shute JK, Rimmer SJ, Akerman CL, Church MK, Holgate ST. Studies of the cellular mechanisms for the generation of superoxide by guinea pig eosinophils and its dissociation from granule peroxidase release. Biochem Pharmacol 1990; 40:2013–2020.
59. Kroegel C, Giembycz MA, Barnes PJ. Staurosporine inhibits protein kinase C and prevents phorbol ester mediated PAF receptor desensitization in guinea pig eosinophils. Eur Respir J 1989; 2:668S.
60. Zavioco GB, Halenda SP, Shaafi RI, Feinstein MB. Phorbol myristate acetate inhibits thrombin-stimulated Ca^{2+} mobilization and phosphoinositol 4,5,bisphosphate hydrolysis in human platelets. Proc Natl Acad Sci USA 1985; 82:3859–3862.
61. Nishizuka Y. The molecular heterogeneity of protein kinase C and its implications for cellular regulation. Nature 1988; 334:661–665.
62. Thompson NT, Bonser RW, Garland LG. Receptor-coupled phospholipase D and its inhibition. Trends Pharmacol Sci 1991; 12:404–407.
63. Billah MM, Anthes JE. The regulation and cellular functions of phosphatidylcholine hydrolysis. Biochem J 1990; 269:281–291.

64. Uings IJ, Thomson NT, Randall RW, et al. Tyrosine phosphorylation is involved in receptor coupling to phospholipase D but not phospholipase C in the human neutrophil. Biochem J 1991; 281:597–600.
65. Bauldry SA, McCall CE, Cousart SL, Bass DA. Tumor necrosis factor-α priming of phospholipase A_2 activation in human neutrophils. J Immunol 1991; 146:1277–1285.
66. Minnicozzi M, Anthes JC, Siegel MI, Billah MM, Egan RW. Activation of phospholipase D in normodense human eosinophils. Biochem Biophys Res Commun 1990; 170:540–547.
67. Dent G, Barnes PJ. Platelet activating factor stimulates a pertussis toxin-sensitive GTPase activity in guinea pig eosinophil membranes. Br J Pharmacol 1991; 104:86P.
68. Gomperts BD. G_E: a GTP-binding protein mediating exocytosis. Annu Rev Physiol 1990; 52:591–606.
69. Cromwell O, Bennett JP, Hide I, Kay AB, Gomperts BD. Mechanisms of granule enzyme secretion from permealised guinea-pig eosinophils. Dependence on Ca^{2+} and guanine nucleotides. J Immunol 1991; 147:1905–1911.
70. Jong EC, Henderson WR, Klebanoff SJ. Bactericidal activity of eosinophil peroxidase. J Immunol 1980; 124:1378–1382.
71. Agosti JM, Altman LC, Ayars GH, Loegering DA, Gleich GJ, Klebanoff SJ. The injurious effect of eosinophil peroxidase hydrogen peroxide and halides on pneumocytes in vitro. J Allergy Clin Immunol 1987; 79:496–504.
72. Barnes PJ. Reactive oxygen species and airway inflammation. Free Rad Biol Med 1990; 9:235–243.
73. Silberstein DS, David JR. The regulation of human eosinophil function by cytokines. Immunol Today 1987; 8:380–385.
74. Sanjar S, Smith D, Kings MA, Morley J. Pretreatment with rh-GMCSF, but not rh-IL3, enhances PAF-induced eosinophil accumulation in guinea-pig airways. Br J Pharmacol 1990; 100:399–400.
75. Chanez P, Dent G, Yukawa T, Barnes PJ, Chung KF. Generation of oxygen free radicals from blood eosinophils from asthma patients after stimulation with PAF or phorbol ester. Eur Respir J 1990; 3:1002–1007.
76. Carlson M, Makansson L, Petersen C, Stalenheim G, Venge P. Secretion of granule proteins from eosinophils and neutrophils is increased in asthma. J Allergy Clin Immunol 1991; 87:27–33.
77. Yazdanbakhsh M, Eckman CM, Koenderman L, Verhoeven AJ, Roos D. Eosinophils do respond to fMLP. Blood 1987; 70:379–383.
78. Lowman MA, Benyon RC, Church MK. Characterization of neuropeptide-induced histamine release from human dispersed skin mast cells. Br J Pharmacol 1988; 95: 121–130.
79. Mousli M, Bueb J-L, Bronner C, Rouot B, Landry Y. G protein activation: a receptor-independent mode of action for cationic amphilic neuropeptides and venom peptides. Trends Pharmacol Sci 1990; 11:358–361.
80. Bach MK, Brasher JR. FMLP is a potent activator of guinea-pig eosinophils but its activity is dependent on the poor overnight in vitro culture of the cells (facilitation). Immunology 1992; 75:680–687.

81. Arai K, Lee F, Miyajima A, Miyataki S, Arai N, Yokota T. Cytokines: coordinators of immune and inflammatory responses. Annu Rev Biochem 1990; 59:783–836.
82. Shepherd VL. Cytokine receptors in lung. Am J Respir Cell Mol Biol 1991; 5: 403–410.
83. Muegge K, Durum SK. Cytokines and transcription factors. Cytokine 1990; 2:1–8.
84. Barnes PJ, Chung KF. Questions about inhaled β2-agonists in asthma. Trends Pharmacol Sci 1992; 13:20–23.
85. Yukawa T, Ukena D, Chanez P, Dent G, Chung KF, Barnes PJ. Beta-adrenergic receptors on eosinophils: binding and functional studies. Am Rev Respir Dis 1990; 141:1446–1552.
86. Giembycz MA, Rabe KF, Dent G, Perkins RS, Ravenall S, Barnes PJ. Stimulation of thromboxane biosynthesis by leukotriene B_4 in guinea-pig eosinophils: effect of phosphodiesterase inhibitors, β-adrenoceptor agonists and lipophilic cyclic nucleotide analogues (abstr). Br J Pharmacol. In press.
87. Kita H, Abu-Ghazaleh RI, Gleich GJ, Abraham RT. Regulation of Ig-induced eosinophil degranulation by adenosine-3′5′-cyclic monophosphate. J Immunol 1991; 146:2712–2718.
88. Tecoma ES, Motulsky HJ, Traynor AE, Oman GM, Muller H, Sklar LA. Transient catecholamine modulation of neutrophil activation: kinetic and intracellular aspects of isoproterenol actions. J Leuk Biol 1986; 40:629–644.
89. Barnes PJ, Dent G, Rabe KF, Giembycz MA. Modulation of guinea-pig eosinophil function by a selective β-adrenoceptor agonist procaterol. Proc ICACI Meeting, Kyoto, 1991.
90. Rabe KF, Giembycz M, Dent G, Evans PM, Barnes PJ. β_2-Adrenoceptor agonists and respiratory burst activity in guinea pig and human eosinophils. Fund Clin Pharmacol 1992; 5:402.
91. Kraan J, Koeter GH, Van der Mark TW, Sluiter HJ, De Vries K. Changes in bronchial hyperreactivity induced by 4 weeks of treatment with antiasthmatic drugs in patients with allergic asthma: a comparison between budesonide and terbutaline. J Allergy Clin Immunol 1985; 76:628–636.
92. Kerrebijn KF, Von Essen-Zandvliet EEM, Neijens HJ. Effect of long-term treatment with inhaled corticosteroids and beta-agonists on bronchial responsiveness in asthmatic children. J Allergy Clin Immunol 1987; 79:653–659.
93. Sears MR, Taylor DR, Print CG, et al. Regular inhaled beta-agonist treatment in bronchial asthma. Lancet 1990; 336:1391–1396.
94. Van Schayck CP, Graafsma SJ, Visch MB, Dompeling E, van Weel C, Herwaarden CLA. Increased bronchial hyperresponsiveness after inhaling salbutamol during 1 year is not caused by subsensitization to salbuterol. J Allergy Clin Immunol 1990; 86:736–800.
95. Collins S, Caron MG, Lefkowitz RJ. β-Adrenergic receptors in hamster smooth muscle cells are transcriptionally regulated by glucocorticoids. J Biol Chem 1988; 263:9067–9070.
96. Hadcock JR, Wang HY, Malbon CC. Agonist-induced destabilization of β-adrenergic receptor mRNA: attenuation of glucocorticoid-induced up-regulation of β-adrenergic receptors. J Biol Chem 1989; 264:19928–19933.

97. Mak JCW, Adcock I, Barnes PJ. Dexamethasone increases β_2-adrenoceptor gene expression in human lung (abstr). Am Rev Respir Dis. In press.
98. Yukawa T, Kroegel C, Dent G, Chanez P, Ukena D, Barnes PJ. Effect of theophylline and adenosine on eosinophil function. Am Rev Respir Dis 1989; 140:327–333.
99. Dent G, Giembycz MA, Rabe KF, Barnes PJ. Inhibition of guinea pig eosinophil cyclic nucleotide phosphodiesterase activity and opsonized zymosan-stimulated respiratory burst by type IV phosphodiesterase inhibitors. Br J Pharmacol 1991; 103: 1339–1346.
100. Souness JC, Carter CM, Diocee BH, Hassal GA, Wood LJ, Turner NC. Characterization of guinea-pig eosinophil phosphodiesterase activity. Assessment of its involvement in regulatory superoxide generation. Biochem Pharmacol 1991; 42: 937–945.
101. Schrier DJ, Imre RM. The effects of adenosine antagonists on human neutrophil function. J Immunol 1986; 137:3284–3289.
102. Beavo JA, Reifsnyder DH. Primary sequence of cyclic nucleotide phosphodiesterase isoenzymes and the design of selective inhibitors. Trends Pharmacol Sci 1990; 11: 150–155.
103. Nicholson CD, Challiss RAJ, Shahid M. Differential modulation of tissue function and therapeutic potential of selective inhibitors of cyclic nucleotide phosphodiesterase isoenzymes. Trends Pharmacol Sci 1991; 12:19–27.
104. Dent G, Evans PM, Chung KF, Barnes PJ. Zardaverine inhibits respiratory burst activity in human eosinophils. Am Rev Respir Dis 1990; 141:A392.
105. Torphy TJ, Undem RJ. Phosphodiesterase inhibitors: new opportunities for the treatment of asthma. Thorax 1991; 46:499–503.
106. Giembycz MA. Could selective cyclic nucleotide phosphodiesterase inhibitors render bronchodilator therapy redundant in the treatment of bronchial asthma? Biochem Pharmacol. In press.
107. Sanjar S, Aoki S, Kristersson A, Smith D, Morley J. Antigen challenge induces pulmonary eosinophil accumulation and airway hyperreactivity in sensitized guinea pigs: the effect of anti-asthma drugs. Br J Pharmacol 1990; 99:679–686.
108. Sturton G, Norman P. Prostanoid receptors on human PMN and eosinophils. Am Rev Respir Dis 1991; 143:A641.
109. Barnes PJ. A new approach to asthma therapy. N Engl J Med 1989; 321:1517–1527.
110. Barnes PJ. Effect of corticosteroids on airway hyperresponsiveness. Am Rev Respir Dis 1990; 141:S162–S165.
111. Baigelman W, Chodosh S, Pizzuto D, Cupples LA. Sputum and blood eosinophils during corticosteroid treatment of acute exacerbations of asthma. Am J Med 1983; 75:929–936.
112. O'Connor BJ, Evans PM, Ridge SM, Fuller RW, Barnes PJ. Effect of an inhaled steroid (budesonide) on indirect airway responsiveness and eosinophils in asthma. Am Rev Respir Dis 1991; 143:A22.
113. Slavick FT, Abboud CN, Brennan JK, Lichtman M. Modulation in vitro eosinophil progenitors by hydrocortisone: role of accessory cells and interleukins. Blood 1985; 66:1072–1079.
114. Lamas AM, Leon OG, Schleimer RP. Glucocorticoids inhibit eosinophil responses

to granulocyte-macrophage colony-stimulating factor. J Immunol 1991; 147: 254–259.
115. Tai PC, Sun L, Spry CJF. Effects of IL-5, granulocyte/macrophage colony-stimulating factor (GM-CSF) and IL-3 on the survival of human blood eosinophils in vitro. Clin Exp Immunol 1991; 85:312–316.
116. Dahl R, Venge P, Fredens K. Eosinophils. In: Barnes PJ, Rodger IW, Thomson NC, eds. Asthma: Basic Mechanisms and Clinical Management. London: Academic Press, 1988:115–129.
117. Kita H, Abu-Ghazaleh R, Sanderson CJ, Gleich GJ. Effect of steroids on immunoglobulin-induced eosinophil degranulation. J Allergy Clin Immunol 1991; 87:70–77.
118. Evans PM, Barnes PJ, Chung KF. Effect of corticosteroids on human eosinophils in vitro. Eur J Pharmacol 1990; 3:160S.
119. Bruijnzeel PLB, Warringa RAJ, Kok PTM, Kreukniet J. Inhibition of neutrophil and eosinophil induced chemotaxis by nedocromil sodium. Br J Pharmacol 1990; 99: 798–802.
120. Burke LA, Crea AEG, Wilkinson JRW, Arm JP, Spur BW, Lee TH. Comparison of the generation of platelet activating factor and leukotriene C_4 in human eosinophils stimulated with unopsonized zymosan and the calcium ionophore A23178: the effects of nedocromil sodium. J Allergy Clin Immunol 1990; 85:26–35.
121. Chung KF, Barnes PJ. Loop diuretics and asthma. Pulm Pharmacol 1992; 5:1–7.
122. O'Connor BJ, Chung KF, Chen-Wordsell YM, Fuller RW, Barnes PJ. Effect of inhaled furosemide and bumetamide on adenosine 5′-monophosphate and sodium metabisulphite-induced bronchoconstriction. Am Rev Respir Dis 1991; 143:1329–1333.
123. Perkins RS, Dent G, Chung KF, Barnes PJ. Effects of anion transport inhibitors and chloride ions on eosinophil respiratory burst activity. Biochem Pharmacol 1992; 43: 2480–2482.
124. Shult PA, Graziano FM, Wallow IH, Busse WW. Comparison of superoxide generation and luminol-dependent chemiluminescence with eosinophils and neutrophils from normal individuals. J Lab Clin Med 1985; 106:638–645.

DISCUSSION (Speaker: P. J. Barnes)

Kay: You seem to be intermingling data from eosinophils obtained from the guinea pig and humans. Can you justify this? Also, could you summarize the main differences between eosinophils and neutrophils in terms of the various activation pathways you have described?

Barnes: We have found it necessary to use guinea pig peritoneal eosinophils for the studies that involve detailed biochemical characterization, because the number and purity of cells is a limiting factor. We have carried out limited studies on human eosinophils (often obtained from hypereosinophilic donors) to confirm some of our animal studies, but the number of molecular pharmacology studies that can be conducted on human blood eosinophils is limited. There are similarities and differences between human and guinea pig eosinophils; PAF activates eosinophils of both species but is more consistently effective in human cells, whereas LTB_4 is a consistent activator of guinea pig but not human cells. Eosinophils from both species are activated by phorbol esters and opsonized zymosan. As regards modulatory mechanisms, both guinea pig and human eosinophils have β-receptors and adenosine A_2-receptors and both express type IV phosphodiesterase.

Egan: I compliment you on recognizing the importance of phospholipase D (PLD) in guinea pig eosinophil activation. Your observations are very similar to studies we have published using human eosinophils. For PLD to generate diacylglycerol, the phosphatidic acid must be diphosphorylated by the phosphatidate phosphohydrolase, which can be inhibited by 200 μM propranolol. Propranolol can, therefore, be used to more fully define the role of PLD in your eosinophils. What was the readout for PLD? PDBU?

Barnes: We have not specifically investigated the effect of propranolol on PLD generation by eosinophils, but high concentrations of propranolol have several nonspecific effects that may make interpretation difficult. We have used the phorbol ester phorbol dibutyrate to activate PLD, but have also used LTB_4. We have used the $[^3H]$butan-1-ol transphosphatidylation method to quantify PLD activation.

Venge: I have a comment as to your findings of an anion inhibitory effect of steroids on eosinophil functions. The conclusion today is that these cells are basically steroid-resistant, but I think it may be a little more complicated than that. Thus we find that the chemotactic response of eosinophils is completely resistant to any effect of glucocorticosteroids (GCS). However, when it comes to degranulation, it seems to be more variable. Thus, some preparations from animals show a significant inhibition by GCS, whereas others seem to be completely resistant. We do not know or understand the reason for this variability and my question is: do you see a similar variation in your degranulation experiments?

Barnes: We have carried out relatively few studies on the effects of corticosteroids on human eosinophils and have demonstrated only a small inhibitory effect of superoxide anion release. We also measured EDN, as an index of degranulation, but found that steroids had little effect. I agree with you that sometimes greater inhibitory effects are seen and the state of prior activation or priming of eosinophils may be important in determining their responsiveness to activating stimuli. It is also possible that synthesis of cytokines may be inhibited by steroids in eosinophils, but such studies have not yet been reported.

Morley: With reference to the comment of Per Venge, we routinely study the activation of eosinophils (superoxide generation to PMA and FMLP), using highly purified eosinophils from allergic or nonallergic subjects. In our experience, GCS have no inhibitory effect, and we see no variability in this outcome.

Barnes: It is important to recognize that steroids may have effects on eosinophils other than inhibition of superoxide anion generation. Steroids clearly have some direct effects on eosinophils. For example, they prevent the effects of cytokines such as IL-5 and GM-CSF in sustaining eosinophil survival, so it can be assumed that these eosinophils have glucocorticoid receptors.

Kita: We are using human peripheral blood normodense eosinophils and EDN release as readout. We experience some differences from your data.

1. We cannot observe EDN release from PAF-stimulated eosinophils.
2. PMA induced EDN release from eosinophils, in contrast to your inhibitory effects.

3. When Ig-coated beads were used as stimuli, EDN release was inhibited up to 40% by β-agonists and phosphodiesterase inhibitor.

Could you comment on these differences?

Barnes:

1. There may well be differences in the activation and modulation of eosinophils depending on their prior history. Most of our studies in human blood eosinophils have been carried out on eosinophils from hypereosinophilic patients; it is possible that these cells have a different responsiveness.
2. Phorbol esters also activate human and guinea pig eosinophils in our studies. The results I presented refer to an inhibitory effect of phorbol esters on the calcium transient, suggesting that protein kinase C may be involved in terminating responses to inflammatory stimuli, in addition to stimulating the oxidative burst. It is possible that different isoenzymes of protein kinase C are involved.
3. We also find an inhibitory effect of β-agonists on human eosinophils, but it is important to recognize that this response is tachyphylactic. Although we might expect to see a greater effect in the presence of PDE inhibitors, we have not often observed this. This has suggested to us that β-receptors may be inhibiting eosinophils via some mechanism other than via a rise in intracellular cyclic AMP, such as direct opening of an ion channel, as has been reported recently in several other cell types.

Durham: You mentioned that PGE_2 inhibits eosinophil production of thromboxane. Is there a specific receptor for PGE_2? Are other eosinophil functions altered? Does PGE_2 have potential as an anti-inflammatory drug?

Barnes: PGE_2 is effective in inhibiting eosinophil activation and presumably works via an EP-receptor. Subtypes of EP-receptors have now been recognized, and there is some evidence that an EP_2-receptor may be involved. PGE_2 may have a modulatory role on airway inflammation and may have the capacity to inhibit eosinophils, macrophages, lymphocytes, and other inflammatory cells. This may be worthy of further investigation now that EP-receptor subtype selective agonists have been developed.

Denburg: Is there any information on the modulation of cytokine release by eosinophils by the pharmacological agents you described?

Barnes: I am not aware of such studies, but I believe that this is an important area for future investigation. The effect of corticosteroids on eosinophil GM-CSF, TNF-α, and IL-5 synthesis would be of particular interest.

Townley: We have reported that formoterol blocks the late reactions in guinea pig and also blocks the influx of eosinophils in the late reaction. We have extended these studies to show that eosinophils from antigen-challenged guinea pigs are primed for superoxide production following stimulation with PAF, FMLP, or PMA. Formoterol given 15 min prior to antigen challenge was very effective in preventing superoxide production by PAF and FMLP eosinophils, but not by PMA-stimulated eosinophils. Dr. Barnes, in light of your studies of postreceptor eosinophil activation, could you interpret these effects of β-agonists on eosinophils? Obviously these are important findings and may help to further understand how β-agonists may inhibit one pathway and not another.

Barnes: The differences in the effects of a β-agonist on different activating stimuli may help to elucidate their mode of action. If β-agonists inhibit receptor-mediated eosinophil activation, but not phorbol ester–mediated activation, this suggests that they are blocking the steps prior to activation of PKC. We have some evidence that β-agonists reduce the calcium transient in guinea pig eosinophils and may therefore inhibit either calcium influx or the release of Ca^{2+} from intracellular stores. You have highlighted that there may be differences between different activating stimuli, and it is important to test the modulatory action of drugs against different activating stimuli and using different readouts of eosinophil function (mediator release, degranulation, cytokine production, survival, etc.).

17

Transgenic Experiments with Interleukin-5

Colin J. Sanderson
Western Australian Research Institute for Child Health, Perth, Australia

Malcolm Strath and Ian Mudway
National Institute for Medical Research, London, England

Lindsay A. Dent
University of South Australia, Adelaide, Australia

I. INTRODUCTION

Three important features of eosinophilia provide information about the mechanism of control of eosinophil production. First, eosinophilia is under the control of T lymphocytes (T cells). Thus it is the consequence of an immune response. Second, increases in eosinophil numbers are frequently observed independently of increases in other blood leukocytes. This biological specificity implies a mechanism of control that is independent from the mechanisms controlling the production of the other leukocytes. Third, eosinophilia is observed in a restricted number of diseases, which indicates that the immune system is able to distinguish these particular types of antigenic challenge from the majority that do not induce eosinophils. This suggests that a subset of T cells is involved.

II. PRODUCTION OF EOSINOPHILS IN VITRO

A. Mouse Cell Cultures

Simple liquid culture systems give high levels of eosinophil production and formed the basis for a sensitive assay for eosinophil differentiation factors (1,2). This assay led to identification of a cytokine first called eosinophil differentiation factor and subsequently termed interleukin-5 (IL-5) (3–6). In contrast, in the classic agar-colony assay, only a small number of eosinophil colonies are produced from bone marrow from normal mice in the presence of IL-5, and this number is not increased significantly when bone marrow from mice undergoing eosinophilia as a consequence of infection by *Mesocestoides corti* is used.

The low efficiency of the colony assay for murine eosinophil precursors has limited the study of eosinophil hemopoiesis. Although some attempts have been made to use the colony assay, none of these has been entirely satisfactory. In our own work it has only been possible to estimate relative numbers by comparison of the number of eosinophils produced in liquid cultures. Thus, under similar conditions a low production of eosinophils indicates a few precursors, and a high production indicates relatively more precursors (7,8). Using the liquid assay with IL-5, there is at least a 100-fold increase in eosinophils produced from bone marrow from eosinophilic mice compared to normal mice (3). A more realistic comparison can be made by a limiting dilution assay in which the number of bone marrow cells required to produce a fixed number of eosinophils is used to estimate relative precursor frequency. Another approach is to compare the dose of IL-5 required to produce a fixed number of eosinophils, as the sensitivity toward IL-5 is proportional to the potential to produce eosinophils (9,10).

In the mouse system, IL-5 induces the specific and lineage-restricted production of eosinophils in liquid bone marrow cultures (3). In contrast, both granulocyte-macrophage colony-stimulating factor (GM-CSF) and IL-3 induce the production of neutrophils, macrophages, and eosinophils (11). The microenvironment in the liquid bone marrow cultures allows production of neutrophils for extended periods of time without the addition of exogenous factors (12). No eosinophils are observed under these conditions. However, addition of IL-5 results in the induction of eosinophil production, which peaks at about 3 weeks and continues until about 6–8 weeks. This transient production suggests that IL-5 is incapable of stimulating the production of eosinophil precursors, at least in these bone marrow cultures (3). This led us to suggest that IL-5 is a late-acting factor in eosinophilopoiesis (4–6).

A similar conclusion was based on the observation that IL-5 induced no eosinophil colonies from spleens from mice treated with 5-fluorouracil (5-FU), in which only primitive stem cells survive. However, G-CSF and IL-3 did support colony growth, and addition of IL-5 induced eosinophils in a proportion of the colonies formed. This was interpreted as evidence that IL-5 promoted only the terminal differentiation of eosinophils (13). In a similar approach, it was demonstrated that IL-3 and GM-CSF, but not G-CSF induced a large increase in IL-5-responsive eosinophil precursors from normal bone marrow. Both IL-1 and IL-3 were necessary to induce precursors when bone marrow cells from mice treated with 5-FU were used (14), again suggesting that IL-5 is a late-acting factor.

B. Human Cell Cultures

In contrast to studies in the mouse, cultures of human bone marrow in semisolid agar give significant numbers of eosinophil colonies in the presence of IL-5. However, both IL-3 and GM-CSF stimulate production of a greater number of eosinophil colonies than IL-5 (15). As all the cells in these colonies are morphologically mature and there are no obvious differences in colony size, this surprising result suggests that either IL-3 or GM-CSF is capable of inducing eosinophilia without the action of IL-5. In addition, it suggests that there may be a large pool of eosinophil precursors that are unresponsive to IL-5. In attempts to clarify these results, similar experiments were carried out with human bone marrow in liquid cultures (where the total number of eosinophils produced, rather than the number of colonies, can be assessed). The results from these experiments are markedly different from those obtained in the colony assay, with IL-5 inducing a larger number of eosinophils than IL-3 or GM-CSF. Thus these two in vitro assay systems lead to two contradictory interpretations of the role of IL-5.

When human bone marrow cells were incubated in liquid culture with IL-3 or GM-CSF, there was a significant increase in eosinophil precursors, as assessed by colony formation in semisolid agar cultures with IL-5. In contrast, IL-1, IL-6, and IL-5 had no effect (16). These experiments were carried out with total mononuclear cells from the bone marrow, so the lack of effect of IL-1 and IL-6 might have been due to production of these cytokines by other cells. However, when experiments with highly enriched populations of progenitor cells were performed, similar results were obtained (17).

The studies with human cells are consistent with the experiments with mouse cells, suggesting that IL-5 is a late-acting factor, and that eosinophil production requires other factors to cause the differentiation of eosinophil precursors from primitive stem cells. This apparent requirement for a cytokine network in the development of eosinophilia is paradoxical, because all the cytokines that seem to be active in the early stages of the differentiation pathway are powerful growth factors for other lineages, yet eosinophilia can occur in vivo with little if any change in the numbers of other leukocytes. Experiments in vivo clearly indicate the central role of IL-5 in eosinophilia and indicate that experiments in vitro may be misleading because the culture conditions do not accurately reproduce the microenvironment of the bone marrow. Thus while it appears from experiments in vitro that IL-5 is only a late-acting factor, experiments in vivo indicate that IL-5 induces the full pathway of eosinophil differentiation.

C. Other Activities

IL-5 is an activating factor for human eosinophils (18,19) and increases their survival in vitro (20). This latter activity provides a simple and sensitive assay for human IL-5. IL-5 has been shown to be active on basophils (21), suggesting a close relationship between the eosinophil and basophil lineages. In the mouse, IL-5 is a well-characterized B-cell growth factor (22), although human IL-5 has no activity in analogous assays on human B cells (23). There is no explanation for this intriguing species difference.

III. EXPERIMENTS IN VIVO

A. Reconstitution with Cells Expressing IL-5

A retroviral vector containing the IL-5 cDNA coding sequence was used to produce infectious retrovirus carrying the cytokine gene. This was used to infect hemopoietic stem cells from the fetal liver of mice treated with 5-FU (24).

Mice carrying hemopoietic cells infected with the retrovirus carrying the IL-5 gene produced high-level eosinophilia for at least 12 months. There were excess numbers of eosinophils in the bone marrow, spleen, liver, lung, and gut. No changes were detected in conventional B-cell populations, although there was an increase in Ly-1$^+$ B cells. In contrast to the animals expressing IL-3 or GM-CSF using this system, the IL-5 mice remained healthy, suggesting that expansion of these populations of cells had no pathological consequences (24).

B. IL-5 Transgenic Mice

As IL-5 is normally expressed only after antigen stimulation, in a subpopulation of T cells, the production of transgenic mice in which IL-5 is produced constitutively should provide a constant source of IL-5 in a physiologically relevant manner. Two different approaches have been used to produce transgenic mice expressing IL-5. First, the IL-5 gene was ligated to the human CD2 locus control region (LCR) to give constitutive expression of IL-5 by all T cells (9). Second, the IL-5 cDNA was ligated to the mouse metallothionein promoter to give a low-level constitutive production of IL-5 from spleen, liver, kidney, and bone marrow. Treatment with heavy-metal ions induced expression of the metallothionein promoter and increased the serum IL-5 levels by about fivefold (25).

The CD2 antigen is constitutively expressed on all T cells. Expression of human CD2 in transgenic mice was shown to be T cell specific, integration site independent, and the levels of expression were proportional to the transgene copy number (26). The DNA sequences controlling this tissue specificity (the CD2 LCR) were identified in the 3′ region of the CD2 gene and were shown to confer a similar expression pattern as CD2 on other reporter genes (27).

For the production of transgenic mice, the construct used consisted of a 2-kb fragment containing the hCD2 LCR ligated to a 10-kb fragment containing the mouse IL-5 gene (Fig. 1). This construct contains about 3 kb of 5′ flanking

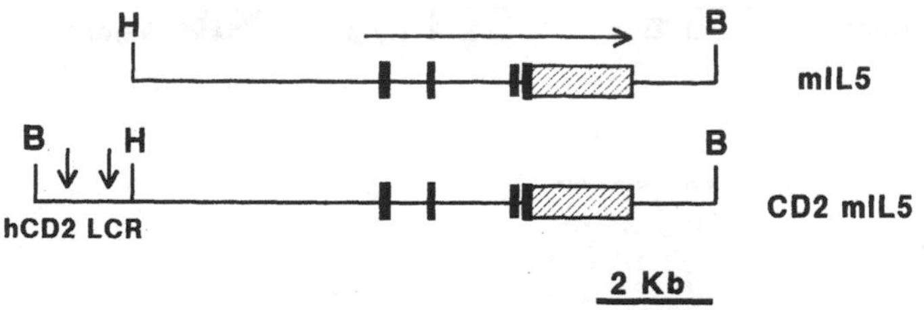

Figure 1 Construct used for the generation of CD2-IL-5 transgenic mice (9). The upper panel shows the structure of the mouse IL-5 gene, boxes indicate exons, and the hatched part shows the 3′ untranslated region. (H) HindIII and (B) BamHI restriction enzyme sites are indicated. The large arrow indicates the direction of transcription. The lower panel shows the construct used. A 2-kb BamHI-to-HindIII fragment from the human CD2 gene was ligated to the mouse DNA. The vertical arrows indicate two DNAse-hypersensitive sites identified in the locus control region (LCR) (27). (Adapted from Ref. 9.)

sequence and about 2 kb of 3′ flanking sequence, so the IL-5 gene is expressed from its own promoter. Two types of expression occur: first, a low-level constitutive expression due to the CD2 LCR, and second, inducible expression when the T cells are stimulated with antigen due to the IL-5 controlling sequences. We have produced four lines of transgenic mice using this construct, and all lines showed high-level eosinophil production that was proportional to the transgene copy number. These animals have detectable levels of IL-5 in the serum and show a profound and lifelong eosinophilia, with large numbers of eosinophils in the blood, spleen, bone marrow, lung, and gut wall (9). Data for two of these IL-5 transgenic lines are shown in Figures 2 and 3.

Infection of these mice with influenza virus by the intranasal route resulted in antihemagglutinin antibody of IgM and IgG1 isotypes, but the levels were the same in transgenic mice as nontransgenic mice. No antibody was detected of other isotypes in either group (Fig. 4).

The inducibility of the CD2-IL-5 was confirmed by demonstrating that spleen cells in culture produced a low level of IL-5 production, but this was increased nearly 10-fold when the cells were stimulated with concanavalin A. Infection of these transgenic mice with *M. corti*, itself a potent inducer of eosinophilia, increased serum IL-5 from about 40 units to over 1000 units. However, the number of eosinophils in the blood, bone marrow, and spleen decreased during the period corresponding to high-level serum IL-5 (Fig. 5). This was shown to be due to a decrease in eosinophil precursors. Therefore, it is likely that some form of control is operating to prevent the overproduction of eosinophils (10).

The effect of infection on total immunoglobulin levels is shown in Figure 6. Although levels of IgG2a, IgG3, and IgM were significantly increased in the

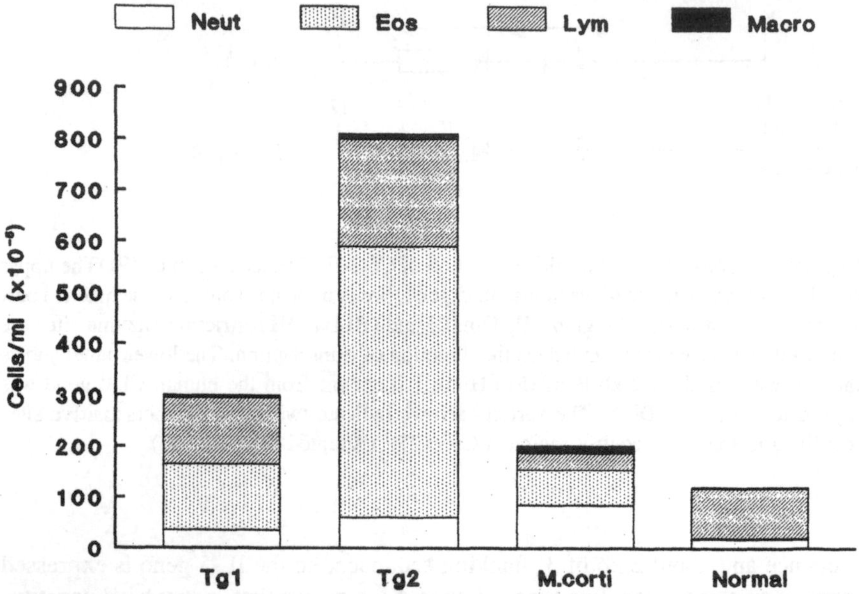

Figure 2 Blood leukocytes observed in CD2-IL-5 transgenic mice expressing about eight copies of the transgene (Tg1) and 30 copies (Tg2), compared to mice 18 days after infection with *M. corti* and normal mice. The number of eosinophils is higher in the transgenic line carrying the higher number of transgenes. (Adapted from Ref. 9.)

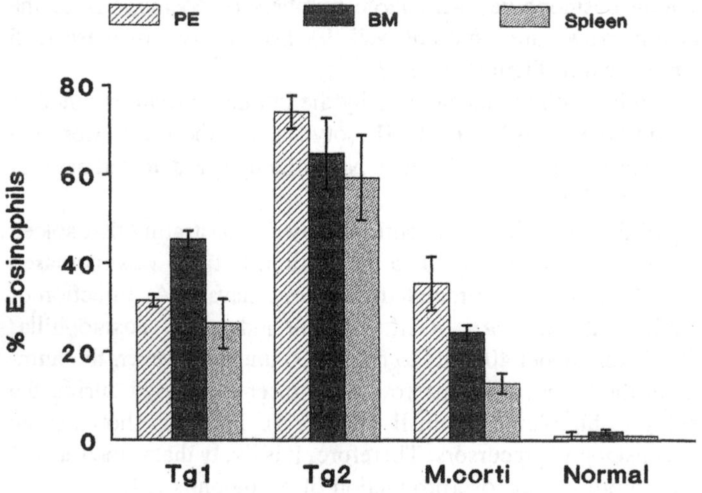

Figure 3 Distribution of eosinophils in peritoneal exudate (PE), bone marrow (BM), and spleen of transgenic mice, mice infected with *M. corti*, and normal mice. See Figure 2 for details. (From Ref. 9.)

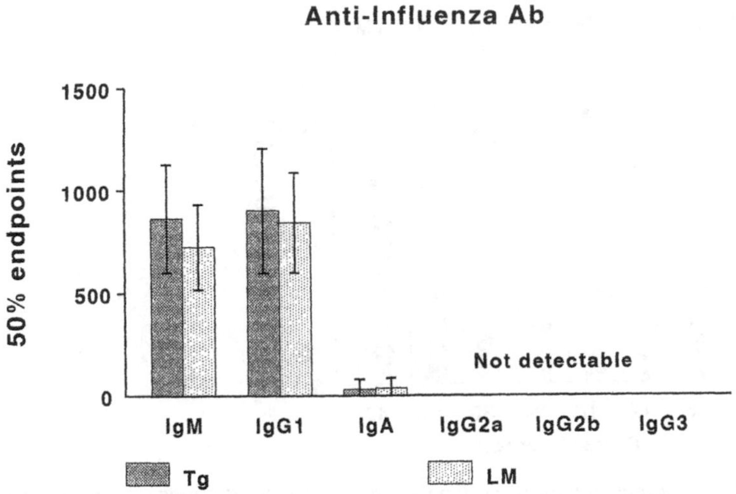

Figure 4 Anti-influenza antibody expressed as a serum dilution giving a 50% end-point, in transgenic (Tg) and nontransgenic littermates (LM). No differences were observed between the two groups. Serum antibody was assayed at 18 days after intranasal inoculation of influenza virus (strain X31). Assay: The wells of a microplate were coated with antigen by dispensing 50 μl of purified influenza hemagglutinin (1 unit/ml) and incubating overnight, followed by washing; 60 μl of serum dilutions was added and incubated 40 min and then washed; 50 μl of biotin-labeled goat antimouse isotype antibody (Southern Biotechnology Associates) was added and incubated 40 min before washing; 50 μl of horseradish peroxidase–labeled Strepavidin was added and incubated 40 min before washing. Finally, 50 μl of peroxidase substrate (tetramethyl benzidine HCl) solution was added and the plates read in a plate reader.

uninfected transgenic animals compared with nontransgenic animals, this effect was small. There was increased production of all isotypes during the course of infection, but the changes were the same in both transgenic and control groups. Specific antibody to parasite antigens showed a slight increase in IgM in the transgenics compared to nontransgenics, but no differences were observed between the two groups in any of the other isotypes (Fig. 7).

The other approach using a metallothionein promoter to produce IL-5 transgenic mice provided some different insights into the biological activity of IL-5. The construct consisted of a truncated murine IL-5 cDNA flanked by rabbit β-globin splice and polyadenylation sites, ligated to the mouse metallothionein promoter, which is inducible by heavy-metal ions. Of two transgenic lines produced, one carried about five copies of the transgene and produced low and variable levels of eosinophils; but the other carried about 40 copies and gave high-level eosinophilia, which was further increased after injection of cadmium (25).

In contrast to the CD2-IL-5 mice, the MT-1-IL-5 mice showed, in addition to

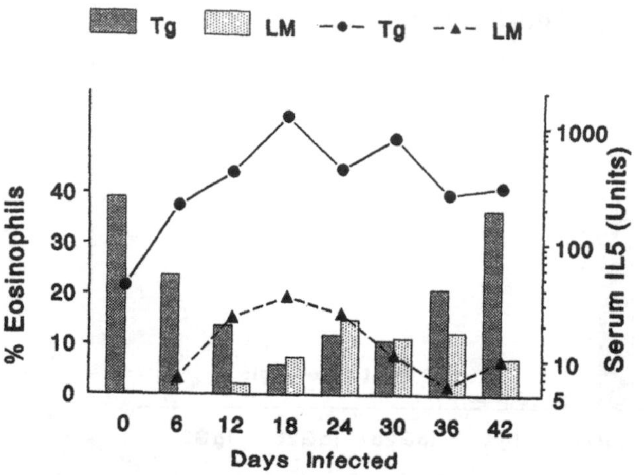

Figure 5 Effect of infection by *M. corti* on serum IL-5 levels. (●) Transgenic mice (Tg); (▲) nontransgenic littermates (LM) and eosinophil numbers in the spleen. Eosinophils in the spleen of transgenic mice (Tg) are shown in dark bars, and those in the spleen of nontransgenic littermates are shown in light bars. The level of IL-5 in the littermates rises to a maximum of about 50 units, while the levels in the transgenics rises to about 1000 units. The littermates show a peak eosinophil response slightly later than the peak IL-5 response, while the transgenic mice show a decrease in eosinophil numbers, which corresponds approximately to the high levels of IL-5. (From Ref. 10.)

the spleen and peritoneum, significant numbers of eosinophils in the liver, where transgene expression would also be expected. This suggests that eosinophils migrate from the hemopoietic tissues, where they are produced, into tissues where IL-5 is expressed. This demonstrates that IL-5 is involved in the localization of eosinophils to the tissue where it is being produced.

The MT-1-IL-5 mice showed increased IgM and IgA, as well as an accumulation of Ly-1^+ cells in the spleen after cadmium administration.

C. Administration of Anti-IL-5 Antibody

Mice infected with *Nippostrongylus brasiliensis* develop eosinophilia and increased levels of IgE. However, when they were treated with an anti-IL-5 antibody, no eosinophils were observed (28) and the numbers of eosinophil precursors present in the bone marrow were also depressed (28). Treatment with anti-IL-4 antibody inhibited the development of IgE response, but treatment with anti-IL-5 antibody had no effect on IgG responses in vivo, suggesting IL-5 may have little if any activity in the generation of antibody responses (29). Similarly, treatment of mice with anti-IL-5 antibody during an infection with *Schistosoma mansoni* (30) or the nematodes *Heligmosomoides polygyrus* (31) and *Strongyloides venezuelensis*

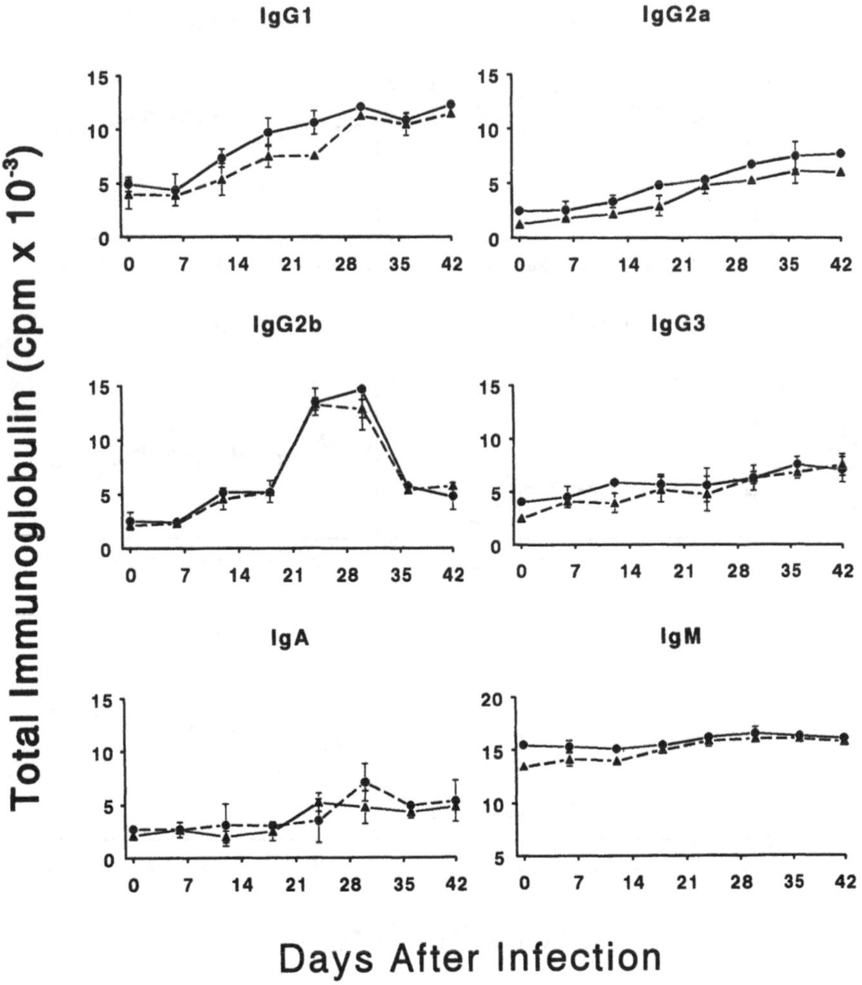

Figure 6 Total immunoglobulin levels during an infection of transgenic (●) and non-transgenic littermates (▲) with *M. corti*. Where error bars (±1 SD) are not shown, they are smaller than the symbols. There is significantly higher IgG2a, IgG3, and IgM in the transgenic mice before infection, but there is no difference in immunoglobulin levels between transgenic and normal mice. Assay: Serum dilutions were added to the wells of a microplate and incubated for 1 hr before washing; 25 μl of biotin-labeled goat antimouse isotype antibody (Southern Biotechnology Associates) was added and incubated 1 h before washing; 25 μl of ^{125}I-Streptavidin was added and incubated for 1 h before washing. The wells were then cut out and counted in a gamma counter.

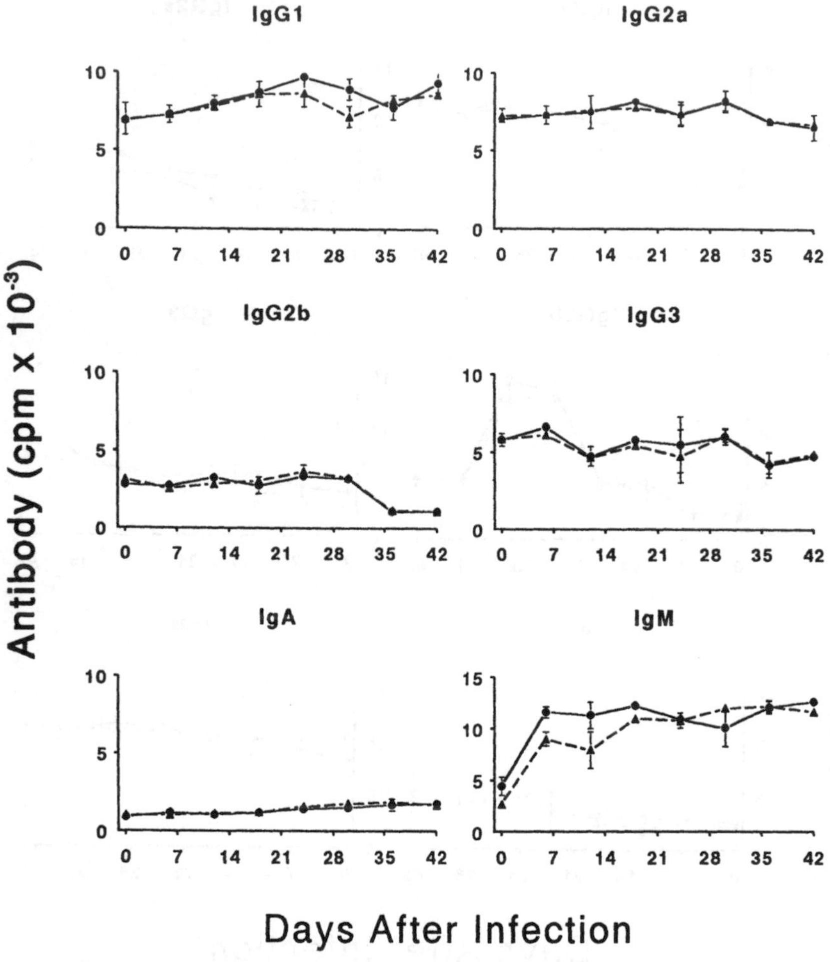

Figure 7 Specific antibody to *M. corti* antigens during an infection of transgenic (●) and nontransgenic littermates (▲) with *M. corti*. Where error bars (±1 SD) are not shown, they are smaller than the symbols. There was a significantly higher level of IgM antibody (background) reacting to the antigen in uninfected transgenics (day 0), but there was no difference between the two groups in antibody production of any isotype. Assay: *M. corti* antigen was prepared by homogenizing the tetrathyridia in PBS and then clarifying the suspension by centrifugation. This crude antigen was used to coat the wells of a microplate by adding 25 μl to each well, incubating for 1 h then washing. The assay for antibody was carried out as described in Figure 4. There was a significant background (counts bound in the absence of serum), which has not been subtracted from the data.

(32) totally blocked the development of eosinophilia. These experiments demonstrate the essential role that IL-5 performs in the control of eosinophilia in such parasite infections. In addition, they show that the apparent redundancy seen in vitro, where both IL-3 and GM-CSF are also able to induce eosinophil production, does not operate in these infections.

IV. SUMMARY AND CONCLUSION

The development of eosinophilia in transgenic mice expressing IL-5 suggests that IL-5 expression is sufficient to induce the full pathway of eosinophil differentiation. In contrast to experiments in vitro where other cytokines appear to be necessary for the development of eosinophil precursors from more undifferentiated stem cells, these experiments in vivo suggest that IL-5 is capable of inducing the production of precursors from stem cells.

Experiments with neutralizing anti-IL-5 antibody also indicate that IL-5 is sufficient for the production of eosinophils and indicate that IL-3 and GM-CSF are not involved in this model. These results all indicate a unique role for IL-5 in the regulation of eosinophilia. Interestingly, despite the massive, long-lasting eosinophilia in both models of IL-5 transgenic mice, they remained apparently normal. This illustrates that increased numbers of eosinophils are not themselves harmful. The tissue damage seen in allergic reactions and other diseases must be due to agents that trigger the eosinophils to degranulate, such as antibody-antigen complexes.

In view of the potential for eosinophils to cause tissue damage, it is interesting that production of very high levels of IL-5 appears to suppress the production of eosinophils. This suggests some form of control to prevent their overproduction. Although the mechanism of this control is not understood, it is possible that high levels of IL-5 lead to expression of the soluble form (rather than the membrane form) of the IL-5R.

The CD2-IL-5 transgenic mice showed a slight increase in IgG2a, IgG3, and IgM, but there was no effect on specific antibody production after infection with *M. corti* or influenza virus. The MT-1-IL-5 mice had increased levels of IgM and IgA immunoglobulin. Treatment with anti-IL-5 antibody had no effect on antibody responses, although treatment with anti-IL-4 inhibited the IgE response. While the activity of IL-5 on antibody production in vivo in the mouse is not yet clear, there is not yet any convincing evidence that IL-5 has activity on human B cells (23).

REFERENCES

1. Strath M, Warren DJ, Sanderson CJ. Detection of eosinophils using an eosinophil peroxidase assay. Its use as an assay for eosinophil differentiation factors. J Immunol Methods 1985; 83:209–215.

2. Strath M, Clutterbuck EJ, Sanderson CJ. Production of human and murine eosinophils in vitro and assay for eosinophil differentiation factors. In: Pollard JW, Walker JM, eds. Methods in Molecular Biology. Vol 5, Animal Cell Culture. Clifton, NJ: Humana Press, 1990:361–378.
3. Sanderson CJ, Warren DJ, Strath M. Identification of a lymphokine that stimulates eosinophil differentiation in vitro. Its relationship to IL3, and functional properties of eosinophils produced in cultures. J Exp Med 1985; 162:60–74.
4. Sanderson CJ, Campbell HD, Young IG. Molecular and cellular biology of eosinophil differentiation factor (interleukin-5) and its effects on human and mouse B cells. Immunol Rev 1988; 102:29–50.
5. Sanderson CJ. Eosinophil differentiation factor (interleukin-5). In: Dexter TM, Garland JM, Testa NG, eds. Colony-Stimulating Factors: Molecular and Cellular Biology. New York: Marcel Dekker, 1990:231–256.
6. Sanderson CJ. Interleukin-5. In: Thomson AW, ed. Immunology and Molecular Biology of Cytokines. London: Academic Press, 1991:149–168.
7. Warren DJ, Sanderson CJ. Production of a T cell hybrid producing a lymphokine stimulating eosinophil differentiation. Immunology 1985; 54:615–623.
8. Strath M, Sanderson CJ. Detection of eosinophil differentiation factor and its relationship to eosinophilia in *Mesocestoides corti*–infected mice. Exp Hematol 1986; 14: 16–20.
9. Dent LA, Strath M, Mellor AL, Sanderson CJ. Eosinophilia in transgenic mice expressing interleukin 5. J Exp Med 1990; 172:1425–1431.
10. Strath M, Dent LA, Sanderson CJ. Infection of IL5 transgenic mice with *Mesocestoides corti* induces very high levels of IL5 but depressed production of eosinophils. Exp Hematol 1992; 20:229–234.
11. Campbell HD, Sanderson CJ, Wang Y, Hort Y, Martinson ME, Tucker WQ, Stellwagen A, Strath M, Young IG. Isolation, structure and expression of cDNA and genomic clones for murine eosinophil differentiation factor. Comparison with other eosinophilopoietic lymphokines and identity with interleukin-5. Eur J Biochem 1988; 174:345–352.
12. Dexter TM, Allen TD, Lajtha LG. Conditions controlling the proliferation of hemopoietic stem cells in culture. J Cell Physiol 1977; 91:335–344.
13. Yamaguchi Y, Suda T, Suda J, Eguchi M, Miura Y, Harada N, Tominaga A, Takatsu K. Purified interleukin 5 supports the terminal differentiation and proliferation of murine eosinophilic precursors. J Exp Med 1988; 167:43–56.
14. Warren DJ, Moore MA. Synergism among interleukin 1, interleukin 3, and interleukin 5 in the production of eosinophils from primitive hemopoietic stem cells. J Immunol 1988; 140:94–99.
15. Clutterbuck EJ, Hirst EM, Sanderson CJ. Human interleukin-5 (IL-5) regulates the production of eosinophils in human bone marrow cultures: comparison and interaction with IL-1, IL-3, IL-6, and GM-CSF. Blood 1989; 73:1504–1512.
16. Clutterbuck EJ, Sanderson CJ. The regulation of human eosinophil precursor production by cytokines: a comparison of rhIL1, rhIL3, rhIL4, rhIL6 and GM-CSF. Blood 1990; 75:1774–1779.
17. Lu L, Lin ZH, Shen RN, Warren DJ, Leemhuis T, Broxmeyer HE. Influence of

interleukins 3, 5, and 6 on the growth of eosinophil progenitors in highly enriched human bone marrow in the absence of serum. Exp Hematol 1990; 18:1180–1186.
18. Lopez AF, Begley CG, Williamson DJ, Warren DJ, Vadas MA, Sanderson CJ. Murine eosinophil differentiation factor. An eosinophil-specific colony stimulating factor with activity for human cells. J Exp Med 1986; 163:1085–1099.
19. Lopez AF, Sanderson CJ, Gamble JR, Campbell HD, Young IG, Vadas MA. Recombinant human interleukin 5 is a selective activator of human eosinophil function. J Exp Med 1988; 167:219–224.
20. Begley CG, Lopez AF, Nicola NA, Warren DJ, Vadas MA, Sanderson CJ, Metcalf D. Purified colony stimulating factors enhance the survival of human neutrophils and eosinophils in vitro: a rapid and sensitive microassay for colony stimulating factors. Blood 1986; 68:162–166.
21. Denburg JA. Basophil and mast cell lineages in vitro and in vivo. Blood 1992; 79: 846–860.
22. Kinashi T, Harada N, Severinson E, Tanabe T, Sideras P, Konishi M, Azuma C, Tominaga A, Bergstedt-Lindqvist S, Takahashi M, Matsuda F, Yaoita Y, Takatsu K, Honjo T. Cloning of a complementary DNA encoding T cell replacing factor and identity with B cell growth factor II. Nature 1986; 324:70–73.
23. Clutterbuck E, Shields JG, Gordon J, Smith SH, Boyd A, Callard RE, Campbell HD, Young IG, Sanderson CJ. Recombinant human interleukin-5 is an eosinophil differentiation factor but has no activity in standard human B cell growth factor assays. Eur J Immunol 1987; 17:1743–1750.
24. Vaux DL, Lalor PA, Cory S, Johnson GR. In vivo expression of interleukin 5 induces an eosinophilia and expanded Ly-1B lineage populations. Intern Immunol 1990; 2: 965–971.
25. Tominaga A, Takaki S, Koyama N, Katoh S, Matsumoto R, Migita M, Hitoshi Y, Hosoya Y, Yamauchi S, Kanai Y, Miyazaki J-I, Usuku G, Yamamura K-I, Takatsu K. Transgenic mice expressing a B cell growth and differentiation factor gene (interleukin 5) develop eosinophilia and autoantibody production. J Exp Med 1991; 173: 429–437.
26. Lang GD, Wotton MJ, Owen WA, Sewell MH, Brown DY, Mason MJ, Crumpton, Kiousis D. The structure of the human CD2 gene and its expression in transgenic mice. EMBO J 1988; 7:1675–1682.
27. Greaves DR, Wilson FD, Lang G, Kiousis D. Human CD2 3′-flanking sequences confer high-level, T cell-specific, position-independent gene expression in transgenic mice. Cell 1989; 56:979–986.
28. Coffman RL, Seymour BW, Hudak S, Jackson J, Rennick D. Antibody to interleukin-5 inhibits helminth-induced eosinophilia in mice. Science 1989; 245: 308–310.
29. Finkelman FD, Holmes J, Katona IM, Urban JF, Beckmann MP, Park LS, Schooley KA, Coffman RL, Mosmann TR, Paul WE. Lymphocyte control of in vivo immunoglobulin isotype selection. Annu Rev Immunol 1990; 8:303–333.
30. Sher A, Coffman RL, Hieny S, Scott P, Cheever AW. Interleukin 5 is required for the blood and tissue eosinophilia but not granuloma formation induced by infection with *Schistosoma mansoni*. Proc Natl Acad Sci USA 1990; 87:61–65.

31. Urban JF Jr, Katona IM, Paul WE, Finkelman FD. Interleukin 4 is important in protective immunity to a gastrointestinal nematode infection in mice. Proc Natl Acad Sci USA 1991; 88:5513–5517.
32. Korenaga M, Hitoshi Y, Yamaguchi N, Sato Y, Takatsu K, Tada I. The role of interleukin-5 in protective immunity to *Strongyloides venezuelensis* infection in mice. Immunology 1991; 72:502–507.

DISCUSSION (Speaker: C. Sanderson)

Galli: Would you care to comment on your observation that transgenic mice with large numbers of tissue eosinophils appear healthy and do not exhibit damage to tissue infiltrated with eosinophils? Also, do transgenic mice with high levels of IL-5 differ from normal mice in the kinetics of eosinophil production, half-life of eosinophils in the blood, or eosinophil survival?

Sanderson: Our transgenic mice (Tg1) show no obvious pathological effects and appear physically normal. Dr. Takatsu also reported that his IL-5 transgenic mice were normal (1). Furthermore, mice reconstituted with bone marrow expressing IL-5 had high-level eosinophilia but remained normal (2). This is in contrast to transgenic mice expressing GM-CSF, which have severe pathological effects (3), and reconstitution experiments with GM-CSF (4) and with IL-3 (5), both of which result in fatal pathological effects. I suggest that eosinophils are not themselves harmful, but require a signal to cause degranulation before tissue damage occurs.

Spry: Could you outline your views on why IL-5 seems less effective in inducing cell division in immature eosinophils and their precursors in vitro, compared to their striking actions in vivo on all stages in their development?

Sanderson: The most plausible explanation is that we do not reproduce the microenvironment of the bone marrow in tissue culture. The production of eosinophil precursors must depend on interactions with stromal cells or extracellular matrix as well as IL-5. It is worth noting that in semisolid colony assays IL-5 is inferior to IL-3 and GM-CSF in the production of eosinophils. However, in liquid cultures where cell interactions can occur, IL-5 is superior in the production of eosinophils (6). This suggests that even in the later stages of the pathway, IL-5 requires a specialized microenvironment.

Ackerman: You show a profound eosinopenia that develops in the parasite-infected IL-5 transgenics. What happens to the eosinophils? Do they interact with the cestode or is there a down-regulation of eosinophil production via some unknown feedback regulation in these mice?

Sanderson: The data point to a decreased production of eosinophils as the number of eosinophil precursors in the bone marrow is markedly reduced (7). However, the mechanism of this feedback control is not known.

Townley: Does the route of administration of the organism affect eosinophilia? Are eosinophils present in lung?

Sanderson: We administer the parasite into the peritoneum and have not tested other routes. The transgenic mice have increased numbers of eosinophils in the lung, but we have not studied the effect of infection on this.

Schleimer: Does the construct have the promoter region for IL-5? If so, is it your concept that the huge increase in IL-5 in the infected, transgenic animals results from induction of the transgene by the endogenous *trans*-activating factors?

Sanderson: Yes. The effect of the CD2 LCR is to produce a constitutive low-level expression of IL-5 from the transgenes. In addition, the transgenes retain the ability to be induced by activation of the T-cell receptor. Thus stimulation with concanavalin A in vitro gives increased IL-5 production by spleen cells, and infection with the parasite gives these very high serum levels of IL-5. Both sets of data suggest that all eight of the transgenes are inducible.

Weller: Since helminthic parasites have evoked many mechanisms to survive within the infected host, I wonder if there is any data to indicate that the eosinopenia observed in *Mesocestoides*-infected IL-5 transgenic mice might actually be mediated by the parasite? With infections with varying numbers of parasites, is there a parasite dose response in the suppression of eosinophilia?

Sanderson: This would seem unlikely since the parasite infection is continuing and actually expanding at the time when eosinophil levels return to the normal high level. Infection of nontransgenic mice results in a peak eosinophilia at about 3 weeks; this is the time of eosinopenia in the transgenics, suggesting that the parasite itself is not directly mediating the eosinopenia.

Busse: Have you had the opportunity to evaluate the function of the eosinophils in the IL-5 transgenic mice?

Sanderson: No, this work is still to be undertaken.

Ackerman: Dr. Takatsu's transgenic, IL-5-producing mice express IL-5 in liver and skeletal muscle, and eosinophils evidently develop in these tissues as well. If, as you suggest, the IL-5 transgenics have increased numbers of circulating eosinophil progenitors, is it not possible that they simply proliferate and differentiate if they end up by chance in other tissues expressing the cytokine? Or do you believe this requires IL-5-mediated chemotaxis of the progenitors?

Sanderson: The spleen becomes the major source of eosinophils in both the transgenic mice and parasite-infected normal mice. Thus it seems likely that eosinophil progenitors do migrate from the bone marrow to other tissues. However, I think it is the migration of mature eosinophils rather than the precursors that accounts for most of the eosinophils infiltrating peripheral tissues.

Vadas: Could you comment on the hypothesized reasons for increased numbers of eosinophil precursors in IL-5 transgenics and the drop in their numbers after infection.

Sanderson: The mechanism of control is an interesting point and I emphasize that I can only speculate about it at present. I was interested to note that in the cloning of the IL-5 receptor α chain (8; and see Chapter 18, this volume) the soluble form of the receptor was the major transcript in eosinophils. As these eosinophils are produced in the presence of IL-5, it seems possible that IL-5 may be involved in controlling whether an active membrane-bound or inactive soluble form of the receptor is produced. Thus high levels of IL-5 may cause a switch toward the inactive form of the receptor and hence reduce the ability of eosinophils to respond to IL-5. This would provide a mechanism for stimulating the production of eosinophils as well as a feedback control to prevent their overproduction, both driven by IL-5.

Denburg: IL-5 is a good stimulator of human peripheral blood colony-forming cells. Perhaps circulating progenitors are particularly responsive to IL-5, and at a level comparable to IL-3 and GM-CSF responsiveness.

Sanderson: Yes, experiments in vivo indicate that eosinophil progenitors are particularly responsive to IL-5. However, I suggest that IL-3 and GM-CSF are relatively poor inducers of eosinophilia. It must be remembered that the optimum concentration of IL-3 or GM-CSF for eosinophil production is 10-fold higher than that required for neutrophil or macrophage production (6). Furthermore, administration of either IL-3 or GM-CSF in vivo, both experimentally and in the clinic, shows a rather poor production of eosinophils compared to neutrophils (9).

Coffman: Do you think that IL-5 induces commitment of a multipotential precursor to eosinophil differentiation and, if so, how do you reconcile this with the emerging view that IL-3, GM-CSF, and IL-5 act through a common signal-transducing chain?

Sanderson: This is an interesting question for which there is not yet a definitive answer. I agree that the mature eosinophil appears to respond equally to all three cytokines, and this is consistent with a common signal-transducing chain. However, there does seem to be a difference at the level of the progenitor cell, which is uniquely sensitive to IL-5. Indeed, the biological specificity of eosinophilia suggests there must be a difference in activity between these cytokines. I think that our knowledge of the receptor complex is incomplete and that a mechanism that will allow the cell to distinguish between these cytokines will be found.

Spry: The great increase in blood eosinophil counts that occurs in many diseases could come about from one or perhaps two additional cell divisions in the eosinophil differentiation pathway. Equally, a large fall could be due to just one less division from, say, six divisions to five, as the number of cells produced is exponential ($\log_2$). Do you think this could account for the alterations you have seen in the transgenic mice with and without infections?

Sanderson: Clearly, there are two possible ways to control eosinophil numbers: first, as you indicate, by restricting the number of divisions of a fixed number of precursors; second, by altering the number of precursors. In these experiments we found that the number of precursors was markedly decreased, suggesting the second control mechanism. However, we cannot rule out that the first mechanism is also operating.

Moqbel: Is the hypereosinophilic syndrome (HES) experiment equivalent to your IL-5 transgenic mice?

Sanderson: There is evidence for the production of IL-5 in HES (10).

DISCUSSION REFERENCES

1. Tominaga A, Takaki S, Koyama N, Katoh S, Matsumoto R, Migita M, Hitoshi Y, Hosoya Y, Yamauchi S, Kanai Y, Miyazaki J-I, Usuku G, Yamamura K-I, Takatsu K. Transgenic mice expressing a B cell growth and differentiation factor gene (interleukin 5) develop eosinophilia and autoantibody production. J Exp Med 1991; 173: 429–437.
2. Vaux DL, Lalor PA, Cory S, Johnson GR. In vivo expression of interleukin 5 induces an eosinophilia and expanded Ly-1B lineage populations. Intern Immunol 1990; 2: 965–971.
3. Elliott MJ, Strasser A, Metcalf D. Selective up-regulation of macrophage function in granulocyte-macrophage colony-stimulating factor transgenic mice. J Immunol 1991; 147:2957–2963.
4. Johnson GR, Gonda TJ, Metcalf D, Hariharan IK, Cory S. A lethal myeloproliferative syndrome in mice transplanted with bone marrow cells infected with a retrovirus expressing granulocyte macrophage-colony stimulating factor. EMBO J 1989; 8: 441–448.
5. Chang JM, Metcalf D, Lang RA, Gonda JJ, Johnson GR. Nonneoplastic hematopoietic myeloproliferative syndrome induced by dysregulated multi-CSF (IL-3) expression. Blood 1989; 73:1487–1497.
6. Clutterbuck EJ, Hirst EM, Sanderson CJ. Human interleukin-5 (IL-5) regulates the production of eosinophils in human bone marrow cultures: comparison and interaction with IL-1, IL-3, IL-6, and GM-CSF. Blood 1989; 73:1504–1512.
7. Strath M, Dent LA, Sanderson CJ. Infection of IL-5 transgenic mice with *Mesocestoides corti* induces very high levels of IL-5 but depressed production of eosinophils. Exp Hematol 1992; 20:229–234.
8. Tavernier J, Devos R, Cornelis S, Tuypens, Van der Heyden J, Fiers W, Plaetinck G. A human high affinity interleukin-5 receptor (IL-5R) is composed of an IL-5 specific α chain and a β chain shared with the receptor for GM-CSF. Cell 1991; 66:1175–1184.
9. Sanderson CJ. Interleukin-5, eosinophils, and disease. Blood 1992; 79:3101–3109.
10. Owen WF, Rothenberg ME, Petersen J, Weller P, Silberstein DS, Sheffer AL, Stevens RL, Soberman RJ, Austen KF. Interleukin 5 and phenotypically altered eosinophils in the blood of patients with the idiopathic hypereosinophilic syndrome. J Exp Med 1989; 170:343–348.

18

Interleukin-5 Receptor

Satoshi Takaki, Yoshiyuki Murata, Akira Tominaga, and Kiyoshi Takatsu
Kumamoto University Medical School, Kumamoto and University of Tokyo, Tokyo, Japan

I. INTRODUCTION

The production of eosinophils can be influenced by various cytokines, including interleukins (IL-5 and IL-3) and granulocyte-macrophage colony-stimulating factor (GM-CSF) (1). There is ample evidence that IL-5 plays a crucial role in both in vitro and in vivo assay systems in eosinophilopoiesis in a lineage-specific manner (2–10). IL-5 is a T-cell-derived glycoprotein that stimulates eosinophil colony formation and growth and differentiation of eosinophils. Murine IL-5 (mIL-5), but not human IL-5 (hIL-5) controls the differentiation of B cells into antibody-secreting cells as well as acting to induce the growth of Ly-1(CD5)-positive B cells and B-cell tumors (11–15). The production of IL-5 transgenic mice also revealed that aberrant expression of the IL-5 gene induces the full pathway of eosinophil differentiation and polyclonal B-cell activation (16,17).

A key question regarding the action of IL-5 in responsive cells has been the molecular mechanism of signal transduction after IL-5-binding to the functional IL-5 receptor (IL-5R). Cloning of cDNA for IL-5 and production of anti-IL-5 MAb enabled us to yield large quantities of purified IL-5 and have made it possible to examine binding sites for IL-5 (18). Binding studies revealed that like other cytokines, IL-5 interacts with target cells with biphasic equilibrium binding kinetics, reflecting two classes of binding sites with high and low affinity on B cells and eosinophils. The IL-5 signals can be transduced through the high-affinity

IL-5R that consists of two different polypeptide chains: α and β (19). Here we will summarize recent advances of IL-5R research studies of structure, physiological functions, and unique mode of receptor-mediated signaling.

II. STRUCTURE OF IL-5 RECEPTOR

A. Molecular Basis of a High-Affinity IL-5 Receptor

IL-5 mediates its function by binding to receptors (IL-5R) expressed mainly on naturally activated B cells, eosinophils, as well as IL-5-dependent B-cell lines. Binding studies revealed that in mouse cells, mIL-5-responsive cell types express relatively small numbers (~100) of high-affinity mIL-5R (K_d of ~150 pM) and large numbers (~1000) of low-affinity mIL-5R (K_d of ~30 nM) (20,21). The order of mIL-5 responsiveness and numbers of high-affinity mIL-5R were in good agreement. The rough coincidence between the biological dose-response and the high-affinity binding curve of IL-5 suggested that the biological response appears to be proportional to the extent of IL-5 binding to the high-affinity sites. Human freshly prepared eosinophils and eosinophilic cell lines were shown to express only high-affinity IL-5-binding sites (K_d of 170–330 pM and 260–380 sites/cell) (22–24).

Chemical cross-linking experiments of mIL-5R on a murine B-cell line with mIL-5 revealed two cross-linked bands of approximately 100 kDa and 160 kDa both of which were specific for IL-5 (21). We therefore hypothesized a two-chain model for the mIL-5R. The intensity of the cross-linked band of 160 kDa was correlated with the number of the high-affinity mIL-5R. The two different polypeptide chains of about 60 kDa and 120–130 kDa, by subtracting M_r of IL-5, appeared to be involved in the formation of the high-affinity mIL-5R. In contrast, cross-linking studies of hIL-5 revealed the presence of a single polypeptide chain (55–60 kDa) on eosinophils (22) and two polypeptide chains (60 kDa and 130 kDa) on eosinophilic sublines of the promyelocytic leukemia HL-60 cells (24).

We and others prepared anti-mIL-5R monoclonal antibodies (MAbs) H7 and R52.120 that recognize the 60-kDa protein and the 130-kDa protein, respectively (25,26). H7 MAb specifically inhibited IL-5 binding to target cells, while R52.120 MAb partially inhibited IL-5- and IL-3-induced proliferation of a pre-B-cell line that responds to mIL-5 and mIL-3 as well, strongly suggesting that the 60-kDa protein is the IL-5-binding protein. Hereafter, we will refer to the 60-kDa protein and the 130-kDa protein as the mIL-5Rα and mIL-5Rβ, respectively. Based on the surface-staining analysis, the mIL-5Rα was shown to be expressed on IL-5-dependent early B-cell lines and more than 70% of peritoneal B cells most of which are Ly-1(CD5)$^+$ and 4–8% of splenic B cells most of which are Ly-1$^-$ (27). Peritoneal eosinophils were shown to express mIL-5Rα. In contrast, the mIL-5Rβ recognized by R52.120 MAb was expressed not only in the IL-5-dependent cell line, but also in the IL-3-dependent cell line (28). When anti-mIL-5R MAb was administered intraperitoneally into the IL-5 transgenic mice, the number of recognizable eosinophils in peripheral blood dropped to normal levels within 5 days (29). These effects and kinetics were similar to those induced by treatment with anti-mIL-5 MAb, which was reported by Coffman et al. (9).

B. Molecular Structure of IL-5-Binding Protein (IL-5Rα)

1. Membrane-Anchored IL-5Rα

We have isolated cDNA clones encoding a mIL-5Rα by expression screening of a library prepared from a murine IL-5-dependent early B-cell line, Y16, in the mammalian expression vector CDM8 (30). A cDNA library was expressed in COS7 cells and screened by panning with the use of anti-mIL-5Rα MAb (H7). The analysis of amino acid sequence deduced from nucleotide sequences of a cDNA demonstrated that the mIL-5Rα is a type I membrane protein with a putative extracellular domain consisting of 415 amino acids (M_r 45,284), including an amino-terminal signal peptide (20 amino acids), a glycosylated extracellular domain (322 amino acids), a single transmembrane segment (22 amino acids), and a cytoplasmic tail (54 amino acids). The extracellular domain of the mIL-5Rα consisted of three domains with fibronectin type III motif. Each domain consists of ~100 amino acid residues. The second domain contains a particular spacing of four cysteine residues and the third domain contains the tryptophan-serine-(X)-tryptophan-serine (WSxWS) motif located close to the transmembrane domain, which are characteristics of a set of receptors for cytokines, growth hormone, and prolactin. The cytoplasmic domain of the mIL-5Rα does not contain the consensus sequences for either a tyrosine kinase domain or a catalytic domain of protein kinases. It has regions rich in proline following the transmembrane domain that is well conserved between mIL-5Rα and hIL-5Rα as well as among receptors for IL-3, GM-CSF, prolactin, and growth hormone. The proline-rich region may be involved in the association with mIL-5Rβ or with other, unidentified signaling apparatus. Furthermore, most of the intracytoplasmic domain of the mIL-5Rα is homologous to a part of the actin-binding domain of human β-spectrin (31), suggesting that the intracytoplasmic domain of the IL-5R may interact with actin. Northern blot analysis showed that two species of mRNAs (5.0 kb and 5.8 kb) were detected in cell lines that display binding sites for mIL-5. Overall, the level of mRNA expression in particular cells correlated well with the number of low-affinity mIL-5R. The relative ratio of 5.0 kb and 5.8 kb mRNA varied among mIL-5R-bearing cells.

The full length of the hIL-5Rα cDNA was isolated from cDNA libraries of peripheral blood eosinophils by plaque hybridization techniques using mIL-5Rα cDNA as a probe (32). The entire nucleotide sequence of hIL-5Rα cDNA showed considerable similarity to coding sequence of the mIL-5Rα. The hIL-5Rα has about 70% amino acid sequence homology with the mIL-5Rα and retains features common to a cytokine receptor superfamily. A hIL-5Rα cDNA encodes a glycoprotein of 420 amino acids (M_r 47,670) with an amino-terminal hydrophobic region (20 amino acids), a glycosylated extracellular domain (324 amino acids), a transmembrane domain (21 amino acids), and a cytoplasmic domain (55 amino acids). RNA blot analysis of human cells demonstrated two transcripts (5.3 and 1.4 kb). Both of these were expressed in normal human eosinophils and in erythroleukemic cell line TF-1 (32). Although 1.4-kb mRNA was specifically expressed in the cells that have hIL-5Rs and hybridized with the hIL-5Rα cDNA probe, it is too short to code for the entire 1260-bp open reading frame. This size of hIL-5Rα mRNA may code for a soluble form of hIL-5Rα.

COS7 cells transfected with the mIL-5Rα cDNA expressed a 60-kDa protein, which bound IL-5 with a single class of affinity (low affinity, K_d of 2–10 nM) (30). Interestingly, however, the IL-5 nonbinding but IL-3-dependent cell line FDC-P1 bound IL-5 with both high (K_d of 30 pM) and low affinity (K_d of 6 nM) and acquired responsiveness to IL-5 for proliferation when the mIL-5Rα cDNA was transfected (30). These results indicated that mIL-5Rα-negative FDC-P1 parental cells constitutively express the additional IL-5R component (mIL-5Rβ), which confers high-affinity binding when mIL-5Rα is expressed. COS7 cells transfected with the hIL-5α cDNA expressed a 60-kDa protein and bound IL-5 with a single class of affinity (K_d of ~300 pM) (32). Both K_d values and estimated molecular sizes of recombinant hIL-5Rα were similar to those observed in human normal eosinophils (22). Despite their homology (about 70% in terms of amino acid sequence), there is a significant difference in the affinity for IL-5-binding between hIL-5Rα and mIL-5Rα. In contrast to the mIL-5Rα, hIL-5Rα expressed on COS cells has much higher affinity. Therefore, the contribution of the β chain to the increase of affinity for the IL-5-binding is much less in case of hIL-5R than in case of mIL-5R.

2. *Soluble IL-5Rα*

In the process of the molecular cloning of both mIL-5Rα and hIL-5Rα, we also isolated several cDNA clones that lack the coding regions for transmembrane and cytoplasmic domains (30,32). Analysis of the gene amplification of the mIL-5Rα cDNA by PCR technique revealed that spleen cells, peritoneal exudate cells, and IL-5-dependent cell lines expressed transcripts corresponding to both membrane-bound and soluble mIL-5Rα (30,34). The transcripts for the membrane-bound mIL-5Rα appeared to be expressed most abundantly among transcripts for the mIL-5Rα. Transcripts for soluble mIL-5Rα may be produced by an alternative splicing of the mature mRNA for mIL-5Rα. COS7 cells transfected with soluble mIL-5Rα cDNA secreted a 50-kDa protein that can inhibit the IL-5-binding to

IL-5-responding B-cell line Y16 (31). This 50-kDa soluble mIL-5Rα showed a little antagonistic effect, if any, on IL-5-induced proliferation of Y16 (Kikuchi, Takaki, Takatsu, et al., submitted). We established an ELISA system for soluble mIL-5Rα and detected it in sera from autoimmune-prone mice.

The major form of the hIL-5Rα mRNA present in differentiated eosinophils encodes a soluble hIL-5Rα, suggesting a regulatory role of the soluble receptors. Actually, it was reported that soluble hIL-5Rα inhibited the hIL-5-driven proliferation and differentiation of eosinophils from human cord-blood culture (33). Functional differences of soluble hIL-5Rα and soluble mIL-5Rα on IL-5 activity may be due to the higher affinity of hIL-5Rα for IL-5-binding than that of mIL-5Rα.

C. Molecular Structure of a β Subunit of IL-5R (IL-5Rβ)

Anti-mIL-5R MAb (R52.120) established by Rolink et al. (26) inhibits IL-5-induced proliferation by down-regulation of the number and K_d of the high-affinity mIL-5R (28). R52.120 MAb also inhibited IL-3-driven proliferation of FDC-P1 (35). Moreover, anti-mIL-3R (anti-Aic-2) MAb (36) reacted with all five IL-5-responsive cell lines and partially down-regulated IL-5-binding sites (35). It was clearly shown that anti-Aic-2 MAb recognizes both the low-affinity mIL-3R (AIC2A protein) (38) and its homologous protein (AIC2B) (39), both of which belong to a cytokine receptor superfamily. Both R52.120 and anti-Aic-2 MAbs immunoprecipitated similar doublet membrane proteins of 130 kDa/140 kDa, whereas they did not react with mIL-5α (28). We then hypothesized that the 130-kDa/140-kDa protein (either AIC2A or AIC2B) recognized by R52.120 MAb or anti-mIL-3R mAb can associate with mIL-5Rα for the formation of the high-affinity mIL-5R. This was proved by the following experimental results. First, the R52 protein turned out to be identical to AIC2 protein (35,37) that was also recognized by anti-Aic-2 MAb. Second, the high-affinity mIL-5R was reconstructed on an L-cell transfectant coexpressing mIL-5Rα and the AIC2B protein, whereas only the low-affinity mIL-5R was detected on a transfectant coexpressing the AIC2A protein and mIL-5Rα (35,40). Chemical cross-linking experiments revealed that the AIC2B protein did not bind IL-5 by itself, but was cross-linked with IL-5 in the presence of mIL-5Rα. The dissociation of IL-5 from the high-affinity mIL-5Rαβ was much slower than from the low-affinity mIL-5Rα, whereas association kinetics of IL-5 to both mIL-5Rαβ and mIL-5Rα were almost similar (Mita, Takaki, Takatsu, et al., submitted). The mIL-5Rβ may therefore stabilize the binding of IL-5 to the mIL-5Rα. It was demonstrated that the expression of hGM-CSFRα together with the AIC2B protein in the murine IL-2-dependent T-cell line CTLL allows it to respond to hGM-CSF (41). Thus, the AIC2B protein is not only the β chain of mIL-5R, but also the β chain of mGM-

CSFR. Recently, mIL-3Rα cDNA was isolated (42) and mIL-3Rα bound IL-3 with low affinity and formed high-affinity mIL-3R with not only the AIC2B protein, but also the AIC2A protein, indicating that the AIC2B protein is also a β subunit of mIL-3R.

Recombinant hIL-5Rα expressed in COS7 cells showed relatively high-affinity IL-5 binding (K_d of ~300 pM). The COS7 cells expressing hIL-5Rα do not respond to IL-5. However, FDC-P1 cells transfected with the hIL-5Rα cDNA expressed high-affinity hIL-5R and acquired hIL-5 responsiveness for their growth (32), indicating that the hIL-5Rα chain may associate with mIL-5Rβ chain (AIC2B protein). These results further support the notion that functional hIL-5R is also composed of hIL-5-binding protein (hIL-5Rα) and an associated molecule (hIL-5Rβ). In the human system, only one AIC2B homolog, KH97, was shown to be a β chain of hGM-CSFR and hIL-3R (43,44). Recently, Tavernier et al. reported the reconstitution of the high-affinity hIL-5R by human-mouse chimeric α chain that has mouse transmembrane and cytoplasmic domains and hGM-CSFRβ chain (KH97) (33). COS cells transfected with the chimeric membrane-anchored hIL-5Rα and KH97 cDNAs showed two cross-linked bands (60 kDa and 130 kDa) with IL-5. We also carried out cotransfection experiments with intact hIL-5Rα and the KH97 protein cDNAs into mouse CTLL and asked whether transfectants bind IL-5 with high affinity and whether they respond to hIL-5 for their growth. CTLL cells expressing both hIL-5Rα and the KH97 protein showed similar binding affinity to transfectants of hIL-5Rα cDNA alone. However, these transfectants showed two cross-linked complexes (60 kDa and 130 kDa) and acquired responsiveness to hIL-5 for proliferation (Takaki, Miyajima, Takatsu, et al., submitted). Taking all these results together, the KH97 protein is the β chain of hIL-5R, indicating that hIL-5R, hIL-3R, and hGM-CSFR share the same β chain, the KH97 protein.

III. FUTURE PERSPECTIVES

Human and mouse IL-5 predominantly function as an eosinophilopoietic factor. In addition, mIL-5 induces not only early B-cell development of Ly-1$^+$ lineage, but also differentiation of mature B cells into Ig-secreting cells. Murine IL-5-dependent pro-B-cell lines also proliferate in response to IL-3 and differentiate into macrophages in response to GM-CSF (45), indicating that they express at least IL-5R, IL-3R, and GM-CSFR on their surface. Each of these receptors consists of a ligand-specific α subunit and a common β subunit. This shared β chain does not have any binding ability by itself to each of the respective cytokines (IL-5, IL-3, or GM-CSF). The IL-5Rα specifically binds IL-5 with low affinity and associates with a common β subunit in the presence of IL-5, resulting in the formation of the high-affinity IL-5R.

IL-5, IL-3, and GM-CSF have some different functions, although they have a common biological activity on eosinophils. However, it is unclear how a unique signal of IL-5 is differently transduced from that of GM-CSF or IL-3. Two interpretations are likely. First, signals generated by these cytokines may be equivalent, and different functions of these cytokines are due to the developmental stages of cells expressing the receptors. Alternatively, different signal-transducing molecules, which generate a specific signal for each cytokine, are associated with the respective ligand-binding subunit (α chain) of each receptor. These possibilities are not mutually exclusive. It is noteworthy that IL-5, IL-3, and GM-CSF induce phosphorylation of a similar set of proteins (46,47). Recently, Takaki and Takatsu demonstrated that FDC-P1 cells transfected with the mIL-5Rβ cDNA and truncated mIL-5Rα cDNA, which lacks a cytoplasmic domain, expressed high-affinity mIL-5R, but failed to respond to IL-5 for proliferation (Takaki S, Takatsu K, submitted). We infer from these results that the cytoplasmic domain of mIL-5Rα plays a role in the signal transduction in conjunction with a β subunit. Murine IL-5Rα may couple with an IL-5-specific signaling pathway.

Multiple cytokines including IL-5 control cells in any one lineage and are active on cells of more than one lineage. This redundancy and pleiotropy of cytokines are explained by two processes. The first is interaction at the ligand-receptor level. Individual cells in each lineage simultaneously display receptors for more than one growth factor. The second process is cross-talk at the intracytoplasmic level for the signal transduction cascade. Different cytokine receptors may use the same second messenger. Accumulating evidence suggests that PI turnover, Ca^{2+} mobilization, and activation of A-kinase or C-kinase appear unlikely to be essential for IL-5-induced differentiation signals. Tyrosine phosphorylation is usually one of the earliest biochemical events in the signal transduction cascade induced by growth factors such as EGF, PDGF, FGF, and CSF-1, in which their ligand-binding activity is on the same molecule of their enzymatic activity. Like other cytokine receptors, IL-5Rαβ lacks intrinsic kinase activity in the intracellular signal transduction portion. However, predominant tyrosine phosphorylation of four different proteins (p140, p92, p53, and p45) is reported within 5 min after exposure to IL-5 (46), suggesting that tyrosine kinases are involved in the signal-transducing pathways of IL-5. Similar sets of protein phosphorylations are reported upon stimulation with IL-3 and GM-CSF (47). Thus, intracellular signals induced by at least IL-5, IL-3, and GM-CSF are likely to use the same or similar pathways. However, a signal-transducing pathway specifically associated with IL-5 may be operating as mentioned above.

We have clearly shown the shared usage of a common β subunit among receptors for IL-3, IL-5, and GM-CSF. These cytokines certainly have a redundant function in the production of eosinophils. Hence, the following model of receptor systems for differentiation of B cells and eosinophils can be considered. The putative signal-transducing subunit (β chain) is constitutively expressed at an immature, namely IL-3-responsive, stage, and the ligand-binding subunit (α chain) for IL-5, IL-3, or GM-CSF is expressed at defined stages during development, creating cells responsive to each differentiation factor. As described (1),

IL-5, IL-3, and GM-CSF have a highly overlapping set of biological activities on eosinophils. However, the activity of hIL-5 in the human system is mainly on eosinophils and basophils. An attractive explanation is the existence of an IL-5-specific signaling. Studying the gene expression of IL-5Rα and the signal-transducing machinery of IL-5R will provide an understanding of the mechanisms involved in eosinophil differentiation.

IV. SUMMARY

IL-5 is a T-cell-derived glycoprotein that stimulates eosinophil production and activation. mIL-5, but not hIL-5 is also active on preactivated B cells to induce differentiation into Ig-producing cells. There are two classes of the mIL-5 binding sites, namely high- and low-affinity receptors, on eosinophils and B cells. The IL-5 signal can be transduced through the high-affinity IL-5 receptors and consist of two different polypeptide chains; α and β. We isolated the mIL-5 receptor α chain (mIL-5Rα) cDNA that encodes the 60-kDa IL-5-binding protein with low affinity. The hIL-5R cDNA, encoding the 60-kDa hIL-5-binding protein with high affinity, was also isolated by plaque hybridization using the mIL-5Rα cDNA as a probe. Both mIL-5Rα and hIL-5Rα are membrane-penetrated glycoproteins that retain features common to a cytokine receptor superfamily. cDNAs encoding soluble mIL-5Rα and hIL-5Rα were also isolated. The β chain (the 130-kDa protein) of mIL-5R was identified as the protein (AIC2B) that is a common β chain between IL-3R and GM-CSFR. The β chain itself does not bind mIL-5, but forms high-affinity receptor with mIL-5Rα. High-affinity mIL-5Rαβ expressed on a T-cell line transmitted a growth signal in response to IL-5. We also identified hIL-5Rβ chain as the KH97 protein that is also a common β subunit between IL-3 and GM-CSF receptors. The role of the IL-5Rα chain in IL-5 signal transduction is discussed.

ACKNOWLEDGMENT

We thank all our collaborators for their individual work, particularly Drs. Yasumich Hitoshi, Eiichiro Sonoda, Masahiro Migita, Seiji Mita, Naoto Yamaguchi, Toshio Kitamura, and Atushi Miyajima. This work was supported in part by a grant-in-aid for Special Project Research, Cancer Bioscience, from the Ministry of Education, Science, and Culture, Japan.

REFERENCES

1. Spry CJF. Eosinophils. Oxford: Oxford Medical Publications, 1988.
2. Sanderson CJ, Campbell HD, Young IG. Molecular and cellular biology of eosinophil differentiation factor (interleukin-5) and its effects on human and mouse B cells. Immunol Rev 1988; 102:29–50.

3. Yokota T, Coffman RL, Hagiwara H, Rennick DM, Takebe Y, Yokota K, Gemmell L, Shrader B, Yang G, Meyerson P, Luh J, Hoy P, Pène J, Briere F, Spits H, Banchereau J, De Vries J, Lee FD, Arai N, Arai K-I. Isolation and characterization of lymphokine cDNA clones encoding mouse and human IgA-enhancing factor and eosinophil colony-stimulating factor activities: relationship to interleukin 5. Proc Natl Acad Sci USA 1987; 84:7388–7392.
4. Yamaguchi Y, Suda T, Suda J, Eguchi M, Miura Y, Harada N, Tominaga A, Takatsu K. Purified interleukin 5 supports the terminal proliferation and differentiation of murine eosinophilic precursors. J Exp Med 1988; 167:45–53.
5. Yamaguchi Y, Hayashi Y, Sugama Y, Miura Y, Kasahara T, Kitamura S, Torisu M, Mita S, Tominaga A, Takatsu K, Suda T. Highly purified murine interleukin 5 (IL-5) stimulates eosinophil function and prolongs in vitro survival. IL-5 as an eosinophil chemotactic factor. J Exp Med 1988; 167:1737–1742.
6. Yamaguchi Y, Suda T, Nakano K, Shinozaki H, Miura Y, Hitoshi Y, Tominaga A, Takatsu K, Kasahara T. In vitro and in vivo induction of IL-5 mRNA by IL-2 that causes eosinophilia in mice. J Immunol 1990; 145:873–877.
7. Owen WF, Rothenberg ME, Petersen J, Weller PF, Silberstein D, Sheffer AL, Stevens RL, Soberman RJ, Austen KF. Interleukin 5 and phenotypically altered eosinophils in the blood of patients with the idiopathic hypereosinophilic syndrome. J Exp Med 1989; 170:343–348.
8. Owen WF, Peterson J, Sheff DM, Folkerth RD, Anderson RJ, Sheffer AL, Austen KF. Hypodense eosinophilia and interleukin 5 activity in the blood of patients with the eosinophilia-myalgia syndrome. Proc Natl Acad Sci USA 1990; 87:8647–8651.
9. Coffman RL, Seymour BWP, Hudak S, Jackson I, Rennick D. Antibody to interleukin 5 inhibits helminth induced eosinophilia. Science 1990; 245:308–310.
10. Limaye AP, Abrams JS, Silver JE, Ottensen EA, Nutman TB. Regulation of parasite-induced eosinophilia selectively increased interleukin-5 production in helminth-infected patients. J Exp Med 1990; 172:399–402.
11. Takatsu K, Tominaga A, Harada N, Mita S, Matsumoto M, Takahashi T, Kikuchi Y, Takahashi T, Yamaguchi N. T cell-replacing factor (TRF)/interleukin 5 (IL-5): molecular and functional properties. Immunol Rev 1988; 102:107–136.
12. Kinashi T, Harada N, Severinson E, Tanabe T, Sideras P, Konishi M, Azuma C, Tominaga A, Bergstedt-Lindqvist S, Takahashi M, Matsuda F, Yaoita Y, Takatsu K, Honjo T. Cloning of complementary DNA encoding T-cell replacing factor and identity with B-cell growth factor II. Nature 1986; 324:70–73.
13. Azuma C, Tanabe T, Konishi M, Kinashi T, Noma T, Matsuda F, Yaoita Y, Takatsu K, Hammerstrom L, Smith CIE, Severinson E, Honjo T. Cloning of cDNA for human T-cell replacing factor (interleukin-5) and comparison with the murine homologue. Nucleic Acids Res 1986; 14:9149–9158.
14. Tominaga A, Mita S, Kikuchi Y, Hitoshi Y, Takatsu K, Nishikawa S-I, Ogawa M. Establishment of IL-5 dependent early B cell lines by long-term bone marrow cultures. Growth Factors 1989; 1:135–146.
15. Sonoda E, Matsumoto R, Hitoshi Y, Mita S, Ishii N, Sugimoto M, Araki S, Tominaga A, Yamaguchi N, Takatsu K. Transforming Growth Factor β induces IgA production and acts additively with interleukin 5 for IgA production. J Exp Med 1989; 170:1415–1420.

16. Tominaga A, Takaki S, Koyama N, Katoh S, Matsumoto R, Migita M, Hitoshi Y, Hosoya Y, Yamaguchi Y, Kanai Y, Miyazaki J-I, Usuku G, Yamamura K-I, Takatsu K. Transgenic mice expressing a B cell growth and differentiation factor (interleukin 5) gene develop eosinophilia and autoantibody production. J Exp Med 1991; 173: 429–437.
17. Dent LA, Strath M, Mellor AL, Sanderson CJ. Eosinophilia in transgenic mice expressing interleukin 5. J Exp Med 1990; 172:1425–1431.
18. Honjo T, Takatsu K. Interleukin 5. In: Sporn MB, Roberts AB, eds. Handbook of Experimental Pharmacology. Vol 95/I. Peptide Growth Factors and Their Receptors I. New York: Springer-Verlag; 1990:609–632.
19. Takatsu K. Interleukin 5 and its receptor. Microbiol Immunol 1991; 35:865–878.
20. Mita S, Harada N, Naomi S, Hitoshi Y, Sakamoto K, Akagi M, Tominaga A, Takatsu K. Receptors for T-cell-replacing factor (TRF)/interleukin 5 (IL-5); specificity, quantitation and its implication. J Exp Med 1988; 167:863–878.
21. Mita S, Tominaga A, Hitoshi Y, Honjo T, Sakamoto K, Akagi M, Kikuchi Y, Takatsu K. Characterization of high-affinity receptors for interleukin 5 (IL-5) on IL-5 dependent cell lines. Proc Natl Acad Sci USA 1989; 86:2311–2315.
22. Migita M, Yamaguchi N, Mita S, Higuchi S, Hitoshi Y, Yoshida Y, Tomonaga M, Matsuda I, Tominaga A, Takatsu K. Characterization of the human IL-5 receptors on eosinophils. Cell Immunol 1991; 133:484–497.
23. Chihara J, Plumas J, Gruart V, Tavernier J, Prin L, Capron A, Capron M. Characterization of a receptor for interleukin 5 on human eosinophils: variable expression and induction by granulocyte/macrophage colony-stimulating factor. J Exp Med 1990; 172:1347–1351.
24. Plaetinck G, Van der Heyden J, Tavernier J, Fache I, Tuypens T, Fischkoff S, Fiers W, Devos R. Characterization of interleukin 5 receptors on eosinophilic sublines from human promyelocytic leukemia (HL-60) cells. J Exp Med 1990; 172:683–691.
25. Yamaguchi N, Hitoshi Y, Mita S, Hosoya Y, Kikuchi Y, Tominaga A, Takatsu K. Characterization of the murine IL-5 receptors by using a monoclonal antibody. Int Immunol 1990; 2:181–187.
26. Rolink AG, Melchers F, Palacios R. Monoclonal antibodies reactive with the mouse interleukin 5 receptor. J Exp Med 1989; 69:1693–1705.
27. Hitoshi Y, Yamaguchi N, Mita S, Sonoda E, Takaki S, Tominaga A, Takatsu K. Distribution of IL-5 receptor-positive B cells: expression of IL-5 receptor on Ly-1(CD5)$^+$ B cells. J Immunol 1990; 144:4218–4225.
28. Mita S, Takaki S, Hitoshi Y, Rolink AG, Tominaga A, Yamaguchi N, Takatsu K. Molecular characterization of the β chain of the murine interleukin 5 receptor. Int Immunol 1991; 3:665–672.
29. Hitoshi Y, Yamaguchi N, Korenaga M, Mita S, Tominaga A, Takatsu K. In vivo administration of antibody to murine IL-5 receptor inhibits eosinophilia of IL-5 transgenic mice. Int Immunol 1991; 3:135–139.
30. Takaki S, Tominaga A, Hitoshi Y, Mita S, Sonoda E, Yamaguchi N, Takatsu K. Molecular cloning and expression of the murine interleukin 5 receptor. EMBO J 1990; 9:4367–4374.
31. Yamaguchi N, Hitoshi Y, Takaki S, Murata Y, Migita M, Kamiya T, Minowada J,

Tominaga A, Takatsu K. Murine interleukin 5 receptor isolated by immunoaffinity chromatography: comparison of determined N-terminal sequence and deduced primary sequence from cDNA and implication of a role of the intracytoplasmic domain. Int Immunol 1991; 3:889–898.
32. Murata Y, Takaki S, Migita M, Kikuchi Y, Tominaga A, Takatsu K. Molecular cloning and expression of the human interleukin 5 receptor. J Exp Med 1992; 175:341–351.
33. Tavernier J, Devos R, Cornelis S, Tuypens T, Van der Heyden J, Fiers W, Plaetinck G. A human high affinity interleukin-5 receptor (IL-5R) is composed of an IL-5-specific α chain and a β chain shared with the receptor for GM-CSF. Cell 1991; 66:1175–1184.
34. Takatsu K, Takaki S, Hitoshi Y, Katoh S, Yamaguchi N, Tominaga A. IL-5 and its receptor system. In: CD5 B cells in development and disease. Ann NY Acad Sci 1992; 651:241–258.
35. Takaki S, Mita S, Kitamura T, Yonehara S, Yamaguchi N, Tominaga A, Miyajima A, Takatsu K. Identification of the second subunit of the murine interleukin 5 receptor: interleukin 3 receptor-like protein, AIC2B is a component of the high-affinity interleukin 5 receptor. EMBO J 1991; 10:2833–2838.
36. Yonehara S, Ishii A, Yonehara M, Koyasu S, Miyajima A, Scheurs J, Arai K, Yahara I. Identification of a cell surface 105 kD protein (Aic-2 antigen) which binds interleukin-3. Int Immunol 1990; 2:143–150.
37. Devos R, Vandekerckhove J, Rolink AG, Plaetinck G, Van der Heyden J, Fiers W, Tavernier J. Amino acid sequence of a mouse interleukin 5 receptor protein reveals homology with a mouse interleukin 3 receptor protein. Eur J Immunol 1990; 21: 1315–1317.
38. Itoh N, Yonehara S, Scheurs J, Gorman DM, Muramatsu K, Ishii A, Yahara I, Arai K, Miyajima A. Cloning of an interleukin-3 receptor gene: a member of a distinct receptor gene family. Science 1989; 247:324–327.
39. Gorman DM, Itoh N, Kitamura T, Scheurs J, Yonehara S, Yahara I, Arai K, Miyajima A. Cloning and expression of a gene encoding an interleukin 3 receptor-like protein: identification of another member of the cytokine receptor gene family. Proc Natl Acad Sci USA 1990; 87:5459–5463.
40. Devos R, Plaetinck G, Van der Heyden J, Cornelis S, Vandekerkhove J, Fiers W, Tavernier J. Molecular basis of a high affinity murine interleukin-5 receptor. EMBO J 1991; 10:2133–2137.
41. Kitamura T, Hayashida K, Sakamaki K, Yokota T, Arai K, Miyajima A. Reconstitution of functional receptors for human granulocyte/macrophage colony-stimulating factor (GM-CSF): evidence that the protein encoded by the AIC2B cDNA is a subunit of murine GM-CSF receptor. Proc Natl Acad Sci USA 1991; 88:5082–5086.
42. Hara T, Miyajima A. Two distinct functional high affinity receptors for mouse interleukin-3 (IL-3). EMBO J 1992; 11:1875–1884.
43. Hayashida K, Kitamura T, Gorman DM, Arai K-I, Yokota T, Miyajima A. Molecular cloning of a second subunit of the receptor for human granulocyte-macrophage colony-stimulating factor (GM-CSF): reconstitution of a high-affinity GM-CSF receptor. Proc Natl Acad Sci USA 1990; 87:9655–9659.
44. Kitamura T, Sato N, Arai K-I, Miyajima A. Expression cloning of the human IL-3

receptor cDNA reveals a shared β subunit for the human IL-3 and GM-CSF receptors. Cell 1991; 66:1165–1174.

45. Katoh S, Tominaga A, Kudo A, Takatsu K. Conversion of Ly-1-positive B-lineage cells into Ly-1-positive macrophages in long-term bone marrow cultures. Dev Immunol 1991; 1:135–142.
46. Murata Y, Yamaguchi N, Hitoshi Y, Tominaga A, Takatsu K. Interleukin 5 and interleukin 3 induce serine and tyrosine phosphorylations of several cellular proteins in an interleukin 5-dependent cell line. Biochem Biophys Res Commun 1990; 173: 1102–1107.
47. Kanakura Y, Druker B, Cannistra SA, Furukawa Y, Torimoto Y, Griffin JD. Signal transduction of the human granulocyte-macrophage colony-stimulating factor and interleukin-3 receptors involves tyrosine phosphorylation of a common set of cytoplasmic proteins. Blood 1990; 76:706–715.

DISCUSSION (Speaker: K. Takatsu)

Lopez: We found that the IL-5 receptor of eosinophils is of high affinity (approx. 100 pM) and is recognized not only by IL-5 but also by IL-3 and GM-CSF. Dr. Woodcock in the laboratory showed in cross-linking experiments that ^{125}I-IL-5 associated with two molecular weight species of 55,000 MW and 130,000 MW, and only the latter was competed by IL-3 and GM-CSF. These experiments on human eosinophils together with the recent cloning of the IL-5, IL-3, and GM-CSF receptors by Devos, Takatsu, Nicola, and the DNAX groups indicate that these receptors express a common chain (β chain) and unique specific chains (α chains). We have proposed that this feature is responsible for the common biological effects of IL-5, IL-3, and GM-CSF on human eosinophils. That this is probably the case is further suggested by the creation of a GM-CSF mutated molecule (GM-CSF Arg^{21}), which was defective at binding to the common β chain but recognized the GM-CSF receptor α chain as well as native GM-CSF. This mutated GM-CSF molecule exhibited a 300-fold decrease in stimulatory activity in multiple assays of eosinophil activation, thus linking the common eosinophil β chain to function and strongly suggesting that the β chain of this receptor complex is responsible for the common biological activities of IL-5, IL-3, and GM-CSF on mature human eosinophil function.

M. Capron: I was surprised by the absence of mRNA encoding IL-5 receptor in your EOL-3 cells since we have shown the IL-5 dependent differentiation of EOL-3 cells.

Takatsu: I presented the data obtained by using our EOL-3 cells. Our EOL-3 cells showed neither detectable numbers of IL-5-binding sites nor mRNA expression by Northern blot analysis. Differences in results between your group and ours may be due to differences of sublines of EOL-3 cells or differences of stimulation. We used unstimulated EOL-3 cells.

Devos: Some years ago we published that on mouse IL-5-dependent cell lines, mIL-5 and mIL-3 could not compete with each other's binding. Now this makes sense, as there are two β chains in the murine system—one that is used only by mIL-5 (the AIC2-B molecule) and one that is used only by mIL-3 (the AIC2-A molecule). Recently, however, Miyajima (DNAX) showed that mIL-3 can both use AIC2-A and AIC2-B for its receptor. Did you observe this also?

Although most of the mRNA for hIL-5Rα present in human eosinophils and in HL60 cells potentially codes for a soluble form, we have no evidence that this mRNA is translated. Do you have evidence for that? What about in the mouse?

Takatsu: We also carried out the cross-competition experiment of radiolabeled mIL-5-binding with mIL-5 and mIL-3 and found that mIL-5, but not mIL-3, did compete with radiolabeled mIL-5. Our results were the same as yours.

Concerning soluble hIL-5Rα, the mRNA expression for soluble hIL-5Rα was abundant in normal human eosinophils. We have not yet checked the translation of soluble hIL-5α mRNA. We will do such experiments soon. In the mouse, we established an ELISA system to detect soluble mIL-5Rα and found that soluble mIL-5Rα was detected in sera of mice bearing mIL-5R-positive murine chronic B-cell leukemia, BCL1 and murine myeloma, MOPC104E, and in sera from the aged autoimmune-prone mice (Kikuchi Y, Takaki S, Takatsu K, submitted).

Venge: Do you think there could be a functional cross-reactivity between the three cytokines IL-5, IL-3, and GM-CSF, which is related to the common β chain of the receptors? The reason for asking this is that Lena Hakansson in my group has shown that all three cytokines will prime the chemotactic and chemokinetic response of human eosinophils and that, much to our surprise, the neutrophil is also primed by the same cytokines, although they seem to lack the proper receptors. It should be said that the IL-5 effect required about 10-fold higher concentrations in the neutrophil system, but that it was inhibited by neutralizing IL-5 antibodies.

Takatsu: IL-3, IL-5, and GM-CSF act on eosinophil precursors for their proliferation and differentiation. Furthermore, in the mouse system, our Ly-1-positive early B-cell lines, established by long-term bone marrow culture, responded to both IL-3 and IL-5 for their proliferation and to GM-CSF for conversion to Ly-1-positive macrophages. I think these results indicate that IL-5, IL-3 and GM-CSF have a functional cross-reactivity.

Concerning your experiments using neutrophils, it is particularly interesting that IL-5 primed neutrophils for the chemotactic and chemokinetic response. There may be two interpretations for your observation. First, neutrophils may express on their surface a very small number of hIL-5Rα that was usually undetectable by IL-5-binding assay. Second, there may be unidentified additional components expressed in neutrophils that might bind hIL-5 with the β chain for hIL-5R. We do not have any evidence at this moment for the existence of the third component. Since Dr. Toshio Sugamura, of Tohoku University, Sendai, Japan, has identified the γ chain of hIL-2R, we have started collaborative work to examine whether the same γ chain of hIL-2R might operate the hIL-5R system for signal transduction.

Devos: A comment on the previous question. We have evidence both in the mouse and in the human system that additional components are required besides the α and β chains to reconstitute a high-affinity IL-5 receptor.

Gleich: Were pathological changes found in the transgenic mice at sites where eosinophils had migrated, and if not, was eosinophil degranulation evident?

Takatsu: We only found pathological changes around muscle in the transgenic mice at sites where eosinophils had migrated. Severe derangement of sarcolemma and disappearance of striation was observed. However, it was difficult to see eosinophil degranulation technically with our effort.

Barnes: Do you have any information about regulation of the IL-5 receptor α-chain or common β-chain gene expression? In particular, does IL-3, IL-5, or GM-CSF have any effects on mRNA levels and do glucocorticosteroids have any inhibitory effect?

Takatsu: This is a very important question. We have just finished the cloning of the genomic mIL-5Rα DNA and started to examine carefully the regulatory regions of the gene. We do not know at present whether IL-3, IL-5 or GM-CSF has any effects on the expression of IL-5Rα mRNA or whether glucocorticoids have any inhibitory effect on the expression. We will have the answer soon.

Hansel: Could you comment on the 3D association of the varying IL-3R, IL-5R, and GM-CSF α chains with their common β chain? Is this (1) a dynamic situation with transient α-β interactions prior to ligand binding, or (2) are there stable α-β dimers, or (3) complexes of multiple α-β units?

Takatsu: I would say that complexes of the common β chain with each α chain are not preformed prior to ligand binding. IL-5Rα binds IL-5 at first with low affinity and then they bind with the common β chain. These views are supported by cross-competition experiments of IL-5-binding with IL-3 or GM-CSF that were presented by Dr. Lopez.

Vadas: What is the role of the β chain if signal transduction by receptor is abolished by deletion of the cytoplasmic domain of the α chain? Can antibodies replace the need for β chain?

Takatsu: There are two possibilities to account for our experimental results. One of them is that the cytoplasmic domain of mIL-5Rα may be required to associate with the common β chain for signal transduction by which IL-5-signaling is mainly coupled with the common β chain. Another possibility is that there is a signaling machinery coupled to the IL-5Rα chain and that signaling may be different from the signaling through the common β chain.

19

The Control of Differentiation and Function of the Th2 Subset of CD4+ T Cells

Robert L. Coffman
DNAX Research Institute, Palo Alto, California

I. INTRODUCTION

It has been only 7 years since Tim Mosmann and I proposed subdivision of CD4+ T cells based on patterns of cytokine production and suggested that the Th2 subset made the right combination of cytokines to coordinately stimulate production of the effector cells and molecules most characteristic of allergic responses: IgE, eosinophils, and mucosal mast cells (Fig. 1) (1). Since then, considerable evidence has accumulated, in humans and in murine models, to support this view. Thus, IL-4 has been shown to induce switching to IgE in mouse B cells in vitro (2), and primary IgE responses are absent in mice treated with anti-IL-4 antibodies (3) or mice with genetic defects in IL-4 production (4) or in the IL-4 structural gene itself (5). IL-5, like IL-4, is produced by Th2, but not Th1, cells in the mouse and appears to be required for enhanced eosinophil production in response to challenge with helminth parasites (6) or protein antigens (Pailler C, Coffman RL, unpublished). The combination of IL-3, IL-4, and IL-10 provides for optimum growth and development of mucosal mast cells in vitro (7), and the importance of IL-3 and IL-4 in parasite-induced mastocytosis has been shown in mice (8). There is now considerable evidence in humans that T cells with a Th2 cytokine pattern exist and that they stimulate essentially the same set of effector functions (9). Allergen-specific Th clones and populations from atopic donors secrete high levels of IL-4 and IL-5, but little or no IFN-γ (10,11). In vitro, the effects of IL-4 on IgE

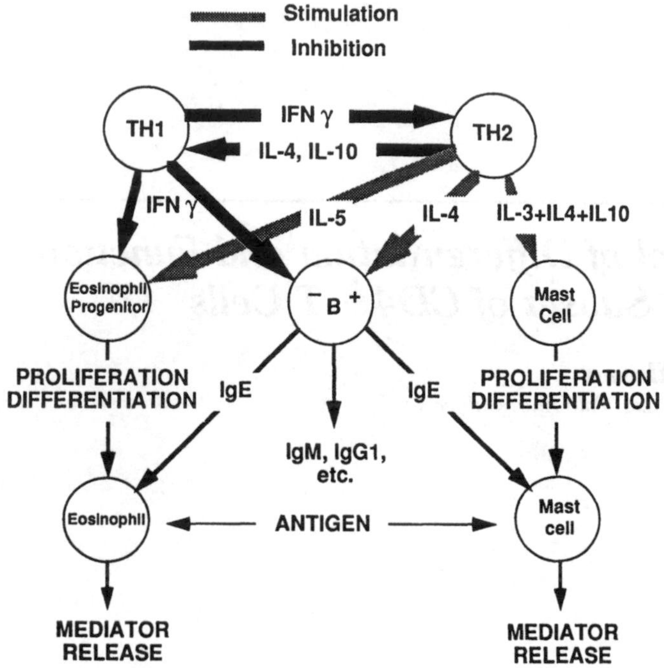

Figure 1 The central features of the regulation of allergic responses by Th2 cells.

switching (12) and of IL-5 on eosinophil differentiation (13,14) are essentially as described in the mouse. There are also many reports of elevations in IL-4 and/or IL-5 production in patients with asthma (15–17), allergic diseases (10,18), and parasite infections (19).

The counterpart to the Th2 subset is the Th1 subset, which regulates a very different set of effector functions centered around macrophage activation and T-cell-mediated immunity. A central theme of T-cell immunoregulation is the dynamic interplay between these two subsets, an interplay mediated largely, if not entirely, by the cytokines produced by each subset. The interactions between subsets are primarily inhibitory and act at multiple stages in the differentiation and function of the Th subsets:

1. The differentiation of Th1 and Th2 cells from a common Th precursor
2. Proliferation and/or cytokine production by differentiated Th cells
3. Inhibition of the effector functions stimulated by the opposite subset

II. THE CONTROL OF Th SUBSET DIFFERENTIATION

Several lines of evidence suggest, but have not formally proven, that Th1 and Th2 cells arise from a common postthymic precursor in a process driven by antigenic stimulation (20). Although many factors have been suggested to influence this differentiation branchpoint, only two have been demonstrated in several experimental systems, both in vitro and, more important, in vivo. These two factors are IL-4 and IFN-γ. The evidence for the role of these cytokines in Th subset differentiation in vivo comes primarily from studies of *Leishmania major* infection in mice. Cutaneous infection of most inbred mouse strains with this protozoan parasite leads to a strongly polarized Th1 response, which gradually resolves the infection and confers immunity to reinfection (21,22). Similar infection of BALB/c mice, however, leads to a predominant Th2 response, which does not control parasite replication and leads to a fatal, systemic infection. This Th2 response in BALB/c mice can be converted into a Th1 response by injecting animals with anti-IL-4 antibodies within the first week or two of infection (23). Neutralization of IL-4 after the second week, however, has no effect on the development of the Th2 response or on the ultimate course of disease. More direct evidence that IL-4 acts to induce Th2 differentiation in the early stages of a Th response comes from analysis of cytokines produced in the earliest detectable in vitro recall responses from the draining lymph nodes of infected animals. By days 3–4 postinfection, draining lymph nodes from BALB/c mice produce substantial levels of IL-4 and IL-5 and relatively modest levels of IFN-γ, whereas draining lymph nodes from anti-IL-4-treated mice produce high IFN-γ and virtually no IL-4 (23). Similarly, injection of IL-4 into resistant C3H mice converts this early response from a Th1 to a Th2 pattern. Thus, the presence of IL-4 during the initial stages of the Th response to *L. major* leads to differentiation of the responding T cells predominantly into Th2 cells. A completely analogous body of evidence demonstrates the parallel role of IFN-γ in the differentiation of Th1 responses in resistant mouse strains (24,25).

IFN-γ and IL-4 have similar abilities to induce Th1 and Th2 differentiation, respectively, in short-term in vitro culture systems in which primary T cells are stimulated and expanded by antigen or mitogen (26,27). A third cytokine, TGF-β, is also active in these systems, acting as a significant inducer of Th1 differentiation in vitro (27). Whether these factors stimulate the appropriate differentiation or inhibit the inappropriate one is not clear, nor is it known whether this initial IL-4 and IFN-γ comes from the responding T cells themselves or from other T or non-T cells. We have suggested in the case of the Th1 response to *L. major* infection that NK cells are an important source of the IFN-γ that induces Th1 differentiation (28). There is some evidence that there may be additional factors that lead to polarization of Th responses. In mice infected with the helminth parasite *Nippostrongylus brasiliensis*, treatment with anti-IL-4 antibody does not inhibit IL-5-dependent eosinophilia (6) or the generation of IL-4 and IL-5 producing Th cells (Seymour BWP, Coffman RL, unpublished).

III. RECIPROCAL REGULATION OF Th1 AND Th2 CELLS

It has long been recognized that there is a reciprocal relationship between antibody production and cell-mediated immunity (see 29 for review of the older literature). Antigen doses, forms, or routes of administration that stimulate optimum production of T-dependent Ig production are usually accompanied by little or no delayed-type hypersensitivity (DTH), and vice versa, despite the fact that both B-cell help and DTH are properties of CD4+ T cells. Furthermore, there is evidence that the dominance of one type of response can be accompanied by active suppression of the other type of response by "suppressor cells" that are either CD4+ (30) or CD8+ (31). We now recognize that DTH and B-cell help are mediated by the Th1 and Th2 subsets, respectively, and these older observations suggest that each of these subsets produces factors that inhibit the growth or function of the other. The experiments of Gajewski, Fitch, and colleagues demonstrated quite clearly that IFN-γ (a Th1 product) is a potent inhibitor of proliferation, but not function, of Th2 clones (32).

The search for a parallel factor mediating the Th2 inhibition of Th1 cells led to the discovery of IL-10 by Fiorentino, Mosmann, and colleagues. In contrast to IFN-γ, the IL-10 inhibits cytokine production by, rather than the proliferation of, Th1 clones (33,34). This inhibition appears to be indirect, with IL-10 acting primarily on antigen-presenting cells (APC), rather than on the T cells themselves (33,35). Some APC, such as macrophages, respond well to IL-10, whereas others, such as B cells, do not. This means that the degree of inhibition of Th2 function by IL-10 in an intact organ or animal will depend on the proportion of antigen presented by different APC.

In the initial experiments that defined the activities of IL-10, long-term clones were used as the source of Th1 cells. In such cultures, IL-4 did not inhibit the production of IFN-γ or other Th1 cytokines (33). In recent experiments with highly polarized CD4+ Th1 populations taken directly from *L. major*–infected mice, we have found that IL-4 can be as potent an inhibitor of IFN-γ production as IL-10 and the two together are often more inhibitory than either alone (Powrie F, Coffman RL, submitted). The ability of IL-4 to inhibit depends on the nature of the APC. When antigen is presented on highly enriched macrophages, IL-10 causes a five- to 10-fold inhibition of IFN-γ production, whereas IL-4 has no effect. If antigen is presented on T-depleted spleen cells, however, either IL-4 or IL-10 alone causes a three- to fourfold inhibition and the combination leads to a five- to 10-fold reduction in IFN-γ. Thus, both of these Th2-specific cytokines can act as inhibitors of function, but not proliferation, of the Th1 subset, although the relative importance of each may vary.

IV. THE INHIBITION OF Th2 EFFECTOR FUNCTIONS

At present, virtually all of the Th1-mediated inhibition of Th2-stimulated effector functions can be explained by IFN-γ. At relatively low concentrations, IFN-γ is a specific and complete inhibitor of B-cell responses to IL-4, especially switch recombination to IgE and (in mouse) IgG1 (36). At higher concentrations, IFN-γ is a more general inhibitor of T-cell-mediated B-cell activation and differentiation (37,38). IFN-γ can also act as an inhibitor of IL-5-stimulated eosinophilia in mice infected with helminth parasites and can increase the severity of the infection (39). Interestingly, IFN-α has similar activity in inhibiting IgE production (40,41) and parasite-induced eosinophilia (39; and Seymour BWP, Coffman RL, in preparation). This is of particular interest from a therapeutic standpoint since IFN-α is approved for clinical use worldwide for a variety of diseases and has been shown to inhibit IgE (42) and eosinophilia (43) in limited clinical trials in humans.

V. SUMMARY

The picture that emerges from the studies summarized here is that Th2 cells regulate, at many levels, a specialized subset of immune effector functions. The secreted products of Th2 cells stimulate antibody, especially IgE, production and promote growth and differentiation of the cell types that utilize IgE to confer specificity to their effector functions. In addition, IL-4 promotes the differentiation of new Th2 cells and, along with IL-10, inhibits cytokine production by Th1 cells. This inhibition of the development and function of the Th1 subset is critical, because it prevents the production of IFN-γ, which is a potent inhibitor of both Th2 growth and differentiation and of many of the effector functions stimulated by IL-4 and IL-5.

ACKNOWLEDGMENT

The DNAX Research Institute is supported by the Schering-Plough Corporation.

REFERENCES

1. Mosmann TR, Coffman RL. Two types of mouse helper T cell clone: implications for immune regulation. Immunol Today 1987; 8:223–227.
2. Rothman P, Lutzker S, Cook W, Coffman R, Alt FW. Mitogen plus interleukin 4

induction of C epsilon transcripts in B lymphoid cells. J Exp Med 1988; 168:2385–2389.

3. Finkelman FD, Katona IM, Urban JF Jr, Snapper CM, Ohara J, Paul WE. Suppression of in vivo polyclonal IgE responses by monoclonal antibody to the lymphokine B-cell stimulatory factor 1. Proc Natl Acad Sci USA 1986; 83:9675–9678.
4. Savelkoul HF, Seymour BW, Sullivan L, Coffman RL. IL-4 can correct defective IgE production in SJA/9 mice. J Immunol 1991; 146:1801–1805.
5. Kuhn R, Rajewsky K, Muller W. Generation and analysis of interleukin-4 deficient mice. Science 1991; 254:707–710.
6. Coffman RL, Seymour BWP, Hudak S, Jackson J, Rennick D. Antibody to interleukin-5 inhibits helminth-induced eosinophilia in mice. Science 1989; 245:308–310.
7. Thompson-Snipes L, Dhar V, Bond MW, Mosmann TR, Moore KW, Rennick DM. Interleukin 10: a novel stimulatory factor for mast cells and their progenitors. J Exp Med 1991; 173:507–510.
8. Madden KB, Urban JF Jr, Ziltener HJ, Schrader JW, Finkelman FD, Katona IM. Antibodies to IL-3 and IL-4 suppress helminth-induced intestinal mastocytosis. J Immunol 1991; 147:1387–1391.
9. Romagnani S. Human TH1 and TH2 subsets: doubt no more. Immunol Today 1991; 12:256–257.
10. Parronchi P, Macchia D, Piccinni MP, Biswas P, Simonelli C, Maggi E, Ricci M, Ansari AA, Romagnani S. Allergen- and bacterial antigen-specific T-cell clones established from atopic donors show a different profile of cytokine production. Proc Natl Acad Sci USA 1991; 88:4538–4542.
11. Wierenga EA, Snoek M, Bos JD, Jansen HM, Kapsenberg ML. Comparison of diversity and function of house dust mite-specific T lymphocyte clones from atopic and non-atopic donors. Eur J Immunol 1990; 20:1519–1526.
12. Gauchat JF, Lebman DA, Coffman RL, Gascan H, de Vries JE. Structure and expression of germline epsilon transcripts in human B cells induced by interleukin 4 to switch to IgE production. J Exp Med 1990; 172:463–473.
13. Lopez AF, Sanderson CJ, Gamble JR, Campbell HD, Young IG, Vadas MA. Recombinant human interleukin 5 is a selective activator of human eosinophil function. J Exp Med 1988; 167:219–224.
14. Clutterbuck EJ, Hirst EM, Sanderson CJ. Human interleukin-5 (IL-5) regulates the production of eosinophils in human bone marrow cultures: comparison and interaction with IL-1, IL-3, IL-6, and GMCSF. Blood 1989; 73:1504–1512.
15. Hamid Q, Azzawi M, Ying S, Moqbel R, Wardlaw AJ, Corrigan CJ, Bradley B, Durham SR, Collins JV, Jeffery PK, et al. Expression of mRNA for interleukin-5 in mucosal bronchial biopsies from asthma. J Clin Invest 1991; 87:1541–1546.
16. Robinson DS, Hamid Q, Ying S, Tsicopoulos A, Barkans J, Bentley AM, Corrigan C, Durham SR, Kay AB. Predominant TH2-like bronchoalveolar T-lymphocyte population in atopic asthma. N Engl J Med 1992; 326:298–304.
17. Sedgwick JB, Calhoun WJ, Gleich GJ, Kita H, Abrams JS, Schwartz LB, Volovitz B, Ben-Yaakov M, Busse WW. Immediate and late airway response of allergic rhinitis patients to segmental antigen challenge. Characterization of eosinophil and mast cell mediators. Am Rev Respir Dis 1991; 144:1274–1281.

18. Maggi E, Parronchi P, Manetti R, Simonelli C, Piccinni MP, Rugiu FS, De Carli M, Ricci M, Romagnani S. Reciprocal regulatory effects of IFN-gamma and IL-4 on the in vitro development of human Th1 and Th2 clones. J Immunol 1992; 148:2142–2147.
19. Limaye AP, Abrams JS, Silver JE, Awadzi K, Francis HF, Ottesen EA, Nutman TB. Interleukin-5 and the posttreatment eosinophilia in patients with onchocerciasis. J Clin Invest 1991; 88:1418–1421.
20. Rocken M, Saurat JH, Hauser C. A common precursor for CD4+ T cells producing IL-2 or IL-4. J Immunol 1992; 148:1031–1036.
21. Heinzel FP, Sadick MD, Holaday BJ, Coffman RL, Locksley RM. Reciprocal expression of interferon gamma or IL4 during the resolution or progression of murine leishmaniasis. Evidence for expansion of distinct helper T cell subsets. J Exp Med 1989; 169:59–72.
22. Scott P, Pearce E, Cheever AW, Coffman RL, Sher A. Role of cytokines and CD4+ T-cell subsets in the regulation of parasite immunity and disease. Immunol Rev 1989; 112:161–182.
23. Chatelain R, Varkila K, Coffman RL. IL-4 induces a Th2 response in *Leishmania-major*–infected mice. J Immunol 1992; 148:1182–1187.
24. Belosevic M, Finbloom DS, Van Der Meide PH, Slayter MV, Nacy CA. Administration of monoclonal anti-IFN-gamma antibodies in vivo abrogates natural resistance of C3H/HeN mice to infection with *Leishmania major*. J Immunol 1989; 143:266–274.
25. Scott P. IFN-γ modulates the early development of Th1 and Th2 responses in a murine model of cutaneous leishmaniasis. J Immunol 1991; 147:3149–3155.
26. Le Gros G, Ben-Sasson SZ, Seder R, Finkelman FD, Paul WE. Generation of interleukin 4 (IL-4)-producing cells in vivo and in vitro: IL-2 and IL-4 are required for in vitro generation of IL-4-producing cells. J Exp Med 1990; 172:921–929.
27. Swain SL, Bradley LM, Croft M, Tonkonogy S, Atkins G, Weinberg AD, Duncan DD, Hedrick SM, Dutton RW, Huston G. Helper T-cell subsets: phenotype, function and the role of lymphokines in regulating their development. Immunol Rev 1991; 123: 115–144.
28. Coffman RL, Varkila K, Scott P, Chatelain R. The role of cytokines in the differentiation of CD4+ T cell subsets in vivo. Immunol Rev 1991; 123:189–207.
29. Parish CR. The relationship between humoral and cell-mediated immunity. Transplant Rev 1972; 13:35–66.
30. Liew FY. Functional heterogeneity of CD4+ T cells in leishmaniasis. Immunol Today 1989; 10:40–45.
31. Tuttosi S, Bretscher PA. Antigen-specific CD8+ T cells switch the immune response induced by antigen from an IgG to a cell-mediated mode. J Immunol 1992; 148: 397–403.
32. Gajewski TF, Schell SR, Nau G, Fitch FW. Regulation of T-cell activation: differences among T-cell subsets. Immunol Rev 1989; 111:79–110.
33. Fiorentino DF, Bond MW, Mosmann TR. Two types of mouse T helper cell IV: TH2 clones secrete a factor that inhibits cytokine production by TH1 clones. J Exp Med 1989; 170:2081–2095.
34. Moore KW, Vieira P, Fiorentino DF, Trounstine ML, Khan TA, Mosmann TR.

Homology of cytokine synthesis inhibitory factor (IL-10) to the Epstein-Barr virus gene BCRFI. Science 1990; 248:1230–1234.
35. Fiorentino DF, Zlotnik A, Vieira P, Mosmann TR, Howard M, Moore KW, O'Garra A. IL-10 acts on the antigen-presenting cell to inhibit cytokine production by Th1 cells. J Immunol 1991; 146:3444–3451.
36. Coffman RL, Ohara J, Bond MW, Carty J, Zlotnik A, Paul WE. B cell stimulatory factor 1 enhances the IgE response of lipopolysaccharide-activated B cells. J Immunol 1986; 136:4538–4541.
37. Coffman RL, Seymour BW, Lebman DA, Hiraki DD, Christiansen JA, Shrader B, Cherwinski HM, Savelkoul HF, Finkelman FD, Bond MW, Mosmann TR. The role of helper T cell products in mouse B cell differentiation and isotype regulation. Immunol Rev 1988; 102:5–28.
38. Reynolds DS, Boom WH, Abbas AK. Inhibition of B lymphocyte activation by interferon-gamma. J Immunol 1987; 139:767–773.
39. Urban JF, Madden KB, Svetic A, Cheever A, Trotta PP, Gause WC, Katona IM, Finkelman FD. The importance of Th2 cytokines in protective immunity to nematodes. Immunol Rev 1992; 127:205–220.
40. Burd PR, Freeman GJ, Wilson SD, Berman M, DeKruyff R, Billings PR, Dorf ME. Cloning and characterization of a novel T cell activation gene. J Immunol 1987; 139: 3126–3131.
41. Finkelman FD, Svetic A, Gresser I, Snapper C, Holmes J, Trotta PP, Katona IM, Gause WC. Regulation by interferon alpha of immunoglobulin isotype selection and lymphokine production in mice. J Exp Med 1991; 174:1179–1188.
42. Souillet G, Rousset F, de Vries JE. Alpha-interferon treatment of patient with hyper IgE syndrome. Lancet 1989; 1:1384.
43. Zielinski RM, Lawrence WD. Interferon-alpha for the hypereosinophilic syndrome. Ann Intern Med 1990; 113:716–718.

DISCUSSION (Speaker: R. Coffman)

Egan: In three species we have demonstrated that a monoclonal antibody to IL-5 blocks eosinophil infiltration into the lung. In sensitized guinea pigs challenged with ovalbumin, eosinophils infiltrate into the lungs within 24 h. This 10-fold increase in eosinophils can be blocked back to baseline with 100 μg of the TRFK-5 antibody that is specific for IL-5. This is seen in both lavage and tissue, has an ED_{50} of 20 μg/kg, and is dependent on active antibody. In the guinea pig, it is difficult to measure consistent pulmonary hyperreactivity and there are very few immunological reagents. We have, therefore, turned to doing some of our studies in mice and monkeys. The mice are sensitized to and challenged with ovalbumin. As with the guinea pig, there is extensive allergic eosinophilia in the mouse that is blocked by low levels of the TRFK-5 antibody. This model should be extremely useful, and we plan to do pulmonary mechanisms in the mouse and to study the connection between eosinophilia and pulmonary function.

Finally, we have used monkeys as a species whose biology should be most similar to humans. In these studies each monkey serves as its own control. The monkey is instrumented and the pulmonary function and BAL are evaluated. The monkey is then challenged with *Ascaris* antigen, to which it is naturally sensitive, and 24 h later it is reevaluated. The monkey shows an *Ascaris*-dependent eosinophilia that is sensitive to 0.3 mg/kg i.v. of the TRFK-5 antibody.

We are fortunate with IL-5 to have a cytokine and cytokine antibodies that cross species, indicating that IL-5 is most likely important in humans and indicating that anti-IL-5 reagents should be useful therapeutics for pulmonary inflammation.

Gleich: A comment and two questions. Concerning IFN-α, we have treated several patients with the hypereosinophilic syndrome with IFN-α and in one case we were impressed by a dramatic change in the clinical course of the disease with reversal of disease severity. Concerning the switch between Th1 and Th2, can you substitute IFN-γ injection, possibly using a minipump, to convert Th2 to Th1 instead of antibody to IL-4? And does anti-IL-10 have any effect on the Th1-Th2 switching?

Coffman: For reasons we do not understand, we have had little success converting Th2 to Th1 responses by administering IFN-γ, even in our murine *Leishmania* infections where we know that IFN-γ plays a decisive role in the development of Th1 responses. IL-10, so far, does not seem to directly influence Th-subset differentiation, but can do so indirectly in certain situations by inhibiting IFN-γ production.

Denburg: What happens to the mast cell response in the IL-4-deleted or IFN-γ-treated mice?

Coffman: We have not studied this at all.

Ackerman: Could you comment on the irreversibility of Th1 to Th2, or vice versa, of the human T-cell clones characterized by Dr. Romagnani concerning the functions of IFN-γ. Are populations of T cells, as opposed to cloned lines, selected by exposure to IFN-γ?

Coffman: Rodrigo Correa-Oliveira in my laboratory has done a series of experiments transferring polarized Th1 and Th2 populations into *scid* mice and changing their polarity by altering the relative levels of IL-4 and IFN-γ. In a similar system, Holaday et al. (1) can reverse the phenotype of T-cell lines, but not of clones derived from those lines. These results, along with those you mentioned from Dr. Romagnani, lead me to think that individual Th cells become irreversibly committed to one cytokine pattern, but populations of committed cells can change when conditions lead to preferential inhibition or expansion of one Th type. Definitive experiments on this question, clearly, have yet to be done.

Durham: We have confirmed that in atopic subjects allergen provocation in the skin, nose, and lung results in a dominant Th2 pattern of cytokine mRNA expression during late responses. Corticosteroid therapy in asthma resulted in a switch in the bronchoalveolar lavage to a Th1 pattern, whereas specific immunotherapy for rhinitis resulted in an additional Th1 component after cutaneous allergen provocation.

Vadas: Did the BALB/c mice susceptible to leishmaniasis and having a Th2-dominant response have eosinophilia?

Coffman: BALB/c mice infected with *L. major* do not have elevated eosinophil numbers, nor do they make much IL-5, despite having high IgE levels and substantial IL-4 production. This contrasts with typical helminth infection, which gives similar elevations in IgE and IL-4, but substantial IL-5-mediated eosinophilia.

Wardlaw: Is there any difference in the in vitro inhibitory effects of corticosteroids in Th1 and Th2 cells?

What is known about the genetic basis for the difference between BALB/c mice and the resistant mouse strain in their response (i.e., Th2 vs. Th1) to leishmaniasis infection?

Coffman: We have not studied corticosteroids at all, but others, notably Raymond Daynes and his colleagues, have. Much of the difference between resistant and susceptible mice is accounted for by the *Scl-1* locus, although several other genes can have an influence. Nothing is known about the action of this gene.

Galli: Perhaps these comments can serve to introduce the next part of the general discussion. Do the IL-4 knockout mice express any abnormality in the course of infection with *N. brasiliensis*? The fact that the selective deletion of a single cytokine, mediator, or cell type does not detectably influence the expression of a given biological response tells us that the deleted element does not have an essential function in those aspects of the response that have been measured. However, in many biological responses, two or more elements may have similar or overlapping roles. In such cases, all these elements may have to be eliminated or their effects antagonized before the expression of the response can be significantly altered. Thus, if deletion of IL-4, IL-5, eosinophils, and mast cells has no discernible effect on a biological response, this does not rule out the possibility that the deleted element contributes to the response. But the finding does show that the deleted element is not the only element responsible for the aspect of the response that has been analyzed.

Coffman: I quite agree with your comment. With any of these "deletion" analyses, whether done with antibodies or by genetic means, one must be very clear about just what they do and do not say about the role of the deleted component in the complex process being studied. With regard to your specific question, the strain of *N. brasiliensis* we use is not adapted to productively infect mice, and the worms are expelled from the gut rapidly, even in *scid* mice. One needs to use a more chronic and immunologically regulated parasite to answer the question you posed.

Butterworth: I agree entirely that most of the elegant experiments that have been carried out in *murine* models of helminth infection do not support the hypothesis that eosinophils are involved in protective immunity, although a recent series of experiments by Drs. Else and Grenus on *Trichuris muris* indicate an association between Th2 responses and protection. However, I would argue that there are marked differences in the mechanisms of protective immunity between different host species. Several years ago, Monique Capron and colleagues demonstrated by adoptive transfer experiments that eosinophils were able to mediate antibody-dependent protection against *Schistosoma mansoni* infection in the rat. More recently, several studies in human schistosomiasis have investigated the relationship between an age-dependent resistance to reinfection after treatment and various immune responses: these include work by Hagan and colleagues on *S. haematobium* in The Gambia, and on *S. mansoni* by Dessein and colleagues in Brazil and by ourselves in Kenya. Dr. Kagan has reported an association between eosinophils and resistance to reinfection in children. All three groups have demonstrated a correlation between IgE responses against worm antigens and resistance to reinfection. In our own studies we have shown that the correlation of immunity with IgE responses is restricted to a particular category of worm antigens; and that it is not a fortuitous consequence of the fact that both immunity and IgE responses are age-dependent, since an effect of IgE can still be demonstrated after contrasting for age in multiple regression or logistic regression analysis. In contrast, antibodies that block eosinophil-dependent killing of schistosomula (IgM and IgG2) are correlated with susceptibility.

These findings indicate a role for Th2 responses in protection and suggest indirectly that the eosinophil-mediated antibody-dependent killing that we have previously described might be important. To test this further, we have examined cytokine production in a further cohort of schistosome-infected individuals, before and after treatment (Roberts M et al., unpublished data). The correlations with levels of reinfection after treatment are not yet available. However, inspection of the raw data shows that, when peripheral blood mononuclear cells are cultured with worm antigens, very few patients make IFN-γ, whereas the majority make greater or lesser amounts of IL-5, increasing in an age-dependent manner. IL-4 assays have not yet been done. Thus, we have *no* evidence for an association of Th1 responses with immunity, as is seen in the mouse model: instead, all our data so far are compatible with a role for Th2 responses.

Moqbel: I would like to comment on the possible role of eosinophils in protection against helminths. Many studies have shown that the process of acquired immunity to helminths and worm elimination is very complex and multifactorial, which includes both specific and nonspecific inflammatory responses part of which involves eosinophils. In enteral parasitic models, it is most likely that eosinophils contribute to worm elimination by changing the environment of the gut in an inimical fashion to the parasite rather than by direct cytotoxicity. Expelled worms are found live. Only in hyperimmune animals have eosinophils been seen in juxtaposition to dead larvae. I also think that some models of parasite infection may reflect more of a Th1 (IFN-γ-dependent, macrophage-mediated) response rather than a predominant Th2 (IL-4, IL-5, eosinophil-mediated) response. Finally, I think that the timing of the anti-IL-5 treatment may be concise in clarifying the possible protective role of eosinophils.

Silberstein: The failure of anti-IL-5 to affect resistance in *Trichinella spiralis* infection should not be taken as an indication that eosinophils do not protect against some helminths. *Trichinella* is probably a poor model to test this hypothesis. While it is possible to get eosinophils to kill migrating larvae in vitro, this is probably not important in vivo. The main component of resistance is directed at adult forms of the parasite, which exhibit reduced residence time and reduced fecundity. Eosinophils probably have only a minor involvement in this effect.

Gleich: Two comments. First, in the discussion of Th2 responses and immunity to *T. muris*, Else et al. noted that an IgA monoclonal antibody to the parasite conferred immunity, underscoring the importance of the IgA isotype in resistance. And, second, in the guinea pig tick model antibasophil and antieosinophil sera abolished immunity to the tick, pointing to the importance of these cells to resistance in that situation.

DISCUSSION REFERENCE

1. Holaday BJ, Sadick MD, Wang Z-E, Reiner SL, Heinzel FP, Parslow TG, Locksley RM. Reconstitution of *Leishmania* immunity in severe combined immunodeficient mice using Th1- and Th2-like cell lines. J Immunol 1991; 147:1653–1658.

20

Eosinophils in a Guinea Pig Model of Allergic Airways Disease

M. G. Campos, M-C. Seminario, J. K. Shute, T. C. Hunt, S. T. Holgate, and M. K. Church
Southampton General Hospital, Southampton, England

I. INTRODUCTION

Increased numbers of eosinophils in peripheral blood (1), sputum (2), lavage fluid (3), and bronchial tissue (4) in asthmatic subjects have led to the theory that the eosinophil is a major effector cell (5,6), or even a causative cell (7,8), in the chronic inflammatory process that underlies human bronchial asthma. However, evidence, although strong, is circumstantial and several questions still remain. In attempts to answer these questions, guinea pig models of allergen-induced airway inflammation have been developed to study the relationship between airway eosinophilia, the generation of eosinophil mediators and airway function and hyperreactivity, and the mediators modulating those responses (9–12).

Our group has developed a model of allergic airways disease in which pulmonary function in conscious free-breathing guinea pigs, sensitized and challenged with aerosolized ovalbumin, is assessed by whole-body plethysmography (13). In this model three phases of airways obstruction were observed: an early response peaking at 2 h, a late response peaking at 17 h, and a further late response at 72 h. The late response was accompanied by a 13-fold rise in neutrophils and a fourfold rise in eosinophils recovered by bronchoalveolar lavage (BAL) at 17 h. By 72 h, the BAL content of neutrophils had returned to near normal, whereas eosinophil numbers had risen to 6.7-fold above baseline (13). Despite the fact that these findings share similarities with those of human asthma, pulmonary function was variable enough as to consider this model as not being reproducible.

In the process of solving this problem, the relative deposition of particles within the lungs and nose of guinea pigs has been assessed, since the upper airways may represent the largest resistance component within the airways (14) and is a major contributor to allergen-induced alteration in airways resistance in guinea pigs (15). Thus, a radioaerosol of technicium 99m in diethylenetriaminepentaacetic acid with mass median diameter of 1.5 μm was generated using a Wright nebulizer and was inhaled by free-breathing guinea pigs. Aerosol distribution and activity was measured with a gamma camera and analyzed with a VAX computer. Thirty percent of the aerosol delivered to the animals was retained in the airways, of which 80% was deposited in the nose and only 12% in the lungs (16).

II. SENSITIZATION AND CHALLENGE THROUGH ENDOTRACHEAL INTUBATION

To overcome the problem of aerosol deposition in the nasal turbinates when the aerosol is inhaled via the nose, a model has been developed that allows aerosol delivery directly to the lower airways through an endotracheal tube introduced into the trachea of sedated animals with the aid of a modified laryngoscope and an introducer. On three separate occasions, animals are sedated and intubated to perform a sensitization and challenge protocol as follows: exposure of guinea pigs for 3 min to aerosolized 1% ovalbumin in 0.9% sterile sodium chloride on two occasions separated by 7 days. Antigen challenge consists of 2% ovalbumin inhalation, again through the endotracheal tube, for 5 min under the protection of mepyramine maleate (17).

Following endotracheal intubation, 88% of aerosol retained is observed within the lungs, with only a small proportion in the trachea or on the fur around the nose (16). Thus, aerosol exposure after endotracheal intubation offers a direct route to the lower airways, which more closely mimics experimental exposure to allergen in human studies, and provides a suitable model of allergic airways disease as deducted from pulmonary function and cellular infiltration studies.

III. PULMONARY FUNCTION

Pulmonary function was assessed through an automated whole-body plethysmograph (18) both before challenge and at different time points after saline and ovalbumin provocation over a 24-h period. Fifteen minutes after ovalbumin challenge, specific airways conductance fell to levels $-74.7 \pm 2.6\%$ from baseline ($p < 0.05$). Airways conductance levels remained significantly lower than in controls for up to 2 h ($-39.6 \pm 6.6\%$; $p < 0.05$). After this time, levels

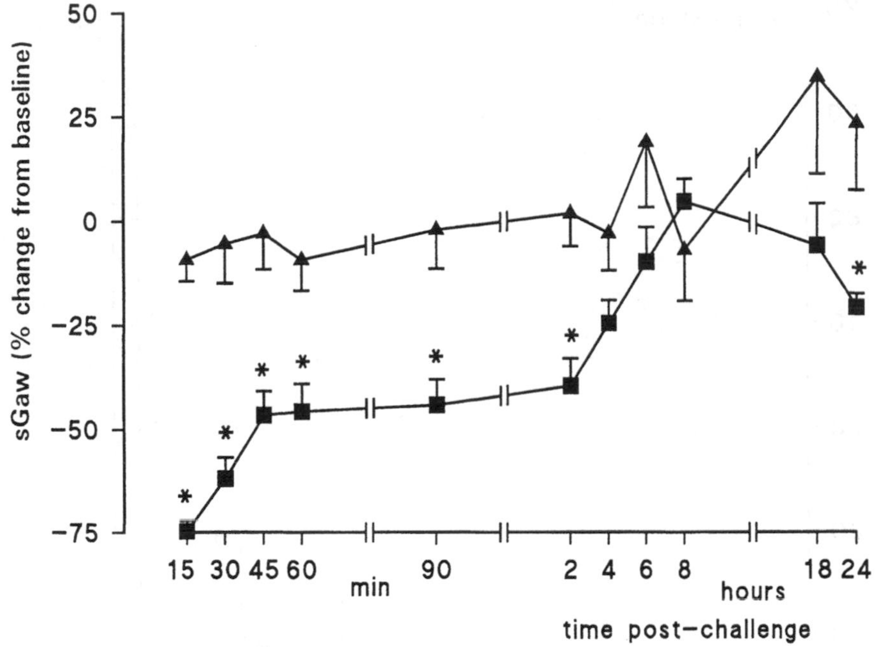

Figure 1 Specific airways conductance (sGaw) at different times after challenge with either 2% ovalbumin (squares) or saline (triangles) in sensitized guinea pigs by endotracheal exposure. Results are expressed as mean ± SEM; $*p < 0.05$, $n \geqslant 7$.

remained close to baseline up to 24 h when a second significant fall was observed ($-20.66 \pm 3.3\%$; $p < 0.05$) (Fig. 1). All the changes observed were accompanied by changes in airways resistance, whereas thoracic gas volume levels only differed from controls during the immediate response. It is noteworthy that 92% of the animals manifested both responses at 15 min and 24 h after challenge, whereas in other guinea pig models 40% or less animals were dual responders (12,19).

IV. CELLULAR INFILTRATION

The differential counts of cells recovered by bronchoalveolar lavage from guinea pigs sensitized and challenged with ovalbumin by the endotracheal procedure showed a gradual eosinophilia with continuous rising of eosinophil numbers, which reached a statistical significance at 24 h, when compared with saline-challenged animals (5.97×10^6 vs. 1.23×10^6 cells/ml; $p < 0.05$), constituting 35% of the total leukocyte population (Fig. 2). Neutrophil numbers were signifi-

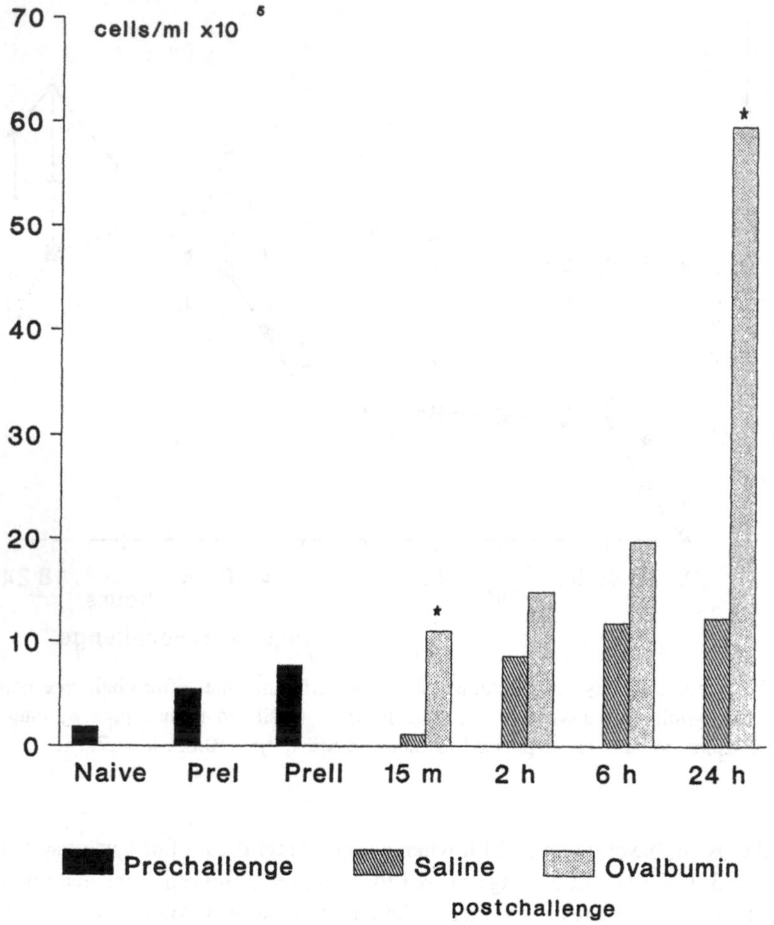

Figure 2 Effect of ovalbumin sensitization and challenge on the eosinophil numbers recovered in bronchoalveolar lavage from guinea pigs. Results are expressed as medians $\times 10^6$; $*p < 0.05$, $n \geq 7$.

cantly increased following allergen challenge at every time point studied, achieving a peak at 6 h, 4.97×10^6 versus 0.70×10^6 in saline-challenged animals ($p < 0.05$), constituting 47% of the total leukocyte population. Mononuclear cells were significantly elevated at 2 h 24 h after challenge.

The observation of a predominant neutrophilia in the first hours after challenge, followed by a slower but pronounced eosinophilia, resembles temporally the findings observed in human BAL studied after allergen challenge by bronchoscopy (20,21).

V. CHARACTERISTICS OF EOSINOPHILS RECOVERED FROM BRONCHOALVEOLAR LAVAGE FLUID

Accumulation of eosinophils in BAL fluid (BALF) in association with the late phase of the bronchoconstrictor response to allergen bronchoprovocation may indicate a causative role for these cells. However, in addition to an increase in cell numbers, it is also necessary to show signs of activation. Eosinophils in BAL from asthmatic patients differ from normal eosinophils in that they show degranulation and loss of granule cationic proteins (22) and are recognized as hypodense eosinophils (23). Morphological studies indicating partial degranulation suggest that hypodense eosinophils are activated. This is supported by the observations that hypodense eosinophils display increased oxygen consumption, deoxyglucose uptake, cytotoxicity against schistosomula targets, and leukotriene C_4 production (29). Hypodense eosinophils also have increased expression of receptors, including low-affinity IgG ($Fc_{\gamma}RII$) and IgE ($Fc_{\epsilon}RII$) and complement receptors (29). It has been further suggested by Capron (30) that only hypodense eosinophils possess IgE receptors that enable them to respond to anti-IgE by release of eosinophil peroxidase and major basic protein (MBP).

A. Density Profiles of BALF and Blood Eosinophils in Guinea Pigs

We have investigated the density of eosinophils in the BALF of naïve guinea pigs and after allergen sensitization and challenge to determine the density and morphological changes that accompany changes in allergic status (31). We define hypodense eosinophils as having density <1.080 g/ml, normodense as 1.080–1.096 g/ml, and hyperdense eosinophils as >1.096 g/ml when separated on continuous gradients of Percoll. This is based on the observation that eosinophils elicited in the peritoneal cavity of guinea pigs in response to weekly injections of horse serum show a mean peak density of 1.088 ± 0.001 g/ml, with more than 95% eosinophils within the density range 1.080–1.096 g/ml. In contrast, BALF eosinophils from naïve, sensitized, and challenged animals show a heterogeneous density profile consisting of a mixture of hypodense, normodense, and hyperdense eosinophils in all BALF samples studied. A similar heterogeneity is seen in the density profile of blood eosinophils. This latter profile was not significantly changed in sensitized and challenged animals compared with naïve guinea pigs, in respect to the percentage of total eosinophils present in blood or the proportion of each density subgroup. The density profile of BALF eosinophils was, however, influenced by sensitization to ovalbumin and subsequent challenge with this allergen. After sensitization, the total number of eosinophils in BALF increased significantly, with further increases occurring after allergen challenge. At 72 h after allergen challenge, eosinophils comprised 51.9% of all BALF cells. Sensiti-

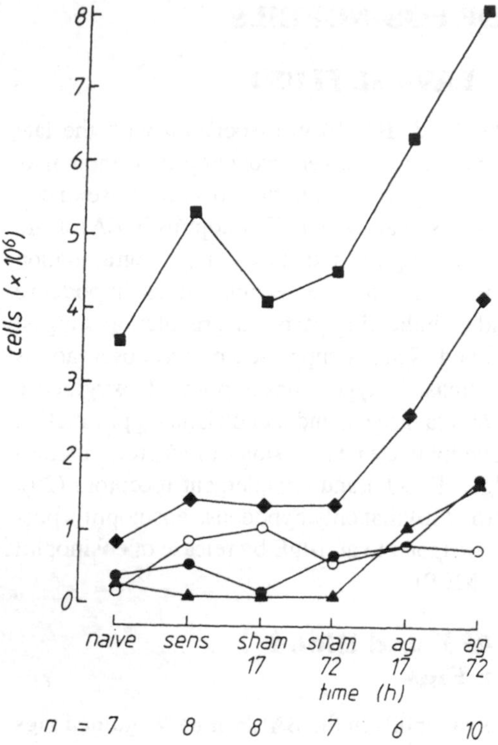

Figure 3 Absolute numbers of all cells and eosinophils in bronchoalveolar lavage fluid (BALF) in naïve, sensitized unchallenged, sensitized 17 and 72 h after saline challenge, and sensitized 17 and 72 h after allergen (ag) challenge animals. (Squares) Total cells in BALF; (diamonds) total eosinophils in BALF; (open circles) normodense eosinophils 1.080–1.096; (triangles) hyperdense eosinophils > 1.096.

zation by two exposures to ovalbumin also resulted in a significant increase in the proportion of hypodense eosinophils (as a percentage of all eosinophils) to 63.3 ± 8.7% from 25.3 ± 5.9% in naïve animals. The total number of hypodense eosinophils did not alter significantly on subsequent allergen challenge. A significant decrease in the proportion of hypodense eosinophils after allergen challenge reflected an increase in the number and relative proportion of hyperdense cells. As the numbers of hypodense eosinophils in BALF after allergen challenge did not differ significantly from values at the time of sensitization some 10 days earlier, it appears that these cells represent a relatively constant population of eosinophils with prolonged viability.

Following sensitization, a decrease in the mean maximal density of eosinophils was demonstrated, consistent with the observed increase in total eosinophil numbers as well as the absolute number and percentage of hypodense eosinophils in BALF. As no significant decrease in the percentage or numbers of circulating hypodense eosinophils was measured, we may hypothesize that the inflammatory process occurring during sensitization results in local cytokine production and eosinophil activation. A recent study (32) in the same animal model has demonstrated an increase in CD8+ T cells in the bronchial mucosa during sensitization. T lymphocytes are a source of cytokines, including granulocyte/macrophage colony-stimulating factor, interleukin-3, and interleukin-5. This group of cytokines is capable of promoting cell growth, differentiation, viability, and functional activation of eosinophils, rendering them hypodense (33). This would provide a mechanism for activation of normodense eosinophils in the bronchial mucosa and the appearance of hypodense eosinophils in BALF.

Our findings indicate, therefore, that in sensitized animals, many of the eosinophils present in BALF are hypodense, and after allergen challenge a massive eosinophil influx occurs with cells of the same density distribution found in blood.

B. Morphological Analysis of BALF Eosinophils

Electron microscopic analysis of eosinophils obtained 72 h after allergen challenge of sensitized guinea pigs reveals that granular volume is lower in hypodense than in normal cells. As the total number of granules per cell remains unchanged, this indicates a reduction (47%) in mean granular area in hypodense eosinophils. When compared with normodense eosinophils, the major qualitative differences observed in hypodense cells are in the prevalence of granules with light core(s) and the reduction in granules with more than one core per granule, indicative of major basic protein (MBP) release. Additionally, flow cytometric analysis revealed an increase in the mean diameter of hypodense eosinophils compared to normal cells. An increase in cell volume and a reduced granule content are therefore the major factors contributing to the reduced density of hypodense eosinophils.

Hyperdense cells are morphologically distinct from normodense cells. They have an increased vacuolar volume, an increased number of granules in which the core is of equal density to the matrix, and a reduced number of dark core granules, which are the classically described eosinophil granules. These are viable, functional cells that may develop in response to the local environment; their role, however, remains unclear.

C. Functional Analysis of Normodense and Hypodense BALF Eosinophils

Hypodense (<1.080 g/ml) eosinophils and all other eosinophils (>1.080 g/ml) were separated from BALF of guinea pig obtained 72 h after allergen challenge.

Table 1 Superoxide Generation by BALF Eosinophils

Agonist	Concentration (M)	nmol O_2^-/min/10^7 cells Hypodense	Normodense
PAF	10^{-6}	137.3 ± 7.8	42.0 ± 3.2
PMA	1.6×10^{-9}	70.2 ± 3.4	487 ± 7.2
A23187	2.5×10^{-6}	22.5 ± 3.7	6.8 ± 2.5

Superoxide production by hypodense eosinophils in response to PMA, calcium ionophore A23187, and PAF is significantly greater than that of normodense cells, either from BALF or those obtained by peritoneal lavage of animals sensitized with horse serum (34) (Table 1).

While normodense cells in BALF respond to the same degree as peritoneal cells, it is evident that the hypodense population is primed for enhanced function. It is conceivable, therefore, that the same as yet unidentified cytokines responsible for the decrease in eosinophil density in sensitized guinea pig pulmonary tissue also prime these cells for increased secretory responses. The study of cellular events in the guinea pig model of allergen-induced inflammation therefore continues to be relevant.

VI. GUINEA PIG EOSINOPHIL MAJOR BASIC PROTEIN

MBP, which accounts for 50% of the total granule protein of eosinophils (24), has been identified by immunofluorescence in necrotic areas inside the bronchial wall of patients with bronchial asthma (25). Elevated concentrations of MBP have been measured in sputum and BALF from asthmatics (26,27), suggesting eosinophil activation and secretion of granule proteins.

We have raised monoclonal and polyclonal antibodies to guinea pig eosinophil MBP and developed specific staining of eosinophils in cytospin preparations and lung tissue sections, as well as a quantitative enzyme-linked immunosorbent assay (ELISA).

A. Antibody Production

MBP was purified from guinea pig peritoneal eosinophils by a method modified from Gleich et al. (24). Confirmation of the identity of the MBP was obtained by the fact that it formed a single band on SDS-PAGE with a molecular weight of approximately 14.5–15 kDa, relative to markers of known molecular weight. This agrees with the molecular weight of approximately 14 kDa estimated from the amino acid sequence of guinea pig eosinophil MBP (28). The MBP also reacted specifically in Ouchterlony and radial immunodiffusion assays with a rabbit polyclonal anti-MBP antiserum (provided by G. J. Gleich).

The purified MBP was used to raise our own polyclonal antisera in rabbits, and also to produce five different murine monoclonals specific for guinea pig MBP. Isotype analysis demonstrated that all five antibodies were of the IgG_1 subclass. Testing in an indirect ELISA demonstrated that all five monoclonal antibodies bound to purified reduced and alkylated MBP coated onto polystyrene microtiter plates. Their reaction with the purified MBP was further confirmed by immunoblotting. When the five monoclonals were tested in combination in an additive ELISA, there was no, or relatively little, additive effect, indicating that the five monoclonals were specific for closely related or identical epitopes on the MBP molecule.

Our polyclonal antisera were capable of specifically staining eosinophils in cytospin preparations by an indirect immunofluorescence technique and could be used for qualitative and quantitative detection of reduced and alkylate MBP in Ouchterlony and radial immunodiffusion assays respectively.

B. Immunocytochemistry

Only one of the monoclonal antibodies (designated 8A12) was found to be suitable for use in immunocytochemistry. This stained eosinophils strongly and specifically in cytospin preparations, frozen sections, or wax-embedded sections (after protease digestion). Either indirect immunofluorescence or alkaline phosphatase–anti-alkaline phosphatase techniques could be used to visualize the staining. This resulted in strong granular staining of the eosinophils and proved to be an extremely sensitive means of localizing this cell type in tissue sections. The 8A12 monoclonal was also capable of localizing MBP deposited extracellularly by eosinophils at sites of allergic inflammation. Staining with this antibody was specific for eosinophils and did not stain other guinea pig leukocyte types.

C. Antigen Capture ELISA

Microtiter plates were coated with antibody 8A12 (2 μg/ml in coating buffer) overnight at 4°C. Standard MBP was applied in concentrations ranging from 0.1 to 10,000 ng/ml with overnight incubation at 4°C. Biotinylated antibody 8D12 was added and incubated for 2 h at room temperature, followed by avidin-biotin peroxidase complex for 30 min. Color was developed using *O*-phenylenediamine dihydrochloride, and absorbance was read at 490 nm after 30 min.

Using this assay protocol described above, a reproducible standard curve over the range of 1–10,000 ng/ml MBP was obtained. The assay had a minimum detection limit of 2 ng/ml MBP, with accurate measurement of MBP levels from 10 to 1000 ng/ml. There was no cross-reactivity with guinea pig eosinophil peroxidase (EPO) or with human MBP or EPO.

D. Quantitation of MBP in Bronchoalveolar Lavage

Measurement of MBP by ELISA in BALF from guinea pigs sensitized and challenge with ovalbumin by endotracheal exposure showed a significant increase only at 24 h after challenge (1643 ± 224 ng/ml). The levels observed at this time were also significantly higher than those measured in BALF from guinea pigs sensitized and challenged by free-breathing exposure (318 ± 142 ng/ml) (Fig. 4).

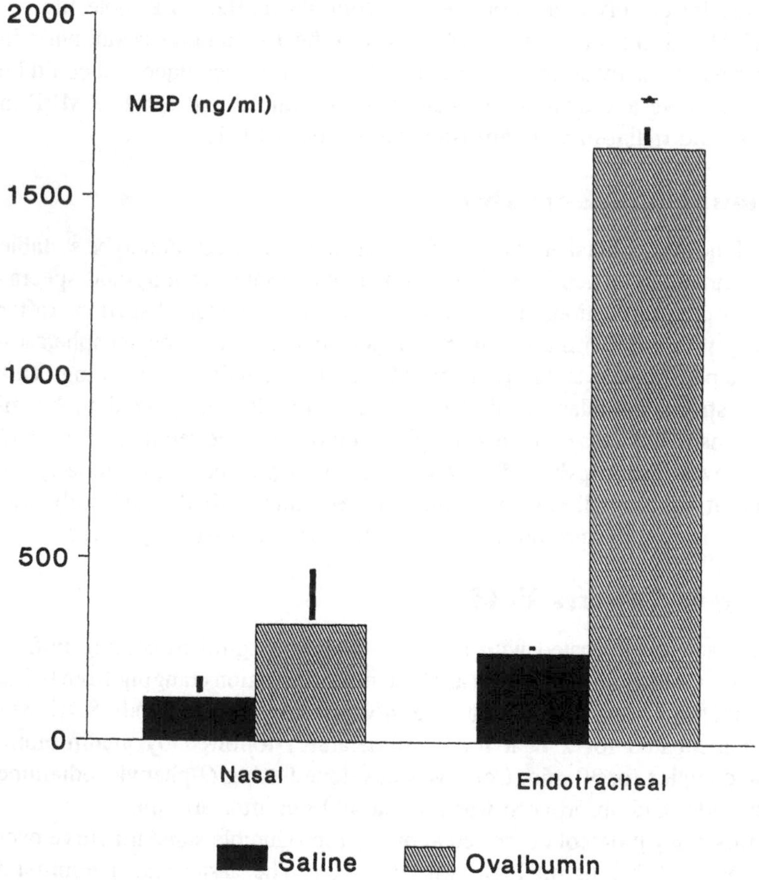

Figure 4 Concentration of major basic protein in bronchoalveolar lavage fluid according to exposure method 24 h postchallenge. Data are expressed as mean ± SEM; $*p < 0.01$, $n \geqslant 6$.

VII. SUMMARY AND CONCLUSIONS

Our experiments in guinea pigs demonstrate that, following delivery of allergen to the airways via an endotracheal tube, there is an influx of eosinophils into the airways, which then become activated to release at least MBP. This activation observed at 24 h may be related to the decline in airways function at this time. Studies of airway pathology and bronchial hyperreactivity at this time are necessary to extend our present studies to obtain relationships between eosinophil influx, activation, and airways pathology and function.

REFERENCES

1. Durham SR, Loegering DA, Dunnette S, Gleich GJ, Kay AB. Blood eosinophils and eosinophil-derived proteins in allergic asthma. J Allergy Clin Immunol 1989; 84: 931–936.
2. Bousquet J, Chanez P, Lacoste HY, Barneon G, Ghavanian N, Enander I, Venge P, Ahlstedt S, Simony-Lafontaine J, Godard P, Michel FB. Eosinophilic inflammation in asthma. N Engl J Med 1990; 323:1033–1039.
3. Diaz P, Gonzalez MC, Galleguillos FR, Ancic P, Cromwell P, Shepherd D, Durham SR, Gleich GJ, Kay AB. Leukocytes and mediators in bronchoalveolar lavage during allergen-induced late-phase asthmatic reactions. Am Rev Respir Dis 1989; 139:1383–1389.
4. Djukanovic R, Wilson JW, Britten KM, Wilson SJ, Walls AF, Roche WR, Howarth PH, Holgate ST. Quantitation of mast cells and eosinophils in the bronchial mucosa of symptomatic atopic asthmatics and healthy control subjects using immunohistochemistry. Am Rev Respir Dis 1990; 142:863–871.
5. Kay AB. Eosinophils as effector cells in immunity and hypersensitivity disorders. Clin Exp Immunol 1985; 62:1–12.
6. Wardlaw AJ, Kay AB. The role of the eosinophil in the pathogenesis of asthma. Allergy 1987; 42:321–335.
7. Gleich GJ, Hamann KJ. Eosinophils and eosinophil granule proteins in asthma: the eosinophil hypothesis. In: Kobayashi S, Bellani JA, eds. Advances in Asthmology. Amsterdam: Elsevier, 1991:195–200.
8. Venge P, Hakansson L. Current understanding of the role of the eosinophil granulocyte in asthma. Clin Exp Allergy 1991; 21 (Suppl 3):31–7.
9. Gulbenkian AR, Fernandez X, Kreutner W, Minnicozzi M, Watnick AS, Kung T, Egan RW. Anaphylactic challenge causes eosinophil accumulation in bronchoalveolar lavage fluid of guinea pigs. Am Rev Respir Dis 1990; 142:680–685.
10. Sanjar S, Aoki S, Kristersson A, Smith D, Morley J. Antigen challenge induces pulmonary airway eosinophil accumulation and airway hyperreactivity in sensitized guinea pigs: the effect of anti-asthma drugs. Br J Pharmacol 1990; 99:679–686.
11. Richards IM, Griffin RL, Oostveen JA, Morris J, Wishka DG, Dunn CJ. Effect of the selective leukotriene B_4 antagonist U-75302 on antigen-induced bronchopulmonary eosinophilia in sensitized guinea pigs. Am Rev Respir Dis 1989; 140:1712–1716.
12. Iijima H, Ishii M, Yamaguchi K, Chao CL, Kimura K, Shimura S, Shindoh Y, Inoue

H, Mue S, Takishima T. Bronchoalveolar lavage and histologic characterization of late asthmatic response in guinea pigs. Am Rev Respir Dis 1987; 136:922–929.

13. Hutson PA, Church MK, Clay TP, Miller P, Holgate ST. Early and late phase bronchoconstriction after allergen challenge of non anaesthetized guinea pigs. 1. The association of disordered airway physiology to leukocyte infiltration. Am Rev Respir Dis 1988; 137:548–557.
14. Holroyde MC, Norris AA. The effect of ozone on reactivity of upper and lower airways in guinea pigs. Br J Pharmacol 1988; 94:938–946.
15. Johns K, Sorkness R, Graziano F, Castleman W, Lemanske RF. Contribution of upper airways to antigen-induced late airway obstructive responses in guinea pigs. Am Rev Respir Dis 1990; 142:138–142.
16. Varley JG, Perring S, Seminario M-C, Fleming J, Holgate ST, Church MK. Aerosol deposition in the guinea pig. A comparison of two methods of exposure. Am Rev Respir Dis 1991; 143:A708.
17. Seminario M-C, Varley JG, Holgate ST, Church MK. Aerosol challenge through an endotracheal tube in guinea pigs. Eur Respir J 1991; 4:595s.
18. Varley JG, Heath JR, Bacon R, Holgate ST, Church MK. Reproducibility of a new automated body plethysmograph. Am Rev Respir Dis 1991; 143:A353.
19. Varley JG, Heath JF, Bacon R, Holgate ST, Church MK. Pulmonary function values using a new guinea pig body plethysmograph. Am Rev Respir Dis 1991; 143:A354.
20. Fabbri LM, Boschetto P, Zocca E, Milani G, Pivirotto F, Plebani M, Burlina A, Licata B, Mapp CE. Bronchoalveolar neutrophilia during late asthmatic reactions induced by toluene diisocyanate. Am Rev Respir Dis 1987; 136:36–42.
21. Metzger WJ, Richerson HB, Worden K, Monick M, Hunninghake GW. Bronchoalveolar lavage of allergic asthmatic patients following allergen provocation. Chest 1986; 89:477–483.
22. Metzger WJ, Zavala D, Richerson HB, Moseley P, Iwamota P, Monick M, Sjoerdsma K, Hunninghake GW. Local allergen challenge and bronchoalveolar lavage of allergic asthmatic lungs. Description of the model and local airway inflammation. Am Rev Respir Dis 1987; 135:433–440.
23. Frick WE, Sedgwick JB, Busse WW. The appearance of hypodense eosinophils in antigen-dependent late phase asthma. Am Rev Respir Dis 1989; 139:1401–1406.
24. Gleich GJ, Loegering DA, Maldonado JE. Identification of a major basic protein in guinea pig eosinophil granules. J Exp Med 1973; 137:1459–1471.
25. Filley WV, Kephart GM, Holley LE, Gleich GJ. Identification by immunofluorescence of eosinophil granule major basic protein in lung tissue of patients with bronchial asthma. Lancet 1982; 2:11–16.
26. Frigas E, Loegering DA, Solly GO, Farrow GM, Gleich GJ. Elevated levels of the eosinophil granule major basic protein in the sputum of patients with bronchial asthma. Mayo Clin Proc 1981; 56:345–353.
27. Wardlaw AJ, Dunnette S, Gleich GJ, Collins JV, Kay AB. Eosinophils and mast cells in bronchoalveolar lavage in subjects with mild asthma. Am Rev Respir Dis 1988; 137:62–69.
28. Aoki I, Shindoh Y, Nishida T, Nakai S, Hong YM, Mio M, Saito T, Tasaka K.

Sequencing and cloning of the cDNA of guinea pig eosinophil major basic protein. FEBS Lett 1991; 279:330–334.
29. Weller PF. The immunobiology of eosinophils. N Engl J Med 1991; 324:1110–1118.
30. Capron M. Eosinophils: receptors and mediators in hypersensitivity. Clin Exp Allergy 1989; (Suppl 1):3–8.
31. Rimmer SJ, Akerman CL, Hunt TC, Church MK, Holgate ST, Shute JK. Density profile of bronchoalveolar lavage eosinophils in the guinea pig model of allergen-induced late-phase allergic responses. Am J Respir Cell Mol Biol 1992; 6:340–348.
32. Frew AJ, Moqbel R, Azzawi M, Hartnell A, Barkans J, Jeffery PK, Kay AB, Scheper RJ, Varley J, Church MK, Holgate ST. T-lymphocytes and eosinophils in allergen-induced late-phase asthmatic reactions in the guinea pig. Am Rev Respir Dis 1990; 141:407–413.
33. Owen WF. Cytokine regulation of eosinophil inflammatory disease. Allergy Clin Immunol News 1991; 3:85–89.
34. Shute JK, Rimmer SJ, Ackerman CL, Church MK, Holgate ST. Studies of cellular mechanisms for the generation of superoxide by guinea pig eosinophils and its dissociation from granule peroxidase release. Biochem Pharmacol 1990; 40:2013–2021.

DISCUSSION (Speaker: M. K. Church)

Makino: Guinea pigs sensitized by antigen inhalation developed immediate- and late-phase airway narrowing associated with prolonged eosinophil infiltration in the airway after antigen inhalation challenge. However, guinea pigs sensitized by intraperitoneal injection of the antigen did not develop late-phase airway narrowing by antigen inhalation challenge, though they showed similar eosinophilia in the airway. Do you have any suggestion about the discrepancy of eosinophil infiltration and airway responses between two different models of the route of sensitization?

Church: We have not studied airways hyperreactivity in this model, but I can give you some information on the effect of different routes of sensitization on IgG production and sensitization of the tracheal smooth muscle as assessed by the Schultz-Dale response in vitro. Intraperitoneal sensitization gave high circulating levels of IgG and a good Schultz-Dale response. However, when we sensitized the animals by the intratracheal route, we had a strong Schultz-Dale response in the absence of circulating IgG. This strongly suggests a local production of immunoglobulins in the airways.

Morley: It may be of interest if I comment on our experience using anesthetized ventilated animals sensitized either actively or passively for measurement of airway hyperreactivity in relationship to eosinophil accumulation. We utilize lower concentrations of ovalbumin (0.1–0.01%) to induce airway hyperreactivity and eosinophilia that is maximal. Thus, 0.1% ovalbumin consistently induces pronounced airway hyperreactivity and eosinophilia, whereas 0.01% evokes maximal eosinophilia usually without evident airway hyperreactivity (to histamine). This finding questions the presumption that eosinophilia in the airways is the determinant of allergic airway hyperreactivity. Our doubts are reinforced by two further observations. First, in animals passively sensitized by IgE, infusion of low doses of antigen, insufficient to effect overt airway obstruction, leads to pronounced airway hyperreactivity to LTC_4, histamine, bradykinin, and $PGD_{2\alpha}$, yet there is no associated eosinophil accumulation at this time (first hour). Second, administration of cyclosporin A as a single dose prior to inhalation of allergen can suffice to abolish allergic eosinophilia, yet acute administration of cyclosporin A fails to influence acute allergic bronchospasm or associated increased reactivity to histamine.

These findings can be accommodated by proposing that acute airway hyperreactivity and allergic inflammation are separate events, a viewpoint that predicts the occurrence of drugs that will inhibit expression or development of airway hyperreactivity and accumulation or activation of eosinophils in the airways separately or together. Thus, potassium channel openers and sympathomimetics suppress expression of airway hyperreactivity without influencing airway eosinophilia; cyclosporin A inhibits fully allergic eosinophilia in the airways yet does not influence acute allergic bronchospasm or allergic hyperreactivity (histamine); PDE isoenzyme inhibitors (types III and IV) may inhibit airway hyperreactivity and airway eosinophilia separately or together according to the pattern of inhibition selected. I might add in passing that in all these drug categories our selection has been primarily based on selection by reference to whole-animal pharmacology using these two techniques rather than by reference to mechanistic approaches. Hence, our experience is not inconsistent with the failure of Professor Church to demonstrate an association between airway obstruction and MBP release as an index of eosinophil activation.

Church: I would be interested to measure MBP in your antigen challenge experiments.

Lee: In the recent model of guinea pig asthma, have you measured bronchial hyperreactivity?

Church: We have not measured bronchial hyperreactivity as yet.

Egan: I sympathize with your difficulties generating consistent hyperreactivity in guinea pigs. We have had similar experiences, and using pertussis and $Al(OH)_3$ during sensitization, rather than only albumin, we find eosinophilia and more extensive and consistent hyperreactivity. We are now focusing more on other species, including mice and monkeys.

Gundel: In your new guinea pig model, do you see damage to the bronchial epithelium due to the high levels of MBP?

Church: We have not yet studied this in detail.

M. Capron: In relation to the sticky properties of MBP, have you looked at the presence of other granule proteins such as EPO, which is easy to titrate owing to its enzymatic properties?

Church: No, we have not.

Wardlaw: The justification for animal models is that they can tell us something conclusive about human disease, in this case asthma, that cannot be revealed by human studies. Notwithstanding the fundamental species differences between guinea pigs and humans, the difficulties you have had with your model seem to suggest that the guinea pig is inappropriate as a model for asthma. More fundamentally, is there anything we have learned about asthma from guinea pig studies that has not been shown in human studies?

Church: No single animal model is predictive of human asthma. They are only models and as such can only be used to examine experimental hypotheses and to perform experiments that are not ethically possible in humans. Also, models are necessary to examine newly synthesized drugs.

Egan: In response to Dr. Wardlaw's comment, it is important to bear in mind that animal models that simulate the desired human disease are essential to drug discovery projects. In several studies, eosinophil peroxidase enzyme activity is being used to indicate the presence of eosinophils. It is important to keep in mind that, as with other peroxidases, EPO self-destructs as it catalyzes its reaction. Therefore, the amount of EPO could be dramatically underestimated. Immunological methods are much more reliable for detection of EPO.

Vadas: Could you comment on the wide error and lack of correlation between hyperreactivity and MBP levels?

Church: My data were from an experiment we performed very recently. This work needs to be confirmed before publication.

Townley: Are you implying that in your original challenge studies, the pulmonary function response was due to a nasal reflex? I would think this surprising since in awake sensitized guinea pigs, challenge with histamine or methacholine or allergen results in wheezing, severe dyspnea, and death by all these challenges. The lungs become markedly hyperinflated, and in antigen challenge a marked eosinophilia is noted in the bronchi, as well as damage to epithelium.

Church: One of the greatest problems with the majority of smaller mammals is that they are obligate nose breathers. Thus in conscious animals, measurement of airways function involves the nose and the lungs in series. From our studies it also appears that there are reflexes, in both directions, between the nose and the lungs. These need to be examined in more detail before your question can be answered definitively.

Durham: In your intubated guinea pig model was there any relationship between cellular infiltration and the late asthmatic cellular infiltration and the late asthmatic response as represented by total gas volume rather than airways conductance measurements?

Church: In our new model there is no classic late reaction using our present allergen loads.

21

Asthma, Eosinophils, and Interleukin-5

A. Barry Kay, Qutayba Hamid, Douglas S. Robinson, Andrew M. Bentley, Anne Tsicopoulos, Sun Ying, Redwan Moqbel, Christopher J. Corrigan, and Stephen R. Durham
National Heart and Lung Institute, London, England

I. INTRODUCTION

Asthma is characterized by infiltration of the bronchial mucosa with large numbers of activated eosinophils and the presence of elevated concentrations of eosinophil-derived proteins such as major basic protein (1) and the eosinophil cationic protein (2). The degree of the eosinophilia has been shown to correlate with the severity of airways hyperresponsiveness (1,3). In a primate model of asthma, inhibition of the eosinophilic response to allergen substantially blocked the development of airways hyperresponsiveness (4). Eosinophil-derived mediators have potential for producing many of the pathological features of asthma. For example, epithelial shedding might possibly occur through the cytotoxic effects of secreted eosinophil granule proteins (5) and mucus hypersecretion and bronchoconstriction possibly through the release of platelet-activating factor (PAF) and leukotriene C_4 (6). Recent studies have also suggested a role for T lymphocytes in asthma. T-lymphocyte infiltration is a feature of the late-phase response to allergen in atopic individuals in both the skin (7) and lung (8). Increased numbers of activated T cells and concentrations of their products have been observed in the peripheral blood of acute, severe asthmatics (9). By specific immunostaining of bronchial mucosal biopsies obtained via the fiberoptic bronchoscope, we demonstrated a significant increase in numbers of interleukin-2 (IL-2) receptor-bearing cells in the airways of mild, steady-state asthmatics (10). This was associated with an elevation in the numbers of EG2+ cells (EG2 is a monoclonal antibody that recognizes the secreted form of the eosinophilic cationic protein) (11).

It has been known for several years that T lymphocytes play a central role in eosinophil production and function through the release of soluble mediators. A number of cytokines with selective actions on eosinophils have now been sequenced and cloned. One of the most important is interleukin-5 (IL-5), which promotes terminal differentiation of the committed eosinophil precursor (12) as well as enhancing the effector capacity of mature eosinophils (13). IL-5 also prolongs the survival of eosinophils in vitro (14). Furthermore, antibodies against IL-5 ablate the eosinophilic response to helminthic infection in mouse (15), and an IL-5-like substance has been found in the serum of patients with hypereosinophilia (16). IL-5 increases eosinophil, but not neutrophil, adhesion to vascular endothelium (17) and also primes eosinophils for enhanced locomotory responsiveness to PAF and other chemoattractants (18). Release of IL-5 at the site of allergic inflammation could explain, in part, the specific eosinophil accumulation seen in these conditions. Similarly, release of IL-5, in and around the bronchial mucosa, by activated T lymphocytes could lead to the specific recruitment of eosinophils, enhanced eosinophil cytotoxicity, and prolonged eosinophil survival.

For these reasons we have attempted to determine whether IL-5 is associated with, and possibly regulates, the asthma process. Several approaches have been used. These include measurements of serum concentrations of IL-5 in chronic severe asthma and the identification of messenger RNA (mRNA) for IL-5 in bronchial biopsies and bronchoalveolar lavage (BAL) cells in patients with milder disease. We have been able to investigate steady-state asthma, provoked asthma in the clinical laboratory, and asthmatics before and after treatment with corticosteroids.

II. EXPRESSION OF mRNA FOR INTERLEUKIN-5 IN MUCOSAL BRONCHIAL BIOPSIES IN ASTHMA

Identification of mRNA for IL-5 in the bronchi of asthmatics might provide important evidence of local IL-5 generation as well as emphasizing the possible link between eosinophils and T lymphocytes in this disease. We therefore used the technique of in situ hybridization using an IL-5 complementary RNA probe to investigate the expression of IL-5 mRNA and the pattern of distribution of IL-5-producing cells in bronchial tissue obtained from biopsies of asthmatics and normal individuals (19). We also attempted to relate the expression of IL-5 mRNA to the severity of the disease and the degree of infiltration of the airway mucosa with eosinophils and activated T lymphocytes.

Bronchial biopsies were obtained from 10 asthmatics and nine nonatopic normal controls. A radiolabeled cRNA probe was prepared from an IL-5 cDNA and hybridized to permeabilized sections. These were washed extensively before processing for autoradiography. An IL-5-producing T-cell clone derived from a patient with the hyper-IgE syndrome was used as a positive control. As a negative control, sections were also treated with a "sense" IL-5 probe. Specific hybridization signals for IL-5 mRNA were demonstrated within the bronchial mucosa in six of the 10 asthmatic subjects. Cells exhibiting hybridization signals were located beneath the epithelial basement membrane. In contrast, there was no hybridization in the control group. No hybridization was observed with the sense probe.

The six IL-5 mRNA-positive asthmatics tended to have more severe disease than the negative asthmatics, as assessed by symptoms and lung function, and showed a significant increase in the degree of infiltration of the bronchial mucosa by secreting (EG2+) eosinophils and activated (CD25+) T lymphocytes. Within the subjects who showed positive IL-5 mRNA, there was a correlation between the numbers of IL-5 mRNA+ cells and the number of CD25+ and EG2+ cells and total eosinophil count.

Therefore, this study provided evidence for the cellular localization of IL-5 mRNA in the bronchial mucosa of asthmatics and supported the concept that this cytokine regulates eosinophil function in bronchial asthma.

III. T-CELL ORIGIN OF IL-5 mRNA IN MILD ATOPIC ASTHMA

To examine whether mRNA for IL-5 was expressed by T cells, cytocentrifuge preparations of BAL cells from three subjects with asthma were fixed in 4% paraformaldehyde and washed in 15% sucrose in phosphate-buffered saline (20). Cells were incubated simultaneously with IL-5 cRNA probes labeled with uridine triphosphate-biotin and with monoclonal antibody to CD3 directly conjugated to fluorescein isothiocyanate. The conditions for in situ hybridization were as previously described (21), and positive hybridization of probe to cytokine mRNA was detected using streptavidin–Texas red staining. Controls were IgG1-fluorescein isothiocyanate with sense probes and RNase pretreatment. Cells were quantified by fluorescence microscopy, and the percentages of cells expressing both CD3 and cytokine mRNA as well as the percentages of singly labeled cells were evaluated by counting at least 200 positive cells.

Dual fluorescence for CD3 and mRNA for IL-5 in BAL cytospin preparations showed that a mean of 91% of the cells positive for IL-5 mRNA were positive for CD3 (89%, 92%, and 91% in the three subjects), and 58% of the CD3+ cells expressed IL-5 mRNA.

To examine further whether T cells were the source of IL-5 mRNA, BAL cells from five subjects with asthma were incubated with immunomagnetic beads covalently bound to monoclonal antibody to CD2 in a ratio of three beads to one lymphocyte for 20 min at 4°C. Cells were separated with a magnetic cell separator.

Table 1 Percentages of Total BAL Cells on Cytocentrifuge Preparations from Five Subjects with Asthma Identified as Lymphocytes on May Grünwald Giemsa Staining and as IL-4 and IL-5 mRNA-Positive by *In Situ* Hybridization

		% mRNA+[a]	
Cell	% lymphocytes[a]	IL-4	IL-5
Unseparated	13 (8–14)	7 (4–8)	7 (6–11)
CD2 positive	70 (62–85)	42 (39–62)	59 (37–70)
CD2 negative	3 (1–4)	1 (0–1)	2 (1–4)

[a]Medians with range.
Bronchoalveolar lavage samples were analyzed after immunomagnetic selection of CD2+ cells (CD positive), depletion of CD2+ cells (CD2 negative), or no separation procedure (unseparated).

CD2+ cells were washed four times in phosphate-buffered saline with 0.1% bovine serum albumin, and cytospin preparations were made from both positively and negatively separated cells as well as from unseparated cells for differential counts and in situ hybridization for IL-5 mRNA. Hybridization was detected and quantified in cells as a percentage of BAL cells in positively or negatively separated and unseparated samples. These results were compared with the percentages of lymphocytes on cytospin slides as determined on May Grünwald Giemsa staining (Table 1). There was a clear association between cytokine mRNA expression and CD2+ cells. Binding of anti-CD2-coated beads to mRNA+ cells in cytospin preparations identified these cells as T lymphocytes.

IV. IL-5 mRNA AND THE LATE-PHASE ASTHMATIC REACTION

Inhalation challenge of atopic asthmatic subjects with allergen provokes an immediate or early asthmatic response detectable within minutes as a decrease in forced expiratory volume in 1 s (FEV_1). This may be followed by a second fall in FEV_1 between 4 and 8 h after allergen exposure, which may persist for 3–8 h and is associated with an increase in nonspecific bronchial responsiveness that may persist for several days (22–24). While the early asthmatic response appears to be dependent on IgE triggering of mediator release from mast cells resulting in bronchoconstriction and airway edema (25,26), the late asthmatic response is associated with an influx of inflammatory cells, particularly eosinophil leukocytes (2,8,27), into the bronchial mucosa. Eosinophil infiltration of the bronchial mucosa is a feature of the airways of patients dying of asthma (28) and is present in bronchial biopsies (29) and BAL (1,30) of patients with mild disease. Thus, although not entirely reproducing the pattern of natural exposure, allergen challenge may provide a model for investigating allergen-induced inflammatory events in atopic asthma (31).

Participation of T lymphocytes in allergen-induced asthmatic responses was suggested by the finding of lymphocyte and eosinophil infiltration in BAL after local allergen challenge (8) and changes in CD4 and CD8 T cells in both peripheral blood and BAL after allergen challenge (32–34). We, therefore, hypothesized that allergen inhalation challenge of atopic asthmatics results in local activation of IL-5-producing lymphocytes with subsequent eosinophilia and that this in turn contributes to the associated changes in airway caliber and hyperresponsiveness.

To test this hypothesis we examined the expression of IL-5 mRNA in BAL cells (35) and bronchial biopsies (36) in sensitized atopic asthmatics 24 h after inhalation challenge with either allergen or diluent control. Compared with diluent, there were significant increases after allergen challenge in the numbers of cells expressing mRNA for IL-5 from both BAL and biopsies. There were also increases in eosinophils in both bronchial wash and BAL after allergen but not when compared to diluent challenge. Close associations were observed between the numbers of CD25+ BAL CD4+ T cells after allergen challenge, cells expressing IL-5 mRNA, and eosinophils. There was also a correlation between the numbers of cells expressing mRNA for IL-3 and IL-5 and BAL eosinophils on the allergen day. BAL and bronchial wash eosinophilia also closely correlated with maximal late fall in FEV_1 after allergen challenge. Similarly, the numbers of EG2+ cells in biopsies and mRNA IL-5+ cells also correlated. We concluded, therefore, that IL-5 may contribute to late asthmatic responses in the airway by mechanisms that include eosinophil accumulation (Fig. 1).

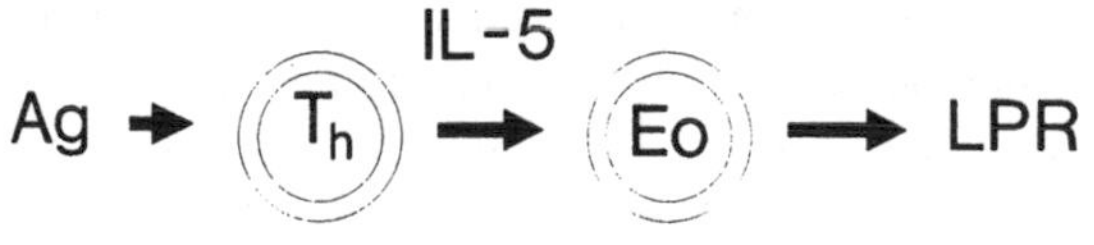

CD4/CD25+% vs. IL-5 mRNA+	r=0.70,p=0.008
CD4/CD25+% vs. Eosinophils%	r=0.62,p=0.02
IL-5 mRNA+ vs. Eosinophils%	r=0.60,p=0.03
Eosinophils% vs. LPR	r=0.73,p=0.005

Figure 1 Hypothesized mechanism of eosinophil activation contributing to late asthmatic responses. Allergen challenge activates Th2-like CD4+ T cells, leading to IL-5 production. Correlations between these variables in BAL obtained 24 h after allergen challenge of 13 atopic asthmatic subjects are shown.

V. THE EFFECT OF CORTICOSTEROIDS ON IL-5 SYNTHESIS AND SECRETION IN ASTHMA

The beneficial effects of corticosteroids in the treatment of asthma are well documented (37), and the use of inhaled corticosteroids in chronic asthma is widely accepted (38,39). It has been suggested that corticosteroids mediate their beneficial role in the reversal of airway obstruction and bronchial hyperreactivity through an effect on T-cell-associated inflammatory processes (40). However, the mechanism of action of corticosteroids in asthma remains unclear.

We hypothesized that inhibition of T-cell cytokine production (particularly IL-5) in vivo may underlie, at least in part, the anti-inflammatory actions of corticosteroids in chronic asthma. To examine the effect of prednisolone on IL-5 gene expression in symptomatic chronic asthma, we performed fiberoptic bronchoscopy with BAL and endobronchial biopsies before and after 2 weeks of therapy in a double-blind, placebo-controlled parallel group study (41). To assess the clinical relevance of cellular changes, the response to treatment was followed closely by measurements of lung function and bronchial responsiveness.

Clinical improvement in the patients receiving prednisolone was shown by decreases in airflow obstruction and in bronchial responsiveness to inhaled methacholine, which were not seen in patients receiving placebo. Between-group comparison showed a significant fall in numbers of BAL cells per 1000 with positive in situ hybridization signals for mRNA for IL-5 with prednisolone treatment. There was also a reduction in BAL eosinophils in prednisolone-treated patients compared with those receiving placebo. Immunohistology of bronchial mucosal biopsies revealed a significant decrease in the numbers of T cells (CD3+) and EG2+ eosinophils in those patients receiving prednisolone, together with a significant reduction in numbers of tryptase only (MC_T) but not tryptase/chymase-positive (MC_{TC}) mast cells by double sequential immunostaining (Table 2).

Finally, we also obtained evidence that in chronic severe asthma IL-5 was detectable in the serum and decreased after treatment with prednisolone (42). Peripheral blood mononuclear cells and serum were obtained, on two occasions, from 15 asthmatic patients who required oral glucocorticoid therapy for moderate to severe disease exacerbations. Samples were obtained immediately before commencement of oral glucocorticoids (day 1) and again after 7 days of treatment (day 7). Samples were also isolated on two occasions 7 days apart from a group of seven untreated normal volunteers. Serum concentrations of IL-5 were measured using an ELISA technique. IL-5 was detectable in the serum of eight of the asthmatic patients on day 1, but in none of these patients on day 7. Serum IL-5 was undetectable in all the control subjects on both occasions. The numbers of activated (CD4+/CD25+) T helper cells were also elevated in the asthmatics and decreased after treatment with prednisolone. These observations are consistent with the hypothesis that exacerbations of asthma are associated with activation

Table 2 Summary of Findings in BAL and Endobronchial Biopsies in a Double-Blind Study of Prednisolone Treatment in Symptomatic Asthma

Bronchoalveolar lavage	
↓ Eosinophils	$p < 0.05$
↓ IL-4 and IL-5 mRNA+ cells	$p < 0.01$
↑ IFN-γ mRNA+ cells	$p < 0.01$
Bronchial biopsies	
↓ CD3+ cells	$p < 0.05$
↓ EG2+ eosinophils	$p < 0.05$
↓ MC_T mast cells	$p < 0.05$

There were significant decreases in BAL eosinophils and cells positive for IL-4 and IL-5 mRNA together with an increase in cells expressing interferon-γ mRNA when subjects receiving prednisolone were compared with those receiving placebo. Bronchial biopsies showed significant reductions in CD3, EG2, and tryptase-positive cells after prednisolone treatment.

of CD4 T lymphocytes that secrete IL-5, and that glucocorticoid therapy results in reduction of the activation status of these cells concomitant with inhibition of IL-5 secretion.

VI. SUMMARY AND CONCLUSIONS

Eosinophils are implicated as major proinflammatory cells in the asthma process. A number of eosinophil functions are regulated by IL-5. These include terminal differentiation of the committed eosinophil precursor, increased adhesiveness of eosinophil to vascular endothelium, and survival of eosinophils in vitro. Here we present evidence that IL-5 mRNA transcripts are detectable in T lymphocytes in ongoing steady-state asthma as well as asthma provoked by inhalational challenge (i.e., the late-phase reactions). Furthermore, the numbers of mRNA+ cells in BAL in asthma and circulating concentrations of serum IL-5 decrease after corticosteroid therapy. Taken together, these results suggest that IL-5 plays a critical role in eosinophil-induced bronchial inflammation in asthma. They also suggest that agents that selectively target IL-5 may be of value in the treatment of this disease.

REFERENCES

1. Wardlaw AJ, Dunnette S, Gleich GJ, Collins JV, Kay AB. Eosinophils and mast cells in bronchoalveolar lavage in subjects with mild asthma. Relationship to bronchial hyperreactivity. Am Rev Respir Dis 1988; 137:62–69.

2. De Monchy JGR, Kauffman HK, Venge P, Koeter GH, Jansen HM, Sluiter HJ, De Vries K. Bronchoalveolar eosinophilia during allergen-induced late asthmatic reactions. Am Rev Respir Dis 1985; 131:373–376.
3. Durham SR, Kay AB. Eosinophils, bronchial hyperreactivity and late-phase asthmatic reactions. Clin Allergy 1985; 15:411–418.
4. Wegner CD, Gundel RH, Reilly P, Haynes N, Letts GL, Rothlein R. Intercellular adhesion molecule-1 (ICAM-1) in the pathogenesis of asthma. Science 1990; 247: 456–459.
5. Gleich GJ, Frigas E, Loegering DA, Wassom DL, Steinmuller D. Cytotoxic properties of the eosinophil major basic protein. J Immunol 1979; 123:2925–2927.
6. Holgate ST, Abraham WM, Barnes PJ, Lee TH. Pharmacology and treatment. In: Holgate ST, Howell JBL, Burney PGJ, Drazen JM, Hargreave FE, Kay AB, Kerrebijn KF, Reid LM, eds. The Role of Inflammatory Processes in Airway Hyperresponsiveness. Oxford: Blackwell, 1989:179–221.
7. Frew AJ, Kay AB. The relationship between infiltrating CD4+ lymphocytes, activated eosinophils and the magnitude of the allergen-induced late phase cutaneous reaction in man. J Immunol 1988; 141:4158–4164.
8. Metzger WJ, Zavala D, Richerson HB, Moseley P, Iwamota P, Monick M, Sjoerdsma K, Hunninghake GW. Local allergen challenge and bronchoalveolar lavage of allergic asthmatic lungs. Description of the model and local airway inflammation. Am Rev Respir Dis 1987; 135:433–440.
9. Corrigan CJ, Hartnell A, Kay AB. T-lymphocyte activation in acute severe asthma. Lancet 1988; 1:1129–1132.
10. Azzawi M, Bradley B, Jeffery PK, Frew AJ, Wardlaw AJ, Assoufi B, Collins JV, Durham SR, Knowles GK, Kay AB. Identification of activated T lymphocytes and eosinophils in bronchial biopsies in stable atopic asthma. Am Rev Respir Dis 1990; 142:1407–1413.
11. Tai P-C, Spry CJF, Peterson C, Venge P, Olsson I. Monoclonal antibodies distinguish between storage and secreted forms of eosinophil cationic peptide. Nature 1984; 309: 182–184.
12. Clutterbuck EJ, Hirst EMA, Sanderson CJ. Human interleukin-5 (IL-5) regulates the production of eosinophils in human bone marrow cultures: comparison and interaction with IL-1, IL-3, IL-6 and GM-CSF. Blood 1988; 73:1504–1513.
13. Lopez AF, Sanderson CJ, Gamble JR, Campbell HR, Young IG, Vadas MA. Recombinant human interleukin-5 is a selective activator of human eosinophil function. J Exp Med 1988; 167:219–224.
14. Yamaguchi Y, Hayashi Y, Sugama Y, Miura Y, Kasahara T, Kitamura S, Torisu M, Mita S, Tominaga A, Takatsu K, Suda T. Highly purified murine interleukin-5 (IL-5) stimulates eosinophil function and prolongs in vitro survival. J Exp Med 1988; 167: 1737–1742.
15. Coffman LR, Seymour WB, Hudak S, Jackson J, Rennick D. Antibody to interleukin 5 inhibits helminth-induced eosinophilia in mice. Science 1989; 245:308–310.
16. Owen WF, Rothenberg ME, Petersen J, Weller PF, Silberstein D, Sheffer AL, Stevens RL, Soberman RJ, Austen KF. Interleukin-5 and phenotypically altered eosinophils in the blood of patients with the idiopathic hypereosinophilic syndrome. J Exp Med 1989; 170:343–348.

17. Walsh GM, Hartnell A, Wardlaw AJ, Kurihara K, Sanderson CJ, Kay AB. IL-5 enhances the in vitro adhesion of human eosinophils but not neutrophils, in a leukocyte integrin (CD11/18)-dependent manner. Immunology 1990; 71:258–265.
18. Sehmi R, Wardlaw AJ, Cromwell O, Kurihara K, Waltmann P, Kay AB. Interleukin-5 (IL-5) selectively enhances the chemotactic response of eosinophils obtained from normal but not eosinophilic subjects. Blood 1992; 79:2952–2959.
19. Hamid Q, Azzawi M, Sun Ying, Moqbel R, Wardlaw AJ, Corrigan CJ, Bradley B, Durham SR, Collins JV, Jeffery PK, Quint DJ, Kay AB. Expression of mRNA for interleukin-5 in mucosal bronchial biopsies from asthma. J Clin Invest 1991; 87: 1541–1546.
20. Robinson DS, Hamid Q, Sun Ying, Tsicopoulos A, Barkans J, Bentley AM, Corrigan CJ, Durham SR, Kay AB. Predominant T_{H2}-type bronchoalveolar lavage T-lymphocyte population in atopic asthma. N Engl J Med 1992; 326:298–304.
21. Giaid A. Non-isotopic RNA probes. Histochemistry 1989; 93:191–196.
22. Booij-Nord H, Orie NGM, deVries K. Immediate and late bronchial obstructive reactions to inhalation of house dust and protective effect of disodium cromoglycate and prednisolone. J Allergy Clin Immunol 1971; 48:344–353.
23. Robertson DG, Kerrigan AT, Hargreave FE, Chalmers R, Dolovich J. Late asthmatic responses induced by ragweed pollen allergens. J Allergy Clin Immunol 1974; 54: 244–254.
24. Cartier A, Thomson NC, Frith PA, Roberts R, Hargreave FE. Allergen-induced increase in bronchial responsiveness to histamine: relationship to the late asthmatic response and change in airway caliber. J Allergy Clin Immunol 1982; 70:170–177.
25. Liu MC, Hubbard WC, Proud D, Stealey BA, Galli SJ, Kagey-Sobotka A, Bleeker ER, Lichtenstein LM. Immediate and late inflammatory responses to ragweed antigen challenge of the peripheral airways in allergic asthmatics. Am Rev Respir Dis 1991; 144:51–58.
26. Casale TB, Wood D, Richerson HB, Zehr B, Zavala D, Hunninghake GW. Direct evidence of a role for mast cells in the pathogenesis of antigen-induced bronchoconstriction. J Clin Invest 1987; 80:1507–1511.
27. Diaz P, Gonzalez C, Galleguillos FR, Ancic P, Cromwell O, Shepherd D, Durham SR, Gleich GJ, Kay AB. Leukocytes and mediators in bronchoalveolar lavage during allergen-induced late-phase asthmatic reactions. Am Rev Respir Dis 1989; 139:1383–1389.
28. Dunnill MS. The pathology of asthma. In: Middleton E, Reed CE, Ellis EF, eds. Allergy; Principles and Practice. St. Louis; Mosby, 1978:678–686.
29. Bradley BL, Azzawi M, Jacobson M, Assoufi B, Collins JV, Irani A-M, Schwartz LB, Durham SR, Jeffery PK, Kay AB. Eosinophils, T-lymphocytes, mast cells, neutrophils and macrophages in bronchial biopsy specimens from atopic subjects with asthma: comparison with biopsy specimens from atopic subjects without asthma and normal control subjects and relationship to bronchial hyperresponsiveness. J Allergy Clin Immunol 1991; 88:661–674.
30. Bousquet J, Chanez P, Lacoste JY, Barneon G, Ghavanian N, Enander I, Venge P, Ahlstedt S, Simony-Lafontaine J, Godard P, Michel FB. Eosinophilic inflammation in asthma. N Engl J Med 1990; 323:1033–1039.
31. O'Byrne PM, Dolovich J, Hargreave FE. Late asthmatic responses. State of art. Am Rev Respir Dis 1987; 136:740–751.

32. Gonzalez MC, Diaz P, Galleguillos FR, Ancic P, Cromwell O, Kay AB. Allergen-induced recruitment of bronchoalveolar helper (OKT4) and suppressor (OKT8) T cells in asthma. Am Rev Respir Dis 1987; 136:600–604.
33. Gerblich AA, Campbell AE, Schuyler MR. Changes in T lymphocyte subpopulations after antigenic bronchial provocation in asthmatics. N Engl J Med 1984; 310:1349–1352.
34. Gerblich AA, Salik H, Schuyler MR. Dynamic T cell changes in peripheral blood and bronchoalveolar lavage after antigen bronchoprovocation in asthmatics. Am Rev Respir Dis 1991; 143:533–537.
35. Robinson DS, Hamid Q, Sun Ying, Bentley AM, Barkans J, Durham SR, Kay AB. Activated T helper cells and interleukin-5 gene expression in bronchoalveolar lavage from atopic asthma. Relationship to symptoms and bronchial responsiveness. Thorax 1993; 48:26–32.
36. Bentley AM, Qiu Meng, Robinson DS, Hamid Q, Kay AB, Durham SR. Increases in activated T lymphocytes, eosinophils and cytokine messenger RNA expression for IL-5 and GM-CSF in bronchial biopsies after allergen inhalation challenge in atopic asthmatics. Am J Respir Cell Mol Biol 1993; 8:35–42.
37. Walsh SD, Grant IWB. Corticosteroids in the treatment of chronic asthma. Br Med J 1966; 2:796.
38. British Thoracic Society. Guidelines for management of asthma in adults. Statement by the research unit of the Royal College of Physicians of London, King's Fund Centre, National Asthma Campaign. Br Med J 1990; 142:434–457.
39. Expert Panel Report Guidelines for Diagnosis and Management of Asthma. Bethesda, MD: National Heart, Lung and Blood Institute Information Center, 1991.
40. Reed CE. Aerosol steroids as primary treatment of mild asthma. N Engl J Med 1991; 325:425–426.
41. Robinson DS, Hamid Q, Sun Ying, Bentley AM, Assoufi B, North J, Qui Meng, Durham SR, Kay AB. Prednisolone treatment in bronchial asthma. Clinical improvement is accompanied by reduction in bronchoalveolar lavage eosinophilia and modulation of IL-4, IL-5 and IFN-gamma cytokine gene expression. Am Rev Respir Dis. In press.
42. Corrigan CJ, Haczku A, Gemou-Engesaeth V, Doi S, Kikuchi Y, Takatsu K, Durham SR, Kay AB. CD4 T-lymphocyte activation in asthma is accompanied by increased serum concentrations of interleukin-5: effect of glucocorticoid therapy. Am Rev Respir Dis 1993; 147:540–547.

DISCUSSION (Speaker: A. B. Kay)

Gleich: Is it possible that the CD2 selection of T cells induced mRNA for IL-4 and IL-5? Are all of the IL-5/IL-4 hybridizing cells CD2-positive T cells?

Kay: It is unlikely that T cells were activated by CD2 selection. The separation was performed in the cold and completed within an hour or so. All the IL-5/IL-4 hybridizing cells were CD2 positive. We have shown this not only by the magnetic bead technique, but also by combined immunofluorescence and in situ hybridization.

Busse: Have you retrieved lymphocytes from the lavage for culture and measured IL-5 secretion from either spontaneous release or following antigen activation?

Kay: These experiments are ongoing and we have no results to report at present.

Makino: In electron microscopic examination of the density of eosinophils and lymphocytes, the damage of the bronchial epithelium and airway responsiveness was not different between atopic and nonatopic asthma. Do you think Th2 and eosinophil relationship is present in nonatopic asthma as in atopic asthma?

Kay: The Th2 cytokine profile has only been found in atopic asthma.

Hansel: Christoph Walker and colleagues have studied cytokine patterns in BAL from intrinsic (nonatopic) and extrinsic (atopic) asthmatics. BAL from intrinsic asthmatics contained increased amounts of IL-5, IL-2, and IFN-γ. In contrast, BAL from extrinsic asthmatics contained increased IL-5 and IL-4. These results have been published (1).

Galli: Do you, or others, have information about the course of allergic asthma or other allergic disorders in patients who experience T-cell deficits as a result of HIV infection?

Kay: Although there are some anecdotal reports of "allergic breakthrough" in HIV infection, I do not think there is any solid data. A proper prospective study would be required to answer your question satisfactorily, but this would be very difficult.

Venge: Did you distinguish between CD45RO and CD45RA cells in your BALs after allergen challenge?

Kay: Almost all the T cells in BAL from normal as well as asthmatics are CD45RO positive (2).

Ackerman: Could you comment on your results with cyclosporin A treatment of asthma with regard to changes in the T-cell populations and cytokine profiles you see in the lung?

Kay: Our cyclosporin A trial (3) was in chronic steroid-dependent asthma. Thus we were unable to sample from the lung. The results of the serum cytokine measurements are currently being analyzed.

Bochner: You presented data showing detectable levels of IL-5 in the serum of some, but not all, of your more severe asthmatics. Can you comment whether there were any associated characteristics that distinguished these products from those without detectable IL-5 levels (e.g., degree of peripheral blood eosinophilia, airways physiology, sinus disease, etc.)?

Kay: There were no obvious differences between the serum IL-5-positive and IL-5-negative individuals.

Townley: You show prednisone increased IFN-γ. Is it known from your studies or other studies that prednisone increases IFN-γ? Do asthmatics have decreased baseline IFN-γ? Is there any comparison of IFN-γ levels in normals versus asthmatics?

Durham: We have no comparison of IFN-γ between normals and asthmatics. I am not aware of studies of effect of prednisone on IFN-γ and we were somewhat surprised by the findings.

Hansel: It is interesting that IL-5 levels are raised in about half of chronic severe asthmatics. Have you measured levels of cytokines such as GM-CSF, IFN-γ, IL-2, and IL-3 in these sera?

Kay: In a previous study we showed that IFN-γ and IL-2R were elevated in the serum of chronic asthmatics. We do not yet have the results of GM-CSF or IL-3.

Coffman: The data you presented on the role of IL-5 in chemotaxis as well as data from earlier presentations about eosinophil homology do not seem to predict a decisive role for IL-5 in short-term, antigen-stimulated eosinophil infiltration. Yet the data from Dr. Egan's group show almost total inhibition of eosinophils in BAL after 24 h in mouse, monkey, and guinea pig. Can you comment on this apparent discrepancy? On the other hand, the later influx of eosinophils (i.e., within hours) may have an IL-5 component.

Gleich: These are elegant studies on the pathophysiology of asthma. I have two comments, and a question. First, Ohnishi et al. in our laboratory have measured elevated concentrations of IL-5 in BAL fluids of patients with chronic severe asthma, in keeping with your findings. Second, we also detected a potent inhibitor of cytokine-mediated survival in BAL fluids that appears to be lidocaine. Finally, you seem to provide an alternative to the dogma that the mast cell is central to allergic inflammation, your data pointing to an important role for the CD4 T cell. Can you comment?

Kay: Our scheme, which is just a hypothesis, emphasizes the role of the CD4 T cell in chronic allergic inflammation. So far we have no evidence that mast cells participate in ongoing allergic inflammation, although I appreciate that they probably contribute to the late-phase reaction that follows a one-off antigen challenge.

Spry: Does IL-5 affect neurotransmitter release/smooth muscle? Is asthma a disease of the microenvironment or macroenvironment?

Kay: Asthma is a disease of bronchial epithelium. It affects the bronchial epithelial cells, the underlying extracellular matrix proteins, blood vessels, mucus glands, and smooth muscle. It is associated with an intense infiltration of eosinophils and mononuclear cells. The initial lesion or principal molecular target is unknown.

DISCUSSION REFERENCES

1. Walker C, Bode E, Boer L, Hansel TT, Blaser K, Virchow J-C Jr. Allergic and non-allergic asthmatics have different patterns of T cell activation and cytokine production in peripheral blood and bronchoalveolar lavage. Am Rev Respir Dis 1992; 146: 109–115.
2. Robinson DS, Bentley AM, Hartnell A, Kay AB, Durham SR. Activated memory T helper cells in bronchoalveolar lavage from atopic asthmatics. Relationship to asthma symptoms, lung function and bronchial responsiveness. Thorax 1993; 48:26–32.
3. Alexander AG, Barnes NC, Kay AB. Trial of cyclosporin A in corticosteroid-dependent chronic severe asthma. Lancet 1992; 339:324–328.

22

Eosinophils and Late-Phase Reactions in Primates

Robert H. Gundel, Craig D. Wegner, and L. Gordon Letts
Boehringer Ingelheim Pharmaceuticals, Inc., Ridgefield, Connecticut

I. INTRODUCTION

Allergen inhalation by allergic asthmatics can result in the occurrence of an immediate, acute bronchoconstriction (acute response) that peaks 10–20 min and resolves 1–2 h after allergen exposure. A subgroup of asthmatics also experiences a second, late-phase broncho-obstruction (late-phase response) that is characterized in part by a slow onset, usually beginning 4–5 h after initial allergen exposure and persisting up to 24 h or more (1,2). The acute response is thought to be the result of the release of bronchoactive and vasoactive mediators from resident airway cells resulting in a rapid smooth muscle constriction, mucus secretion, and edema formation. In contrast, the time course and severity of the late-phase response are associated with the recruitment of inflammatory cells into the lungs (3,4). These observations suggest an important role for infiltrating (nonresident) cells that can synthesize and release preformed and/or newly synthesized inflammatory mediators capable of contributing to the late-phase obstructive response. In addition, the similarity of these features with those that characterize the airway inflammation found in asthma suggests that clinical asthma may be the result of repeated or superimposed late-phase reactions. The efficacy of corticosteroid in both late-phase reactions and clinical asthma further supports this contention.

Numerous cell types have been implicated in the pathogenesis of the late-phase response, including eosinophils, platelets, and neutrophils (5–7). Each of these cells is well-equipped and quite capable of altering airway function by the release of biologically active, preformed granule-associated mediators as well as newly synthesized membrane-derived lipid mediators, including platelet-activating factor (PAF), leukotrienes, and prostaglandins. However, studies of cell influx and late-phase airway responses in both humans and animal models of asthma have produced conflicting results. For example, de Monchy and colleagues (4) reported increased numbers of eosinophils and elevated levels of eosinophil cationic protein ECP bronchoalveolar lavage (BAL) fluid during allergen-induced late-phase responses in allergic asthmatics. In contrast, Diaz and co-workers (5) demonstrated a mixed leukocyte infiltration including increased numbers of both neutrophils and eosinophils recovered by BAL. Finally, Page (7) suggested a major role for platelets in antigen-induced responses and asthma. Thus, the specific role of resident and newly infiltrating cells in the development of the late-phase asthmatic response remains unclear.

We have developed a primate model of allergic asthma in order to examine the role of inflammatory cells and mediators in the pathogenesis of asthma and the late-phase response. The animals utilized in these studies are adult male cynomolgus monkeys (*Macaca fascicularis*) that were selected based on a naturally occurring respiratory sensitivity to inhaled *Ascaris suum* extract. The methods of assessing airway cellular composition by BAL or biopsy, administration of inhaled antigen, and measurements of pulmonary function by oscillatory mechanics have been described in detail elsewhere (8,9).

II. EXPERIMENTAL RESULTS

A. Acute and Late-Phase Responses

Early on, a study was performed to investigate functional airway responses and airway cellular composition before and after a single antigen challenge (10). In this study, pulmonary function was assessed prior to and at several time points out to 24 h after inhaled antigen. Airway cellular composition (assessed by BAL) was determined prior to and out to 14 days after antigen challenge. Based on the functional airway response to inhaled antigen, two groups of animals were distinguished: those that had an isolated immediate bronchoconstriction response (single responders) that peaked at 10–15 min and resolved 2 h after antigen challenge and those that had a similar immediate response followed by a late-phase response (dual responders) that peaked 6–8 h postchallenge (Fig. 1). Analysis of airway cellular composition revealed that dual-responder monkeys, relative to single responders, had elevated numbers of eosinophils and eosinophil peroxidase (EPO) activity in BAL fluid prior to antigen challenge (i.e., a baseline, persistent eosinophilic inflammation). Six hours after antigen challenge, concurrent with the

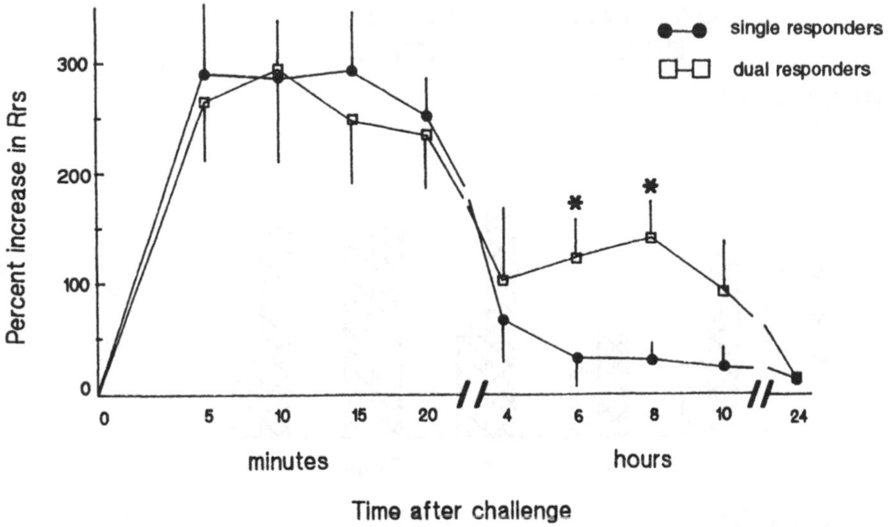

Figure 1 Increases in respiratory system resistance (Rrs) during the acute and late-phase airway obstruction response following inhaled antigen in single- and dual-responder monkeys. The degree of bronchoconstriction in single- and dual-responder monkeys was similar; however, only the dual responders had an increase in Rrs 6–8 h after antigen inhalation. *Statistical significance between groups, $p < 0.05$, $n = 6$ per group. Data are the mean ± SEM.

late-phase airway obstruction, the number of eosinophils recovered by BAL was decreased; however, the concentration of EPO in BAL fluid was significantly increased, suggesting that the eosinophils were activated and degranulating in the airways (Figs. 2 and 3, respectively). In contrast, there were no significant changes in the number of eosinophils or BAL EPO concentrations 6 h after antigen challenge in single-responder monkeys. Neutrophils, which were very low before antigen challenge in both single and dual responders, were increased in both 6 h after antigen challenge; however, the magnitude of the neutrophil influx was significantly greater in dual-responder primates (Fig. 2). In addition, the magnitude of the neutrophil influx 6 h after antigen inhalation significantly correlated with the magnitude of late-phase airway obstruction, suggesting a possible effector cell role for the neutrophil ($r = 0.60$, $p < 0.01$).

Studies examining airway responsiveness to inhaled methacholine revealed that the airways of dual-responder monkeys are hyperresponsive compared to naïve (unchallenged) or single-responder monkeys (Fig. 4). Multiple antigen inhalations will induce eosinophilic inflammation and increase airway responsiveness in naïve or single-responder monkeys to a degree similar to that of dual responders

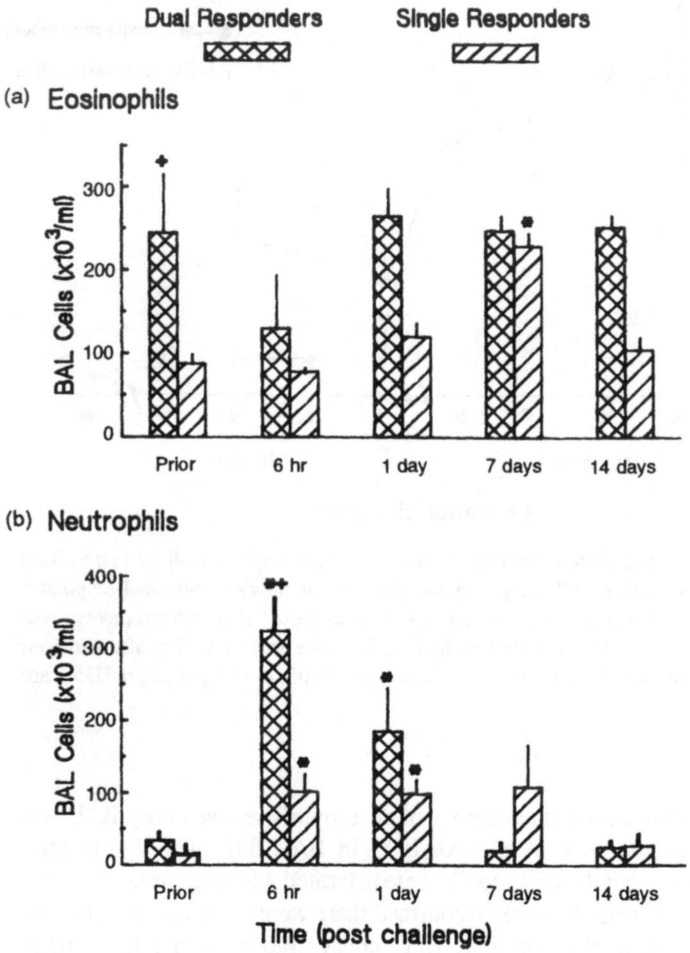

Figure 2 (a) Changes in airway eosinophils before and after inhaled antigen in single- and dual-responder monkeys. Dual responders had a significantly higher number of eosinophils before antigen than single-responder monkeys. BAL eosinophils remained elevated in dual-responder monkeys. BAL eosinophils in single responders were significantly increased 7 days after challenge and returned to baseline at day 14. (b) Changes in airway neutrophils before and after inhaled antigen in single- and dual-responder monkeys. Both single and dual responders had a significant increase in neutrophils 6 h and 1 day after challenge. The magnitude of the neutrophil influx in the dual responders was significantly greater than that occurring in single responders 6 h after challenge. *Statistical significance from baseline (prior to antigen challenge) within each group. +Statistical significance between groups. $p < 0.05$, $n = 6$ per group. Data are the mean ± SEM.

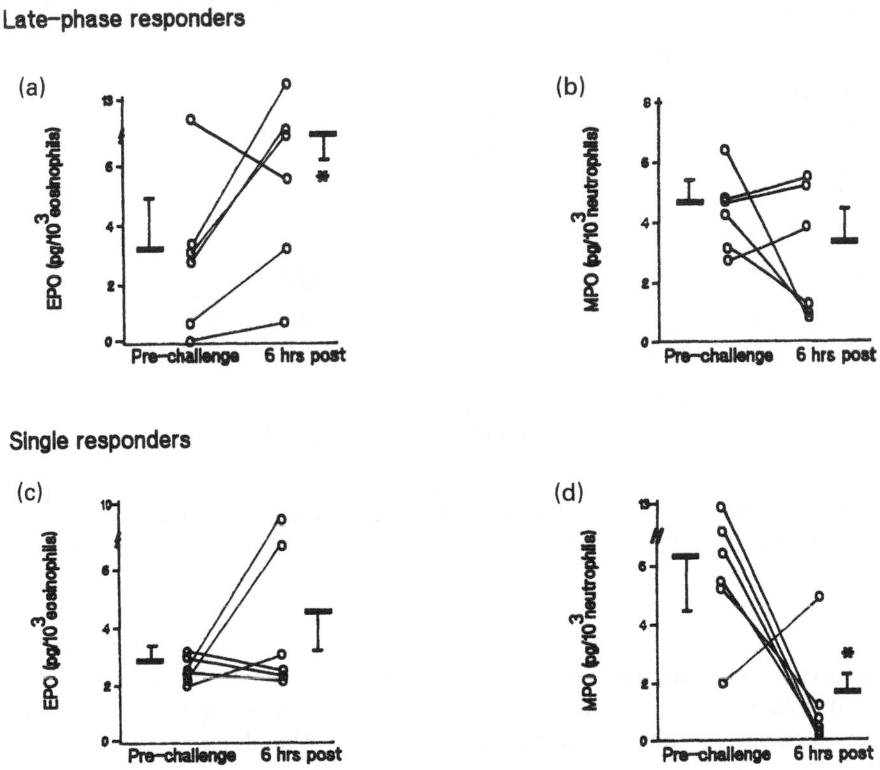

Figure 3 Changes in (a) BAL EPO, (b) BAL MPO in dual-responder monkeys, and (c) BAL EPO, (d) BAL MPO in single-responder monkeys. Basal levels of EPO were similar between single- and dual-responder monkeys; however, only the dual-responder monkeys had a significant increase in BAL EPO 6 h after antigen inhalation. There were no significant changes in BAL MPO concentrations in dual responders 6 h after challenge, but the levels of BAL MPO were significantly decreased in single-responder monkeys. *Statistical significance between prechallenge values and 6 h values.

(8,9). Therefore, it is likely that the chronic airway eosinophilia and airway hyperresponsiveness found in dual-responder monkeys contributes to the late-phase airway response.

Thus, we are able to distinguish between single and dual responders based on the resident airway population as well as the functional airway response to inhaled antigen and responsiveness to inhaled methacholine as follows. Dual-responder primates, in contrast to single responders, have a baseline, chronic airway eosinophilia that is associated with airway hyperresponsiveness before antigen

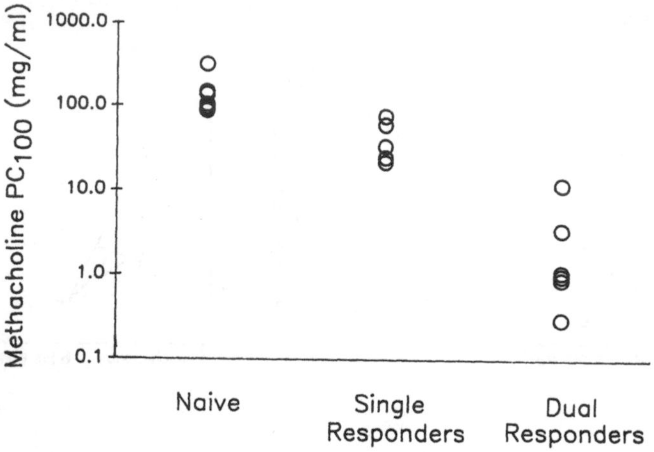

Figure 4 Airway responsiveness to inhaled methacholine in naïve (unchallenged, uninflamed airways) single- and late-phase-responder monkeys. Late-phase monkeys are hyperresponsive to inhaled methacholine compared to both naïve and single-responder monkeys.

inhalation. The late-phase airway obstruction is associated with an increase in eosinophil-derived proteins in BAL and the development of an acute airway neutrophilia superimposed on the chronic airway eosinophilia.

B. Acute Airway Inflammation and Late-Phase Airway Obstruction

Having characterized dual responders in terms of airway function and resident and infiltrating cell populations after antigen challenge, we next began to investigate the role of each cell in the pathogenesis of the late-phase response.

Antigen inhalation results in the release of preformed and newly synthesized mediators that contribute to the acute bronchoconstriction and neutrophil infiltration into the lungs (Table 1) (11). Our suggestion of a possible effector cell role for newly recruited neutrophils in the late-phase response is based on several lines of evidence. We have shown that a single aerosol treatment with a direct 5-lipoxygenase inhibitor (BI-L-0239) or a translocation inhibitor (MK886) before antigen inhalation results in a dose-related inhibition of the late-phase response (Fig. 5) (12). The efficacy of these compounds in blocking the late-phase response parallels their ability to inhibit the production and release of leukotrienes in the lungs during the acute response (Table 2). In addition, inhibition of leukotriene biosynthesis in the lungs blocks the acute influx of neutrophils into the airways associated with the late-phase response (12). These data suggest that the production and release of

Table 1 Mediators Recovered in Monkey BAL Fluid

Challenge	n	Histamine (μg)	i-LTC_4 (ng)	i-PGD_2 (ng)	i-TXB_2 (ng)	i-PGE_2 (pg)	i-LTB_4 (ng)
Baseline	12	1.0 ± 1.5	1.5 ± 0.7	2.4 ± 0.9	4.2 ± 0.4	BDL	BDL
PBS	6	2.1 ± 0.8	2.5 ± 0.5	1.9 ± 1.2	5.3 ± 1.7	BDL	BDL
Ovalbumin	6	3.7 ± 1.3	2.9 ± 0.6	2.6 ± 1.0	3.6 ± 0.9	BDL	BDL
Ascaris						BDL	BDL
10 min	12	25.9 ± 5.5*	41.6 ± 12.7*	25.9 ± 5.5*	13.0 ± 4.0	BDL	BDL
20 min	12	19.9 ± 4.2*	41.5 ± 14.3*	29.6 ± 9.0*	11.9 ± 4.9	BDL	BDL

*$p < 0.05$.
Data values are the mean ± SEM.
BDL = below detectable limits of the assay.

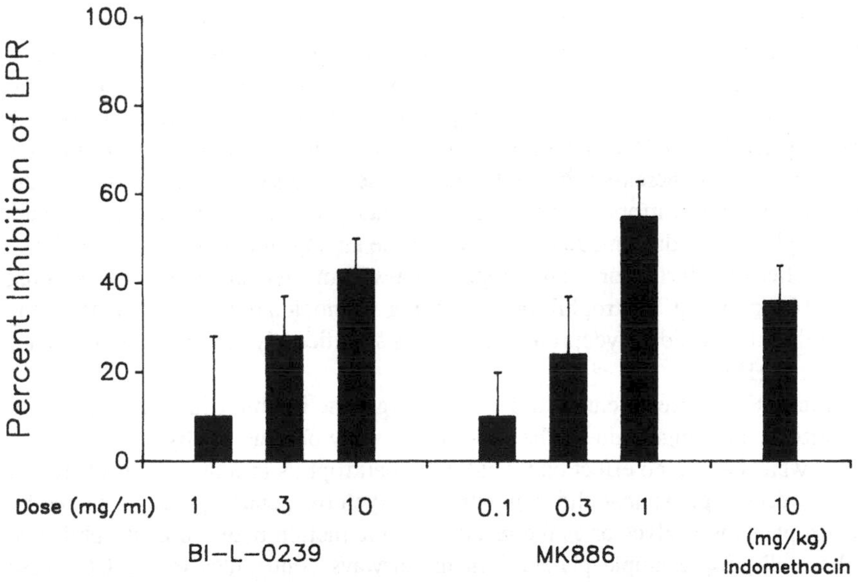

Figure 5 The effects of various treatments on antigen-induced late-phase airway obstruction in monkeys. Pretreatment with aerosolized BI-L-0239 (1–10 mg/ml) or MK886 (0.1–1 mg/ml) administered 10 min before antigen challenge resulted in a dose-related inhibition of the late-phase response. Indomethacin, administered intraperitoneally at a dose of 10 mg/kg 24, 4, and 1 h before antigen challenge reduced the late-phase response by approximately 40%. (n = 3–6 animals per group).

Table 2 Inhibition of Antigen-Induced Mediator Release

Compound	*n*	Dose	Percent inhibition	
			i-LTC4	i-PGD2
BI-L-0239	6	1 mg/ml	35 ± 74	0
	5	3 mg/ml	61 ± 17	30 ± 12
	6	10 mg/ml	73 ± 12	52 ± 14
MK886	5	0.1 mg/ml	12 ± 6	0
	6	0.3 mg/ml	41 ± 9	31 ± 13
	6	1.0 mg/ml	78 ± 12	59 ± 19

leukotrienes (LTB_4?) during the acute-phase response (immediately after antigen inhalation) can contribute to, as well as trigger, events leading to the acute neutrophil infiltration and activation in the lungs. Once in the lungs, activated neutrophils may contribute to the late-phase airway obstruction via products of de novo synthesis (i.e., lipid mediators, oxidative products following respiratory burst) that can affect airway and vascular smooth muscle, resulting in a prolonged "obstructive" response (13). Several groups have demonstrated that eosinophil granule products (i.e., major basic protein, MBP) are capable of activating neutrophils (14,15). We have shown that the levels of eosinophil-derived proteins are significantly increased during the late-phase response. Thus, exposure of the newly recruited neutrophils to eosinophil granule proteins may act to stimulate the neutrophils to produce mediators that influence airway function. Further evidence of an effector cell role for the neutrophil comes from a recent study demonstrating that blocking the neutrophil influx with a monoclonal antibody against the endothelial adhesion glycoprotein E-selectin significantly reduced the late-phase response (16).

Interestingly, treatment with the cyclooxygenase inhibitor indomethacin prior to antigen challenge reduces the late-phase airway obstruction by approximately 40% while having no effect on the influx of neutrophils into the airways (Fig. 5). These data suggest that the generation of cyclooxygenase products either by neutrophils themselves or as the result of an interaction between neutrophils and other cells, for example platelets, in the airways contributes to the late-phase airway obstruction. Preliminary evidence from our laboratory suggests, in fact, that the number of platelets present in BAL fluid is increased 6 h after antigen inhalation, indicating an association between platelet infiltration and the late-phase response. In addition, several studies have demonstrated that interactions between platelets and neutrophils (i.e., substrate sharing) can result in enhanced production of lipid mediators and other products of de novo synthesis. These data support the notion that newly recruited neutrophils infiltrate into the lung and come in contact with mediators (i.e., MBP) and platelets that trigger a further activation of the neutrophils and synthesis of vasoactive and bronchoactive substances, resulting in airway smooth muscle constriction and edema in lung tissue.

C. Chronic Airway Inflammation and Late-Phase Obstruction

As described above, dual-responder primates have an existing, baseline chronic airway eosinophilia. In addition to airway eosinophilia, dual responders have increased numbers of activated T lymphocytes (identified in BAL fluid), and the airways are hyperresponsive to inhaled methacholine. The levels of eosinophil-derived proteins present in BAL fluid, before antigen inhalation, are higher in dual-responder primates than in single responders, suggesting that the eosinophils are activated and degranulating in the airways. It may be that the accumulation of activated eosinophils in the airways is the result of T-lymphocyte activation and cytokine release. This leads to the development of airway hyperresponsiveness and contributes to the late-phase airway obstruction after antigen inhalation. Evidence to support this contention comes from studies where we have demonstrated that multiple antigen inhalations to monkeys with normal airway cellular composition and airway responsiveness induce a profound and selective eosinophilia that is associated with damage to the airway epithelium and a striking increase in airway responsiveness (8,9,17). Other studies have shown that a direct intratracheal instillation of purified human MBP results in a transient bronchoconstriction and a 10-fold increase in airway responsiveness (18). Thus, eosinophil granule proteins have direct effects on airway function resulting in bronchoconstriction and increases in airway responsiveness.

Recently we examined the effects of treatment with systemic or inhaled corticosteroids on the airway eosinophilia and airway hyperresponsiveness in late-phase responder primates (19). In a vehicle-controlled, cross-over study design, six primates were treated with systemic dexamethasone (0.2 mg/kg, i.m.) or vehicle once a day for 7 consecutive days. Dexamethasone significantly reduced the total number of leukocytes and the numbers of eosinophils recovered by BAL, the amount of EPO in BAL fluid was significantly decreased, and there was a substantial reduction in airway hyperresponsiveness in five of the six animals (Figs. 6 and 7, respectively). In addition, dexamethasone treatment resulted in a significant inhibition of both the acute and late-phase airway response after antigen challenge as well as the increase in BAL EPO concentrations and the acute influx of neutrophils into the airways (Fig. 8). Similar results were seen in a study with inhaled beclomethasone (4 mg/ml) administered b.i.d. for 7 consecutive days. In this study, the number of activated T lymphocytes (CD25+) in BAL and in the peripheral blood was also reduced after inhaled beclomethasone

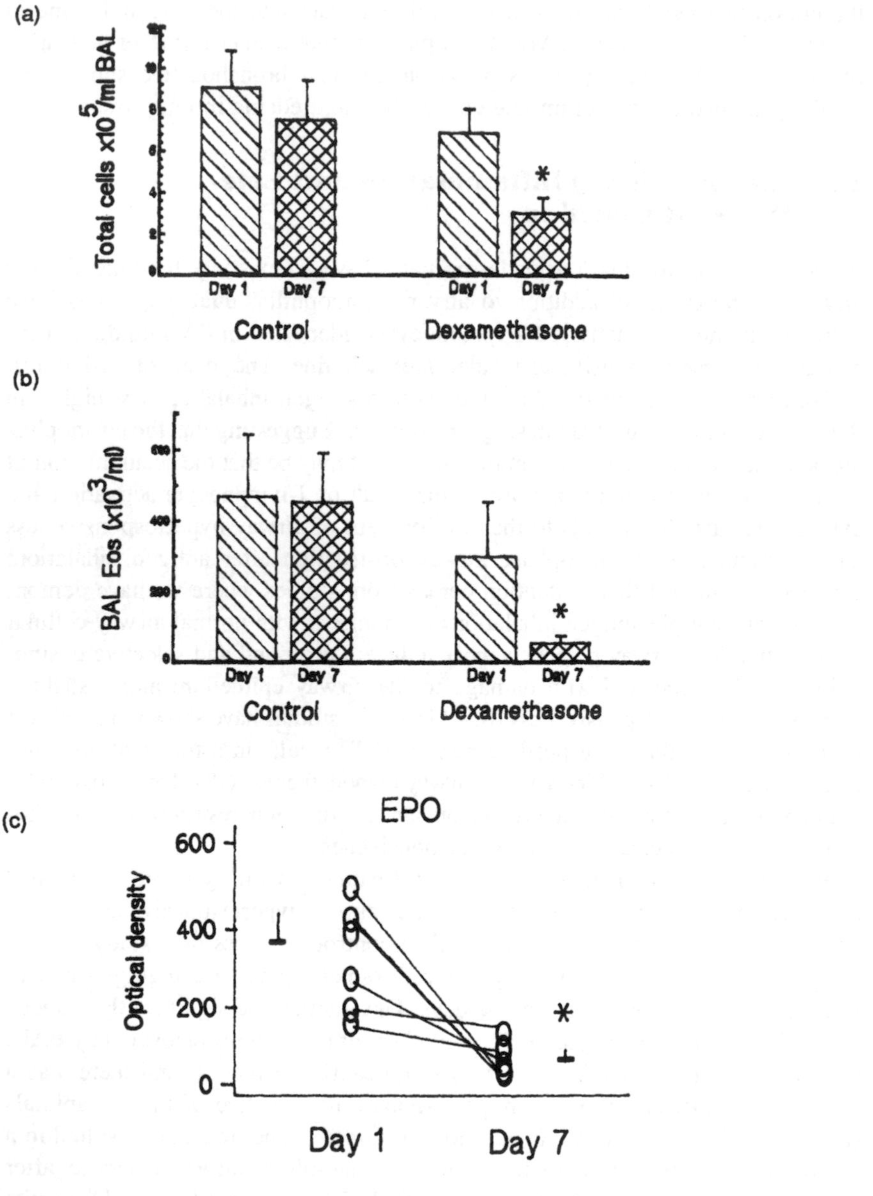

Figure 6 (a) Total number of leukocytes recovered by BAL before and 7 days after treatment with vehicle (control) or dexamethasone. Dexamethasone treatment resulted in a significant decrease in the number of leukocytes recovered by BAL. (b) Number of eosinophils recovered by BAL before and 7 days after dexamethasone or vehicle treatment. Dexamethasone treatment significantly reduced the number of eosinophils recovered

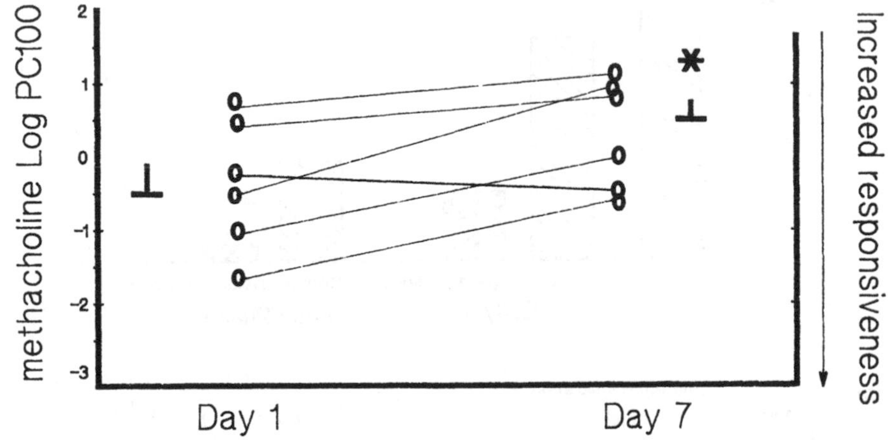

Figure 7 Effects of treatment with dexamethasone on airway responsiveness to inhaled methacholine. Dexamethasone treatment significantly reduced (increased PC100 values) airway responsiveness. Data for individual animals are shown. Solid bars represent the mean ± SEM of the group. *Statistical significance from day 1, $p < 0.05$.

treatment. Thus, treatment with corticosteroids reduces the baseline airway eosinophilia, lowers the number of activated eosinophils and T lymphocytes, and reduces the airway responsiveness to inhaled methacholine. Corticosteroid treatment also significantly reduces the acute and late-phase airway obstruction after antigen challenge.

III. DISCUSSION

Features typical of the pathology of asthma include an intense infiltration of eosinophils and deposition of eosinophil-derived proteins in and around the airway mucosa (20). Areas of epithelial damage or desquamation are also common in asthma as well as airway edema and mucus hypersecretion resulting in considerable plugging of the peripheral airways (21,22). Studies from our laboratory with a primate model of allergic asthma have demonstrated that before antigen inhalation, dual-responder monkeys have higher numbers of activated eosinophils and T lymphocytes present in the airways relative to single-responder monkeys. Antigen

---

by BAL. (c) Amount of EPO measured in BAL fluid obtained before and 7 days after treatment with vehicle or dexamethasone. Dexamethasone treatment significantly reduced the level of BAL EPO. *Statistical significance from day 1 of the study. $p < 0.05$, $n = 6$, Data are the mean ± SEM.

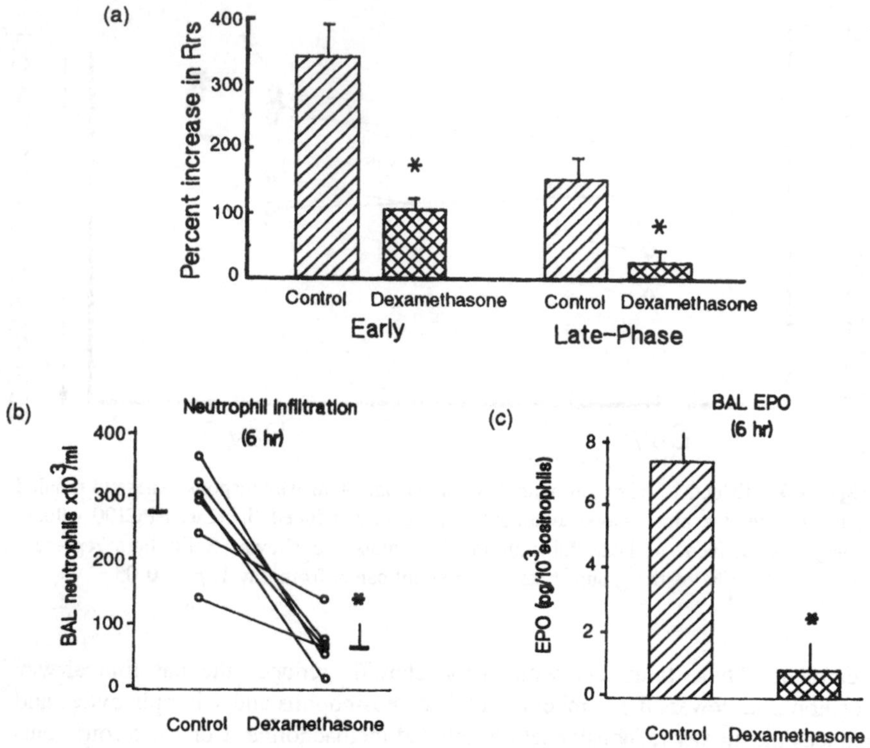

Figure 8 (a) Increases in Rrs during the acute and late-phase response to inhaled antigen in dual responders during control and dexamethasone treatment studies. Dexamethasone treatment caused a significant reduction in both the acute and late-phase airway responses. (b) Effects of dexamethasone treatment on the acute infiltration of neutrophils associated with the late-phase response. Dexamethasone treatment significantly reduced the number of neutrophils infiltrating into the airways 6 h after inhaled antigen compared to cross-over vehicle control studies. (c) Effects of dexamethasone treatment on BAL levels of EPO 6 h after inhaled antigen. Dexamethasone treatment significantly reduced the levels of EPO in BAL fluid recovered during the late-phase response. *Statistical significance from control study values. $p < 0.05$, $n = 6$. Data are the mean ± SEM.

inhalation induces a transient neutrophil influx in both single and dual responders; however, the magnitude of the neutrophil influx is much greater in dual responders. In addition to a temporal relationship between neutrophil infiltration and the late-phase response, there is a modest, but significant correlation between the magnitude of the neutrophil influx and the magnitude of the late-phase response, suggesting an effector cell role for the neutrophil. The level of BAL eosinophil-derived EPO is also increased 6 h after antigen challenge during the late-phase response. Thus, our results implicate possible roles for both the eosinophil and neutrophil in the development of the late-phase response.

The role of inflammatory cell influx into the airways and late-phase airway obstruction has been examined in humans as well as in animal models of asthma, with conflicting results. For example, de Monchy and colleagues (4) have reported a selective eosinophil influx into the airways and elevated levels of ECP in BAL fluid concurrent with the onset of late-phase airway obstruction in asthmatics. On the other hand, Diaz and co-workers (5) have shown a substantial neutrophil influx in addition to eosinophil infiltration, while Page (7) contends that platelet infiltration and activation plays a major role. Our results with the primate asthma model suggest involvement of eosinophils, neutrophils, and perhaps platelets in the late-phase response. Similar to the results of de Monchy et al. (4), we have observed a direct relationship between eosinophil activation (increased BAL EPO) and the late-phase response. However, the presence of large numbers of eosinophils in the airways before antigen inhalation in late-phase responders differs from the findings of other studies of the late-phase response in mild asthmatics and studies with other animal models (23–25). This difference may be attributed to several factors. For example, the monkeys utilized in our experiments are studied biweekly (i.e., receive an inhaled antigen challenge every 14 days), have inflamed airways, and, thus, may be more similar to symptomatic patients with spontaneous asthma than asymptomatic mild asthmatics. Wardlaw and colleagues (26) demonstrated that while both symptomatic and asymptomatic patients with spontaneous asthma (no antigen challenge) had significantly higher numbers of eosinophils and MBP concentrations in BAL fluid compared to normal subjects, symptomatic patients had significantly greater numbers of eosinophils and MBP compared to the asymptomatic group.

For many years the association between blood and sputum eosinophilia with bronchial asthma has been recognized. Eosinophil MBP has been identified on lung tissue taken from patients who have died from asthma (20). More recent studies have demonstrated an association between blood eosinophilia and dual asthmatic responses as well as a direct correlation between blood eosinophilia and the degree of nonspecific airway hyperresponsiveness (6). In addition, MBP and other eosinophil granule proteins have been shown to alter airway smooth muscle function in vitro and in vivo (18,25). Our results demonstrate that eosinophil activation and degranulation is occurring to a certain degree prior to antigen inhalation and that degranulation is enhanced after challenge during the late-phase airway obstructive response. It is well known that eosinophils can also generate membrane-derived lipid mediators (i.e., LTC_4 and PAF) that have been purported to play a role in the pathogenesis of asthma and may contribute to the late-phase response (27,28). We suggest that, in this primate model of allergic asthma, the increased number of eosinophils in the airways, in addition to contributing to the late-phase response, may have the potential to prime the airways for the occurrence of late-phase obstruction. Our studies with corticosteroids support this hypothesis. Chronic treatment with either dexamethasone or beclomethasone reduced the number of eosinophils in the airways as well as the level of eosinophil-derived proteins recovered in BAL fluid. In addition, this treatment regime resulted in a significant inhibition of both the acute and late-phase airway responses.

Perhaps a more important question is what is responsible for, or what is "driving," the selective eosinophilia found in asthma. Recent work suggests that the T lymphocyte, in particular the Th2 subtype, may be a major contributor to the development of the eosinophilia in allergic asthma. In support of this hypothesis, Frew and Kay (29) demonstrated an increase in the number of T cells in allergen-induced cutaneous late responses. In addition, the number of CD4+ T lymphocytes correlated significantly with the number of activated eosinophils infiltrating into the allergen injection site. In situ hybridization studies demonstrated a strong signal for IL-3, IL-4, IL-5, and GM-CSF, with little or no IFN (Th2 pattern). These observations have been extended by the examination of both biopsy tissue and airway cells recovered by BAL from the lungs of asthmatic patients. A similar pattern was observed in terms of the type of T lymphocytes found in asthmatics and the correlation between activated T cells (CD25+) and the number of activated eosinophils (30). Perhaps the most striking evidence for a role for the T lymphocyte comes from a study by Alexander and colleagues (31) demonstrating a significant improvement in lung function parameters following treatment with cyclosporin A.

We have also demonstrated the development of an acute, transient neutrophil inflammation, superimposed onto the chronic eosinophilic inflammation, that is temporally related to the late-phase airway obstruction. Earlier work by several investigators has suggested a role for the neutrophil in asthma and late-phase airway responses. For example, increased levels of high-molecular-weight neutrophil chemotactic activity (HMW-NCA) have been detected in the blood of asthmatic patients after antigen-induced acute and late-phase reactions (32,33). More recently, these observations have been confirmed in a study with severe asthmatics (34). In other studies in our laboratory, we have shown that blocking the neutrophil influx significantly inhibits the late-phase airway response (16). These data support the hypothesis of an effector cell role for the neutrophil in the late-phase response. The neutrophil is a cell quite capable of altering airway function. For instance, neutrophils generate lipid mediators such as prostaglandins and PAF in response to many different stimuli including MBP. Neutrophils may also contribute to the late-phase response by generating toxic oxygen species that can alter airway smooth function and damage airway epithelium and vascular permeability, leading to bronchial obstruction.

IV. SUMMARY AND CONCLUSION

We have demonstrated that before antigen inhalation, dual-responder monkeys have elevated numbers of activated eosinophils and T lymphocytes in the airways when compared to single responders. In addition, dual responders have a large influx of neutrophils into the airways concurrent with the late-phase airway obstruction response that occurs 6–8 h after antigen inhalation. Chronic treatment with corticosteroids reduced the number of resident eosinophils and activated T lymphocytes in the lung and BAL levels of eosinophil-derived proteins as well as blocking the acute and late-phase response. Therefore, we suggest that the eosinophil, under the influence of T lymphocytes, may have a priming action on the airways, making them more responsive in terms of the development of an "acute" inflammatory response (neutrophil influx) and subsequent late-phase obstruction after antigen inhalation. However, the critical number of resident eosinophils required for priming may differ among animal species as well as in humans.

ACKNOWLEDGMENTS

The authors acknowledge the excellent technical assistance of Ms. Carol A. Torcellini, Mr. Donald Sousa, and Ms. Carol Stearns with this work.

REFERENCES

1. O'Byrne PM, Dolovich J, Hargreave FE. Late asthmatic responses. Am Rev Respir Dis 1987; 136:740–748.
2. Larsen GL. The pulmonary late-phase response. Hosp Pract 1987; 22:155–169.
3. Marsh WR, Irvin CG, Murphy KR, Behrens BL, Larsen GL. Increases in airway reactivity to histamine and inflammatory cells in bronchoalveolar lavage after the late asthmatic response in an animal model. Am Rev Respir Dis 1985; 135:875–879.
4. de Monchy JGR, Kaufman HR, Venge P, Koeter GH, Jansen HM, Sluiter HJ, de Vries K. Bronchoalveolar eosinophilia during allergen-induced late asthmatic reactions. Am Rev Respir Dis 1985; 131:373–376.
5. Diaz P, Gonzalez MC, Galleguillos FR, Ancic P, Cromwell O, Shepherd D, Durham SR, Gleich GJ, Kay AB. Leukocytes and mediators in bronchoalveolar lavage during allergen-induced late-phase asthmatic reactions. Am Rev Respir Dis 1989; 139:1383–1389.
6. Durham SR, Kay AB. Eosinophils, bronchial hyperreactivity and late-phase asthmatic reactions. Clin Allergy 1985; 15:411–418.
7. Page CP. Platelets and asthma. Ann NY Acad Sci 1991; 629:38–47.
8. Gundel RH, Gerritsen ME, Wegner CD. Antigen-coated sepharose beads induce airway eosinophilia and airway hyperresponsiveness in cynomolgus monkeys. Am Rev Respir Dis 1989; 140:629–633.
9. Gundel RH, Gerritsen ME, Gleich GJ, Wegner CD. Repeated antigen inhalation

results in a prolonged airway eosinophilia and airway hyperresponsiveness in primates. J Appl Physiol 1990; 68:779–786.
10. Gundel RH, Wegner CD, Letts LG. Antigen-induced acute and late-phase responses in primates. Am Rev Respir Dis 1992; 146:369–373.
11. Gundel RH, Kinkade P, Torcellini CA, Clarke CC, Watrous J, Desai S, Homon CA, Farina PR, Wegner CD. Antigen-induced mediator release in primates. Am Rev Respir Dis 1991; 144:76–82.
12. Gundel RH, Torcellini CA, Clarke CC, Desai SN, Lazer ES, Wegner CD. The effects of a 5-lipoxygenase inhibitor on antigen-induced mediator release, late-phase bronchoconstriction and cell infiltrates in primates. Adv Prost Throm Leuko Res 1990; 21:457–460.
13. Ward PA, Till GO, Kunkel R, Bearchamp C. Evidence for a role of hydroxyl radical in complement and neutrophil dependent tissue injury. J Clin Invest 1983; 72:789–793.
14. Moy JN, Gleich GJ, Thomas LL. Noncytotoxic activation of neutrophils by eosinophil granule major basic protein: effect on superoxide anion generation and lysosomal enzyme release. J Immunol 1989; 143:952–955.
15. Rohrbach MS, Wheatley CL, Slifman NR, Gleich GJ. Activation of platelets by eosinophil granule proteins. J Exp Med 1990; 172:1272–1274.
16. Gundel RH, Wegner CD, Torcellini CA, Clarke CC, Rothlein R, Haynes N, Smith CW, Letts LG. Endothelial-leukocyte adhesion molecule-1 mediates antigen-induced acute airway inflammation and late-phase airway obstruction in monkeys. J Clin Invest 1991; 88:1407–1411.
17. Wegner CD, Rothlein R, Gundel RH. Adhesion molecules in the pathogenesis of asthma. Agents Actions 1991; 34:529–544.
18. Gundel RH, Letts LG, Gleich GJ. Human eosinophil major basic protein induces airway constriction and airway hyperresponsiveness in primates. J Clin Invest 1991; 87:1470–1473.
19. Gundel RH, Wegner CD, Torcellini CA, Letts LG. The role of intercellular adhesion molecule-1 in chronic airway inflammation. Clin Exp Allergy 1992; 22:569–575.
20. Filley WV, Holley KE, Kephart GM, Gleich GJ. Identification by immunofluorescence of eosinophil granule major basic protein in lung tissue of patients with bronchial asthma. Lancet 1982; 2:11-16.
21. Ryley HC, Brogan TD. Variation in the composition of sputum in chronic chest diseases. Br J Exp Pathol 1968; 49:25–33.
22. Brogan RD, Ryley HC, Neale L, Yassa L. Soluble proteins of bronchopulmonary secretions from patients with cystic fibrosis, asthma and bronchitis. Thorax 1975; 30: 72–79.
23. Rossi GA, Crimi E, Lantero S, Gianiorio P, Oddera S, Crimi P, Brusasco V. Late-phase asthmatic reactions to inhaled allergen is associated with early recruitment of eosinophils in the airways. Am Rev Respir Dis 1991; 144:379–383.
24. Murphy KR, Wilson MC, Irvin CG, Glezen LS, Marsh WR, Haslett C, Henson PM, Larsen GL. The requirement for polymorphonuclear leukocytes in the late asthmatic response and heightened airways reactivity in an animal model. Am Rev Respir Dis 1986; 134:62–68.
25. Abraham WM, Sielczak MW, Wanner A, et al. Cellular markers of inflammation in

the airways of sheep with and without allergen-induced late-phase responses. Am Rev Respir Dis 1988; 138:1565–1571.
26. Wardlaw AJ, Dunnette S, Gleich GJ, Collins JV, Kay AB. Eosinophils and mast cells in bronchoalveolar lavage in mild asthma: Relationship to bronchial hyperreactivity. Am Rev Respir Dis 1988; 137:62–69.
27. Shaw RJ, Walsh GM, Cromwell O, Moqbel R, Spry CJF, Kay AB. Activated human eosinophils generate SRS-A leukotrienes following physiological (IgG-dependent) stimulation. Nature 1985; 316:150–152.
28. Lee TC, Lenihan DJ, Malone B, Ruddy LL, Wasserman SI. Increased biosynthesis of platelet activating factor in activated human eosinophils. J Biol Chem 1984; 259: 5520–5530.
29. Frew AJ, Kay AB. The relationship between infiltrating CD4+ lymphocytes, activated eosinophils and the magnitude of the allergen-induced late phase cutaneous reaction in man. J Immunol 1988; 141:4158–4164.
30. Hamid Q, Azzawi M, Sun Ying, Moqbel R, Wardlaw AJ, Corrigan CJ, Bradley B, Durham SR, Collins JV, Jeffery PK, Quint DJ, Kay AB. Expression of m-RNA for interleukin5 in mucosal bronchial biopsies from asthmatics. J Clin Invest 1991; 87: 1541–1545.
31. Alexander AG, Barnes NC, Kay AB. Trial of cyclosporin A in corticosteroid-dependent chronic severe asthma. Lancet 1992; 339:324–328.
32. Atkins PC, Norman M, Weiner H, Zweiman B. Release of neutrophil chemotactic activity during immediate hypersensitivity reactions in humans. Ann Intern Med 1977; 86:415–418.
33. Nagy L, Lee TH, Kay AB. Neutrophil chemotactic activity in antigen-induced late asthmatic reactions. N Engl J Med 1982; 306:497–501.
34. Buchanan DR, Cromwell O, Kay AB. Neutrophil chemotactic activity in acute severe asthma ("status asthmaticus"). Am Rev Respir Dis 1987; 136:1397–1402.

DISCUSSION (Speaker: R. Gundel)

Gleich: In keeping with your findings of neutrophil involvement in the late-phase response we have studied cases of fatal asthma and have found predominant neutrophil airway inflammation in patients dying acutely of asthma. And a question on the ability of leukocytes to produce LTC_4: can primate neutrophils produce LTC_4? Among human leukocytes several studies suggest that only eosinophils produce LTC_4?

Gundel: As in humans, monkey neutrophils produce LTB_4 and eosinophils produce LTC_4. I discussed the data with leukotriene biosynthesis inhibitors to demonstrate, first, that leukotrienes are produced and released during the immediate response to inhaled antigen and, second, to show that the ability of leukotriene biosynthesis inhibitors to inhibit the late-phase airway obstruction parallels their ability to inhibit the synthesis of leukotrienes during the immediate response. Their activity is also related to their ability to block the acute neutrophil influx into the airways during the late-phase response.

Venge: Your data of high eosinophil numbers in the lung predicting the outcome of an allergen challenge with respect to a late-phase reaction are in keeping with our findings in humans. However, your neutrophil data are very different from our results in humans. Thus, we have never seen any increase in either neutrophil numbers or myeloperoxidase (MPO) levels 2, 6, or 24 h after allergen challenge of allergic asthmatics. Furthermore, we have conducted a treatment study with budesonide with lavages before and after and saw no changes in neutrophil numbers or MPO levels as opposed to eosinophil numbers and ECP levels, which were significantly decreased.

Gundel: Our model is different from other animal models and findings in mild asthmatic subjects in that we are showing a baseline chronic airway eosinophilia before antigen challenge. I think our model may be more closely related to patients with severe asthma rather than nonsymptomatic patients with mild asthma. Our primate model is the only one that shows a chronic baseline eosinophilia with a superimposed "acute" neutrophil inflammation following antigen inhalation. Again, I think this is more closely related to patients with a severe form of asthma. Dexamethasone treatment clearly reduces the number of eosinophils, the level of BAL EPO, and airway hyperresponsiveness in our model as well as inhibiting the late-phase associated acute neutrophil infiltration into the airways.

Konig: One has to be cautious to correlate eosinophil numbers with LTC_4. Obviously, a transcellular activation of cells induced by leukotriene A_4 may occur. It is known that exogenous addition of LTA_4 to platelets generates large amounts of LTC_4. Incubation of lymphocytes with LTA_4 leads to LTB_4 generation. Thus the amounts of LTA_4 and the presence of cells may be crucial for the net results of leukotrienes in BAL.

Gundel: I think that the late-phase airway obstruction could well be due, in part, to the interaction of newly recruited neutrophils and platelets into the airways. This interaction could result in a large production of LTC_4 by the platelets as well as the de novo synthesis of other substances capable of causing airway obstruction by neutrophils.

Schleimer: I noticed that at 7 days after challenge of the single responders, the eosinophil count in BAL was equal to that of the dual responders. It seems that you should be able to test your hypothesis by challenging the single responders at day 7 and looking for a late-phase response. Have you done this experiment and, if so, do you see LPR? When studying cell recruitment, it is always necessary to be sure that endotoxin is not influencing the experiment. Since I am sure it is difficult to purify the *Ascaris* antigen without some endotoxin present, have you assessed whether endotoxin is playing some role in the cell recruitment that you have described?

Gundel: Antigen inhalation 7 days after the first challenge does not produce a consistent late-phase response in single-responder patients. These data demonstrate that just having increased numbers of eosinophils present in the airways is not enough to switch a single-responder monkey to a dual responder. My feeling is that the chronic airway eosinophilia must be present for a long period (possibly years) before animals show a consistent dual response to inhaled antigen. No, we have not done experiments to define the role of endotoxin in the primate late-phase model.

Bochner: Based on in vitro studies from several laboratories, ICAM-1 appears to be important for transendothelial migration. Can you speculate on why the anti-E-selectin monoclonal antibody, but not the anti-ICAM-1 monoclonal antibody, inhibited antigen-induced neutrophil recruitment in the airways in your model?

Schleimer: Clare Doerschuk and John Harlan have shown that the lung is peculiar in that there is a CD18-independent accumulation of neutrophils in response to endotoxin. In light of the failure of ICAM-1 antibody to inhibit neutrophil accumulation, I wonder whether you have tried a CD18 antibody in this model and, if such an antibody fails, what conclusions can be drawn?

Gundel: We were also surprised that anti-ICAM-1 treatment did not reduce the neutrophil influx. However, when we examined the expression of adhesion molecules in the airways before and after antigen challenges, we found that E-selectin was markedly upregulated 6 h after antigen, while low levels of ICAM-1 were expressed both prior to and 6 h after antigen challenge. Thus, the acute neutrophil influx into the airways 6 h after antigen inhalation occurs via an E-selectin-dependent mechanism. We have also examined the effects of an anti-CD18 antibody and found very similar results to our anti-ICAM-1 studies in that there was no effect on the acute neutrophil influx or on the late-phase airway obstruction.

Lee: Have you used antagonists?

Gundel: No, we are currently studying the effects of an LTD_4 receptor antagonist in this model and have plans to test an LTB_4 receptor antagonist in the near future.

Silberstein: Could natural *Ascaris* infection explain differences between early and late-phase reactions?

Gundel: All of the monkeys used in our studies have a naturally occurring respiratory sensitivity to *A. suum* extract. My feeling is that during the years we have studied these animals and performed antigen challenges on a repetitive basis, the two types of responders (single and dual) have developed. Thus, there may be some predisposition of animals to become dual responders, and this may also depend on how often and for how long a period of time the animals are exposed to antigen.

Busse: In human studies, it is difficult to predict who will have a late asthmatic reaction to antigen. In your studies, you indicate that the level of airway responsiveness will predict the late responders. This is not true in humans. Is this a difference between your model and humans or the conditions of sensitization of the primate model?

Gundel: If you have an animal that has chronic airway eosinophilia and also shows very good sensitivity to antigen (i.e., responds to very low doses), there is a very good chance that this animal will be a dual responder. Thus, one can predict dual responders fairly well on this basis. I am not sure the same situation exists in humans.

Kay: In our experience raised numbers of eosinophils in BAL (at baseline) do not predict whether patients develop an antigen-induced single early or dual (early + late phase) response.

Gundel: Again, I think our model represents one of severe asthma. Most, if not all, of the studies in humans are done in mild asymptomatic asthmatics, and this may account for the differences between our model and the results in humans.

Galli: Have you tried assessing the effect on the late-phase reaction of combining treatment with the antibody to E-selectin and the leukotriene synthesis inhibitors, since neither alone appears to completely ablate the response?

Gundel: This is an interesting suggestion. No, we have not done these experiments.

Gleich: Have you challenged your monkeys with endotoxin only?

Gundel: No, I agree that these studies need to be done. It is very likely that our purified *Ascaris* antigen has a fair amount of endotoxin in it.

Vadas: What causes the Th2 cells to migrate and is this the key step?

Kay: This is a critical question. We do not know whether Th2-type cells migrate or whether local T cells acquire a Th2 phenotype as a result of influences from the microenvironment.

Church: In your model there were clear increases in neutrophil numbers and neutrophil products but that does not tell you that the neutrophils are activated. Have you corrected your mediator levels for neutrophil numbers to assess whether the cells are more activated in test animals than controls?

Gundel: Yes, if one normalizes the MPO data for the number of neutrophils, then there is no significant increase 6 h after antigen in dual responders. However, in single responders this results in a significant decrease in MPO levels. Thus, in single responders, neutrophils are recruited to the lungs but they are not activated and degranulating. We are planning to measure other parameters of neutrophil function in vivo as well as isolating and studying these newly recruited neutrophils in vitro.

Egan: To an extent, observations in our cynomolgus monkeys differ from yours, possibly due to the maturity of the colony. We do not see the same biologically relevant neutrophil infiltration that you do. The baseline eosinophils are low and they increase dramatically within 24 h. Even in the absence of neutrophils, we see significant hyperreactivity to aerosol histamine.

Gundel: In our model of antigen-induced airway inflammation and hyperresponsiveness, we use animals with normal airway cellular composition and responsiveness and induce airway eosinophilia and hyperresponsiveness with multiple antigen inhalations. In this model we see no significant neutrophil influx into the airways, but rather, a selective airway eosinophilia develops. In other studies we have induced a selective bronchial neutrophilia with repeated polymyxin B inhalations and saw no effect on airway responsiveness to inhaled methacholine (1). The late-phase model differs in that we are using animals that already have tremendous airway inflammation (eosinophils) and hyperresponsiveness, and with a single antigen challenge we are inducing an acute inflammatory response (neutrophil influx) and the late-phase airway obstructive response.

DISCUSSION REFERENCE

1. Gundel RH, Gerritsen ME, Wegner CD. Polymyxin B–induced bronchial neutrophilia does not alter airway responsiveness to methacholine in cynomolgus monkeys. Clin Exp Allergy 1992; 22:357–363.

23

Eosinophils and New Antihistamines

Christine De Vos
UCB SA Pharma Sector, Braine-l'Alleud, Belgium

I. INTRODUCTION

Models of allergen challenge of the skin, the nose, or the lungs have revealed that such responses consist of an initial triggering of mediator-containing cells, followed by a succession of events beginning with the release of vasoactive, chemotactic, and spasmogenic mediators and followed by a delayed influx of inflammatory cells into the site of tissue damage. Vasoactive and spasmogenic mediators lead to the early-phase reaction, whereas cellular inflammatory influx is believed to be concomitant with the late-phase reaction. This late response is believed to represent a link between anaphylaxis and chronic allergic disease (1).

The recruitment of eosinophils after an allergen challenge is now believed to be a hallmark of allergy. In the skin, the recruitment of eosinophils after a dermal challenge in an atopic subject was described as long as 30 years ago by Eidinger et al. (2) and is now extensively confirmed. During the pollen season, there is a significant increase in mast cell and eosinophil density in the nasal mucosa of allergic patients (3). In the development of late-phase asthma, neutrophils and eosinophils have been identified as important contributors (4). Eosinophils appear to play a direct role in the pathogenesis of asthma, particularly in late-phase reactions (5). Eosinophils may damage epithelial cells in the respiratory tract and amplify the allergic response through an activation of basophil and mast cell histamine release (6,7). Finally, a relationship between the presence of activated eosinophils in the lung and clinical inflammation of the airways has recently been reported (8) and confirmed by different research teams (9–12). These observations implicate the eosinophil as a primary proinflammatory cell.

Thus, the development of agents that inhibit or abolish eosinophil recruitment and/or activation at inflammatory sites could have therapeutic benefits.

This chapter provides an overview of the reported characteristics of well-known H_1 antagonists in various models of allergic eosinophils recruitment and/or activation.

II. ANIMAL PHARMACOLOGY

A. Active Anaphylactic Shock in Guinea Pigs (Table 1)

Twenty-four hours after an active anaphylactic shock induced by inhalation of antigen in conscious guinea pigs sensitized by ovalbumin, a noteworthy bronchial inflammation, characterized by increased numbers of neutrophils, mononuclear cells, and eosinophils in the bronchoalveolar lavage fluid, was observed. Some H_1 antagonists were studied in this model. Mepyramine at a dose of 5 mg/kg, i.p. given just before and 5 h after the allergen challenge was reported to be inactive for eosinophil recruitment (13). However, when the dose was increased (30 mg/kg, p.o., 24 h and 2 h before and 4 h after challenge), significant protection was achieved with mepyramine (14). Chlorpheniramine (0.1 mg/kg, s.c.) provided a small (29% inhibition), nonsignificant protection (15). This negative result may have been due to the small dosage used in this trial. Pretreatment with loratadine (0.4 mg/kg) caused a 43% reduction in eosinophils in the bronchoalveolar lavage (BAL) (16); by contrast, cetirizine (10 mg/kg) did not influence eosinophil influx (17).

Table 1 Active Anaphylactic Shock Induced by Inhalation of Ovalbumin in Conscious Sensitized Guinea Pigs Followed by BAL

Drug	Dosage	Route	Result	Ref.
Mepyramine	5 mg/kg × 2	i.p.	Inactive	13
Mepyramine	30 mg/kg 24 h and 2 h before + 4 h after OA challenge	p.o.	51–74% inhibition	14
Chlorpheniramine	0.1 mg/kg	s.c.	29% inhibition, nonsignificant	15
Loratadine	0.4 mg/kg	?	43% inhibition	16
Cetirizine	10 mg/kg	i.p.	Inactive	17

OA = ovalbumin.

B. PAF-Induced Eosinophil Accumulation in the Lungs (Table 2)

Intravenous, intraperitoneal, or inhalation of platelet-activating factor (PAF) caused a dose-related increase in the recovery of eosinophils from the BAL fluid. Various animal species were used with this challenge (guinea pig, rat, or monkey). Mepyramine, oxatomide, azelastine, and meclizine were reported as inactive in reducing pulmonary eosinophil accumulation (18–20). By contrast, cetirizine (10–30 mg/kg, i.p.) was reported as significantly active in preventing eosinophil accumulation in the rat pleural cavity after PAF challenge (20). Ketotifen significantly reduced the eosinophil infiltration in the airways in the guinea pig as well as in the monkey challenged by PAF (19,21). In monkeys treated with ketotifen, this marked reduction of eosinophil number in the BAL fluid was not accompanied by a reduction of airway sensitivity (22).

C. In Vitro Pharmacology

PAF- and LTB_4-induced migration of guinea pig peritoneal eosinophils was used to study terfenadine in vitro. A full inhibition was observed for 10 μmol/L concentration of terfenadine (23). No other data were reported in this model.

D. Sephadex-Induced Eosinophilia in the Rat

This pharmacological model is reviewed in Ref. 24. Briefly, it was shown that intravenous injections of Sephadex particles induced a blood eosinophilia in

Table 2 PAF-Induced Eosinophil Accumulation in the Lungs

Drug	Dosage	Route	Result	Ref.
Mepyramine[a]	1 mg/kg/day during 7 days	s.c.	Inactive	18
Oxatomide[a]	1 mg/kg/day during 6 days	s.c.	Inactive	19
Azelastine[a]	1 mg/kg/day during 6 days	s.c.	Inactive	19
Ketotifen[a]	0.1 mg/kg/day 1.0 mg/kg/day during 6 days	s.c.	Significant inhibition	19
Cetirizine[b]	10–30 mg/kg	i.p.	Significant inhibition	20
Meclizine[b]	40 mg/kg	i.p.	Inactive	20
Ketotifen[c]	1 mg/kg	i.v.	Significant inhibition	21
Ketotifen[c]	2 mg/kg sustained release	i.m.	Significant inhibition	22

[a]Guinea pig.
[b]Rat
[c]Monkey.

rats. In addition, the rats developed bronchial hyperresponsiveness and eosinophil accumulation in the BAL. Pulmonary eosinophil accumulation was highly reduced by dexamethasone (0.1 mg/kg, p.o., 88% reduction) and isoprenaline (0.1 mg/kg, s.c., 83% reduction). Aminophylline (100 mg/kg, p.o.) and sodium cromoglycate (100 mg/kg, s.c.) were moderately, but significantly, active (43% and 52% reduction, respectively). Ketotifen (20 mg/kg, p.o.) and mepyramine (20 mg/kg, s.c.) were inactive.

Thus animal pharmacology offers various models of eosinophil accumulation allowing the pharmacological characterization of drugs. However, the clinical relevance of these models is still not established.

III. HUMAN PHARMACOLOGY

A. In Vitro Testing (Tables 3 and 4)

The chemotactic response of human isolated eosinophils can be induced using different chemotactic mediators. The pharmacological modulation of this response was assessed using this model. Four H_1 antagonists were studied with a model of PAF- and FMLP-induced chemotaxis of eosinophils (Table 3). Dexchlorpheniramine (0.01–1 μg/ml) demonstrated a poor inhibition (less than 30% inhibition) on both responses. Cetirizine significantly inhibited both responses ($IC_{50} \simeq$ 0.01 μg/ml = 0.02 μmol/L) (25,26).

Loratadine was compared with cetirizine in the same model and demonstrated a similar potency on PAF-induced chemotaxis of eosinophils ($IC_{50} \simeq$ 0.01 μg/ml = 0.026 μmol/L) but a poor inhibition of FMLP-induced chemotaxis (27). Ketotifen (10 μmol/L) significantly inhibited PAF-induced chemotaxis of eosinophils isolated from asthmatic patients or normal subjects. However, inhibition in normal eosinophils (50.2%) was significantly higher than in eosinophils from asthmatics (28.4%) (28).

Table 3 In Vitro Inhibition of Human Eosinophil Chemotaxis by H_1 Antagonists

Drug	Chemotactic agent	Result	Ref.
Dexchlorpheniramine	PAF, FMLP	Poor inhibition	25
Cetirizine	PAF, FMLP	$IC_{50} \leqslant 0.02$ μmol/L	25
	PAF, FMLP	0.02 μmol/L $< IC_{50} <$ 2 μmol/L	26
Loratadine	PAF	$IC_{50} < 0.026$ μmol/L	27
	FMLP	Poor inhibition	27
Ketotifen	PAF	$IC_{50} \geqslant 10$ μmol/L	28

Table 4 In Vitro Inhibition of Human Eosinophil Activation by H_1 Antagonists

Drug	Model	Result	Ref.
Ketotifen	A23187-induced LTC_4 release	23% inhibition at 20 μmol/L	28
	PAF-induced actin polymerization	40% inhibition at 10 μmol/L	33
	Anti-IgG-induced morphological changes	$IC_{50} \simeq 1$ μmol/L	29
	Anti-IgG-induced leukotriene release	±30% inhibition at 1 μmol/L	29
Oxatomide	A23187-induced LTC_4 release	$IC_{50} = 5$ μmol/L	30
Terfenadine	A23187-induced LTC_4 release	$IC_{50} \simeq 20$ μmol/L	31
Azelastine	FMLP-induced chemiluminescence	Significant inhibition at 10 μmol/L	33
	PMA-FMLP-A23187 and zymozan-induced O_2^- generation	PMA: $IC_{50} = 18$ μmol/L FMLP: $IC_{50} = 3$ μmol/L A23187: $IC_{50} = 1.7$ μmol/L Zymozan inactive	34
Cetirizine	PAF-induced enhancement of IgG and C3b rosettes	$IC_{50} = 20$ μmol/L	38
	PAF-induced enhancement of eos toxicity	$IC_{50} = 20$ μmol/L	38
	PAF-induced hyperadherence of eos to plasma-coated glass	Significant inhibition at 2 and 200 μmol/L	38
	IL_1^- and FMLP-induced hyperadherence of eos to human umbilical vein endothelial cells	Significant inhibition at 200 μmol/L	39

eos = eosinophil.

Purified human eosinophils can be activated with various stimuli to release leukotrienes, mainly LTC_4. Ketotifen was reported to inhibit anti-IgG and A23187-induced LTC_4 release at concentrations ranging from 1 to 20 μmol/L (28,29). Oxatomide and terfenadine inhibited A23187-induced LTC_4 release with IC_{50} of 5 and 20 μmol/L, respectively (30,31). PAF and anti-IgG stimulation of human isolated eosinophils can induce morphological changes and actin polymerization, which are inhibited by ketotifen (1 μmol/L and 10 μmol/L, respectively) (29,32). Activated eosinophils produce superoxide anion (O_2^-), which can be measured by chemiluminescence or by biochemical method (reduction of cytochrome C). Azelastine was reported to inhibit superoxide anion production induced by PMA ($IC_{50} = 1.8$ μmol/L), FMLP ($IC_{50} = 3$ μmol/L), and A23187

(IC_{50} = 1.7 μmol/L) (33,34). However, zymozan-induced generation of O_2^- was not reduced by azelastine (34).

Eosinophil activation leads to increased membrane receptor expression (35), enhanced cytotoxicity against appropriately opsonized targets (36), and hyperadherence to cultured endothelial cells (37). Cetirizine inhibited PAF-induced, eosinophil-enhanced rosette formation and complement-dependent cytotoxicity (IC_{50} = 20 μmol/L) (38). Moreover, cetirizine reduced the PAF-induced eosinophil hyperadherence to plasma-coated glass (38) and the IL-1- or FMLP-induced hyperadherence of eosinophil to human umbilical vein endothelial cells (39). The concentration of drug required to give significant inhibition of hyperadherence was somewhat high; however, this activity seemed to be selective for the eosinophil since cetirizine has no effect on neutrophil adhesion in either model (38,39).

B. In Vivo Testing

1. Skin (Table 5)

The skin is obviously easier to investigate in humans than the nose or the lungs. Usually the cutaneous cellular infiltration can be evaluated either by the technique of the Rebuck window (40), which is easy to perform, or by histology of a skin biopsy, which is much more invasive.

The pharmacological modulation of allergic cutaneous cellular infiltration is largely documented showing a powerful inhibition in response to steroid treatment (41,42). When the eosinophil influx is induced by an allergen challenge, older H_1 antagonists such as promethazine (50 mg or 25 mg 4×/day) and chlorpheniramine (4 mg 4×/day) seem to be inactive (2,43), whereas a new-generation H_1 antagonist, i.e., cetirizine (10 or 20 mg/day), was reported by several different research teams to reduce the eosinophil accumulation in the skin (44–49). The results reported by these different authors are rather homogeneous, showing an inhibition ranging from 60 to 83%. All these results were obtained with a Rebuck window technique for the assessment of eosinophil cutaneous accumulation.

However, another series of papers described results obtained using skin biopsies for the assessment of eosinophil influx after cutaneous allergen challenge. With such a technique, neither hydroxyzine (100 mg/day) (50), a first-generation H_1 antagonist, nor astemizole (10 mg/day) (51) and cetirizine (10 mg single intake) (42), which are two H_1 antagonists of the last generation, were able to reduce the eosinophil infiltration. The discrepancy between the results obtained with the Rebuck window technique and the biopsy technique could be due to the mode of action of the tested drug.

A drug may specifically inhibit eosinophil adhesion to vascular endothelium, since a biopsy technique would include eosinophils retained within the microvasculature, whereas skin window techniques rely on migration of eosinophils through the vascular endothelium before enumeration. A valid comparison be-

Table 5 In Vivo Inhibition of Cutaneous Eosinophil Accumulation by H_1 Antagonists

Model	Drug	Dosage	Route	% inhibition	Ref.
Allergen	Promethazine	50 mg 4×/day	p.o.	Inactive	2
	Astemizole	10 mg/day	p.o.	Inactive	51
	Hydroxyzine	100 mg/day	p.o.	Inactive	50
	Terfenadine	?	p.o.	Inactive	43
	Promethazine	25 mg 4×/day	p.o.	Inactive	43
	Chlorpheniramine	4 mg 4×/day	p.o.	Inactive	43
	Cetirizine	20 mg/day	p.o.	Significant inhibition	43
	Cetirizine	10 mg single intake	p.o.	60%	44
	Cetirizine	10 mg/day	p.o.	62–64%	45,46
	Cetirizine	10 mg/day	p.o.	85%	47
	Cetirizine	20 mg/day	p.o.	65–83%	48,49
	Cetirizine	10 mg single intake	p.o.	Tendency to reduction	42
PAF	Cetirizine	10 mg 2×/day	p.o.	60–66%	52
	Loratadine	10 mg single intake	p.o.	Inactive	53
Anti-IgE	Cetirizine	10 mg 2×/day	p.o.	45%	54
	Terfenadine	60 mg 2×/day	p.o.	27% NS	54
Anti-IgA	Cetirizine	10 mg 2×/day	p.o.	Inactive	55
Delayed pressure urticaria	Cetirizine	10 mg 3×/day	p.o.	65%	56,57

NS = not significant.

tween two new H_1 antagonists was performed using Rebuck window technique; a significant inhibition was obtained with cetirizine and no inhibition with terfenadine (43).

This property of cetirizine was demonstrated in other cutaneous models leading to a cutaneous eosinophil accumulation. PAF-induced eosinophil recruitment in the skin was significantly inhibited by cetirizine (10 mg 2×/day) (52), whereas loratadine was inactive (10 mg single intake) (53). Skin challenge with anti-IgE antibodies is characterized by a significant accumulation of eosinophils only in atopic subjects. This phenomenon was significantly inhibited by cetirizine (10 mg 2×/day), while terfenadine (60 mg 2×/day) had a mild, nonsignificant inhibitory effect (54). Anti-IgA antibodies, injected intradermally, as well as anti-IgE antibodies, induced an eosinophil influx only in atopic subjects. However, the anti-

IgA-induced cell infiltration was not reduced by cetirizine (10 mg 2×/day) (55). Finally, cetirizine (10 mg 3×/day) reduced pressure-induced wheals and eosinophil recruitment in patients with delayed pressure urticaria. A 65% reduction of eosinophil infiltration was measured using either the Rebuck window technique or the biopsy technique (56,57).

2. *Nose (Table 6)*

Studies on eosinophil accumulation in the nose are not numerous. During the pollen season, there is a significant increase in mast cell and eosinophil density in the nasal mucosa. Treatment with nedocromil sodium significantly inhibited the accumulation of mast cells but not eosinophils (3).

Nasal allergen provocation induced an influx of eosinophils into the nasal mucosa. Topical glucocorticosteroids abolished the eosinophil accumulation in nasal lavage fluids (58). Eosinophil influx in the nose can be measured with different techniques. Nasal washes and nasal smears are not invasive; nasal brushing and nasal biopsy are more unpleasant and invasive.

After a nasal allergen provocation, performed before the pollen season, the eosinophil influx was not reduced by cetirizine (10 mg/day) or terfenadine (60 mg 2×/day), whereas both compounds reduced nasal symptoms and inhibited the increased nonspecific nasal reactivity (59). There is a discrepancy between this first study and a second one reporting that pretreatment with cetirizine (10 mg 2×) 24 h before challenge was able to minimize or to inhibit completely the onset of eosinophilia in the nose (60).

During the pollen season, nasal smears and nasal biopsies showed an increase of eosinophil tissue density compared with the baseline value before pollen. However, cetirizine treatment (10 mg/day) did not prevent this increase of eosinophils in the nasal mucosa despite significant reduction of rhinitis symptoms compared with a placebo in double-blind studies (61). These results are in contradiction with a study comparing cetirizine (10 mg/day) in seasonal rhinitis

Table 6 In Vivo Inhibition of Nasal Eosinophil Accumulation by H_1 Antagonists

Model	Drug	Dosage	Route	Results	Ref.
Allergen nasal provocation	Terfenadine	60 mg 2×/day	p.o.	No change	59
	Cetirizine	10 mg/day	p.o.	No change	59
	Cetirizine	2 × 10 mg	p.o.	Complete inhibition	60
Seasonal rhinitis	Cetirizine	10 mg/day	p.o.	Inactive	61
	Cetirizine	10 mg/day	p.o.	Better inhibition with cetirizine than with terfenadine or astemizole	62
	Terfenadine	60 mg 2×/day	p.o.		
	Astemizole	10 mg/day	p.o.		

Table 7 In Vivo Inhibition of Lung Eosinophil Accumulation by H_1 Antagonists

Model	Drug	Dosage	Route	Results	Ref.
Allergen inhalation	Cetirizine	15 mg 2×/day	p.o.	Significant reduction	64

and reporting a reduction of nasal eosinophil influx. However, this study was not placebo-controlled (62).

3. *Lungs (Table 7)*

Eosinophil recruitment during the late asthmatic response and eosinophil accumulation in the lung during symptomatic episodes of allergic asthma are now well described in the literature. However, trials attempting to measure the influence of drug treatment on cell accumulation in the lung are rare, probably for technical and ethical reasons. It has been shown that sodium cromoglycate could diminish the eosinophil infiltration in the bronchoalveolar fluid of allergic asthmatic subjects (63).

Thus far, only one study has reported an inhibition of bronchial eosinophil recruitment by the H_1 antagonist cetirizine. Bronchoalveolar lavage was performed 24 h after an allergen inhalation inducing early and late asthmatic responses in allergic asthmatics. These asthmatics received either cetirizine (15 mg 2×/day) or placebo for 8 days before challenge. In the placebo-treated group, the total cell number as well as the total eosinophil number were significantly higher in the BAL fluid than in the cetirizine-treated group despite the fact that late asthmatic response (measured as % drop in FEV_1) was the same in both groups (64).

IV. SUMMARY AND CONCLUSIONS

Eosinophils are now considered to be one of the main cell types involved in allergic inflammation associated with the presence of a late allergic response leading to chronic allergic situations.

It has been known for many years that the early, immediate symptoms of allergy, e.g., urticaria or sneezing and rhinorrhea or ocular pruritus, are generally well controlled with antihistaminic drugs. However, chronic allergies such as perennial rhinitis or asthma are less easy to control. Topical corticosteroids have become the "gold standard" to treat such conditions and are shown to abolish the cellular inflammatory component of chronic allergy.

This overview showed that new-generation antihistamines such as cetirizine and compounds such as azelastine and ketotifen seem to possess not only anti-H_1

properties but also other properties, e.g., inhibition of eosinophil recruitment induced by various allergic and nonallergic challenges in humans or animals as well as eosinophil activation in some in vitro models, which could be assumed to be "antiallergic properties."

The concomitant presence of both anti-H_1 and antiallergic properties on the same molecules could have some additional therapeutic benefit in the treatment of allergic manifestations, switching off the vicious circle of early and late-phase allergic reactions and providing an insight into more basic pathogenetic mechanisms.

REFERENCES

1. Gleich GJ. The late phase of the immunoglobulin mediated reaction: a link between anaphylaxis and common allergic disease. J Allergy Clin Immunol 1982; 70:160–169.
2. Eidinger D, Wilkinson R, Rose B. A study of cellular responses in immune reactions utilizing the skin window technique. J Allergy 1964; 35:77–85.
3. Lozewicz S, Gomez Z, Clague J, Gatland D, Davies RJ. Allergen-induced changes in the nasal mucous membrane in seasonal allergic rhinitis: effect of nedocromil sodium. J Allergy Clin Immunol 1990; 85:125–131.
4. Kay AB. Eosinophils as effector cells in immunity and hypersensitivity disorders. Clin Exp Immunol 1985; 62:1–12.
5. Durham SR, Kay AB. Eosinophils, bronchial hyperreactivity and late phase asthmatic reactions. Clin Allergy 1985; 15:411–418.
6. Frigas E, Gleich GJ. The eosinophil and the pathophysiology of asthma. J Allergy Clin Immunol 1986; 77:527–537.
7. Gleich GJ, Loegering DA. Immunobiology of eosinophils. Annu Rev Immunol 1984; 2:429–460.
8. Bousquet J, Chanez P, Lacoste JY, Barnéon G, Ghavanian N, Enander I, Venge P, Ahlstedt S, Simony-Lafontaine J, Godard P, Michel FB. Eosinophilic inflammation in asthma. N Engl J Med 1990; 323:1033–1039.
9. Broide DH, Gleich GJ, Cuomo AJ, Coburn DA, Federman EC, Schwartz LB, Wasserman SI. Evidence of ongoing mast cell and eosinophil degranulation in symptomatic asthma airway. J Allergy Clin Immunol 1991; 88:637–648.
10. Bousquet J, Chanez P, Lacoste JY, Enander I, Venge P, Peterson C, Ahlstedt S, Michel FB, Godard P. Indirect evidence of bronchial inflammation assessed by titration of inflammatory mediators in BAL fluid of patients with asthma. J Allergy Clin Immunol 1991; 88:649–660.
11. Bradley BL, Azzawi M, Jacobson M, Assoufi B, Collins JV, Irani AMA, Schwartz LB, Durham SR, Jeffery PK, Kay AB. Eosinophils, T-lymphocytes, mast cells, neutrophils and macrophages in bronchial biopsy specimens from atopic subjects with asthma. Comparison with biopsy specimens from atopic subjects without asthma and normal control subjects and relationship to bronchial hyperresponsiveness. J Allergy Clin Immunol 1991; 88:661–674.
12. Walker C, Kaegi MK, Braun P, Blaser K. Activated T cells and eosinophilia in

bronchoalveolar lavages from subjects with asthma correlated with disease severity. J Allergy Clin Immunol 1991; 88:935–942.
13. Tarayre JP, Aliaga M, Barbara M, Tisseyre N, Vieu S, Tisne-Versailles J. Pharmacological modulation of a model of bronchial inflammation after aerosol-induced active anaphylactic shock in conscious guinea pigs. Int J Immunopharmacol 1991; 13: 349–356.
14. Chand N, Hess FG, Nolan K, Diamantis W, McGee J, Sofia RD. Aeroallergen-induced immediate asthmatic responses and late-phase associated pulmonary eosinophilia in the guinea pig: effect of methylprednisolone and mepyramine. Int Arch Allergy Appl Immunol 1990; 91:311–314.
15. Hand JM, Gadaleta PE, Norak L, Ferracone J, Osborn RR. Effect of rolipram, salbutamol, chlorpheniramine and SK&F 106203 on bronchoalveolar lavage cells following antigen challenge in guinea pigs. Pharmacologist 1990; 32:172.
16. Cuss F, Gulbenkian A, Egan R, Danzig M, Kreutner W. Eosinophilia in nasal and bronchial lavage is inhibited by loratadine. Clin Exp Allergy 1990; 20:57.
17. Hutson PA, Holgate ST, Church MK. Late bronchial responses in the guinea pig. In Dorsch W, ed. Late Phase Allergic Reactions. CRC Press, Boca Raton, FL, 1990: 373–384.
18. Sanjar S, Aoki S, Boubekeur K, Chapman ID, Smith D, Kings MA. Eosinophil accumulation in pulmonary airways of guinea pigs induced by exposure to an aerosol of platelet-activating factor: effect of anti-asthma drugs. Br J Pharmacol 1990; 99: 267–272.
19. Sanjar S, Aoki S, Boukebeur K, Burrows L, Colditz I, Chapman I. Inhibition of PAF-induced eosinophil accumulation in pulmonary airways of guinea pigs by anti-asthma drugs. Jpn J Pharmacol 1989; 51:167–172.
20. Martins MA, Pasquale C, Silva P, Pires A, Ruffié C, Rihoux JP, Cordeiro R, Vargaftig B. Interference of cetirizine with the late eosinophil accumulation induced by either PAF or compounds 48/80. Br J Pharmacol 1992; 105:176–180.
21. Arnoux B, Denjean A, Page CP, Nolibe D, Morley J, Benveniste J. Accumulation of platelets and eosinophils in baboon lung following PAF-acether challenge: inhibition by ketotifen. Am Rev Respir Dis 1988; 137:855–860.
22. Amsler B, Hoshiko K, Chapman ID, Morley J. Effects of therapeutic doses of ketotifen on guinea pig and monkey airways. Am Rev Respir Dis 1991; 143:A653.
23. Hidi R, Joseph D, Vargaftig BB, Coeffier E. Interference of terfenadine with migration of guinea pig eosinophils. Schweitz Med Wochenschr 1991; 121:27.
24. Spicer BA, Baker RC, Hatt PA, Laycock SM, Smith H. The effects of drugs on sephadex-induced eosinophilia and lung hyperresponsiveness in the rat. Br J Pharmacol 1990; 101:821–828.
25. De Vos C, Joseph M, Leprevost C, Vorng H, Tomassini M, Capron M, Capron A. Inhibition of human eosinophil chemotaxis and of the IgE-dependent stimulation of human blood platelets by cetirizine. Int Arch Allergy Appl Immunol 1989; 88:212–215.
26. Townley RG, Okada C. Use of cetirizine to investigate non-H_1 effects of second-generation antihistamines. Ann Allergy 1992; 68:190–196.
27. De Vos C. H_1-antagonists and inhibitors of eosinophil accumulation. Clin Exp Allergy 1991; 21:277–281.

28. Sugiyama H, Nabe M, Miyagawa H, Agrawal DK, Townley RG. Effect of ketotifen on calcium ionophore induced LTC_4 release and platelet activating factor induced eosinophil chemotaxis. Am Rev Respir Dis 1990; 141:A874.
29. Kishimoto T, Sato T, Ono T, Takahashi K, Kimura J. Effect of ketotifen on the reactivity of eosinophils with the incubation of anti-IgE. Br J Clin Pract 1990; 44: 226–230.
30. Manabe H, Ohmori K, Tomioka H, Yoshida S. Oxatomide inhibits the release of chemical mediators from human lung tissues and from granulocytes. Int Arch Allergy Appl Immunol 1988; 87:91–97.
31. Nabe M, Agrawal DK, Sarmiento EV, Townley RG. Inhibitory effect of terfenadine on mediator release from human blood basophils and eosinophils. Clin Exp Allergy 1989; 19:515–520.
32. Morita M, Tsuruta S, Mori KJ, Mayumi M, Mikawa H. Ketotifen inhibits PAF-induced actin polymerization in a human eosinophilic leukaemia cell line, Eol-1. Eur Respir J 1990; 3:1173–1178.
33. Kurosawa M, Hanawa K, Kobayashi S, Nakano M. Inhibitory effects of azelastine on superoxide anion generation from activated inflammatory cells measured by a simple chemiluminescence method. Arzneim-Forsch/Drug Res 1990; 40:767–770.
34. Busse W, Randlev B, Sedgwick J. The effect of azelastine on neutrophil and eosinophil generation of superoxide. J Allergy Clin Immunol 1989; 83:400–405.
35. Moqbel R, Walsh GM, MacDonald AJ, Kay AB. The effects of disodium cromoglycate on activation of human eosinophils and neutrophils following reverse (anti-IgE) anaphylaxis. Clin Allergy 1986; 16:73–84.
36. Moqbel R, Walsh GM, Nagakura T, MacDonald AJ, Wardlaw AJ, Iikura Y, Kay AB. The effect of platelet activating factor on IgE binding to, and IgE-dependent biological properties of human eosinophils. Immunology 1990; 70:251–257.
37. Kimani G, Tonnesen MG, Henson PM. Stimulation of eosinophil adherence to human vascular endothelial cells in vitro by platelet activating factor. J Immunol 1988; 140: 3161–3166.
38. Walsh GM, Moqbel R, Hartnell A, Kay AB. Effects of cetirizine on human eosinophil and neutrophil activation in vitro. Int Arch Allergy Appl Immunol 1991; 95:158–162.
39. Kyan-Aung U, Hallsworth M, Haskard D, De Vos C, Lee TH. The effect of cetirizine on the adhesion of human eosinophils and neutrophils to cultured human umbilical vein endothelial cells. J Allergy Clin Immunol 1992; 90:270–272.
40. Rebuck JW, Crowley JH. A method of studying leukocytic functions in vivo. Ann NY Acad Sci 1955; 59:757–805.
41. Dunsky EA, Atkins PC, Zweiman B. Histologic responses in human skin test reactions to ragweed. IV. Effect of a single intravenous injection of steroids. J Allergy Clin Immunol 1977; 59:142–146.
42. Varney V, Gaga M, Frew AJ, De Vos C, Kay AB. The effect of a single oral dose of prednisolone or cetirizine on inflammatory cells infiltrating allergen-induced cutaneous late-phase reactions in atopic subjects. Clin Exp Allergy 1992; 22:43–49.
43. Kagey-Sobotka A, Massey WA, Charlesworth EN, Cooper P, Lichtenstein LM.

Terfenadine blocks both the early and late cutaneous response to antigen. J Allergy Clin Immunol 1992; 89:248.
44. Fadel R, Herpin-Richard N, Rihoux JP, Henocq E. Inhibitor effect of cetirizine 2HCl on eosinophil migration in vivo. Clin Allergy 1987; 7:373–379.
45. Fadel R, Herpin-Richard N, Henocq E, Rihoux JP. Effect of cetirizine on in vivo cutaneous leukocyte migration in allergic rhinitis. J Allergy Clin Immunol 1988; 81:178.
46. Michel L, De Vos C, Rihoux JP, Burtin C, Benveniste J, Dubertret L. Inhibitory effect of oral cetirizine on in vivo antigen induced histamine and PAF-acether release and eosinophil recruitment in human skin. J Allergy Clin Immunol 1988; 82:101–109.
47. Cortada-Macias JM, Perello-Servera R. Cetiricina, su efecto sobre la migracion de eosinophilos. Rev Esp Alergol Immunol Clin 1991; 6:175–180.
48. Charlesworth EN, Kagey-Sobotka A, Norman PS, Lichtenstein LM. Effects of cetirizine on mast cell mediator release and cellular traffic during the cutaneous late phase response. J Allergy Clin Immunol 1989; 83:905–912.
49. Michel L, De Vos C, Dubertret L. Cetirizine effects on the cutaneous allergic reaction in humans. Ann Allergy 1990; 65:512–516.
50. Atkins P, Green GR, Zweiman B. Histologic studies of human skin test responses to ragweed, compound 48/80, and histamine. J Allergy 1973; 51:263–273.
51. Bierman CW, Maxwell D, Rytina E, Emanuel MB, Lee TH. Effect of H_1-receptor blockade on late cutaneous reactions to antigen: a double-blind controlled study. J Allergy Clin Immunol 1991; 87:1013–1019.
52. Fadel R, David B, Herpin-Richard N, Borgnon A, Rassemont R, Rihoux JP. In vivo effects of cetirizine on cutaneous reactivity and eosinophil migration induced by platelet-activating factor (PAF-acether) in man. J Allergy Clin Immunol 1990; 86: 314–320.
53. Fadel R, Herpin-Richard N, Dufresne F, Rihoux JP. Pharmacological modulation by cetirizine and loratadine of antigen and histamine-induced skin weals and flares, and late accumulation of eosinophils. J Int Med Res 1990; 18:366–371.
54. Henocq R, Rihoux JP. Does reverse-type anaphylaxis in healthy subjects mimic a real allergic reaction? Clin Exp Allergy 1990; 20:269–272.
55. Rihoux JP, Melac M, Henocq E. Anti-IgE and anti-IgA-induced eosinophil migration in atopics and healthy volunteers. Clin Exp Allergy 1990; 20:11–18.
56. Kontou-Fili K, Maniatakou G, Demaka P, Gonianakis M, Paleologos G. Therapeutic effects of cetirizine in delayed pressure urticaria. Part 1. Effects on weight tests and skin-window cytology. Ann Allergy 1990; 65:517–519.
57. Kontou-Fili K, Maniatakou G, Paleologos G, Aroni K. Cetirizine inhibits delayed pressure urticaria. Part 2. Skin biopsy findings. Ann Allergy 1990; 65:520–522.
58. Andersson M, Andersson P, Pipkorn U. Eosinophils and eosinophil cationic protein (ECP) in nasal lavages in allergen-induced hyperresponsiveness: effects of topical glucocorticosteroid treatment. Allergy 1989; 44:342–348.
59. Klementson H, Andersson M, Pipkon U. Allergen-induced increase in non specific nasal reactivity is blocked by antihistaminics without a clear-cut relationship to eosinophil influx. J Allergy Clin Immunol 1990; 86:466–472.

60. Leonhardt L, Molitor SJ, Richter K. Delayed reactions and eosinophilia after nasal challenge test with allergens. Schweiz Med Wochenschr 1991; 121:15.
61. Howarth PH, Wilson SJ, Breuster H. The influence of cetirizine on symptom generation and nasal eosinophilia in seasonal allergic rhinitis. J Allergy Clin Immunol 1991; 87:151.
62. Spyropoulou-Vlahou M, Emmanuel B, Tsangarakis E, Stauropoulos-Giokas A. Comparative efficacy of non-sedating antihistamines in seasonal allergic rhinitis. Clin Exp Allergy 1990; 20:56.
63. Diaz P, Galleguillos FR, Gonzalez MC, Pantin CFA, Kay AB. Bronchoalveolar lavage in asthma: the effect of disodium cromoglycate on leukocyte counts, immunoglobulins, and complement. J Allergy Clin Immunol 1984; 74:41–48.
64. Redier H, Chanez P, De Vos C, Rifaï N, Clauzel AM, Michel FB, Godard Ph. Inhibition by cetirizine of the bronchial eosinophil recruitment induced by allergen inhalation challenge in allergic asthmatics. J Allergy Clin Immunol 1992; 90: 215–224.

DISCUSSION (Speaker: C. De Vos)

Kay: Have you been able to show an association between the increase in BAL eosinophils and severity of disease?

De Vos: There is no answer to this question at present.

Lee: Do you have information on cetirizine concerning its structure/function relationship of the anti-H_1 and antieosinophil activities?

De Vos: That's a very good question. For the moment we have no answer to that question.

Busse: The modification of eosinophil function by antihistamine is often compound-specific. In the studies we did to evaluate azelastine, modification of superoxide generation was quite specific to this compound. Therefore, it may be possible to develop profiles on antihistamines based on their ability to modify in vitro cell functions and then attempt to translate these observations to in vivo studies.

M. Capron: Has Barry Kay's laboratory looked at antibody-dependent schistosomula killing in the absence of PAF?

Moqbel: We have performed only complement-dependent eosinophil-mediated cytotoxicity experiments in this study.

Kay: Have you been able to demonstrate that cetirizine, given over a period of time, affects asthma severity and that this is accompanied by a decrease in BAL eosinophil numbers?

De Vos: We completed a study in mild asthmatic children comparing DSCG-cetirizine and placebo in a double-blind parallel study ($n = \pm 325$) (1). Briefly, the results showed a significant decrease of frequency of asthma crisis and use of rescue drug under cetirizine treatment compared to placebo or DSCG. We have no answer concerning eosinophil numbers in either blood or BAL fluid in clinical situations with cetirizine treatment.

Galli: You said that cetirizine blocked the accumulation of eosinophils after allergen challenge when the cells were measured in skin blisters, skin windows, or BAL fluid, but not in skin biopsies of allergen-challenged sites. In the first three cases, eosinophils are counted after they have successfully passed through the vasculature and have migrated through an interstitial compartment. The lack of an effect of cetirizine on the number of eosinophils present in the skin biopsies may have reflected either the inability of the cells to pass through the vessel wall, as you suggested, or the accumulation of these cells in the interstitium, perhaps because of an impairment in their migration. It is usually possible to determine, in histological sections of skin, whether leukocytes are within the blood vessels or within the interstitium. In the skin biopsy survey you mentioned, is it clear to what extent the eosinophils in allergen-challenged sites are intravascular as opposed to extravascular?

De Vos: Cetirizine consistently inhibits eosinophil accumulation in the skin after allergen challenge when the cell recruitment is measured with the Rebuck window technique of counting of cells in chambers or blister fluids. However, neither astemizole nor cetirizine affected eosinophil recruitment when measured on biopsies. One explanation could be that on biopsies, cell enumeration is performed even on cells present in the vascular bed. On the contrary, skin windows or skin blister fluids contain only the cells that crossed the endothelial barrier. Tak Lee found eosinophils in tissues in biopsies of patients treated with astemizole.

Schleimer: Some of the in vitro effects of cetirizine had an IC_{50} of 10^{-12} to 10^{-11} M, whereas others were in the 10^{-6} M range. Do you think that cetirizine is working through a single molecular mechanism and, if so, why are the responses to cetirizine so heterogeneous?

De Vos: There are large differences in IC_{50} for cetirizine on eosinophil function (chemotaxis—IC_{50} ranging from 2×10^{-6} mol/L to 5×10^{-11} mol/L depending on the stimulating agent). I have no answer to explain these differences.

DISCUSSION REFERENCE

1. van de Venne H, Hulhoven R, Arendt C. Cetirizine in perennial atopic asthma. Eur Respir J 1991; 4:525s.

24

Eosinophils and Asthma

Julie B. Sedgwick, William F. Calhoun, and William W. Busse
University of Wisconsin Medical School, Madison, Wisconsin

I. INTRODUCTION

In the past two decades, considerable evidence has accrued to indicate that eosinophils contribute to clinical features of asthma: airway obstruction, bronchial hyperresponsiveness, and airway inflammation. Prior to learning the unique features of eosinophil biology, there was highly suggestive clinical evidence that eosinophils may have a causative association with asthma. Although such a conclusion was largely one of correlation, the findings were quite convincing. For example, when patients died of asthma, airway histology revealed a marked inflammation with eosinophilic infiltration predominant (1). Furthermore, in patients with asthma who died of unrelated causes, airway histology also showed bronchial eosinophilia. Finally, peripheral blood, lavage fluid, and airway mucosal eosinophilia is a clinical association of asthma that parallels the severity of airway obstruction (2,3). The absence of eosinophilia, in either the circulation, sputa, or bronchial tissue, raises the question of whether asthma exists. Thus, even in the absence of an understanding of eosinophil function in asthma, there are data to strongly implicate this cell as an important participant in this pulmonary disease.

In this chapter, we will review data from the last decade that attempt to correlate the increasing, large body of information on eosinophil function and molecular biology with an understanding of when, why, and how this cell contributes to asthma. We will focus on a number of issues, including the association of eosinophil subpopulations in asthma and relationship of eosinophil heterogeneity to the development of airway obstruction, and finally discuss early evidence that airway eosinophils in asthma, as well as allergic airway disease, are distinct from peripheral blood eosinophils, and these unique features may prove essential to their ability to contribute to the airway dysfunction of asthma.

II. THE RELATIONSHIP OF EOSINOPHILS TO CLINICAL ASTHMA

Nearly 20 years ago, Horn and colleagues (2) found a clinical association between the intensity of peripheral blood eosinophilia and severity of asthma, as judged by FEV_1 values. There was an inverse correlation between blood eosinophil counts and airway obstruction: the higher the eosinophilia, the lower the FEV_1. These data have been substantiated and extended in recent investigations. Bousquet and colleagues (3) found that not only was peripheral blood eosinophilia associated with asthma severity, as indicated by Aas scores, but this association was also significant when eosinophils in airway lavage fluid and airway tissues were related to the same index. Moreover, when asthma severity is reduced by corticosteroid therapy, in particular, eosinophils disappear (4). When these associations are linked to what is now known of the eosinophil's ability to generate airway inflammation and bronchial smooth muscle dysfunction, it becomes clinically apparent that these cells are likely important contributors to the manifestations of disease (5).

The early finding that asthma severity was reflected by the intensity of peripheral blood eosinophilia was helpful in directing attention to this cell as being important in this disease. Furthermore, the significance of this association raised the possibility that those events which worsened airway function in asthma would also be reflected in circulating cells. This was not necessarily a new concept and is noted in many clinical situations. Nonetheless, this association presented not only a helpful clinical guide to the diagnosis and treatment of asthma, but the possibility that peripheral blood cells were a potential window for events in the lung.

The use of peripheral blood eosinophils for study of asthma was facilitated by a number of advances. First, methods were developed to isolate and purify peripheral blood eosinophils. With purified eosinophils it became possible to specifically characterize function, identify intracellular constituents and their unique features, and establish cell surface markers and to examine these parameters in relationship to cell function and to clinical disease activity. Second, because the separation methods to isolate eosinophils were density gradients, it became apparent that peripheral eosinophils were a heterogeneous population of cells when evaluated by cell buoyance. The concept of eosinophil heterogeneity, and its relationship to asthma, became an important, but confusing milestone in the study of this cell.

A. Low-Density Eosinophils and Asthma

The interest in low-density eosinophils was spurred on by a number of interesting observations. First, low-density eosinophils were increased in a number of diseases, particularly those with eosinophilia (6). Second, airway lavage eosinophils tended to be of low density; this suggested a phenotypic change of eosinophils in the lung and raised the possibility of an etiological link between this subpopulation and disease activity (7). Furthermore, there was evidence that low-density eosinophils had enhanced functional activity (8). Finally, a number of factors, in particular cytokines, were found that reduced eosinophil density and enhanced selected function (9). Thus, there was convincing evidence that low-density eosinophils were up-regulated cells, and efforts to identify their presence, function, and association with disease appeared important.

Because the low-density eosinophil was felt to be a marker of an up-regulated cell, patients with allergic diseases and asthma were evaluated and the presence of "hypodense" eosinophils quantitated. Fukuda et al. (10) and we (11) were able to demonstrate increased numbers of low-density eosinophils in asthma. Furthermore, we found that the percentage of circulating low-density eosinophils, but not absolute eosinophils, correlated inversely with the severity of asthma as indicated by the degree of airway obstruction with FEV_1 (Fig. 1). These findings suggested

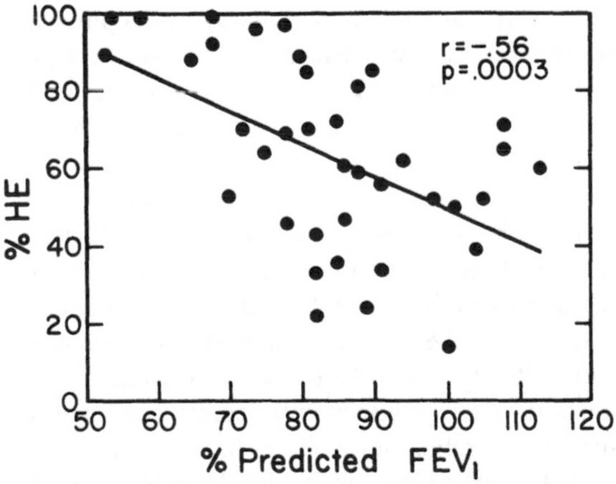

Figure 1 Correlation between percent of low-density eosinophils (HE) in peripheral blood (baseline) and percent of predicted FEV_1 (baseline) in 38 atopic asthmatics. (Reprinted, with permission, from Ref. 12.)

that the percentage of low-density eosinophils was a relatively sensitive marker of asthma severity. Furthermore, because there was a variability in this association from patient to patient, the possibility existed that factors which exacerbated asthma may also be reflected in the number of low-density eosinophils.

To further evaluate this possibility, we determined what influence selected provocative factors of asthma had on the appearance of circulating low-density eosinophils (12). In one study, the effect of antigen challenge was evaluated on the phenotypic features of circulating eosinophils. Patients with allergic asthma were challenged by inhaled antigen. Peripheral blood eosinophils and their density were determined. The pattern of their airway response to antigen was established, i.e., immediate alone or dual (immediate and late), and compared to levels of blood eosinophilia and their cell density. Patients who had an isolated immediate response showed an increase in peripheral eosinophils 24 h after antigen challenge; no change, however, was noted in the percentage of low-density cells. In contrast, patients with an immediate and late response had no increase in total eosinophils, but rather a significant rise in the percentage of low-density circulating cells. Methacholine-induced airway obstruction had no effect on either eosinophil numbers or density.

Thus, there was evidence that the development of the late asthmatic reaction to antigen, presumably a more severe and inflammatory response, was associated with increased numbers of low-density eosinophils. From these data we speculated that the increased numbers of low-density circulating eosinophils represented changes that either occurred in the airway or were the reflected phenotypic changes in circulating eosinophils from the generation of inflammatory messages in the lung. If the latter event was correct, we speculated that the phenotypic conversion to low density occurred prior to recruitment to the airway eosinophils and represented functional up-regulation. Because low-density eosinophils were proposed to have increased inflammatory capabilities, their recruitment to the antigen-challenged site would lead to more severe asthma, i.e., the late asthmatic reaction (LAR). Conversely, if blood eosinophils reflect events in the airway, which is more likely to be the case, the appearance of a greater inflammatory response to antigen, i.e., LAR, would be mirrored in circulating cells. Whatever the final explanation, we speculated that the development of a more intense inflammatory response in the airway was likely to be reflected in circulating cells.

Many factors, including cell density, make up the eosinophil's phenotype. Of particular interest have been the effects of cytokines on eosinophil density. Incubation of eosinophils with a variety of cytokines will decrease cell density. Consequently, we speculated that the generation of cytokines in the airway during antigen challenge may dictate, or at least participate in the development of, a LAR and also have systemic effects. One of these systemic effects could be a change in eosinophil density.

To extend these observations, we evaluated the effect of nocturnal asthma on the appearance of low-density eosinophils (13). A number of mechanisms have been presented to explain the nighttime appearance of airway obstruction in asthma, including changes in circulating catecholamines, cortisol, and possibly signals for inflammation (14). Furthermore, there is evidence that LARs to antigen are more likely to occur at night (15) because cellular constituents of airway inflammation increase at night (16).

In our study, 15 patients with asthma were identified. Five of these patients experienced nighttime episodes of asthma, which we defined by a greater than 15% drop in FEV_1 between 4:00 A.M. and 4:00 P.M. A number of differences were noted in the patients with nighttime asthma. First, patients with nocturnal asthma had increased numbers of peripheral eosinophils at 4 A.M. and 4 P.M. compared to normal subjects and asthma patients without nocturnal asthma (Fig. 2). Second, there was no circadian variation in total eosinophil counts in any of the patient groups. However, there was an increase in the percentage of low-density eosinophils at 4 A.M., but only in patients with nocturnal asthma (Fig. 3). This finding raised the possibility that events that regulate airflow obstruction at night may also contribute to, or be associated with, changes in eosinophil characteristics.

We also evaluated for differences in the function of peripheral blood eosinophils in relationship to the appearance of nocturnal asthma. This was done by measuring in vitro survival. We found that cells obtained at 4 A.M. survived longer in culture. Thus, not only was the appearance of nocturnal asthma associated with an increase

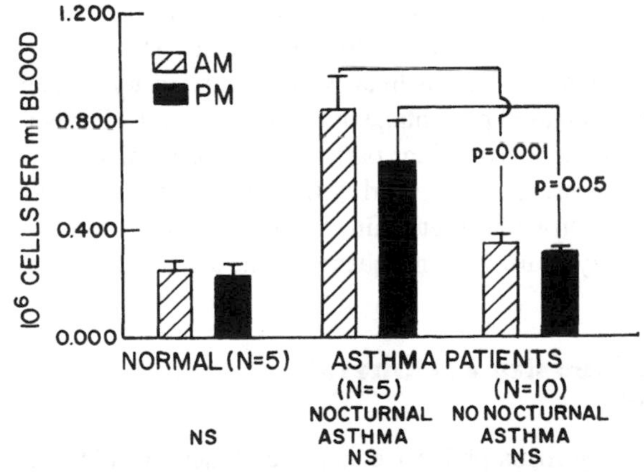

Figure 2 Circulating eosinophil counts at 4:00 A.M. and 4:00 P.M. in normal and asthmatic patients. Values are mean ± SEM. The number of eosinophils was similar A.M. to P.M. in all groups. Subjects with nocturnal asthma had higher circulating eosinophils at each time than did subjects without nocturnal asthma or normal subjects. (Reprinted, with permission, from Ref. 13.)

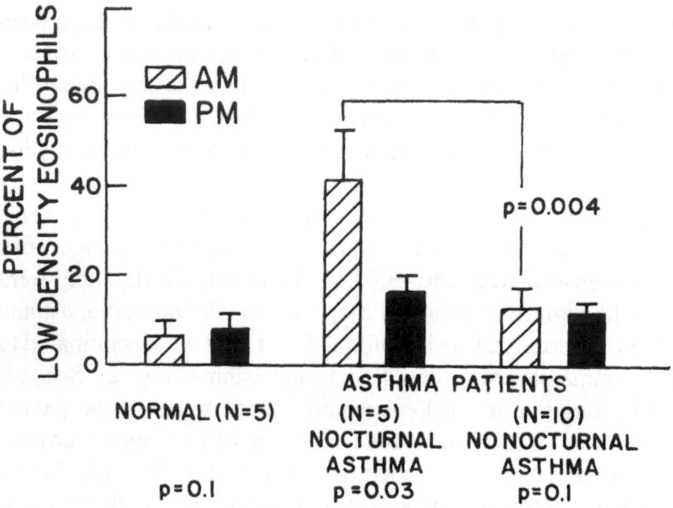

Figure 3 The percentage (mean ± SEM) of low-density eosinophils (≤1.090 g/ml) at 4:00 A.M. and 4:00 P.M. in normal and asthmatic patients. The difference between A.M. and P.M. values is indicated for each subject group. (Reprinted, with permission, from Ref. 13.)

in low-density eosinophils, but these cells had altered function, i.e., enhanced survival. When these data were configured into a three-dimensional concept, a striking relationship was noted between the number of eosinophils, their survival in culture, and the tendency for the FEV_1 to fall at night (Fig. 4).

From these data, we have evidence that asthma is characterized by greater numbers of eosinophils and an increased percentage of low-density cells. Furthermore, factors that exacerbate airway obstruction, particularly those which cause increased bronchial responsiveness, are associated with changes in circulating eosinophil density. However, it is difficult to determine from these studies whether the changes in eosinophil phenotype are the consequence of the asthma exacerbation or the cause of the process.

B. Function of Low-Density Peripheral Blood Eosinophils

To determine whether low-density eosinophils have increased function and therefore may reflect features of cells present in the airway, low- and normal-density eosinophils were collected from seven patients with asthma and activated with the phorbol ester PMA (17). Although it was found that eosinophils from patients with asthma generated more superoxide than cells from normal individuals, no striking differences were noted in the eosinophil response when evaluated by cell density.

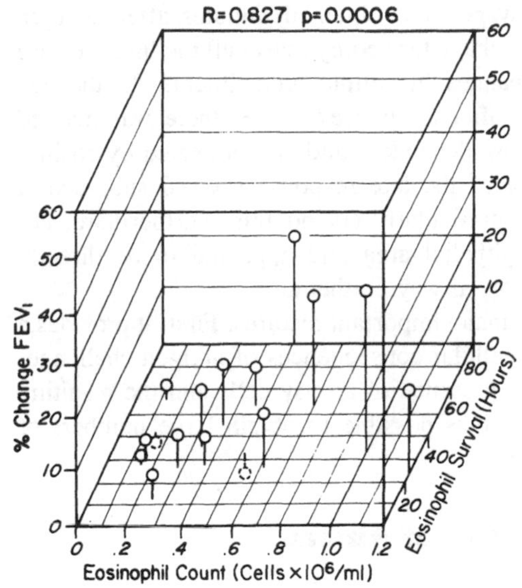

Figure 4 Correlation between eosinophil numbers, in vitro eosinophil survival, and the percent change in FEV_1 A.M. to P.M. Data are calculated by multiple linear regression. (Reprinted, with permission, from Ref. 13.)

Furthermore, when eosinophils were activated by the chemotactic peptide f-met-leu-phe (fMLP), cells of lower density, and from asthma patients, had a greater response than their normal dense counterpart. The differences in response between eosinophils of different density were not striking. From these data, we concluded that although low-density eosinophils are increased in asthma, functional differences between cells of different density are not dramatic when superoxide generation is evaluated. Furthermore, it was very difficult to repetitively obtain large numbers of peripheral blood low-density eosinophils for such studies. Therefore, both from a feasibility aspect and on an informative basis, other approaches to investigation were necessary.

III. AIRWAY EOSINOPHILS IN ALLERGIC DISEASE AND ASTHMA

From the evolving data it became apparent that further study of eosinophils in asthma, particularly in relationship to potential for airway injury, required cells from the lung. To accomplish this, the technique of segmental antigen challenge was developed and used to evaluate the airway response to antigen (18). With this procedure, antigen is introduced in the airway and the consequence of an immediate response measured in lavage fluid obtained minutes after allergen exposure. The immediate response is characterized by mast cell mediator release (histamine and tryptase) but no changes in cellular constituents. In the late response, which we measured 48 h after antigen exposure, there was marked cellular influx, which was antigen dose dependent and predominated by eosinophils. Biopsies obtained at the time of the late response showed submucosal infiltration with inflammatory cells, particularly eosinophils. Furthermore, eosinophils were present in the subepithelial area and appeared to be directly associated with, or causing damage to, airway epithelium.

This technique has proven to have many important features. First, it provides a means to evaluate the mediator and cellular consequences of antigen challenge. Second, accompanied by biopsy, the presence of airway cells and the resulting pathology can be compared. Finally, it is possible to obtain large numbers of eosinophils for further in vitro study.

A. Functional Characteristics of Airway Eosinophils

Large numbers of eosinophils can be isolated from airway lavage fluid obtained 48 h after antigen challenge. Furthermore, it is possible to purify these cells and isolate them by different density. Using this approach, we were able to compare the functional response of airway cells, categorized by density, with eosinophils from the peripheral circulation. A number of significant findings were noted. When blood and airway cells are activated by fMLP, airway eosinophils generate significantly more superoxide than do peripheral blood cells. Furthermore, the amount of superoxide generated by airway eosinophils was not influenced by cell density. We also examined the adherence response of airway and blood eosinophils to collagen-coated plates. Spontaneous adherence was similar with all eosinophil preparations. When the eosinophils were activated by fMLP, there was a marked increase in collagen adherence by airway eosinophils. However, as noted with generation of superoxide, eosinophil density had no bearing on this functional response.

To further expand on these observations, the intracellular calcium concentrations were determined in airway and blood eosinophils and the change in intracellular calcium $[Ca^{2+}]_i$ to fMLP evaluated. To accomplish this, eosinophils were loaded with the fluorescent dye INDO-1 and baseline concentrations of $[Ca^{2+}]_i$ determined. Eosinophils were then activated with fMLP and the fluorescence response measured. The increase in intracellular calcium to fMLP was greater in airway eosinophils than in peripheral blood. Moreover, the pattern of $[Ca^{2+}]_i$ increase to fMLP was also different with airway eosinophils. Peripheral blood eosinophils showed an initial fluorescent spike that represented a rise in intracellular calcium to activation; the rise in calcium rapidly returned to baseline. Airway eosinophils had a greater activation spike to fMLP and a more sustained response. When evaluated in the presence of EGTA, to chelate extracellular calcium, there was a slight reduction in the initial fluorescent spike with blood and airway eosinophils. However, the sustained rise in calcium with the BAL eosinophil was lost in the presence of EGTA. These observations suggest that EGTA chelated extracellular calcium, prevented its influx, and thus blocked the sustained rise in calcium. Our preliminary findings suggest that airway eosinophils in asthma may have altered calcium "gating" following receptor activation. Furthermore, it is possible that these changes in signal transduction may contribute to enhanced eosinophil activity found with airway cells.

IV. SUMMARY

Clinical evidence strongly indicates that eosinophils are an important component of asthma. This conclusion is substantiated by knowledge of the eosinophil's ability to cause airway inflammation. Of great interest is the likelihood that airway eosinophils are phenotypically distinct from cells found in circulation. Although initial efforts in this regard have previously focused on cell density, other factors are now emerging that appear to be of greater importance.

REFERENCES

1. Thurlbeck WM, Hogg JC. Pathology of asthma. In: Middleton E Jr, Reed CE, Ellis EF, Adkinson NF Jr, Yunginger JW, eds. Allergy — Principles and Practices. St. Louis: Mosby, 1988:1008–1017.
2. Horn BR, Robin ED, Theodore J, Van Kessel A. Total eosinophil counts in the management of bronchial asthma. N Engl J Med 1975; 292:1152–1155.
3. Bousquet J, Chanez P, Laciste JY, Arnéon G, Ghavanion N, Enander I, Venge P, Ahlstedt S, Simony-Lafontaine J, Godard P, Michel F-B. Eosinophilic inflammation in asthma. N Engl J Med 1990; 323:1033–1039.
4. Gleich GJ. The eosinophil and bronchial asthma: current understanding. J Allergy Clin Immunol 1990; 85:422–436.
5. Baigelman W, Chodosh S, Pizzuto D, Cupples LA. Sputum and blood eosinophils during corticosteroid treatment of acute exacerbations of asthma. Am J Med 1983; 75:929–936.
6. De Simone C, Donelli G, Meli D, Rosati F, Sorice F. Human eosinophils and parasitic diseases. II. Characterization of two cell fractions isolated at different densities. Clin Exp Immunol 1982; 48:249–255.
7. Fukuda T, Gleich GJ. Heterogeneity of human eosinophils. J Allergy Clin Immunol 1989; 83:369–373.
8. Winqvist I, Olofsson T, Olsson I, Persson A, Hallberg T. Altered density metabo-

lism and surface receptors of eosinophils in eosinophilia. Immunology 1982; 47: 531–539.
9. Owen WF Jr, Rothenberg ME, Silberstein DS, Gassen JC, Stevens RL, Austen KF, Soberman RJ. Regulation of human eosinophil viability, density and function by granulocyte/macrophage colony-stimulating factor in the presence of 3T3 fibroblasts. J Exp Med 1987; 166:129–141.
10. Fukuda T, Dunnette SL, Reed CE, Ackerman SJ, Peters MS, Gleich GJ. Increased numbers of hypodense eosinophils in the blood of patients with bronchial asthma. Am Rev Respir Dis 1985; 132:981–985.
11. Shult PA, Lega M, Jadidi S, Vrtis R, Warner T, Graziano FM, Busse WW. The presence of hypodense eosinophils and diminished chemiluminescence response in asthma. J Allergy Clin Immunol 1988; 81:429–437.
12. Frick WE, Sedgwick JB, Busse WW. The appearance of hypodense eosinophils in antigen-dependent late phase asthma. Am Rev Respir Dis 1989; 139:1401–1406.
13. Calhoun WJ, Bates ME, Schrader L, Sedgwick JB, Busse WW. Characteristics of peripheral blood eosinophils in patients with nocturnal asthma. Am Rev Respir Dis 1992; 145:577–581.
14. Barnes PJ. Circadian variations in airway function. Am J Med 1985; 79(Suppl 6A): 5–9.
15. Mohiuddin AA, Martin RJ. Circadian basis of the late asthmatic response. Am Rev Respir Dis 1990; 142:1153–1157.
16. Martin RJ, Cicutto LC, Smith HR, Ballard RD, Szefler SJ. Airway inflammation in nocturnal asthma. Am Rev Respir Dis 1991; 143:351–357.
17. Sedgwick JB, Geiger KM, Busse WW. Superoxide generation by hypodense eosinophils from patients with asthma. Am Rev Respir Dis 1990; 142:120–125.
18. Sedgwick JB, Calhoun WJ, Gleich GJ, Kita H, Abrams JS, Schwartz LB, Volovitz B, Ben-Yaakov M, Busse WW. Immediate and late airway response of allergic rhinitis patients to segmental antigen challenge. Am Rev Respir Dis 1991; 144:1274–1281.

DISCUSSION (Speaker: W. Busse)

Kay: What, in your opinion, is the stimulus for eosinophil degranulation once the cell has left the microcirculation? Is it in contact with extracellular matrix proteins? Is there any sign of activation/degranulation of eosinophils retrieved by BAL using your technique? Are they, for instance, EG2 positive?

Busse: Although we have evidence that eosinophil degranulation occurs in the airways following cell recruitment, as indicated by release of eosinophil granule proteins, we do not know the mechanism of cell activation. There are, however, a number of possibilities, including IgA interaction, an antigen IgE interaction on eosinophils, and secretion of other activating factors yet to be described.

M. Capron: I think that "hypodense" eosinophils from the tissues are much more heterogeneous than hyperdense eosinophils purified from the blood. We found several years ago that the cytotoxic function of rat eosinophils was increased at night (artificial night induced by inversion of the light cycle). Did you study the function of human eosinophils recovered during the day and at night?

Busse: We have not compared the function of eosinophils obtained at 4 A.M. versus 4 P.M. This is an interesting question and raises the possibility that circadian variation in cytokines, or other up-regulating factors, also influence cell function.

Gundel: In terms of eosinophil activation in the lungs, we have shown that pretreatment with an anti-Mac-1 monoclonal antibody to monkeys does not inhibit antigen-induced eosinophil infiltration into the airways. What it does do, however, is block the activation of the eosinophils once they enter into the lung and prevent the induction of airway hyperresponsiveness. Therefore, cell adhesion seems to play a role in the activation and degranulation of cells once they enter into the lung tissue.

Busse: In these experiments we have not had the opportunity to determine the importance of various adhesion molecules on migration of eosinophils into the airways and, moreover, the influence of these cells on airway responsiveness.

Roos: Do the differences in Ca^{2+} relate to the different responses of BAL and blood eosinophils to fMLP? Can you mimic the BAL eosinophil activation by incubating blood eosinophils with cytokines, and so forth?

Busse: Baseline Ca^{2+} concentrations were similar in airway and blood eosinophils. The intracellular Ca^{2+} response did parallel the function response to fMLP; when there was greater generation of O_2^- to fMLP, the intracellular rise in Ca^{2+} was also enhanced. We have not evaluated the effects of cytokines on the Ca^{2+} responses to various activators. However, cytokines, by themselves, do not enhance eosinophil function to the level found with airway eosinophilia.

Makino: Electron microscopic studies of bronchial biopsies from asthmatic patients revealed that the number of eosinophils showed a significant correlation with the frequency of the opening of the tight junction between bronchial epithelial cells. There was also a significant correlation between eosinophils and airway responsiveness to inhaled acetylcholine, suggesting that eosinophils are effector cells for airway hyperresponsiveness in asthma. In addition to eosinophils, there were also elevated numbers of CD4+ cells as well as cells expressing mRNA for IL-5. These studies suggest that CD4+ T cells contribute to eosinophilia in the airway in asthma.

Rihoux: How long after antigen challenge did you measure ICAM-1 expression on epithelial cells of asthmatic patients?

Busse: Forty-eight hours after antigen challenge.

Durham: Twenty-four hours after antigen challenge.

Hansel: Please comment on the influence of dextran sedimentation and Percoll pH and osmolarity on the number of detected hypodense eosinophils.

Busse: The density of eosinophils can be affected by many factors, including dextran exposure, Percoll osmolarity, and solution pH. However, if these conditions are maintained stable within one laboratory, the appearance of eosinophils of varying density is a valid observation.

Roos: It is difficult to compare data between laboratories.

Busse: Although there is variability in various laboratories as to the definition of low density, this variable is quite consistent *within* each laboratory. Therefore, the distribution of eosinophils by density within a laboratory is quite valid. There has been concern that dextran and other reagents can affect eosinophil density; however, when we separated airway eosinophils according to density, the cells were not exposed to dextran and, even under these conditions, there were low-density cells.

25

Eosinophil Granule Proteins in Cutaneous Disease

Kristin M. Leiferman
Mayo Clinic and Mayo Foundation, Rochester, Minnesota

I. INTRODUCTION

Because of its staining characteristics, the eosinophil is a conspicuous cell in cutaneous inflammation, and eosinophils are frequently observed in cutaneous diseases (1). Eosinophils constitute part of the diagnostic pattern in several diseases (2). In addition, although not identifiable as intact cells, eosinophils may be part of the pathogenic process in skin diseases through deposition of their toxic granule proteins in tissue (3).

It is not clear what causes eosinophils to accumulate in skin lesions and, perhaps more important, what causes eosinophils to become activated and discharge their toxic constituents into tissues. Chemoattractants presumably play a role, as do recently defined eosinophil-active cytokines, including interleukin (IL)-3, IL-5, and granulocyte-macrophage colony-stimulating factor (GM-CSF). The information that is known about these factors in skin diseases will be discussed along with the evidence for eosinophil granule protein deposition.

II. SYNDROMES ASSOCIATED WITH URTICARIA AND EDEMA

Urticaria is characterized by the appearance of transient wheals on skin in any part of the body. The wheals are uncircumscribed, slightly elevated, erythematous, usually pruritic edematous areas in upper dermal tissue. Angioedema is similarly a transient erythematous edema, but involving deeper cutaneous and subcutaneous tissues and usually larger areas. Urticaria and angioedema have many causes and associations (4).

A. Urticarial Reactions

Eosinophils are frequently observed in the otherwise nonspecific inflammatory infiltrate of urticaria. Skin biopsy specimens from chronic urticaria have shown extracellular eosinophil granule protein localization in 43–60% of lesions (5,6). The extracellular protein was prominent and deposition occurred in three patterns: (1) around small blood vessel walls, (2) as granules dispersed in the dermis, and (3) as focal or diffuse immunofluorescence of connective tissue fibers (5).

In addition to chronic urticaria, solar urticaria and delayed pressure urticaria also show extracellular localization of eosinophil granule proteins in lesions (7,8, and unpublished observations). Multiple sequential biopsy specimens of wheals elicited by either a solar simulator or a dermographometer have been examined for eosinophil granule protein deposition. In solar urticaria, progressive eosinophil infiltration and deposition of eosinophil granule proteins accompanied the development of the solar wheal (8). Biopsy specimens of normal skin in patients with solar urticaria did not contain eosinophils or eosinophil granule major basic protein (MBP), indicating that their appearance in tissue was associated with the elicitation of a lesion by ultraviolet light. Tissue deposition of MBP occurred as the lesion evolved in time and generally increased with longer irradiation times. It was greatest at 24 h, when the mononuclear cell tissue infiltrate was also greatest. Neutrophil infiltration occurred along with eosinophil infiltration (8). Both solar urticaria and delayed pressure urticaria showed patterns of extracellular eosinophil and neutrophil protein deposition similar to those found in the late-phase reaction. Thus, two physical urticarias, solar urticaria and delayed pressure urticaria, show extracellular deposition of eosinophil and neutrophil granule proteins and mimic the IgE-mediated cutaneous late-phase reaction (9).

B. Edematous Reactions

Episodic angioedema associated with eosinophilia, first described in 1984, is characterized by recurrent angioedema, urticaria, fever, and elevated levels of IgM. During the cyclical flares of this disease, leukocyte counts are increased to 100,000/mm^3 with greater than 90% eosinophils, and patients gain up to 31% of their body weight in edema fluid (10). The disease activity waxes and wanes with the number of peripheral eosinophils. The disease responds favorably to prednisone therapy. MBP is found extracellularly deposited around collagen bundles and blood vessels in edematous tissues. Some granules are abnormal in peripheral blood eosinophils by electron microscopy, and a spectrum of abnormalities exists in dermal eosinophils ranging from alterations of cytoplasmic granules in intact cells to complete disruption of cells with loss of cellular organelles, including granules, into the spaces among collagen bundles (10). Analyses of peripheral blood helper T cells in one patient with episodic angioedema and eosinophilia showed elevated numbers of activated, HLA-DR-positive helper T cells (11). Similarly, immunophenotyping of the dermal infiltrate in lesional skin of one patient with this disease revealed predominance of T-helper cells, many expressing the surface HLA-DR (12). IL-5 levels in patients with the syndrome of episodic angioedema with eosinophilia are increased; peak serum IL-5 levels are detectable prior to peak eosinophil numbers in the peripheral blood (13). In three of four patients studied, IL-5 levels spontaneously dropped during the latter days of the cyclical attacks. During treatment with prednisone, IL-5 levels became undetectable in three of four patients with associated clinical improvement including a fall in eosinophil numbers and weight loss (13).

A variant of episodic angioedema is recurrent facial edema associated with eosinophilia. Peripheral blood eosinophilia occurs during attacks of localized facial edema; levels of both MBP and Charcot-Leyden crystal protein are elevated in peripheral blood during attacks (14).

Administration of IL-2 is consistently associated with peripheral blood eosinophilia (15,16). IL-2 is a growth factor with a central role in T-cell development and immune function (17), but does not induce or enhance eosinophil differentiation (18). Patients with advanced malignancies undergoing treatment with recombinant IL-2 experience a capillary leak syndrome with weight gain and peripheral blood eosinophilia similar to episodic angioedema with eosinophilia (15,16,19). MBP is strikingly elevated in serum and urine; edematous skin from certain of these patients shows extracellular MBP deposition in a perivascular distribution in the dermis (19). Infusions of IL-2 are associated with increased plasma levels of IL-5, with IL-5 immunoreactivity becoming undetectable between IL-2 infusions. Peak eosinophil counts were noted more than 21 days after IL-2 therapy was begun (19). This delay is comparable to the time required for maximal eosinophil colony formation in bone marrow cultures and likely reflects the effect of IL-5 on relatively immature eosinophil progenitors (20). During the course of IL-2 therapy, plasma levels of eosinophil granule MBP increased and this increase preceded peripheral blood eosinophilia by more than a week (19). These findings suggest that IL-5 may promote or facilitate early eosinophil activation and degranulation. Consistent with this, IL-5 was recently found to increase nonspecific release of eosinophil granule proteins and to enhance secretory IgA- and IgG-mediated degranulation, whereas IL-2 alone in similar in vitro experiments demonstrated no such enhancing effects on the release of eosinophil granule proteins (21).

Cutaneous reactions are also common in patients receiving GM-CSF for cancer therapy protocols and include localized edema at sites of injection and generalized cutaneous reactions (22,23). These patients may also experience a capillary leak syndrome along with prominent peripheral blood eosinophilia clinically resembling episodic angioedema with eosinophilia and IL-2 toxicity. Skin biopsies reveal the presence of extracellular MBP deposition in patients with generalized cutaneous reactions; local reactions have not been studied (23).

In loaiasis, a helminth infection with *Loa loa*, localized angioedematous areas (Calabar swellings) and eosinophilia are part of the clinical presentation. Mitogen-stimulated peripheral blood mononuclear cells from infected patients show greater IL-5 production than cells from normal controls, suggesting a mechanism for eosinophil activation and participation in this helminth infection (24). Increased IL-5 levels are also associated with eosinophilia in patients with onchocerciasis.

Wells' syndrome, or eosinophilic cellulitis, is characterized by recurrent cutaneous swellings (25). Lesions begin with prodromal burning or itching followed by redness and swelling, subsequently evolving in a few days into large areas of edema with violaceous borders. Individual lesions persist for days to weeks and gradually change from bright red to brown-red to finally a blue-gray. The lesions may be single or multiple, may occur in any cutaneous location, may recur, and bullae may develop over the surface. Urticarial lesions may also accompany the disease (26). Familial cases have been reported. The edematous and infiltrative plaques of Wells' syndrome histologically are characterized by foci of dermal amorphous eosinophilic material called "flame figures." When examined by immunofluorescence for MBP, the flame figures show bright extracellular staining, suggesting that extensive eosinophil degranulation has occurred (27).

C. The IgE-Mediated Late-Phase Reaction

1. Immunohistological Findings

The wheal-and-flare cutaneous reaction is characteristic of type I, IgE-mediated hypersensitivity. This response develops rapidly and is followed by a late inflammatory reaction at the same site, which becomes apparent 3–4 h after intradermal challenge, peaks at 6–12 h, and resolves by 24–72 h. At the height of the response, the late-phase reaction is characterized by erythema, warmth, edema, pruritus, and tenderness. Pathological examination of the cutaneous late-phase reaction has shown infiltration with mononuclear cells, neutrophils, basophils, and eosinophils, with edema and mast cell degranulation in tissue (28). Immunofluorescence for neutrophil elastase and MBP shows that both neutrophil and eosinophil granule proteins are deposited in evolving late-phase reactions beginning 1–3 h after antigen challenge (9). The extracellular deposition of these proteins is extensive, out of proportion to the number of infiltrating eosinophils and neutrophils, and not detectable by staining with hematoxylin and eosin. Electron microscopy has shown the presence of degenerating eosinophils and free eosinophil granules in the skin, indicating that eosinophils disrupt in tissue during the late-phase reaction, depositing granules and toxic granule proteins in the skin.

Neutrophil elastase deposition and eosinophil granule protein deposition in the late-phase reaction were similar, supporting contributions from both neutrophils and eosinophils to the pathophysiology of the late-phase reaction (9). Histological studies of inflammatory cell infiltrates in the late-phase reaction have shown a predominance of neutrophils at 6–8 h after antigen challenge (28,29). Elevations in serum levels of neutrophil chemoattractants have been observed during the late-phase reaction (30), and a need for polymorphonuclear leukocytes to induce the late-phase reaction in rabbits has been observed (31). In skin chamber models of allergic inflammation, neutrophils were prominent after antigen challenge and the numbers were correlated with the intensity of the mast cell response as measured by mediator release (32). Furthermore, the neutrophils were capable of chemotaxis and phagocytosis and showed increased oxidative metabolism.

Eosinophil and neutrophil granule proteins were deposited in the dermis 1 h after antigen challenge; the deposition was extensive by 3–4 h and comparable to that observed at 8 h. Thus, the clinical expression of the cutaneous late-phase reaction, with tissue swelling maximal at 6–12 h, lags behind the release of eosinophil and neutrophil granule proteins (9). Certain of the clinical expressions of the late-phase reaction, such as the marked cutaneous edema, may be dependent on the release of granule proteins from eosinophils and neutrophils. Previous studies have found that basophils infiltrate the late-phase reaction (28,33) and that basophil degranulation is provoked by MBP at concentrations as low as 5×10^{-7}M (34). Mast cell numbers, as detected by specific stains, are lowest in the late-phase reaction when MBP deposition is maximal (9). This raises the possibility that MBP stimulates basophils and/or mast cells to release histamine, and it is this added wave of mediator release that culminates in the marked edema characteristic of the late-phase reaction.

2. *Evidence for Cytokine Involvement*

In lung as well as skin, immediate hypersensitivity reactions to allergen exposure are followed by late responses. Bronchoalveolar lavage fluid was obtained from allergic rhinitis patients 12 min after segmental antigen bronchoprovocation (immediate response) and 48 h later (late response) to define the nature of the inflammatory process in allergic airway disease (35). The immediate fluid revealed a significant increase in histamine and tryptase but no cellular response. The late fluid showed marked and significant increases in both hypodense and normal-density eosinophils as well as striking elevations of the eosinophil granule proteins MBP, eosinophil-derived neurotoxin (EDN), eosinophil cationic protein (ECP), and eosinophil peroxidase (EPO). Leukotriene (LT)C_4, but not tryptase, concentrations were also consistently elevated in the late lavage samples. Further, the late lavage samples showed a significant increase in IL-5 concentrations, which correlated with eosinophils and eosinophil granule protein levels. This study suggests that eosinophils are attracted to the airway during the late-phase reaction and that IL-5 may produce changes in airway eosinophil density and promote release of granule proteins that cause airway injury (35).

Skin biopsy specimens from 24 h-induced late-phase cutaneous reactions in 14 atopic patients showed cytokine messenger RNA (mRNA) expression for IL-3 (8/14), IL-4 (10/14), IL-5 (11/14), and GM-CSF (13/14) (36). Only 5 of the 14 specimens gave hybridization signals for IL-2 and 0 of the 14 for interferon-gamma (IFN-γ). Specimens from diluent controls gave only occasional weak signals. These results show that cells infiltrating cutaneous late-phase reactions have mRNA for IL-3, IL-4, IL-5, and GM-CSF and support the hypothesis that atopy is associated with preferential activation of cells having a cytokine profile similar to the murine T helper 2 subset (36). Therefore, expression of these cytokines in the cutaneous late-phase reaction may be associated with eosinophil induction and activation.

D. Evidence Linking Eosinophils to Edema

Several pieces of evidence suggest that eosinophils and eosinophil granule proteins contribute to inflammatory edema (37). In rats the number of uterine eosinophils varies more than 100-fold during the estrus cycle, correlating with edema of the uterus (38). Histamine release from basophils and mast cells can be induced by either EPO, in the presence of hydrogen peroxide and a halide, or MBP alone (34). MBP, EPO, ECP, and EDN all induce cutaneous wheal-and-flare reactions when injected into human skin (39). Activated eosinophils preferentially produce LTC_4 and platelet-activating factor (PAF), which are potent agents in the provocation of vasopermeability (38). IL-5, a selective eosinophil activator, is elevated in association with eosinophilia and edema in the syndrome of episodic angioedema (13) and in the IL-2 toxicity syndrome (19). IL-5 production from stimulated peripheral blood mononuclear cells of patients with parasite-induced swelling (*L. loa* infection) is greater than from cells of normal controls, further showing a relationship between eosinophil activation and edema (24).

Studies of purified eosinophils from patients with eosinophilia have shown that eosinophils exist as populations of varying densities, which may reflect their level of activation. Hypodense eosinophils appear to be in an activated state and generate significantly more superoxide and discharge more granule products than normodense cells. Currently, at least three factors are known to influence eosinophil differentiation in the bone marrow and activation in tissue: GM-CSF, IL-3, and IL-5 (40). Both GM-CSF and IL-3 are believed to act by inducing greater numbers of eosinophil precursors, whereas IL-5 acts later in eosinophil differentiation and influences maturation of precursor cells. IL-5 may also be a selective chemoattractant for eosinophils (41,42), whereas GM-CSF is chemotactic for both eosinophils and neutrophils (43). GM-CSF, IL-3, and IL-5 all enhance eosinophil survival (40,42). In addition to eosinophilopoiesis, chemotaxis, and prolongation of eosinophil survival, GM-CSF, IL-3, and IL-5 are also able to "activate" eosinophils as judged by increased surface receptor expression and enhanced function, including respiratory burst activity, production of inflammatory mediators, and cytotoxicity. IL-5 induces increased eosinophil degranulation stimulated by secretory IgA. The conversion of normodense eosinophils to hypodense cells has been accomplished in vitro by exposure to IL-3 and IL-5 or GM-CSF alone or in association with fibroblasts or endothelial cells (42). An early sign of eosinophil activation and incipient degranulation in tissues may be loss of electron density in eosinophil granule cores. Eosinophils in diseased tissues show both normal-appearing granules with intact cores and other granules in the same cell without cores. The loss of granule core density is correlated with MBP deposition outside of cells and may be related to decreased cell density. Eosinophils that localize to tissues are likely hypodense, activated cells capable of provoking edema (44).

Thus, several cutaneous syndromes show an association between edema and eosinophilia. Although the etiological mechanisms are incompletely understood, the ability of eosinophils to elaborate LTC_4 and PAF, the ability of eosinophil granule proteins to induce histamine release from mediator-containing cells, the finding that eosinophil granule proteins induce cutaneous wheal-and-flare reactions in human skin, and the finding of toxic products of eosinophil degranulation in tissues in the absence of many intact tissue eosinophils support the eosinophil's role as a primary participant in the inflammation and edema associated with these diseases.

III. LICHENIFIED DERMATOSES

A. Atopic Dermatitis

Atopic dermatitis is a common, chronic, eczematous skin disease associated with intense pruritus and lichenification (45). Although eosinophils are rarely a prominent feature of the infiltration in atopic dermatitis, extensive deposition of eosinophil granule MBP has been found in lesions of atopic dermatitis but not in unaffected skin (46,47). Other eosinophil granule proteins, including EDN and ECP, are deposited in atopic dermatitis lesions as well, but neutrophil granule proteins (elastase and lactoferrin) are absent or minimally present in atopic dermatitis lesions (48). This is in contrast to the IgE-mediated late-phase reaction, solar urticaria, and delayed-pressure urticaria where eosinophil *and* neutrophil proteins are deposited in lesions. These observations suggest that atopic dermatitis cannot simply be explained as an ongoing IgE-mediated reaction.

Atopic dermatitis is frequently associated with peripheral blood eosinophilia, and blood eosinophil counts roughly correlate with disease severity (49). Density analyses of eosinophils in atopic dermatitis have revealed a high proportion of hypodense cells that normalize with disease improvement (50). The circulating hypodense eosinophils show signs of activation, including morphological changes (51). Hypodense eosinophils, in fact, correlate better with disease activity than do normodense eosinophils. Increased levels of MBP (52) and ECP (53,54) in the peripheral blood do not correlate with peripheral eosinophil counts, establishing further evidence for eosinophil activation in the disease. Neutrophil elastase levels are not elevated in the serum from atopic dermatitis patients, providing evidence for eosinophil but not neutrophil activation in the disease (48).

An association between allergen contact and atopic dermatitis has been demonstrated in eczematous reactions to 48-h patch tests with aeroallergens to which atopic dermatitis patients show immediate hypersensitivity reactions on prick skin testing (55–58). Analysis of the cellular infiltrate in these eczematous reactions demonstrated an influx of eosinophils into the dermis starting 2–6 h after patch testing. Immunostaining suggested that the eosinophils were activated and had lost part of their granule contents. Eosinophils were observed in the epidermis at 24 h; some were in close proximity to Langerhans cells (59). These studies suggest an active role for eosinophils in patch test reactions to inhalant allergens in atopic dermatitis. Also, evidence for extracellular eosinophil granule protein deposition has been found in eczematous reactions in atopic dermatitis patients after provocative food challenges (47). Clearly, eosinophil granule proteins are deposited in atopic dermatitis lesions and other eczematoid eruptions associated with elevated IgE levels, and eosinophil granule proteins are deposited in lesions provoked by IgE antibodies, but the pathogenetic relationship is yet to be defined.

B. Onchocercal Dermatitis

Infection with *Onchocerca volvulus* causes a dermatitis associated with pruritus and lichenification. Microfilariae are found in the dermis. Peripheral blood eosinophilia is present, and eosinophils are seen in the skin infiltrate, but are widely scattered and few. Skin biopsies stained for MBP by immunofluorescence show extensive fibrillar extracellular MBP deposition in the superficial dermis (46). Neutrophil elastase deposition is minimal (unpublished observations). These observations emphasize the similarities between the immunohistological findings in onchocercal dermatitis and atopic dermatitis; prominent fibrillar extracellular deposition of eosinophil granule proteins is found in the absence of neutrophil granule proteins in the lichenified, pruritic lesions of both these disorders (46).

IV. BULLOUS DISORDERS

The prominence of eosinophils in bullous diseases suggests that they are important. Recent studies demonstrate evidence for eosinophil granule protein deposition in blistering lesions of incontinenti pigmenti, an X-linked neurocutaneous disease (60). The eosinophil-rich, pregnancy-related bullous dermatosis pemphigoid gestationis shows extracellular MBP deposition in lesions with prominence at blister edges (61).

Bullous pemphigoid is a subepidermal blistering disease characterized by circulating autoantibodies reactive with antigens in the lamina lucida of the basement membrane zone. Eosinophils are commonly found intermingled with mononuclear cells and neutrophils in the inflammatory infiltrate and in blisters. Several years ago, studies of blister fluid in bullous pemphigoid showed elevations of MBP levels (62). More recent studies have shown that bullous pemphigoid blister fluid induces the conversion of normodense eosinophils to hypodense eosinophils. This is accompanied by morphological changes and enhanced function, suggesting that activation of eosinophils has occurred (63).

Eosinophil granule proteins are prominently deposited in lesions of bullous pemphigoid (64). When lesions of bullous pemphigoid are studied in evolution from erythematous macules to urticarial wheals to blisters to crusted lesions, the most extensive deposition of eosinophil granule proteins is in the prebullous lesions, particularly urticarial lesions (65). Deposition of all the distinctive eosinophil granule proteins is found in the lesions. Considering the toxic activities of these proteins, including to epithelial cells (66), it is likely they are contributing to the damage in the disease. By electron microscopy, free eosinophil granules were observed on both sides of the basement membrane in proximity to hemidesmosomes and in the dermis in prebullous lesions (65). Eosinophil granule protein levels were increased in the blister fluid. Blister fluids also enhanced eosinophil survival when incubated with eosinophils in culture (65). This enhancement was blocked by combinations of antibodies to IL-5, IL-3, and GM-CSF, indicating that IL-5, IL-3, and, to a lesser extent, GM-CSF were all important cytokines in promoting eosinophil survival in the blister fluid (65). These cytokines also likely account for eosinophil accumulation (41,42) and activation by bullous pemphigoid blister fluid (63). These results point to the early participation of eosinophils in bullous pemphigoid, prior to blister development, and suggest that eosinophil activation is important in basement membrane zone damage in bullous pemphigoid.

V. EOSINOPHIL-MEDIATED VASCULITIS

Three cases of small-vessel necrotizing vasculitis with exclusive eosinophil infiltration have recently been identified. These cases represent a novel syndrome with specific clinical findings (67). Several features of this syndrome point to the importance of eosinophil activation in the pathogenesis.

All three patients developed persistent and recurrent pruritic and purpuric papular skin lesions and angioedema of periorbital areas, hands, and feet. Annular erythematous edematous lesions, urticarial plaques, and vesicular lesions occasionally developed. Chronic periodontitis developed in two of the patients without known cause (67). Lesions responded promptly to glucocorticoid therapy but recurred and followed a chronic course. Peripheral blood eosinophilia was variably present (up to 6000/mm^3), and erythrocyte sedimentation rate was elevated. No systemic symptoms accompanied the disease, and the patients have been maintained on "burst" and alternate-day glucocorticoid therapy. Follow-up has been for extended times of 24, 18, and 3 years in these patients (67).

Two or more skin biopsy specimens from each patient showed necrotizing vasculitis of small dermal vessels with fibrinoid necrosis of vessel walls without leukocytoclasis. Eosinophilic infiltration was prominent perivascularly and in the dermis; eosinophils exclusively infiltrated the vessels. Indirect immunofluorescence for eosinophil granule MBP showed marked extracellular MBP deposition in the involved small vessels and surrounding dermis. Indirect immunofluorescence for neutrophil elastase and neutrophil lactoferrin showed rare perivascular cells staining without extracellular localization (67). Electron microscopic examination of an early skin lesion showed adherence of abnormal eosinophils and free eosinophil granules to endothelial cells and walls of small vessels with destruction of endothelial cells. Eosinophils showed loss of cytoplasmic granule cores, loss of cellular organelles, and chromatolysis of nuclei. Endothelial cells showed nuclear pyknosis, swelling and destruction of mitochondria, and disruption of cellular membranes. Degenerated eosinophils and free granules were also found between collagen bundles (67). Immunoperoxidase staining for localization of vascular cell adhesion molecule-1 (VCAM-1) and very late activation antigen-4 (VLA-4) showed staining for VCAM-1 on endothelial cells of involved vessels with adherence of VLA-4-positive eosinophils. Eosinophil survival assays with serum showed prolongation of survival that was inhibited by antibodies to IL-5 but not by antibodies to IL-3 or GM-CSF; this suggests that IL-5 was important in eosinophil proliferation and activation in this disease (67).

This syndrome is characterized by glucocorticoid-responsive, pruritic, purpuric papular and angioedematous lesions that histologically show eosinophilic infiltration and vascular destruction. Affected vessels express vascular cell adhesion molecules that may selectively recruit eosinophils (68). An eosinophil-active cytokine, IL-5, is present in serum, and eosinophil activation is evident in this syndrome with deposition of toxic granule molecules in affected areas. These findings strongly implicate eosinophils in the pathogenesis of this disease.

VI. SUMMARY

Recent studies defining the composition and activities of eosinophils support a role for the eosinophil as a primary participant in several cutaneous disorders. The eosinophil possesses proteins with toxicity toward a variety of cells and can induce activation of cells such as mast cells and basophils with release of histamine and other vasoactive substances. Because they disrupt in tissue and deposit granule products, eosinophils may not be particularly prominent in the tissues. Potent oxidants and toxic cationic proteins probably achieve high concentrations at points of granule deposition, thus mediating damage to tissue. Therefore, the eosinophil is likely important in the pathophysiology of several cutaneous diseases, and its involvement cannot be assessed by simple enumeration of tissue eosinophils. Recent studies are providing important information about recruitment and activation of eosinophils in cutaneous diseases that ultimately may lead to new, more specific therapies.

REFERENCES

1. Leiferman KM. A current perspective on the role of eosinophils in dermatologic diseases. J Am Acad Dermatol 1991; 24:1101–1112.
2. Peters MS. The eosinophil. In: Callen JP, et al, eds. Advances in Dermatology. Vol 2. Chicago: Year Book Medical Publishers, 1987:129–151.
3. Leiferman KM, Gleich GJ. Cutaneous eosinophilic diseases. In: Fitzpatrick TB, et al, eds. Dermatology in General Medicine. 4th ed, St. Louis: McGraw-Hill. In press.
4. Monroe EW. Urticaria and angioedema. In: Sams WM Jr, Lynch PJ, eds. Principles and Practice of Dermatology. New York: Churchill Livingstone, 1990:Chap 44.
5. Peters MS, Schroeter AL, Kephart GM, Gleich GJ. Localization of eosinophil granule major basic protein in chronic urticaria. J Invest Dermatol 1983; 81:39–43.
6. Tai P-C, Spry CJF, Peterson C, Venge P, Olsson I. Monoclonal antibodies distinguish between storage and secreted forms of eosinophil cationic protein. Nature 1984; 309: 182–184.
7. Peters MS, Winkelmann RK, Greaves MW, Kephart GM, Gleich GJ. Extracellular deposition of eosinophil granule major basic protein in pressure urticaria. J Am Acad Dermatol 1986; 16:513–517.
8. Leiferman KM, Norris PG, Murphy GM, Hawk JLM, Winkelmann RK. Evidence for eosinophil degranulation with deposition of granule major basic protein in solar urticaria. J Am Acad Dermatol 1989; 21:75–80.
9. Leiferman KM, Fujisawa T, Holmes-Gray B, Gleich GJ. Extracellular deposition of eosinophil and neutrophil granule proteins in the IgE-mediated cutaneous late phase reaction. Lab Invest 1990; 62:579–589.
10. Gleich GJ, Schroeter AL, Marcoux JP, Sachs MI, O'Connell EJ, Kohler PF. Episodic angioedema with eosinophilia. N Engl J Med 1984; 310:1621–1626.
11. Katzen DR, Leiferman KM, Weller PF, Leung DYM. Hypereosinophilia and recurrent angioedema in a 2½ year old girl. Am J Dis Child 1986; 140:62–68.
12. Wolf C, Pehamberger H, Breyer S, Leiferman KM, Wolff K. Episodic angioedema with eosinophilia. J Am Acad Dermatol 1989; 20:21–27.
13. Butterfield JH, Leiferman KM, Gonchoroff N, Silver JE, Abrams J, Bower J, Gleich GJ. Elevated serum levels of interleukin 5 in patients with the syndrome of episodic angioedema and eosinophilia. Blood 1992; 79:688–692.

14. Songsiridej V, Peters MS, Dor PJ, Ackerman SJ, Gleich GJ, Busse WW. Facial edema and eosinophilia: evidence for eosinophil degranulation. Ann Intern Med 1985; 103: 503–506.
15. Rosenberg SA, Lotze MT, Muul LM, Leitman S, Chang AE, Ettinghausen SE, Matory YL, Skibber JM, Shiloni E, Vetto JT, Seipp CA, Simpson C, Reichert CM. Observations on the systemic administration of autologous lymphokine-activated killer cells and recombinant interleukin-2 to patients with metastatic cancer. N Engl J Med 1985; 313:1485–1492.
16. Lotze MT, Matory YL, Raynor AA, Ettinghausen SE, Vetto JT, Seipp CA, Rosenberg SA. Clinical effects and toxicity of interleukin 2 in patients with cancer. Cancer 1986; 58:2764–2772.
17. Morgan DA, Ruscetti FW, Gallo RG. Selective in vitro growth of T lymphocytes from normal human bone marrows. Science 1976; 193:1007–1008.
18. Thorne KJI, Richardson BA, Tavine J, Williamson DJ, Vadas MA, Butterworth AE. A comparison of eosinophil-activating factor (EAF) with other monokines and lymphokines. Eur J Immunol 1986; 16:1143–1149.
19. van Haelst-Pisani C, Kovach JS, Kita H, Leiferman KM, Gleich GJ, Silver JE, Dennin R, Abrams JS. Administration of IL-2 results in increased plasma concentrations of IL-5 and eosinophilia in patients with cancer. Blood 1991; 78:1538–1544.
20. Clutterbuck EJ, Hirst EMA, Sanderson CJ. Human interleukin-5 (IL-5) regulates the production of eosinophils in human bone marrow cultures: comparison and interaction with IL-1, IL-3, IL-6, and GM-CSF. Blood 1989; 73:1504–1512.
21. Fujisawa T, Abu-Ghazaleh R, Kita H, Sanderson CJ, Gleich GJ. Regulatory effect of cytokines on eosinophil degranulation. J Immunol 1990; 144:642–646.
22. Horn TD, Burke PJ, Karp JE, Hood AF. Intravenous administration of recombinant human granulocyte-macrophage colony-stimulating factor causes a cutaneous eruption. Arch Dermatol 1991; 127:49–52.
23. Mehregan DR, Fransway AF, Edmonson JH, Leiferman KM. Cutaneous reactions to granulocyte-monocyte colony-stimulating factor. Arch Dermatol 1992; 128:1055–1059.
24. Limaye AP, Abrams JS, Silver JE, Ottesen EA, Nutman TB. Regulation of parasite-induced eosinophilia: selectively increased interleukin 5 production in helminth-infected patients. J Exp Med 1990; 172:399–402.
25. Wells GC, Smith NP. Eosinophilic cellulitis. Br J Dermatol 1979; 100:101–109.
26. Dijkstra JWE, Bergfeld WF, Steck WD, Tuthill RJ. Eosinophilic cellulitis associated with urticaria: a report of two cases. J Am Acad Dermatol 1986; 14:32–38.
27. Peters MS, Schroeter AL, Gleich GJ. Immunofluorescence identification of eosinophil granule major basic protein in the flame figures of Wells' syndrome. Br J Dermatol 1983; 109:141–148.
28. Solley GO, Gleich GJ, Jordon RE, Schroeter AL. The late phase of the immediate wheal and flare skin reaction. Its dependence upon IgE antibodies. J Clin Invest 1976; 58:408–420.
29. Dolovich J, Hargreave FE, Chalmers R, Shier KJ, Gauldie J, Bienenstock J. Late cutaneous allergic responses in isolated IgE-dependent reaction. J Allergy Clin Immunol 1973; 52:38–46.

30. Nagy L, Lee TH, Kay AB. Neutrophil chemotactic activity in antigen-induced late asthmatic reactions. N Engl J Med 1982; 306:497–501.
31. Murphy KR, Wilson MC, Irvin CG, Glezen LS, Marsh WR, Haslett C, Henson PM, Larsen GL. The requirement for polymorphonuclear leukocytes in the late asthmatic response and heightened airways reactivity in an animal mode. Am Rev Respir Dis 1985; 134:62–68.
32. Fleetkop PD, Atkins PC, von Allmen C, Valenzano MC, Shalit M, Zweiman B. Cellular inflammatory responses in human allergic skin reactions. J Allergy Clin Immunol 1987; 80:140–146.
33. Charlesworth EN, Hood AF, Soter NA, Kagey-Sobotka A, Norman PS, Lichtenstein LM. The cutaneous late phase response to allergen: mediator release and inflammatory cell infiltration. J Clin Invest 1989; 83:1519–1526.
34. O'Donnell MC, Ackerman SJ, Gleich GJ, Thomas LL. Activation of basophil and mast cell histamine release by eosinophil granule major basic protein. J Exp Med 1983; 157:1981–1991.
35. Sedgwick JB, Calhoun WJ, Gleich GJ, Kita H, Abrams JS, Schwartz LB, Volvitz B, Ben-Yaakov M, Busse WW. Immediate and late airway response of allergic rhinitis patients to segmental antigen challenge: characterization of eosinophil and mast cell mediators. Am Rev Respir Dis 1991; 144:1274–1281.
36. Kay AB, Ying S, Varney V, Gaga M, Durham SR, Moqbel R, Wardlaw AJ, Hamid Q. Messenger RNA expression of the cytokine gene cluster, interleukin 3 (IL-3), IL-4, IL-5 and granulocyte/macrophage colony-stimulating factor, in allergen-induced late-phase cutaneous reactions in atopic subjects. J Exp Med 1991; 173:775–778.
37. Leiferman KM, Peters MS, Gleich GJ. The eosinophil and cutaneous edema. J Am Acad Dermatol 1986; 15:513–517.
38. Gleich GJ, Adolphson CR. The eosinophilic leukocyte: structure and function. Adv Immunol 1986; 39:177–251.
39. Leiferman KM, Loegering DA, Gleich GJ. Production of wheal-and-flare skin reactions by eosinophil granule proteins (abstr). J Invest Dermatol 1984; 82:414.
40. Silberstein DS, Austen KF, Owen WF, Jr. Hemopoietins for eosinophils: glycoprotein hormones that regulate the development of inflammation in eosinophilia-associated disease. Hematol/Oncol Clin North Am 1989; 3:511–533.
41. Yamaguchi Y, Hayashi Y, Sugama Y, Miura Y, Kasahara T, Kitamura S, Torisu M, Mita S, Tominaga A, Takatsu K, Suda T. Highly purified murine interleukin 5 (IL-5) stimulates eosinophil function and prolongs in vitro survival. IL-5 as an eosinophil chemotactic factor. J Exp Med 1988; 167:1737–1742.
42. Rothenberg ME, Petersen J, Stevens RL, Silberstein DS, McKenzie DT, Austen KF, Owen WF. IL-5 dependent conversion of normodense human eosinophils to the hypodense phenotype uses 3T3 fibroblasts for enhanced viability, accelerated hypodensity, and sustained antibody dependent cytotoxicity. J Immunol 1989; 143:2311–2316.
43. Warringa RAJ, Koenderman L, Kok PTM, Kreukniet J, Bruynzeel PLB. Modulation and induction of eosinophil chemotaxis by granulocyte-macrophage colony-stimulating factor and interleukin-3. Blood 1991; 77:2694–2706.

44. Slifman NR, Adolphson CR, Gleich GJ. Eosinophils: Biochemical and cellular aspects. In: Middleton E, Reed CE, Ellis EF, Adkinson NF, Jr., Yunginger JW, eds. Allergy: Principles and Practice. Vol. 1. St. Louis: Mosby, 1988:179–205.
45. Hanifin JM. Atopic dermatitis. J Allergy Clin Immunol 1984; 73:211.
46. Leiferman KM, Ackerman SJ, Sampson HA, Haugen HS, Venencie PY, Gleich GJ. Dermal deposition of eosinophil granule major basic protein in atopic dermatitis. Comparison with onchocerciasis. N Engl J Med 1985; 313:282–285.
47. Leiferman KM. Eosinophils in atopic dermatitis. Allergy 1989; 44:20–26.
48. Ott NL, Gleich GJ, Fujisawa T, Abrams JS, Sur S, Leiferman KM. Failure to detect neutrophil degranulation in patients with atopic dermatitis (AD): implications for pathophysiology (abstr). J Allergy Clin Immunol 1991; 87:234.
49. Uehara M, Izukura R, Sawai T. Blood eosinophilia in atopic dermatitis. Clin Exp Dermatol 1990; 15:264–266.
50. Miyasato M, Iryo K, Kasada M, Tsuda S. Varied density of eosinophils in patients with atopic dermatitis reflecting treatment with anti-allergic drug (abstr). J Invest Dermatol 1988; 90:589.
51. Tsuda S, Miyasato M, Nakama T, Nakano S, Kasada M, Sasai Y. Evidence of eosinophil degranulation in the peripheral circulation of atopic dermatitis (abstr). J Invest Dermatol 1989; 92:534.
52. Wassom DL, Loegering DA, Solley GO, Moore SB, Schooley RT, Fauci AS, Gleich GJ. Elevated serum levels of the eosinophil granule major basic protein in patients with eosinophilia. J Clin Invest 1981; 67:651–661.
53. Jakob T, Hermann K, Ring J. Eosinophil cationic protein in atopic eczema. Arch Dermatol Res 1991; 283:5–6.
54. Kapp A, Czech W, Krutmann J, Schöpf E. Eosinophil cationic protein in sera of patients with atopic dermatitis. J Am Acad Dermatol 1991; 24:555–558.
55. Clark RAF, Adinoff AD. Aeroallergen contact can exacerbate atopic dermatitis: patch tests as a diagnostic tool. J Am Acad Dermatol 1989; 21:863–869.
56. Mitchell EB, Crow J, Chapman MD, Jouhal SS, Pope FM, Platts-Mills TAE. Basophils in allergen-induced patch test sites in atopic dermatitis. Lancet 1982; 1: 127–130.
57. Reitamo S, Visa K, Kähönen K, Käyhkö K, Stubb S, Salo OP. Eczematous reactions in atopic patients caused by epicutaneous testing with inhalant allergens. Br J Dermatol 1986; 114:303–309.
58. Adinoff AD, Tellez P, Clark RAF. Atopic dermatitis and aeroallergen contact sensitivity. J Allergy Clin Immunol 1988; 81:736–742.
59. Bruynzeel-Koomen CAF, Van Wichen DF, Spry CJF, Venge P, Bruynzeel PLB. Active participation of eosinophils in patch test reactions to inhalant allergens in patients with atopic dermatitis. Br J Dermatol 1988; 118:229–238.
60. Thyresson NH, Goldberg NC, Tye MJ, Leiferman KM. Localization of eosinophil granule major basic protein in incontinentia pigmenti. Pediatr Dermatol 1991; 8: 102–106.
61. Scheman AJ, Hordinsky MD, Groth DW, Vercellotti GM, Leiferman KM. Evidence for eosinophil degranulation in the pathogenesis of herpes gestationis. Arch Dermatol 1989; 125:1079–1083.

62. Wintroub BU, Dvorak AM, Mihm MC, Gleich GJ. Bullous pemphigoid: cytotoxic degranulation and release of eosinophil major basic protein (abstr). J Invest Dermatol 1981; 76:310–311.
63. Miyasato M, Tsuda S, Kasada M, Iryo K, Sasai Y. Alteration in the density, morphology, and biological properties of eosinophils produced by bullous pemphigoid blister fluid. Arch Dermatol Res 1989; 281:304–309.
64. Maynard B, Peters MS, Butterfield JH, Leiferman KM. Bullous pemphigoid: eosinophil, neutrophil and mast cell degranulation in lesional tissue (abstr). J Invest Dermatol 1990; 94:553.
65. Borrego L, Maynard B, Peterson E, George T, Iglesias L, Leiferman KM. Extracellular deposition of eosinophil granule proteins precedes blister formation in bullous pemphigoid (abstr). J Invest Dermatol 1992; 98:588.
66. Gleich GJ, Frigas E, Loegering DA, Wassom DL, Steinmuller D. Cytotoxic properties of the eosinophil major basic protein. J Immunol 1979; 123:2925–2927.
67. Chen K-R, Pittelkow MR, Su WPD, Newman W, Leiferman KM. Eosinophilic vasculitis: eosinophil-mediated small vessel necrotizing vasculitis (abstr). J Invest Dermatol 1992; 98:603.
68. Bochner BS, Luscinskas FW, Gimbrone MA, Newman W, Sterbinsky SA, Derse-Anthony CP, Klunk D, Schleimer RP. Adhesion of human basophils, eosinophils and neutrophils to interleukin 1-activated vascular endothelial cells: contributions of endothelial cell adhesion molecules. J Exp Med 1991; 173:1553–1556.

DISCUSSION (Speaker: K. Leiferman)

Kay: Would you care to speculate on the mechanism whereby eosinophils accumulate rapidly and deposit granule proteins?

Leiferman: I do not know why eosinophils accumulate and deposit granule proteins so extensively and so quickly (within an hour) in the IgE-mediated LPR. There are many possible explanations but no direct evidence for a mechanism.

Morley: With reference to the comment by Professor Kay, we have been measuring the kinetics of eosinophil accumulation within the lung of the guinea pig using isotype-labeled cells and observe that the accumulation of eosinophils within the vasculature is maximal within 5 min.

Gleich: The ability to passively transfer skin test reactivity to normal persons and the development of late-phase reactions indicate that the atopic status is not crucial to the development of the late-phase response.

Moqbel: Using immunostaining as a method to detect MBP by either your polyclonal or our monoclonal antibodies raises the question of the contribution of basophils to MBP release in tissue. Second, have you been able to determine the source of released cytokines that supported eosinophil survival?

Leiferman: We do not have an easy way to identify basophils in the specimens we have studied. We can identify a one-to-one correlation with most cells that fluoresce with the eosinophil granule proteins and those identifiable with H&E staining. The amount of MBP in basophils is significantly less than that in eosinophils. We have not colocalized cytokines to areas of eosinophil granule proteins.

Schleimer: The pattern of endothelial adhesion molecule expression you showed in the eosinophilic vasculitis, VCAM-1 but no ELAM-1, is reminiscent of the pattern induced by IL-4 in vitro. Have you assessed whether IL-4 is produced in this syndrome?

Leiferman: We have not studied IL-4 yet.

Durham: The differences you observe between allergen-induced late cutaneous biopsies and atopic eczema may at least in part be explained by the dose, magnitude, and timing of allergen exposure. Have you studied the influence of topical steroids on the immunopathology of atopic eczema?

Leiferman: We have not systematically studied the effects of topical glucocorticoids on eosinophil granule protein deposition in skin disease. In lesions of atopic dermatitis that have minimal or no eosinophil granule protein deposition, review of the cases shows that the patients have used potent topical steroids or taken systemic steroids.

De Vos: Henocq's investigation showed that PAF intradermal injection induced eosinophil infiltration in atopic subjects and mainly neutrophils in normal subjects. However, a recent study by L. Juhlin (Sweden) showed an eosinophil recruitment in normal subjects when PAF dosage was increased. Thus the difference between atopic and asthmatic subjects is probably at the level of the threshold of reaction for PAF stimulation.

Weller: What is the half-life of basic proteins in skin? Is there differential release of granule proteins?

Leiferman: We do not know the half-life of eosinophil granule proteins in skin. We have not injected them and determined their clearance time. We do know in the late-phase reaction that MBP, EDN, and ECP persist to 56 h after antigen challenge, but ongoing deposition may be occurring. We have no evidence for differential release of granule proteins; where one of the granule proteins is found, the others are also localized. The apparent differences in intensity and extent of staining of the different proteins may be related to biochemical properties of the proteins in the skin reactions. It is important to note that EDN and ECP are also in neutrophils and their deposition in skin may be from neutrophils as well as from eosinophils.

Busse: Are there differences between affected and unaffected eczema sites?

Leiferman: We have studied biopsies of eczematous reactions to food challenge from Hugh Sampson. These show prominent MBP deposition. We have not studied unaffected and affected food challenge sites and do not know why eosinophils are activated in affected sites.

Dvorak: Do you know the subcellular location of EDN/ECP in neutrophils? Is it synthesized or internalized by neutrophils?

Leiferman: We do not know where the EDN and ECP are located in neutrophils. It is deposited in lesional areas where elastase is prominent. Furthermore, Dr. Ackerman has found RNA message for EDN and ECP in neutrophils.

Ackerman: Have you studied T-cell subsets and cytokine profiles in the various skin diseases as yet? Have you examined T-cell subsets or cytokine profiles by immunostaining and/or in situ hybridization in any of the skin diseases you have studied? I am curious in particular about the perivascular mononuclear cell infiltrates seen in atopic dermatitis.

Leiferman: We do not have the technology to look at T-cell subsets currently in the formalin-fixed tissues that we have studied and we have not done comparisons with T-cell subsets in frozen tissues. We have looked at cytokine profiles in the diseases I have discussed but not in tissues.

26

Summing Up: Eosinophils 1992

Christopher J. F. Spry, Ming-Shi Li, Li Sun, and Takahiro Satoh
St. George's Hospital Medical School, London, England

I. INTRODUCTION

Forty-nine scientists and clinicians from 11 countries participated in the meeting on which this book is based. There were 24 presentations lasting a total of 8 h, with another 8 h of discussion. Many significant observations and experiments are reported here, and many exciting discoveries no doubt lie ahead for those privileged to work in this field.

II. A BACKWARDS PERSPECTIVE

First, it is worth taking a few moments to recall some of the early pioneers in eosinophil research. Although Paul Ehrlich is rightly credited with defining the eosinophil in 1879, we should not forget that several others before him had seen eosinophils or their products in tissues and blood. One among these unsung pioneers is Dr. C. Robin (his first name was not published!), who first saw crystals of CLC protein in a patient with (probably eosinophilic) leukemia in Paris in 1853 (1), 26 years before Ehrlich's discovery.

After the early work of defining the diseases in which eosinophils were prominent, much of which took place in France and Germany at the end of the 19th century, innovative work on eosinophils switched to the Johns Hopkins Hospital and the Rockefeller Institute, where a number of new findings about eosinophils were made during the next 60 years. It was not until the 1970s that eosinophil research returned to Europe again. This soon resulted in one of the seminal papers from Oxford by Tony Basten and Paul Beeson, in which they showed the T-lymphocyte dependency of eosinophilia in rats (2). As T cells had only been defined 3 years previously, this discovery was quite a coup, especially as we now accept a central regulatory role for T cells in most aspects of eosinophil biology.

Although several of us realized in the early 1970s that the eosinophil growth factor might now be isolated and defined, our hopes were frustrated by technical limitations. As Basten has both Australian and British parentage, it was fitting that Australian and British teams, especially Colin Sanderson's group, should subsequently isolate and clone interleukin-5 (IL-5) in 1987 (3). The cloning of the IL-5 receptor in Japan (4), where IL-5 had also been cloned and sequenced as a B-cell interleukin (5), was another significant milestone whose implications in disease are still being unraveled.

Meantime, in the late 1970s and 1980s, the eosinophil granule proteins had been isolated and defined at the Mayo Clinic and in Sweden, and the broad spectrum of activities of eosinophils in disease had begun to be apparent. Despite this, we have all become aware that some of the generalizations about eosinophil functions that were promoted in the 1970s may not be as relevant to human disease as we hoped. Armed with a healthy skepticism, we are now entering an interesting period in which it is possible to reassess many of the possible roles for eosinophils in disease using (1) monoclonal antibodies, which eliminate IL-5 functions, (2) knockout and transgenic mice, which provide animal models with defined and limited alterations in one or more proteins that we believe to be central to eosinophil function, and (3) recombinant DNA techniques, which define the molecular features of eosinophil constituents and the factors that affect their generation and properties in disease.

Among other important techniques not already mentioned are rapid and reliable methods to isolate blood and tissue eosinophils. These include immunomagnetic beads coated with monoclonal antibodies to purify eosinophils by negative selection from large numbers of contaminating CD16-bearing cells (6).

III. MOLECULES AND DISEASE

You have been brought up to date on the biosynthesis and properties of many of the molecular properties of the eosinophil granule proteins (7), which, with CLC protein, lipid mediators, and reactive oxygen species, give eosinophils their effector capacity. Our group has also begun to work in this area, and we have found further complexity in relation to major basic protein (MBP). We have found a point mutation in the coding region of MBP in HL60 cDNA library, a gift from Dr. David Bentley, Imperial Cancer Research Funds Laboratory, London. We have also shown that the MBP mRNA transcript in these cells is differentially spliced, giving rise to two forms with different 5′ noncoding regions. Both these transcripts are present in eosinophils and placental tissues, but the "new" transcript is present in smaller amounts than the "classical" mRNA described by Barker and colleagues in 1988 (8).

After 18 months' careful work, we have now successfully expressed the ECP, EDN, and MBP genes in *Escherichia coli* using the pGEX vector system from Pharmacia LKB Biochemicals (Fig. 1). We know that other groups are even further

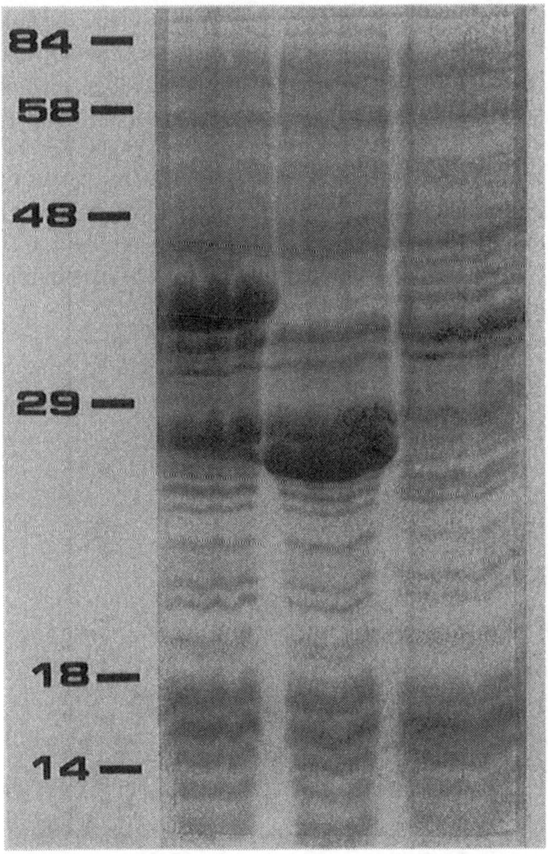

Figure 1 Expression of the mature form of EDN introduced by pGEX into *E. coli*. This 15% SDS-polyacrylamide gel separation of the bacterial proteins stained with Coomassie brilliant blue, shows the GST-mature EDN fusion protein (41 kDa) in lane 1, the pGEX-2T-encoded GST protein (26 kDa) from bacteria in lane 2, and proteins from uninduced bacteria in lane 3.

ahead in this aim to express and study the properties of mutated eosinophil granule proteins to define their active domains and the structural basis for their wide-ranging properties.

Our long-term interest in the hypereosinophilic syndrome (HES) has been greatly illuminated by Monique Capron's work with her colleagues in France, which she outlined here. I hope that many patients with the severe forms of HES will benefit from interferon-α treatment. An open trial of this kind is under way at the Mayo Clinic. We need to find out why this treatment is beneficial and whether it affects long-term prognosis of this disease, which has a half-time survival of over 13 years after diagnosis.

What are the causes of HES? It is already known that some patients have a premalignant disease with some similarities to chronic myeloid leukemia. Others have clonal expansion of T cells, and a few develop a T-cell malignancy several years after the eosinophilia is detected. Others have IL-5 like activity in their sera, and an autocrine production of IL-5 by eosinophils may be a cause for the disease in some patients. We have recently begun to study IL-5 mRNA levels in HES, as we have long championed the hypothesis that HES should not be classified as an autonomous malignant disease, but be treated as a failure of regulation of eosinopoiesis unless there is definite evidence for a clonal defect in eosinophils

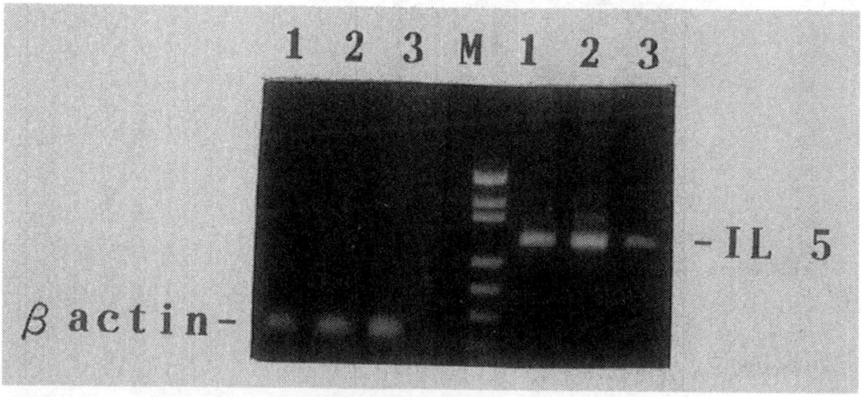

Figure 2 The presence of IL-5 mRNA in unstimulated human bone marrow mononuclear cells shown using a semiquantitative PCR assay with β-actin mRNA as an internal standard. The three lanes on the left show the amplified beta actin, M = molecular weight markers of 489/501, 404, 348, 242, 190, 147, and 110 bp. The three lanes on the right show the IL-5 mRNA transcripts in bone marrow mononuclear cells from the three patients in whom this assay was positive. Eosinophil and mononuclear mRNA samples from other patients were negative.

themselves. Eleven of our patients with HES have been examined so far. None of their blood eosinophils contained detectable amounts of IL-5 mRNA, but peripheral blood and bone marrow mononuclear cells in three patients contained large amounts of IL-5 mRNA (Fig. 2). It was also interesting to find that blood mononuclear cells from patients with marked eosinophilia produced more IL-5 mRNA than cells from normals in response to stimulation with phorbol myristate acetate, suggesting that they were in some way "primed" for an IL-5 response. We have no explanation for why their mononuclear cells were so rich in IL-5 mRNA. Interestingly, two of the three patients with high levels of IL-5 mRNA in their unstimulated blood mononuclear cells had severe forms of the disease. It was also surprising to find that one of them had a clonal chromosome 7q- defect. It is now important for us to pool our clinical and laboratory resources to look further into the molecular defects that give rise to this serious, lifelong syndrome. This may open up several new avenues into the cell biology of eosinophils and the regulation of eosinopoiesis.

IV. PAPERS AND PEOPLE

The pace of eosinophil research has gradually increased in the last 15 years, as shown by the number of papers on eosinophils in the National Library of Medicine's MEDLINE databases. Many of the most prolific authors in this field (Table 1) are represented here. It is a pity that not everyone who has worked in the field could have participated, but as there are over 19,300 authors of the 8,231 eosinophil papers indexed in MEDLINE since 1980, it would have been rather difficult.

Table 1 Authors of >30 Papers on Eosinophils 1980–1992

>30–40	>40–50	>80–90
Dahl R	Loegering DA	Capron M
McLaren DJ	Tai P-C	Capron A
Goetzl EJ	Metcalf D	Spry CJ
Tchernitchin AN	Sanderson CJ	>90–110
Klebanoff SJ	Weller PF	Venge P
Fauci AS	Austen KF	Kay AB
Czarnetzki BM	Ackerman SJ	>190
Ottesen EA	>50–60	Gleich GJ
David JR	Vadas MA	
Moqbel R	Butterworth AE	

V. LOOKING AHEAD

To conclude with some questions that lie at the heart of the topics we have been discussing: What are the evolutionary pressures that have led to the development of eosinophils (e.g., who needs eosinophils)? What is the source of the IL-5 that stimulates eosinophil production in the marrow? What is the molecular basis of the host response that causes IL-5 synthesis and secretion in disease? (Expressed another way: Why do bacterial and viral infections *not* stimulate eosinophil production?) Why are eosinophils not made outside the marrow when there are plenty of circulating eosinophil progenitor cells and IL-5 in disease? Where did the gene for MBP come from, and why is it expressed in X cells in some placental mammals? Why do eosinophils possess not one but two ribonucleases? Is the enzymatic activity of these proteins important in the functions of eosinophils? What are the principal signals given and received by eosinophils in tissues that regulate their functions and induce such varied responses?

ACKNOWLEDGMENTS

We are grateful to the British Heart Foundation, the Wellcome Trust, Sandoz Japan, and the British Medical Research Council for their financial support for our work on eosinophils.

REFERENCES

1. Charcot JM, Robin C. Observation de Leucocythemie. CR Mem Soc Biol 1853; 5:44–50.
2. Basten A, Beeson PB. Mechanisms of eosinophilia. II. Role of the lymphocyte. J Exp Med 1970; 131:1288–1305.
3. Campbell HD, Tucker WQ, Hort Y, Martinson ME, Mayo G, Clutterbuck EJ, Sanderson CJ, Young IG. Molecular cloning, nucleotide sequence, and expression of the gene encoding human eosinophil differentiation factor (interleukin 5). Proc Natl Acad Sci USA 1987; 84:6629–6633.
4. Takaki S, Tominaga A, Hitoshi Y, Mita S, Sonoda E, Yamaguchi N, Takatsu K. Molecular cloning and expression of the murine interleukin-5 receptor. EMBO J 1990; 9:4367–4374.
5. Azuma C, Tanabe T, Konishi M, Kinashi T, Noma T, Matsuda F, Yaoita Y, Takatsu K, Hammarstrom L, Smith CI, Honjo T. Cloning of cDNA for human T-cell replacing factor (interleukin-5) and comparison with the murine homologue. Nucleic Acids Res 1986; 14:9149–9158.
6. Hansel TT, De Vries IJ, Iff T, Rihs S, Wandzilak M, Betz S, Blaser K, Walker C. An improved immunomagnetic procedure for the isolation of highly purified human blood eosinophils. J Immunol Methods 1991; 145:105–110.
7. Hamann KJ, Barker RL, Ten RM, Gleich GJ. The molecular biology of eosinophil granule proteins. Int Arch Allergy Appl Immunol 1991; 94:202–209.
8. Barker RL, Gleich GJ, Pease LR. Acidic precursor revealed in human eosinophil granule major basic protein cDNA. J Exp Med 1988; 168:1493–1498.

Index

(Page numbers in **bold** refer to illustrations)

About the Editors

GERALD J. GLEICH is a Professor of Immunology and Medicine at the Mayo Medical School, Mayo Clinic and Foundation, Rochester, Minnesota. The author of over 300 professional papers and book chapters, Dr. Gleich is a Fellow of the American College of Physicians and the American Association for the Advancement of Science, and a member of the American Academy of Allergy and Immunology, the American Association of Pathologists, and the American Rheumatism Association, among others. His research interests include the function of eosinophils, the structure and function of eosinophil granule proteins, the pathophysiology of allergic inflammatory reactions, and the role of the eosinophil granule major basic protein in human pregnancy. He received the B.A. degree (1953) in chemistry from the University of Michigan, Ann Arbor, and the M.D. degree (1956) from the University of Michigan Medical School, Ann Arbor.

A. BARRY KAY is the Director of and a Professor in the Department of Allergy and Clinical Immunology at the National Heart and Lung Institute, University of London, and an Honorary Consultant Physician at Royal Brompton National Heart and Lung Hospital, London, England. A Fellow of the Royal College of Physicians of Edinburgh, the Royal College of Physicians of London, the American Academy of Allergy and Immunology, the Royal College of Pathologists, and the Royal Society of Edinburgh, Dr. Kay is a member and Past President of the European Academy of Allergology and Clinical Immunology, and a member of several other associations and societies. He received his medical degree (1963) from the University of Edinburgh, Scotland, the B.A. (1965) and the M.A. (1970) degrees in pathology from Jesus College, Cambridge University, the Ph.D. degree (1969) in pathology from Cambridge University, England, and the D.Sc. degree (1976) from the University of Edinburgh, Scotland.

Milton Keynes UK
Ingram Content Group UK Ltd.
UKHW020007071024
449327UK00031B/2684

9 780367 455996